THE UNIVERSITY OF WISCONSIN

Publications in Medieval Science

MARSHALL CLAGETT
General Editor

Campanus of Novara

and Medieval Planetary Theory

Campanus of Novara
and Medieval Planetary Theory

Theorica planetarum

EDITED WITH AN
INTRODUCTION, ENGLISH TRANSLATION,
AND COMMENTARY BY

FRANCIS S. BENJAMIN, JR.

and

G. J. TOOMER

THE UNIVERSITY OF WISCONSIN PRESS
MADISON, MILWAUKEE, AND LONDON
1971

Published 1971
The University of Wisconsin Press
Box 1379, Madison, Wisconsin 53701

The University of Wisconsin Press, Ltd.
70 Great Russell Street, London, WC1B 3BY

First printing

Printed in the Netherlands
Koninklijke Drukkerij G. J. Thieme N.V., Nijmegen

ISBN 0-299-05960-X; LC 78-138057

Contents

Preface

The *Theorica planetarum* is of interest because it is the first detailed account of the Ptolemaic astronomical system, which was the basis of all later medieval cosmology, to be written in the Latin-speaking West. Ptolemy's *Almagest* itself had indeed been available in Latin translation for about a century before this, but its intrinsic difficulty was compounded by the crabbed style of the Latin versions. To make Ptolemy's doctrines generally available, a number of popularizing works *On the Sphere* were composed during the thirteenth century. But Campanus was the first to expound the system in full numerical detail and, at the same time, to exploit a technique (not found in the *Almagest*, but also of Ptolemaic origin) for calculating the exact sizes and distances of the planets. Furthermore, the *Theorica* contains the first known Latin description of an "equatorium," a mechanical device for computing the positions of the heavenly bodies. It achieved remarkable popularity for a technical astronomical work (as is attested by the large number of extant manuscripts). But by the time printing was invented, the type of equatorium Campanus describes had been superseded by more sophisticated devices, and other accounts of the Ptolemaic system, based on the Greek text of the *Almagest*, were available. So the work was never printed, because it was obsolete for the astronomer and too technical for the general public (in contrast, Campanus's popularizing work *On the Sphere* was printed four times in the sixteenth century). The present edition is a belated rectification.

This book, the outgrowth of Benjamin's doctoral studies, is an amalgam of the efforts of both contributors. All the work of finding and assembling manuscript material, establishing a preliminary text, and reading and collating manuscripts was done by Benjamin, who also wrote the introductory sections on Campanus's life and works and on the influence of the *Theorica planetarum*, as well as the resumé of manuscripts and sigla. The translation and commentary and the remaining sections of the introduction are due to Toomer. The final establishment of text and critical apparatus is the result of a joint effort.

We wish to thank the Master and Fellows of Magdalene College, Cambridge, for permission to reproduce figures from the manuscript Pepys 2329. We gratefully acknowledge the help of Robert B. Honeyman, Jr., and of the various libraries and institutions mentioned in the "Manuscripts and Sigla" summary in providing access to or copies of the manuscripts concerned and giving us permission to use the material therein. We acknowledge with thanks a grant from the National Science Foundation subsidizing the costs of publication.

Acknowledgments should also be made to Brown University, Columbia University, Corpus Christi College, Oxford, and Emory University for assistance in a variety of forms. The Carnegie Foundation should be thanked for a grant which permitted the search for and examination of several of the manuscripts *in situ*. It would be impossible to overstate the multiple contributions of the late Lynn Thorndike, who initiated this project as a dissertation topic and saw it through that stage. We also wish to thank Janet Sachs, who typed the final manuscript, Jonathan Sachs, who drew the diagrams, and B. Goldstein and O. Neugebauer for help on various points.

F. S. B.
G. J. T.

Editorial Procedures

The text we print is frankly eclectic. It aims, not to reproduce the original words of the author (an impossible task in the present state of the evidence), but to be readable and intelligible. To this end, we have not hesitated to adopt readings of a minority of manuscripts, and have even, very occasionally, resorted to emendation. No manuscript or group of manuscripts has been followed slavishly, but when there was no reason to choose between equally good alternatives, we have usually adopted the reading of the majority (the α group; see p. xiv, below).[1]

The titles of the subdivisions of the *Theorica* which we have put in the text are those in manuscript *z* (Bologna, Biblioteca Universitaria, 132). They are included merely for illustration, since there is no reason to suppose that they or any of the numerous variants found in other manuscripts were composed by Campanus, nor even that he inserted any titles at all.

Orthography. The spelling in our text is a standardized version of that current in the time of Campanus. Thus, no distinction is made between the consonant *v* and the vowel *u* (such a distinction seems to have been introduced by sixteenth-century printers): we use *u* for the minuscule and *V* for the capital form. The diphthong *æ* of classical Latin is consistently rendered as *e* (in medieval manuscripts it is occasionally, but not consistently, represented by *ę*; however, *æ* is almost never found in them). We have expanded all scribal abbreviations. On the other hand, the punctuation does not attempt to follow medieval practice (which, like the ancient, is more concerned with representing speech rhythm than sense breaks), but is intended solely to aid the reader's understanding. The capitalization and paragraphing, too, are entirely ours and bear no relation to what is found in the manuscripts (most of which have very little of either).

The translation. The translation is intended to elucidate the text rather than to replace it. Thus, it follows the structure of the Latin as closely as is compati-

1. MS *A* was originally used as the master manuscript, except for the dedicatory letter (lines I.4–110), which does not exist in MS *A*; for the dedication, MS *V* was used as master.

ble with intelligibility. However, it has frequently been necessary to add words or phrases to bring out the full sense of the text or to produce idiomatic English. These and all other additions which have nothing corresponding to them in the Latin text are enclosed in square brackets.

Tables. The tables are merely consolidations of the numerous parameters of the parts and motions of the planetary systems mentioned in the body of the text. They are found in nearly all manuscripts (though incomplete in several) scattered through the text at or near the points where the author indicates they should be. For convenience of consultation, we have gathered them all together on pp. 356–63, with cross-references at the appropriate points in the translation text. The format and wording of the tables as we print them are based on manuscript *F* (other manuscripts present numerous but insignificant variations). The numerical entries we print are the correct ones (i.e., those which Campanus himself must have used). We provide no apparatus of the manuscript variants in these numbers. To do so would have required an expenditure of time and space not justified by the result, which would have been largely a repetition of the apparatus concerning the individual parameters at the points where they occur in the text.

Figures. All the figures are numbered consecutively in order of their appearance in the book, but they are in fact of two kinds: (1) drawings based on diagrams actually found in the manuscripts; (2) drawings added by us to illustrate the introduction and commentary. The former are interspersed in the text, as close as possible to the primary discussion of them; to avoid occasional repetition of a particular figure, cross-references have been added in the translation text. Our drawings are intended to aid the reader's comprehension rather than to reproduce their medieval exemplars with minute accuracy. To enable the reader to compare the latter, we have included photographs of four figures from an exceptionally well executed manuscript, Γ.

Critical apparatus. Our critical apparatus is highly selective. There are many manuscripts, and most have large numbers of individual insignificant variations. If we had reported all variants, the size of the apparatus would have swelled to several times that of the text, with little profit to the reader. We have therefore frequently (but not systematically) not reported the following kinds of textual variants:

(1) Omissions or additions of particles: e.g., *etiam, quidem, uero.*

(2) Repetition or omission of a word or phrase occurring earlier in the sentence or sense-group which does not affect the sense: e.g., *linea medii centri et (linea) ueri (centri).*

(3) Different orthography: e.g., *epycicli, epicicli; duo, 2.*

(4) Substitution of a word or phrase by a synonym: e.g., *enim, igitur; scilicet, uidelicet; quilibet, quiuis; geometre, geometrici; que uidere potes, que uideri possunt.*

(5) Substitution of one tense or mood of the same verb by another when the general sense is not affected. For example, in the instructions for making and using the instrument, the future tense is frequently interchanged with the present or the future simple with the future perfect. An example of the substitution of subjunctive for indicative would be *ista sufficiant*, *ista sufficiunt*.

(6) Obvious scribal errors found in one manuscript or group of manuscripts. This applies particularly to the frequent corruptions in numerals.[2]

(7) Obviously later explanatory additions or variations found in one manuscript or group of manuscripts. These are, where we have omitted them, of an extremely trivial nature.

The apparatus is presented in two parts. That on the verso page contains the variants we record to the text as printed. That on the recto page contains those items in the text or margin of individual manuscripts which are clearly not attributable to Campanus himself. These are explanatory additions to the text, marginal notes, and chapter titles (the latter vary so enormously among manuscripts that they clearly do not go back to the author). We have selected the parts of this material which are of some intrinsic interest or illustrate its nature, and have omitted much trivia (e.g., "Nota", with a pointing finger).

Both portions of the apparatus are keyed to the line numbers in the left-hand margin of the Latin text. A superscript "1," "2," etc., attached to the lemma indicates that the reference is to the first, second, etc., occurrence of the word in that line. The lemma is separated from the variants by a colon, and the different lemmata in the same line by a solidus. Occasionally the variant alone is given without any lemma, when the reference is obvious. The principle of reporting variants is generally "negative"; i.e., the absence of a report on what a given manuscript reads indicates that it shares the reading of the text (with the qualifications mentioned above). However, we have given a "positive" report whenever we have emended the text or taken an exceptional reading. When we report a variant peculiar to one manuscript, we record idiosyncratic spellings; but when the variant is shared by several manuscripts, it is reported in the standard spelling, unless all the manuscripts sharing it happen to use a peculiar spelling. A subscript "1" attached to a manuscript siglum (e.g., A_1) means that the reading in question is what the scribe of the manuscript originally wrote (before correction). A subscript "2" (e.g., A_2) indicates that the reading is the work of a second hand (or possibly a correction by the first hand: one cannot always tell the difference). Variants within a long variant passage are treated paren-

2. Clearly, even trivial errors and variations in manuscripts can be significant when one is seeking to establish relationships between manuscripts. With this in mind, we have occasionally recorded which manuscripts share a particular numerical error. But to have done so systematically would have overburdened the apparatus, with, as far as we can judge, no very significant result (the manuscript tradition being too contaminated).

thetically, and if the secondary variant reading involves more than one word, the passage of which it is a variant has been italicized.

Abbreviations used in the apparatus, and occasionally elsewhere, are as follows (note that "*add.*" implies that the word or phrase given follows the lemma in the text):

add.	addit	*rubric.*	rubrica
codd.	codices	*sim.*	similia
mg.	in margine	*tit.*	titulus, titulo
om.	omittit		

Classification of manuscripts. In drawing up a classification, we have separated the prefatory letter of the *Theorica* (section I) from the main body of the work. We have not attempted any classification for the letter, which is represented by far fewer, and in part different, manuscripts than is the main portion. These manuscripts cannot be grouped in the same way as can those for the rest of the work, as an examination of the readings we report will verify.

For the main part of the *Theorica*, we have divided the manuscripts into several groups, designated by lower-case Greek letters (see "Manuscripts and Sigla," pp. 57–125, below). These are not to be taken as "families of manuscripts" in the paleographer's sense: we have not attempted to draw up a stemma (a horrendous prospect, when one considers the number of manuscripts and the degree of contamination, at least among the later ones). But it has become clear to us, while working through the variant readings, that certain groups frequently agree with one another against all or most of the rest. The groups are as follows.

(1) φ, represented by manuscripts *foy*Λ. The peculiar readings and marginal notes of φ are also found in *J*, but *J* can equally well be assigned to other groups (α and both of α's subgroups, β and θ), so it is best regarded as contaminated.

(2) δ, represented by *bDHNVXZ*. These are sometimes joined by *GjL*, which otherwise adhere to the θ group. *H*, too, agrees at times with θ against the rest of δ. *V* is frequently aberrant and in the later part of the work seems to belong to the β group.

(3) α, represented by all other manuscripts. This large and often inhomogeneous group has been partially divided into

(3a) β, consisting of the central group *BhMmOrYz*Γ, sometimes joined by *CdgKkPpqStU*Π, and in the later part of the text by *V*; and

(3b) θ, consisting of the central group *AcFIQsx*Σ, often joined by *eTu*, and sometimes by *aGijLlWw*. *F* and Σ, which are closely related, seem contaminated from β. φ often agrees with θ against β.

The later manuscripts mostly fall into group α, but some cannot be definitely assigned to the β or the θ subgroup: e.g., Ψ agrees now with one, now with the other. In fact, any attempt to subdivide α rigidly is bound to fail because of the contamination problem.

The accompanying diagram (*not* a stemma) illustrates the groupings.

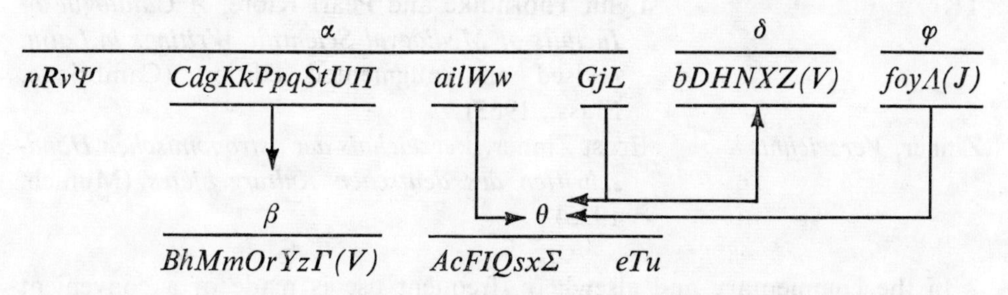

When one of the sigla φδαβθ is attached to a reading in the apparatus, it indicates that all *or a significant number* of the manuscripts in that group share that reading. In the case of subgroups β and θ, assignation of the siglum does not necessarily imply that only the above-mentioned manuscripts in group α share the reading: frequently one or more other manuscripts of group α also have the reading, but it was impracticable to indicate them systematically.

Reference matter. The commentary follows the text and translation on pp. 367–444. It is keyed by superscript reference numbers in the English text, and the entry in the commentary also supplies a cross-reference to the Latin word or phrase and the line numbers.

Bibliographical references (in both the commentary and the introduction) are cited by short title except on first citation (full details will also be found in the consolidated bibliography on pp. 453–65). The following abbreviated references are frequently used.

Alfraganus, Carmody	Alfraganus, *Differentie in quisbusdam collectis scientie astrorum*, edited by F. J. Carmody (Berkeley, 1943)
Duhem, *Système*	Pierre Duhem, *Le Système du monde: Histoire des doctrines cosmologiques de Platon à Copernic*, 10 vols. (Paris, 1913–59)
Gerardus, *Theorica*	Gerardus, *Theorica planetarum*, edited by F. J. Carmody (Berkeley, 1942)
Grosseteste, *De sphaera*	Robert Grosseteste, *De sphaera*, in *Die philosophischen Werke des Robert Grosseteste, Bischofs von Lincoln*, pp. 10–32, edited by Ludwig Baur, Beiträge zur Geschichte der Philosophie des Mittelalters, vol. 9 (Münster i.W., 1912)
Migne, *PL*	J. P. Migne, *Patrologiae cursus completus. Series latina*, 221 vols. (Paris 1844–64)

Ptolemy, Manitius	Ptolemy (Claudius Ptolemäus), *Handbuch der Astronomie*, translated by Karl Manitius, 2 vols. (Leipzig, 1912–13; reprinted 1963)
TKr	Lynn Thorndike and Pearl Kibre, *A Catalogue of Incipits of Mediaeval Scientific Writings in Latin*, revised and augmented edition (Cambridge, Mass., 1963)
Zinner, *Verzeichnis*	Ernst Zinner, *Verzeichnis der astronomischen Handschriften des deutschen Kulturgebietes* (Munich, 1925)

In the commentary and elsewhere, frequent use is made of a convenient symbolism for denoting numbers expressed in the sexagesimal system. According to this notation, for example, 1,44,3;10,16,22 represents $1 \cdot 60^2 + 44 \cdot 60 + 3 + 10 \cdot 60^{-1} + 16 \cdot 60^{-2} + 22 \cdot 60^{-3}$.

The following conventional astronomical symbols are used (usually as subscripts).

⊕	earth	♀	Venus
☾	moon	♂	Mars
⊙	sun	♃	Jupiter
☿	Mercury	♄	Saturn

In addition, the following symbols for the signs of the zodiac will occasionally be found.

♈	Aries (♈0 is the vernal point)	♐	Sagittarius
♊	Gemini	♍	Virgo
♌	Leo		

Indexes are found on pp. 467–90. The first index contains the page references to all manuscripts mentioned in the introduction and commentary. The second is an index of technical terms found in the text (with certain exceptions, as explained on p. 470); this index gives the page and line references for each entry. The third index is a general index.

Appendixes. Appendix A contains a long passage of text found in one group of manuscripts only, which could not conveniently be put in the apparatus. Appendix B is a text of the *Eulogy* of the Emperor Frederick II by Petrus de Vineis which influenced Campanus's prefatory letter to Pope Urban IV. Appendix C is an extract from the astronomical work of Jābir ibn Aflaḥ which seems to have been the inspiration for Campanus's procedure at lines VI.261–340.

Introduction

Campanus of Novara

Life

The mathematician Magister Campanus of Novara may have been born as early as the first decade of the thirteenth century if the first entry-year, 1232, in two sets of astronomical tables ascribed to Campanus[1] can be accepted as their date of composition.[2] Since the tables are not cited prior to 1261[3] and in one

1. See Ernst Zinner, *Verzeichnis der astronomischen Handschriften des deutschen Kulturgebietes* (Munich, 1925), p. 334, no. 10866 (MS Admont, Stiftsbibliothek, F.318 [14th cent.], fols. 2–61), and note on p. 510; p. 334, no. 10868 (MS Lilienfeld, Stiftsbibliothek, 144 [13th cent.], fols. 28–29), and note on pp. 510–11; and Zinner, *Geschichte der Sternkunde von den ersten Anfängen bis zur Gegenwart* (Berlin, 1931), p. 369. The Lilienfeld manuscript has, on fol. 28r, "Tabula ad annos domini ihesu christi expansos" and also "Tabula ad inueniendum annos arabum per annos domini ihesu christi [for the years] 1232, 1260, 1288,…, 1540." On fol. 28v[b]: "Tabula coniunctionis solis et lune secundum medium ipsorum motum in annis domini ihesu christi collectis ad meridiem nauarrie cuius longitudo ab occidente est 30 gradus et 15 minuta, altitudo poli 45 gradus [for the years] 77, 153, 229,…, 1217, 1293, 1369, 1445, 1521, 1597." On fols. 29r[a]–r[c]: "Tempus medie applicationis. [Below] anni collecti [for the years] 622, 650, 679, 698, 1290, 1309."

2. This dating would imply that Campanus was in his mid-twenties at the time of their putative composition. None of his other authentic works suggests any earlier date. His modern biographers include Girolamo Tiraboschi, *Storia della letteratura italiana* (Milan, 1822–26), vol. 4, bk. 2.2, chaps. 8–9, pp. 250–54; P. C. F. Daunou, "Campanus de Novarre, mathematicien," *Histoire littéraire de la France*, vol. 21 (1835), pp. 248–54, 688–89; Pierre Duhem, *Le Système du monde: Histoire des doctrines cosmologiques de Platon à Copernic* (Paris, 1913–59), vol. 3, pp. 317–26, and index; George Sarton, *Introduction to the History of Science* (Baltimore, 1927–47), vol. 2, pt. 2, pp. 985–87; and Giovanni Vacca, "Campano da Novara," *Enciclopedia Italiana di Scienze, Lettere ed Arti*, vol. 8 (Milan, 1930), p. 594. To their resumés I have added the fifteenth-century account of Symon de Phares for the years 924 and 1269 (Ernest Wickersheimer, *Recueil des plus célèbres astrologues et quelques hommes faict par Symon de Phares du temps de Charles VIIIe, publié d'après le manuscrit unique de la Bibliothèque Nationale* [Paris, 1929], pp. 167 and 202) and the sixteenth-century account of Bernardino Baldi ("Vite inedite di matematici italiani," ed. Enrico Narducci, *Bullettino di bibliografia e di storia delle scienze matematiche e fisiche*, vol. 19 [1886], pp. 591–96), as well as my own researches.

3. See note 14, below.

manuscript one set of tables begins with 1260,[4] there may be some other explanation for so early a date in the tables.[5]

Campanus is sometimes given the forename of Johannes, but that name receives no confirmation in the manuscripts of his works which I have seen, nor in the writings or commentaries on his works by astronomers of the fourteenth and fifteenth centuries.[6] Daunou ascribes the first use of this forename to Huet, Bishop of Avranches (1630–1721),[7] but there is at least one citation of Campanus as Johannes as early as 1522.[8] His birthplace was probably Novara in Lombardy.[9]

The contemporary evidence concerning his life, which supplements and corrects the accounts by his modern biographers,[10] begins with an inference from a thirteenth-century manuscript of his edition of Euclid.[11] Just below the colophon of this manuscript in cursive hand is written, "In nomine domini amen. Jacobus Dei gratia Patriarcha Jerusalemitorum omnibus christi fidelibus salutem desideratum." This inscription would date the work between 1255 and 1261, since Jacques Pantaléon was Patriarch of Jerusalem from April 9, 1255, until his elevation to the papacy.[12] The work must in fact have been composed

4. MS London, British Museum, Harley 13, fol. 28r. The date 1254 can be derived from three manuscripts which contain tables, ascribed to Campanus and probably connected with his *Computus*, listing mean conjunctions of the sun and moon for a 19- or 76-year cycle beginning January 1254. These are: Oxford, Bodleian Library, Digby 215 (15th cent.), fol. 93r[a]; Cesena, Biblioteca Malatestiana, Dexter, Pluteus XXVI, Cod. I (13th cent.), fols. 113v–114r; Florence, Biblioteca Nazionale, II.II.67 (15th cent.), fols. 116v–118r. Cf. p. 16, note 59, below.

5. Zinner, *Geschichte der Sternkunde*, p. 369, says (without giving a source) that Campanus computed tables for the new and full moon for the years 1248–79. This provides yet another initial date.

6. C. S. Peirce, in his article "Campanus" (*Science*, n.s., vol. 13 [1901], p. 810), cites fourteenth-century manuscripts of Johannes de Lineriis as if they assigned the name Johannes to Campanus (MSS Oxford, Bodleian Library, Digby 57 and 168; cf. MS Rome, Biblioteca Apostolica Vaticana, Urbinas 1399), but the manuscripts do not confirm Peirce's assertion. Cf. an edition of Lineriis's work by D. J. Price (*Equatorie of the Planetis* [Cambridge, 1955], pp. 188–96), from a Brussels manuscript, where no forename is given. Tiraboschi (*Storia della letteratura*, vol. 4, bk. 2.2, chap. 8, p. 251) says no forename appears in the manuscripts or printed editions of any of Campanus's authentic works.

7. Daunou, "Campanus de Novarre," pp. 688–89.

8. Lynn Thorndike, *A History of Magic and Experimental Science* (New York, 1923–58), vol. 6, p. 355, n. 101.

9. In the manuscripts his name is invariably followed by *nouar[i]ensis*. In the *Theorica planetarum*, whenever he assumes the place of the observer, it is Novara. When cited by later writers in the next several centuries, his name is usually followed by this place designation.

10. See note 2, above.

11. MS New York, Columbia University Library, Plimpton 156, fol. 164v; cited by D. E. Smith, *Rara arithmetica* (Boston and London, 1908), pp. 433–34. This inscription is not in the same hand as that of the main text, but it is the same as that in the diagrams.

12. Conrad Eubel, *Hierarchia catholica Medii Aevi* (Münster, 1913), vol. 1, p. 275.

before May 11, 1259, the date of the earliest known manuscript.[13] In a manuscript work of Petrus Peregrinus, which can be dated from internal evidence as of 1261, the tables of Campanus are already cited.[14]

Several copies of the *Theorica planetarum* are preceded by a dedicatory letter to Pope Urban IV (1261–64) (pp. 128–34, lines I.4–110, of our text). In a letter dated December 21, 1263, at Orvieto, Urban IV confers anew on Magister Campanus the rectorship of the church of Savines in the diocese of Arles.[15] In another letter of February 26, 1264, Urban speaks of a certain Ardizonus, called Flamentus, nephew of Magister Campanus.[16] This Campanus must be the same person as the one mentioned in the preceding letter, for he is likewise closely connected with Ottobonus, cardinal deacon of St. Adrian.[17] In a third

13. MS Florence, Biblioteca Nazionale, Magliabecchi XI, 112, fol. 160r. See J. E. Murdoch, "The Medieval Euclid: Salient Aspects of the Translations of the *Elements* by Adelard of Bath and Campanus of Novara," *Revue de synthèse*, vol. 89, ser. 3 (1968), p. 73, note 18.

14. T. B. Barnabita, "Sopra Pietro Peregrino di Maricourt e la sua epistola de magnete," *Bullettino di bibliografia e di storia delle scienze matematiche e fisiche*, vol. 1 (1868), p. 5. The manuscript cited is Rome, Biblioteca Apostolica Vaticana, Palatinus 1392, fol. 1r. G. Boffito and C. M. D'Eril (*Il trattato dell'astrolabio da Pietro Peregrino di Maricourt: Introduzione e saggio del codice Vaticano Palatino N. 1392 con facsimili* [Florence, 1928], pp. 6–7) date this work prior to 1261. The facsimile, fols. 17r[a]–r[b] reads: "...Et melius est equare ad annum conuenientem primum post bisextum quia tunc inuenimus 365 dies et quartam diem. Propter hoc equauimus solem ad annum [fol. 17r[b]] domini 1261 immediate post bissextum per tabulas quas composuit magister Campanus ad meridiem ciuitatis Nouarie...."

15. Jean Guiraud, *Les Régistres d'Urbain IV (1261–64)* (Paris, 1901–6), vol. 3, p. 428, col. 1, no. 2541: "Orvieto, 21 décembre 1263. Magistro Campano ecclesiam de Saviano, Arelatensis diocesis, ipsi per delegatum Narbonensis archiepiscopi auctoritate apostolica collatam, de novo confert. (REG. 29, fol. 306, n° 1593) 'Magistro Campano, dilecti filii nostri O., s. Adriani diaconi cardinalis. Illo recte rationis.... Datum apud Urbemveterem, xii kalendas januarii, anno tertio.'" (Presumably "capellano" should be added before "dilecti.") We might infer, since the rectorship was conferred anew, that the edition of Euclid's *Elements* may have prompted Campanus's appointment to it for the first time.

16. Guiraud, *Les Régistres d'Urbain IV*, vol. 3, p. 186, col. 1, no. 1356: "Orvieto, 26 février 1264. Praeposito ecclesiae s. Gaudentii Novariensis mandat ut Ardizonum, dictum *Flamentum*, nullum ecclesiasticum beneficium assecutum, nepotem magistri Campani, capellani O., s. Adriani diaconi cardinalis, obtentu ejusdem capellani, in ecclesia s. Julii de Insula Novariensi, in canonicum recipi faciat eique de praebenda provideat. (REG. 29, fol. 119v, n° 406) 'Preposito ecclesie s. Gaudentii Novariensis. Et si ad.... Datum apud Urbemveterem v kalendas martii, anno tertio.'" This Ardizonus is probably the same as the "magistro Arditioni, primicerio ecclesie Mediolanensis," who is mentioned three times in letters of John XXI as his chaplain and collector of the tithe in England. See Léon Cadier, ed., *Les Régistres de Jean XXI, 1276–1277* (Paris, 1898), p. 33, col. 2, no. 103 (15 February 1277); p. 33, col. 2, no. 104 (13 February 1277); and p. 35, col. 2, no. 106 (12 February 1277).

17. See extract in note 15 above. Cf. Guiraud, *Les Régistres d'Urbain IV*, vol. 3, p. 114, col. 2, no. 1125, where a letter of Urban IV confirming an abbacy on a Ribaldus includes a letter of Ottobonus about the same affair listing, among other witnesses,

letter at Orvieto of May 16, 1264, Urban confers a canonicate in the church of Toledo on Magister Campanus, rector of the church of Savines, diocese of Arles, apostolic chaplain.[18] The excerpts Guiraud includes from this letter show that Campanus was well known to and on favorable terms with Urban

---

(Note 17 continued)
"magistro Campano de Nouaria." Ottobonus Fliscus was a nephew of Pope Innocent IV and was made cardinal deacon of St. Adrian's on February 6, 1252. On July 11, 1276, he was elected pope, taking the name of Adrian V, but survived his election by only a little more than a month, dying at Viterbo on August 18, 1276 (Eubel, *Hierarchia catholica*, vol. 1, pp. 7–9). (It is curious that Campanus also died at Viterbo.) Ottobonus was connected with the royal family of England, a fact which explains the favor shown him by Henry III in the way of preferments in Sicily. Although after his uncle's death he does not appear to have been particularly favored by Alexander IV, Urban IV (but he is called "dilecti filii" in the extract in note 15 above), and Gregory X, he stood high with Clement IV, who sent him to England to negotiate peace between the barons and Henry III (1265–68), a task he accomplished extraordinarily well (H. K. Mann, *The Lives of the Popes in the Early Middle Ages* [London, 1925–32], vol. 16, pp. 23–30). Concerning Ottobonus, see Natalie Schöpp, *Papst Hadrian V (Kardinal Ottobuono Fieschi)* (Heidelberg, 1916), and Dorothy M. Williamson, "Some Aspects of the Legation of Cardinal Otto in England, 1237–41," *English Historical Review*, vol. 64 (1949), p. 147, n. 6; F. M. Powicke, *The Thirteenth Century, 1216–1307* (Oxford, 1953), chap. 5 and index; and Powicke, *King Henry III and the Lord Edward: The Community of the Realm in the Thirteenth Century* (Oxford, 1947), pp. 503–50, and index s.v. "Ottobonus."

18. Guiraud, *Les Régistres d'Urbain IV*, vol. 3, p. 259, col. 2, no. 1692: "Orvieto, 16 mai 1264. Magistro Campano, rectori ecclesiae de Saviano, Arelatensis diocesis, capellano apostolico, canonicatum ecclesiae Toletanae confert. (REG. 29, fol. 169v, nº 742) '*Magistro Campano, rectori ecclesie de Saviano, Arelatensis diocesis, capellano nostro*. Fame suavis odor ex merito tue conversationis et vite, dono scientie ac multiplicis probitatis studioque laudabilium actionum hactenus resolutus nonnullas replere dinoscitur regiones, in quibus omnium liberalium artium et aliarum etiam facultatum plenitudinem tuo pectori de super fusam non modico tempore docuisse diceris et exempli doctrina non minus ediffìcasse quam verbi. Odor quoque hujusmodi usque ad apostolicam sedem expansus delectat plurimum sensus nostros, dum per familiarem tui presentiam te famam transcendere meritis experimur, et quodam juris debito sibi vendicat apud nos ut te congruis attollere gratiarum muneribus studeamus, quia licet virtus et probitas sui sint pretio multipliciter approbande, ut tamen ad se plures pertrahant amatores, condignis sunt beneficiis honorande. Hinc est quod nos, attendentes probitatis tue merita ac per hoc volentes te dono specialis gratie confovere, canonicatum ecclesie Tholetane, cum plenitudine juris canonici, ac prebendam cum prestimoniis ejus qui cessit vel decessit, si vacant ibidem ad presens, tibi conferimus, providemus tibi de illis teque de ipsis per nostrum anulum presentialiter investimus. Si vero nulla prebenda vacat ibidem ad presens, prebendam inibi proximo vacaturam et prestimonia cedentis vel decedentis, conferenda tibi, donationi apostolice reservamus ac decernimus irritum et inane si secus de illis a quoquam contigerit attemptari. Non obstante... seu quod in sancte Nutricis Remensis et de Domo Oxole, Novariensis diocesum ecclesiis beneficiatus existis aut quod Tholetana sedes vacat ad presens. Nulli ergo nostre collationis, provisionis, investiture, reservationis et constitutionis etc. Datum apud Urbemveterem, xvii kalendas junii, anno tertio.'" It is not known whether he went to Toledo or enjoyed the benefice *in absentia*.

IV, a fact consistent with the tenor of the dedicatory letter that precedes the *Theorica*. Yet another letter of Urban, dated September 2, 1264, orders the archbishop of Narbonne to sequester the revenues of the church of Savines, which had been unlawfully occupied by another cleric, in favor of Campanus.[19]

In 1267 Roger Bacon, in a work addressed to Pope Clement IV, ranked Campanus among the excellent mathematicians.[20] The next reference to Campanus is dated April 3, 1270, and denotes him as parson of Felmersham, Bedfordshire, England.[21] Since Ottobonus was in England from 1265 to 1268, Campanus may have accompanied him there as his chaplain, and the date of the preferment of Campanus to the Felmersham benefice may therefore lie

19. Guiraud, *Les Régistres d'Urbain IV*, vol. 3, p. 342, col. 2, no. 2121: "Orvieto, 2 septembre 1264. Archiepiscopo Narbonensi mandat ut redditus ecclesiae de Saviniano, Arelatensis diocesis, sequestret. 'Sua nobis dilectus filius magister Campanus, capellanus noster, rector ecclesie de Saviniano, Arelatensis diocesis, petitione monstravit quod nos, in causa que inter ipsum et Vitalem, dictum *Blancum*, clericum ejusdem diocesis, super eo quod dictus clericus predictam ecclesiam occupavit et detinet contra justitiam occupatam, in ipsius capellani prejudicium et gravamen, vertitur, ad sedem apostolicam legitime devoluta, dilectum filium nostrum U., s. Eustachii diaconum cardinalem, dedimus partibus auditorum. Quare prefatus capellanus nobis humiliter supplicavit ut sequestrari fructus ecclesie mandaremus ejusdem. Quocirca fraternitati tue mandamus quatenus, omnes fructus et redditus ecclesie prefate sequestrans, ipsos per te vel per alium colligas et conserves fideliter ad opus illius qui in causa hujusmodi obtinebit. Contradictores etc. usque compescendo. Datum apud Urbemveterem, IIII nonas septembris, anno tertio.'"

20. Roger Bacon, *Opus tertium*, chap. 11, p. 35, in *Opera quaedam hactenus inedita*, ed. J. S. Brewer (London, 1859), vol. 1. See Duhem, *Système*, vol. 3, p. 238 and note 3, p. 317 and note 2. Did this stem from an actual familiarity and respect or did Bacon praise Campanus because he knew that the pope was acquainted with Campanus? I could find no preferments awarded Campanus in the Registers of Clement IV, nor references to him; but, as one of the editors of the papal registers for this period aptly says, the preservation of papal correspondence is very uneven.

21. *Calendar of the Patent Rolls Preserved in the Public Record Office* (London, 1891—), 1266–72, p. 419, where an entry for April 3, 1270, refers to "Master Campanus, parson of Felmersham." If the "Magister Campanus" of the papal registers is identical with Campanus of Novara, then this "Master Campanus," with the same title and connections, also must be Campanus of Novara. The reference to Campanus is a "mandate to the above collectors of the tenth in the bishopric of Lincoln to permit Master Campanus, parson of Felmersham, to be quit of the said tenth in his church for three years of the grant, according to an acquittance made to him by Ottobono, cardinal deacon of St. Adrian's, the late legate; and if they have received anything from him, to restore the same without delay." In 1240 and 1250 the church of Felmersham was held under Bishop Grosseteste by another Robert Grosseteste (F. N. Davis, ed., *Rotuli Roberti Grosseteste, episcopi Lincolniensis, necnon rotulus Henrici di Lexington* [Horncastle, 1914], pp. 330, 332, 336). Since there was a new presentation of the "church of Felmersham in the diocese of Lincoln, Feb. 17, 1297, void by the death of Master Campanus de Novaria, the Pope's chaplain" (*Calendar of the Patent Rolls*, 1292–1301, p. 235), the assumption would be that Campanus obtained the church at some time between 1250 and 1270 and held it until his death.

within those years. However, because Ottobonus was a relative of Henry III and early in his career held extensive benefices from Henry, the award may have been made anytime after 1250. Whether Campanus received this award from Ottobonus himself or from one of the popes cannot be determined.[22] Nonetheless, if Campanus did accompany Ottobonus to England, it would strengthen the possibility of Roger Bacon's knowing him and, therefore, acclaiming him in 1267.

The addition of the Felmersham benefice to the preferments Campanus already held attests to his importance in the eyes of the church. Tiraboschi, Daunou, Duhem, and Sarton identify him with the Campanus who corresponded with Simon of Genoa and to whom the latter dedicated his *Synonyma medicinae* or *Clavis sanationis* (ca. 1288–92), the preface to which Campanus wrote.[23] According to Simon's dedication, Campanus was a chaplain of Nicholas IV (1288–92) and a canon at Paris.[24] Quétif and Échard also mention a letter of Campanus to the Dominican Rainier or Raner da Todi written about the end of the thirteenth century.[25] The next contemporary reference to Campanus is in a letter of Boniface VIII from Anagni, dated September 17, 1296.[26]

22. If the church was not presented to Campanus by Urban IV, it might have been by Clement IV; indeed, such a circumstance could explain the connection between Campanus and Bacon.

23. Tiraboschi, *Storia della letteratura*, vol. 4, bk. 2.2, chap. 8, p. 251; Daunou, "Campanus de Novarre," pp. 248, 250; Daunou, "Simon de Gènes," *Histoire littéraire de la France*, vol. 21 (1835), pp. 241–42, 245–46; Duhem, *Système*, vol. 3, p. 320; Sarton, *Introduction to the History of Science*, vol. 2, p. 1085. Cf. MS Venice, Biblioteca Marciana, VII, 12 (1456; folio): "Simonis Ianuensis, sinonyma medicinae Dedicatio magistro Campano, Nicolai pp. IV cappellano, canonico parisiensi."

24. I could find no mention of Campanus in the Registers of Nicholas IV and have found no earlier mention of Campanus as canon of Paris, so that the time of his preferment to this position must remain unknown. Tiraboschi, *Storia della letteratura*, vol. 4, bk. 2.2, chap. 8, p. 251; Daunou, "Campanus de Novarre," pp. 248, 250; Daunou, "Simon de Gènes," p. 241; Sarton, *Introduction to the History of Science*, vol. 2, p. 985.

25. Jacques Quétif and Jacques Échard, *Scriptores ordinis praedicatorum recensiti*... (Paris, 1719–21), vol. 1, p. 474: "Epistola responsiva D. Campani super quodam dubio de motu octavae Sphaerae. Incipit: Magnae sanctitatis & scientiae religioso viro Fratri Ranero Tudertino de ordine Praedicatorum, Campanus Novariensis de numero peccatorum, orationum suarum cum instantia reverenti deposcit suffragia &c." According to Quétif and Échard, this letter is preserved in codex S. Marci Florence, Arm. IV, Cod. 167, which is now MS Florence, Biblioteca Nazionale, conv. soppr. J.X.40, more completely described in A. A. Björnbo, "Die mathematischen S. Marcohandschriften in Florenz," *Bibliotheca mathematica*, ser. 3, vol. 12 (1912), pp. 201–2.

26. Georges Digard, Maurice Fauçon, Antoine Thomas, and Robert Fawtier, *Les Régistres de Boniface VIII, recueil des bulles de ce pape publiées ou analysées d'après les manuscrits originaux des archives du Vatican*, vol. 1 (Paris, 1907), p. 614, no. 1648: "Anagni 17 septembre 1296. Reservatio beneficiorum magistri Campane de Novaria (fol. 178v). '*Ad perpetuam rei memoriam.* Cum quondam magister Campana [*sic*] de Novaria, canonicus Parisiensis, capellanus noster, nuper

In it Campanus is called not only canon at Paris but also chaplain of Boniface. It is clear from the letter that Campanus had recently died and that Boniface was reserving Campanus's benefices to himself; thus we can set the date of the death of Campanus as 1296, sometime before September 17th.[27] Still another reference to his death may be gleaned from the Viterban archives: for the year 1296, in a citation of one clause in his will, it is stated that Campanus "willed and gave orders for" the construction of "a beautiful and suitable chapel called St. Anna" for the church of the Holy Trinity and that this testament was ratified five years later by Boniface VIII.[28] Evidently Campanus was a relatively affluent man, an inference that his plural benefices would confirm, and important enough for Boniface VIII, in 1301, to see that his will was carried out despite the pope's increasing involvement on the international scene. If Campanus was indeed born in the first decade of the thirteenth century (see note 2, above), then he would have been in or close to his nineties when he died.

In the time of Charles VIII, Symon de Phares compiled a miscellany concerning astrologers and learned men, in which Campanus is identified as two men. In the first mention of "Campanus," under the date 924, Symon describes

---

apud civitatem Viterbiensem, ubi de nostro beneplacito moram traxit, viam sit universe carnis ingressus.... Dat. Anagnie, xv Kal. octobris anno secundo.'"

27. Cf. note 21, above.

28. See P. H. M., "De conventu nostri ordinis Viterbiensi," *Analecta Augustiniana*, vol. 11 (1925–26), p. 230: the author, in a discussion of the Augustinian friars' convent at Viterbo, citing "Arch. hist. Vit., pergam. 22t et 25t," says, "Anno millesimo ducentesimo nonagesimo sexto Magister Campanus, Domini Papae Cappellanus, Canonicus Parisiensis, voluit et mandavit quod fiat (in Ecclesia SS. Trinitatis) una Cappella puchra [*sic*] et decens quae vocatur S. Anna. Quinque annis post, Summus Pontifex Bonifacius VIII ratum habuit et confirmavit testamentum magistri Campani." Since, according to the letter of Boniface, Campanus died at Viterbo, he may have spent his last days at this Augustinian convent and left the chapel in gratitude for care expended upon him by the friars. On May 28, 1320, a miracle is said to have occurred in this chapel in connection with the figure of the Virgin "la quale aveva facta misser Campano" (P. H. M., "De conventu," p. 230). Niccolo della Tuccia, who reported the incident in "Cronache di Viterbo e di altre citta" (in *Cronache e statuti della Città di Viterbo* [Florence, 1872], p. 33), states, "Ma la Vergine misericordisissima, che sta dipinta nella capella di S. Anna.... Il fondatore di quella cappella fu messer Campana castellano di Viterbo." Della Tuccia must have made a mistake in designating Campanus as "castellano" and meant "capellano." There is no record of his being made "chaplain at Viterbo," although he died there.

Yet another reference to Campanus as dead occurs in the archives of the Cathedral at Viterbo (see Pietro Egidio, "L'Archivio della Cattedrale di Viterbo," *Bullettino dell' Istituto Storico Italiano per il Medio evo*, 1906, no. 423, pp. 300–301). In a document concerning a marriage settlement dated July 28, 1298, Campanus is described as "quondam," i.e., "the late": "Anno Domini M.CC.XCVIII., tempore Bonifatii VIII pp., mense iulii, die .XXVIII., ind. .XI. Symonectus qd. Iohannis olim de Navaria, familiaris qd. d. mag. Campani, ad petitionem d. Bonascie qd. Iannocii, sororis Verardi Nicole, confessus est se recepisse a Bonascia pro dote d. Bonascie, future uxoris sue,.CLX. lib. bon. den. papar." etc.

an "excellent and devout teacher...very learned in the science of astrology"[29] who corrected certain errors made by Ptolemy and wrote the *Computum Campani*, a work on the square (quadrant) which begins "Sciendum quod circulus solis habet duas...,"[30] and a work on the equation of the planets.[31] Symon also names two of the many disciples of Campanus, Eugidius, an astrologer at Milan, and Henry de la Forest, a doctor at Pavia, whom I have not further identified.[32] In the second mention of "Campanus," under the date 1269, Symon discusses "Master David Behen, surnamed Campanus," who was in the pay of Urban IV and worked especially on some astrological elections which were aimed to expel the Saracens from the papal territory.[33] Further, Symon says that Campanus foretold the victory of some Swiss over some nobles who wished to oppress them, made some tables for the year 1288,[34] wrote a means of equating

29. Wickersheimer, *Recueil...par Symon de Phares*, pp. 167–68. Since *astrologia* and *astronomia* are virtually interchangeable, "astronomy" may be a better translation.

30. This is the correct *incipit* for Campanus's work on the quadrant (cited in Lynn Thorndike and Pearl Kibre, *A Catalogue of Incipits of Mediaeval Scientific Writings in Latin*, rev. ed. [Cambridge, Mass., 1963], col. 1396).

31. No *incipit* is given, but the work is most likely the *Theorica planetarum*.

32. In view of the margin for error in much of Symon's material, a possible identification of Eugidius might be either Giles (Aegidius) of Lessines (ca. 1230–1304) (Sarton, *Introduction to the History of Science*, vol. 2, p. 946; Lynn Thorndike, *Latin Treatises on Comets between 1238 and 1368 A.D.* [Chicago, 1950], pp. 87–184) or Giles of Rome (ca. 1247–1316) (Sarton, *Introduction to the History of Science*, vol. 2, pp. 922–23). By dating Campanus so early, Symon might have confused Henry de la Forest with Jordanus Nemorarius ("of the wood," "of the grove") (ca. 1150–1237) (Sarton, *Introduction to the History of Science*, vol. 2, pp. 613–14), whose works Campanus seems to have known and may even have commented upon.

33. Wickersheimer, *Recueil...par Symon de Phares*, p. 202: "Environ ce temps fut en fleur à Romme maistre David Behen, surnommé Campanus. Cestui fut stipendié du pape Urbain et travailla grandement aux elections de astrologie, qui furent advisées et prinses pour expolser les Sarrazins de la terre de l'Eglise." Who this David Behen was with whom Phares confuses Campanus I do not know. Moritz Steinschneider ("Vite di matematici arabi tratte da un'opera incdita di Bernardino Baldi," *Bullettino di bibliografia e di storia delle scienze matematiche e fisiche*, vol. 5 [1872], p. 526) mentions in his notes on Jābir "un giudeo David rifiuto la teoria di Giabir avanti l'anno 1247." See also Steinschneider, "Karaitischen Handschriften," *Hebräische Bibliographie*, vol. 11 (1871), p. 43; also Guido Bonati, in his *De astronomia tractatus X* (Basel, 1550), col. 335 (folio), writes, "Illi autem qui fuerunt in tempore meo sicut fuit Hugo Abalugant, Beneguardinus Davidbam, Ioannes Papiensis...." (quoted by Baldassarre Boncompagni, "Della vita e delle opere di Gherardo Cremonese traduttore del secolo duodecimo e di Gherardo da Sabbionetta astronomo del secolo decimoterzo," *Atti dell'Academia pontificia de'Nuovi Lincei*, vol. 4 [1851], p. 449). It is possible that Phares was referring to Henry Bate of Malines, who is called Bethen in some manuscripts, although the first name would militate against that inference (G. Wallerand, *Henri Bate de Malines: Speculum divinorum et quorundam naturalium* [Louvain, 1931], p. 7, note 1; p. 21 and note 44).

34. See Zinner, *Verzeichnis*, note to no. 10866 on p. 510. The tables are probably Campanus's astronomical tables for Novara.

the planets which begins, "Clementissimo patri et piissimo domino Urbano quarto...,"[35] and which is "marvelously well calculated," and gave favorable astrological advice for the expulsion of some nobles who oppressed the people of Bern and Friberg.[36] It seems certain that Symon is referring to the same person in these lives, and his evidence would support the implications of the papal documents concerning the importance of Campanus. Another confirmation is the assessment of Campanus by Bernardino Baldi in his lives of Italian mathematicians: "He was a most famous astrologer in his own time."[37] Francesco Maurolico also cites Campanus as an astrologer.[38]

One more inference that may be drawn from the contemporary documents relates to the numerous instances of his being designated "Magister Campanus." The word "magister" was a title commonly used to describe a professor at one of the universities.[39] Whether Campanus was in fact a professor cannot be proved; the surviving university records from the thirteenth century do not reveal his name. Nonetheless, a possible confirmation of his university affiliation exists in his membership in a "remarkable scientific group," associated with the papal court during the third quarter of the thirteenth century, that included Moerbeke, Witelo, and Johannes Gervasius and perhaps even Thomas Aquinas.[40]

35. The *incipit* is correct for the dedicatory letter to the *Theorica planetarum* (see opening sentence of our text). That Symon read the *Theorica*, if only in part, seems likely from his remarks on Adam (Wickersheimer, *Recueil...par Symon de Phares*, p. 15): "en supposant que ung degre on ciel correspondre en terre 56 mille et demi 8ᵉ [?deux 3ᵉ], comme dit Campanus." Cf. our text, p. 146, line III.52, below.

36. Wickersheimer, *Recueil...par Symon de Phares*, p. 202. Probably the victories of the Swiss and of the peoples of Bern and Friberg are one and the same.

37. Baldi, "Vite inedite," p. 591. Using Campanus's relationship to Urban IV and his references to Robert Grosseteste (Ruperto Linconicie: Baldi, "Vite inedite," p. 595), Baldi, unlike many of his contemporaries, estimated essentially correct dates for Campanus.

38. Francesco Maurolico, "Scritti inediti (BN 7473, fols. 1–16)," ed. Federico Napoli, *Bullettino di bibliografia e di storia delle scienze matematiche e fisiche*, vol. 9 (1876), p. 37. On fol. 13r Maurolico says, "Itaque conflavi super astrorum judiciis quoddam isagogicum compendium ex...Campano [among about thirty others]...aliisque optimis quibusque authoribus congregatum." Maurolico also knew of Campanus's Euclid, *Sphere*, and *Computus maior*. Although he castigates the Euclid and in general accepts the opinion of Regiomontanus, he does say Campanus was most acute (ibid., p. 27: "Quod si perspicacissimi viri, qualem fuisse Campanum nulli dubium est...").

39. See indexes to Hastings Rashdall, *The Universities of Europe in the Middle Ages*, ed. F. M. Powicke and A. B. Emden (Oxford, 1936).

40. Marshall Clagett, *Archimedes in the Middle Ages*, vol. 1, *The Arabo-Latin Tradition* (Madison, Wis., 1964), p. 443, with note 16 citing Martin Grabmann, *Guglielmo di Moerbeke* (Rome, 1946), pp. 56–62.

Works

The works of Campanus of Novara other than the *Theorica planetarum* fall into three categories: those which are undoubtedly genuine, those which are doubtfully so, and those which are probably excerpts from one of the larger works. I shall arrange and discuss them briefly in that order. No attempt will be made to establish a chronological order of his works since the dates of composition for some of them can be only roughly approximate.[41] The earlier biographers list some six or seven works, but the index to the revised Thorndike and Kibre *Catalogue of Incipits*, representing greater attention to the surviving manuscripts and less dependence on printed editions, lists nineteen. (I shall henceforth indicate with the abbreviation TKr and a column number those works which appear in Thorndike and Kibre's *Catalogue*, revised edition.)

(1) Probably the best known and most widely disseminated work of Campanus was his commentary on Euclid's *Elements* (TKr, cols. 1391–92), which went through a possible fourteen printings.[42] Whether Campanus merely

41. See preceding section for some indication of the difficulties in establishing a chronology.

42. The editions and their descriptions are as follows. (i) 1st ed. Venice: Erhardus Ratdolt, 1482 (A. C. Klebs, "Incunabula scientifica et medica," *Osiris*, vol. 4 [1938], p. 134, no. 383.1; Margaret Bingham Stillwell, *Incunabula in American Libraries: A Second Census of Fifteenth-Century Books Owned in the United States, Mexico, and Canada* [New York, 1940], p. 190, no. E86; Tiraboschi, *Storia della letteratura*, vol. 4, p. 749: "Campani Novariensis Comment. in Euclidem. Venetiis, 1472 [*sic*] fol."; Jo. Albert Fabricius, *Bibliotheca latina mediae et infimae aetatis* [Hamburg, 1734–46], vol. 3, p. 97: same incorrect date cited; Pietro Riccardi, *Saggi di una bibliografia Euclidea* [Bologna, 1887], p. 12; Charles Thomas-Stanford, *Early Editions of Euclid's "Elements"* [London, 1926], p. 21). (ii) ?Ulm, 1486 (G. W. Panzer, *Annales typographici ab artis inventae origine ad annum MDXXXVI* [Nuremberg, 1793–1803], vol. 3, p. 536). (iii) Vicenza: Leonardus de Basilea and Gulielmus de Papia, 1491 (Klebs, "Incunabula scientifica," p. 134, no. 383.2; Stillwell, *Incunabula in American Libraries*, p. 190, no. E87; L. A. Cotta, *Museo Novarese* [Milan, 1701], p. 88; Riccardi, *Saggi di una bibliografia*, p. 12; Thomas-Stanford, *Early Editions*, p. 21). (iv) Venice: Joannes Tacuinus, 1505 (Riccardi, *Saggi di una bibliografia*, p. 12; Thomas-Stanford, *Early Editions*, p. 22). (v) Frankfort an der Oder: Ambrosius Lacher de Merspurg, 1506: bks. 1–4 only (Panzer, *Annales typographici*, vol. 7, p. 54, no. 1; Riccardi, *Saggi di una bibliografia*, p. 12; Thomas-Stanford, *Early Editions*, p. 49). (vi) Venice, 1508 (Panzer, *Annales typographici*, vol. 8, p. 375, no. 305; Cotta, *Museo Novarese*, p. 88). (vii) Venice: Paganinus de Paganinis, 1509 (Riccardi, *Saggi di una bibliografia*, p. 13; Thomas-Stanford, *Early Editions*, pp. 22–23). (viii) Venice: Joannes Tacuinus, 1510: reissue of (iv) (Riccardi, *Saggi di una bibliografia*, p. 13; Thomas-Stanford, *Early Editions*, p. 23). (ix) Paris: Henricus Stephanus, 1516 (Fabricius, *Bibliotheca latina*, vol. 3, p. 897; Riccardi, *Saggi di una bibliografia*, p. 13; Thomas-Stanford, *Early Editions*, pp. 6–7: considers 1517 as the more probable date). (x) ?Paris: Simon Colinaeus, 1521 (Riccardi, *Saggi di una bibliografia*, p. 14). (xi) Basel: Johannes Hervagius, 1537 (Cotta, *Museo Novarese*, p. 88; Conrad Gesner, *Bibliotheca universalis, sive catalogus*... [Zürich, 1545], vol. 1, p. 161, and in augmented edition [1583], p. 134, col. 2; Fabricius, *Bibliotheca*

added a commentary to the translation of Adelard of Bath, or revised Adelard's translation in addition to writing the commentary, or composed a new translation with commentary is still a debatable question. The older literature suggests that Campanus had considerable independence from the work of Adelard, but the manuscript evidence is doubtful and disputed,[43] and perhaps it would be better to leave the question open. The date of the work lies between 1255 and 1259 if the colophon of the Plimpton manuscript is correct in ascribing its dedication to Jacques Pantaléon, Patriarch of Jerusalem (later Pope Urban IV).[44]

(2) The *Computus maior* (TKr, cols. 1365, 243–44), a calendrical treatise, was printed twice in the sixteenth century.[45] It is dated to 1268 by a computation of the exact position of the "caput arietis" in Campanus's time according to the theory of Ṯābit's *De motu octaue spere*.[46] Shortened versions of it are found

---

latina, vol. 3, p. 897; Riccardi, *Saggi di una bibliografia*, p. 16; Thomas-Stanford, *Early Editions*, p. 25). (xii) Basel: Johannes Hervagius, 1546 (Fabricius, *Bibliotheca latina*, vol. 3, p. 897; Riccardi, *Saggi di una bibliografia*, p. 17; Thomas-Stanford, *Early Editions*, p. 26). (xiii) ?Basel: Johannes Hervagius, 1557 (Riccardi, *Saggi di una bibliografia*, pp. 19–20). (xiv) Basel: Johannes Hervagius and Bernhardus Brand, 1558 (Tiraboschi, *Storia della letteratura*, vol. 4, p. 749; Fabricius, *Bibliotheca latina*, vol. 3, p. 897; Riccardi, *Saggi di una bibliografia*, p. 20; Thomas-Stanford, *Early Editions*, pp. 27–28).

43. E.g., H. Weissenborn, "Die Übersetzungen des Euklid aus dem Arabischen in das Lateinische durch Adelard von Bath nach zwei Handschriften der königliche Bibliothek in Erfurt," *Zeitschrift für Mathematik und Physik*, hist. sec. vol. 25 (1880), suppl., pp. 141–66; Weissenborn, *Die Übersetzungen des Euklid durch Campano und Zamberti* (Halle, 1882); H. Suter, "Die Mathematiker und Astronomen der Araber und ihre Werke," *Abhandlungen zur Geschichte der mathematischen Wissenschaften mit Einschluss ihrer Anwendungen*, vol. 10 (1900). For the manuscripts, cf. MS Oxford, Corpus Christi College, 234 (15th cent.; 172 fols.): "Euclidis Geometriae Elementorum libri quindecim, cum Theonis commentariis a Campano Latine uersis"; and MS Boncompagni, 298 (end of 16th cent.; 169 fols.) (E. Narducci, *Catalogo di manoscritti ora posseduti da D. Baldassarre Boncompagni*, 2d ed. [Rome, 1892]; now sold and provenance unknown): "Codex XV nitidissimus et fortasse unicus continet Euclidis Elementa mathematica in Libros XV distributa e Greco in Latinum a Hieronymo Campano versa...." The manuscripts are too recent to carry any weight as evidence; the above citation from the second manuscript is, to my knowledge, the only instance in which Campanus is given the forename Hieronymus. See Marshall Clagett, "The Medieval Latin Translation of the *Elements* of Euclid, with Special Emphasis on the Versions of Adelard of Bath," *Isis*, vol. 44 (1953), pp. 16–42, and Clagett, *Archimedes*, vol. 1, *passim*.

44. See above, notes 11 and 13.

45. The printed editions of the *Computus maior* (both accompanied by the *Tractatus de sphera*) are *Sphera mundi nouiter recognita cum commentariis 7 authoribus*, Venice: L. A. de Giunta, 1518; and Venice: Octavianus Scotus, 1518.

46. Campanus's citation of the position appears as follows in MS Florence, Biblioteca Riccardiana, 885 (herein designated *R*), fol. 263v: "Iam autem in hoc nostro tempore factum est capud arietis septentrionale distans a communi sectione parui circuli et equatoris de partibus ipsius parui circuli uersus septentrionem 58 gradibus et 57 minutis fere et propter hoc ipsum capud arietis

in manuscript,[47] and we can deduce from references by Campanus to "computo nostro maiori" that he himself published an epitome.[48] A calendar which does not appear in the printed work accompanies some manuscripts.[49]

(3) The *Tractatus de sphera* (TKr, col. 1770; *incipits* in cols. 302, 1152) was written after the *Computus* and *Theorica*, both of which it cites.[50] It was printed several times in the sixteenth century.[51] One fifteenth-century manuscript at the Bibliothèque Nationale, Paris, erroneously ascribes to Campanus the commentary of Robertus Anglicus on the *Sphere* of John of Sacrobosco.[52]

(4) A *De quadratura circuli* (TKr, col. 136) is attested by Albert of Saxony, a century later, as being a work written by Campanus.[53] It has been printed several times, most recently by Marshall Clagett, who discusses it competently.[54]

(Note 46 continued)
mobilis distat a communi sectione orbis signorum mobilis et equatoris uersus septentrionem de partibus ipsius orbis signorum 9 gradibus 11 minutis et 45 secundis fere." Using the tables in the *De motu* of Ṯābit (or Thebit), one finds that the above figures are derivable from a date in February 1268; however, the date as determined from Campanus's own adaptation of the Toledan tables is March 1268.

47. TKr, index, col. 1770, lists three *incipits* other than the main one (in TKr, col. 1365, for "CUmc" read "CUma").

48. See his letter to Raner da Todi, MS Florence, Biblioteca Nazionale, conv. soppr. J.X.40, fol. 46r, and his *Sphera*, MS Venice, Biblioteca Marciana, VIII, 69, fol. 32v.

49. E.g., MS London, British Museum, Additional 38688, and MS Brussels, Bibliothèque Royale, 1022–47, fol. 148r. A Prague manuscript preserves a calendar derived "ex tabulis Nouariensibus," followed by an incomplete canon, but this seems more likely to have been compiled from the *Computus*: Prague, Universitní knihovna, XIII.C.17 (2292) (14th and 15th cents.), fols. 77r[a]–83v[a]. ([83r[a]] "...iste kalendarius extractus est ex tabulis Nouariensibus quas fecit magister Campanus et est constitutum super meridiem monasterii Riddageshusen prope ciuitatem Brunswik...").

50. See printed edition, Venice: L. A. de Giunta, 1518: fol. 153v[b], chap. 12 (citation of *Computus*); and fol. 154r[a], chap. 13 (citation of *Theorica*). See Duhem, *Système*, vol. 3, pp. 321–22; Lynn Thorndike, ed., *The "Sphere" of Sacrobosco and Its Commentators* (Chicago, 1949), pp. 26–28.

51. Printed editions of the *Tractatus de sphera* are the following (note that the first two were printed together with the *Computus maior*). (i) Venice: L. A. de Giunta, 1518. (ii) Venice: Octavianus Scotus, 1518. (iii) Venice: L. A. de Giunta, 1531, (iv) Mexico: Ioannes Paulus Brissensis, 1557, in *Phisica Speculatio Aedita per R. P. F. Alphonsum a Vera Cruce* (J. G. Icazbalceta, *Bibliografia Mexicana del Siglo XVI* [Mexico, 1954], p. 137; TKr, col. 1152, where the date is mistakenly given as 1554).

52. BN, fonds latin, 10268 (15th cent.), fols. 23r–47v. This commentary is edited in Thorndike, *"Sphere" of Sacrobosco*, pp. 143–98.

53. Sarton, *Introduction to the History of Science*, vol. 2, p. 986.

54. Clagett, *Archimedes*, vol. 1, pp. 581–609, 719. Printed editions of the *De quadratura circuli* are the following. (i) Paris: Guido Mercator, 1495, in Thomas Bradwardine, *Geometria speculativa* (*Gesamtkatalog der Wiegendrucke*, vol. 4, col. 607, no. 5002; Klebs, "Incunabula scientifica," p. 87, no. 208.1; Stillwell, *Incunabula in American Libraries*, p. 110, no. B954; Clagett, *Archimedes*, vol. 1, pp. 586–87). (ii) Venice: Ioan. Bapt. Sessa, 1503 (Cotta, *Museo Novarese*, p. 88: "operetta data in luce da Luca Gaurico, e referita nell'Append. della Margaritta

(5) The *De quadrante* (TKr, cols. 396, 1396, 1405, 1468) was a short popular work, which seems to have some affinities with the *Vetus quadrans*, usually ascribed to Robertus Anglicus. The *incipit* of *De quadrante* is usually "Scire debes" or "Debes scire" (TKr, cols. 396, 1405, and 1770). Where it is not ascribed to Campanus in the manuscripts or appears without ascription it is occasionally assigned to John of Sacrobosco.[55] Daunou claimed that it was printed in *Margarita philosophiae*,[56] but I have not discovered it in any edition of the *Margarita* I have consulted. I have remarked at least forty manuscripts containing this *De quadrante*.

(6) A work on the astrolabe (TKr, col. 1583; *incipit*: "Tres circulos in astrolapsu descriptos...") is also ascribed to Campanus, of which some ten manuscripts remain extant. No printed edition has been discovered.

(7) Campanus also constructed astronomical tables (TKr, cols, 241, 486, 1287) for the meridian of Novara. They were based on the Toledan tables of az-Zarqāl (Arzachel),[57] and have not been printed. We have used the manuscript copy of Campanus's tables that is in Trinity College Library, Dublin, D.4.30, fols. 34v–101r, for comparison with the *Theorica*. Campanus's tables

---

Filosof. e da me veduta presso dell'eruditissimo Semenzi"; Clagett, *Archimedes*, vol. 1, p. 587). (iii) Paris: Volphgangus hopilius et Henricus stephanus, 1503 (Panzer, *Annales typographici*, vol. 7, p. 505, no. 44; Clagett, *Archimedes*, vol. 1, p. 587). (iv) 1507, in Johannes Caesarius Juliacensis, *De quadratura circuli demonstratio ex Campano* (Clagett, *Archimedes*, vol. 1, p. 587). (v) Paris: Jean Petit (with Bradwardine), 1511. (vi) Strasbourg, 1515, in Gregor Reisch, *Margarita philosophica nova...*, ed. Johann Grüniger (Strasbourg, 1515) (Cotta, *Museo Novarese*, p. 88; Fabricius, *Bibliotheca latina*, vol. 3, p. 897; Tiraboschi, *Storia della letteratura*, vol. 4, p. 749, none of whom date the edition), and reprinted in the 1535 and 1583 Basel editions of the *Margarita*. (vii) Paris: Reginald Chandière (with Bradwardine), 1530. (viii) Lyons: G. Rouille, 1559, in Ioan. Buteonis, *De quadratura circuli libri duo*. (ix) Clagett, *Archimedes*, vol. 1, pp. 588–607. Manuscripts other than those cited by Clagett are as follows. (i) Oxford, Bodleian Library, Savile 55 (26123) (1451–54, latin), no. 6, fols. 69–71: "Quadratura circuli edita a [Johanne] Campano [de Novaria]" (included among the writings of Nicholas of Cusa). (ii) Boncompagni, 104 (15th cent.), fols. 35–36 (Narducci, *Catalogo di manoscritti*; now sold and provenance unknown). (iii) New York, Columbia University Library, Plimpton 180 (15th cent.), fols. 27v–31 (formerly Boncompagni, 176 [Narducci, *Catalogo di manoscritti*]). (iv) New York, Columbia University Library, Smith Add. 36 (Western MS 1) (14th cent.), fols. 138v[a]–39r[b] (formerly Voynich 10 [De Ricci, *Census of Medieval and Renaissance Manuscripts in the United States*, vol. 2, p. 1847]). (v) Seitenstetten, Stiftsbibliothek, LXXVII (14th cent.), no. 9 (MS missing).

55. See Thorndike, "*Sphere*" *of Sacrobosco*, p. 3, note 12. The *Vetus quadrans* has been edited by Tannery in Robertus Anglicus, "Le Traité du quadrant," ed. Paul Tannery, *Notices et extraits des manuscrits de la Bibliothèque nationale*, vol. 35 (1897), pp. 561–640; and an edition from most of the manuscripts is now being prepared as a Ph.D. dissertation by Mrs. Nan Britt, one of my graduate students.

56. Daunou, "Campanus de Novarre," p. 252.

57. See above, notes 1, 4, and 14.

for finding the new and full moon probably are excerpted from these astronomical tables for Novara; similarly a set of tables for converting Christian years into Arabic years is probably excerpted from his *Computus*.[58] Another possible date—namely, 1254—for these tables, besides those dates already suggested, may be inferred from MS Rome, Biblioteca Apostolica Vaticana, lat. 3118 (15th cent.), fol. 5v, supralinear.[59]

58. The manuscript preservation of these tables is extremely involved, since they occur in various arrangements. However, there are three canons which appear irregularly with the various arrangements of the tables and which I will attempt to elucidate. The first, beginning "Composui hanc tabulam ad inueniendum diem..." (TKr, col. 241, three variants, and TKr, col. 486 [Florence, Biblioteca Medicea Laurenziana, San Marco 194, 14th cent., fol. 14v]) may confidently be ascribed to Campanus, since it cites the meridian of Novara (the *explicit*, however, cites Paris). The second, "Vt autem annos arabum et menses et per consequens etatem lune ueram..." (TKr, cols. 1613–14; Montpellier, Bibliothèque Municipale, 323 [13th–14th cent.], fols. 157r^a–v^b)/"...fuerint 4 pro illis abiectis addendus est diebus unus dies et tunc minue quod debes et patebit quesitum," is an adaptation from the Toledan tables canon: "Vt autem annos arabum per hanc tabulam sequentem..." (TKr, col. 1614; Madrid, Real Biblioteca del Escorial, O.II.10 [14th cent.], fol. 98r)/"...ex 365 diebus et quarta, menses uero trigenarii 12 non continent ultra 360 dies." In several cases this second canon is followed by considerable additional material. The third canon, "Quia uero perutile est scire annos et menses arabum idcirco ad hoc feci tabulam sequentem qua leuiter reperiuntur..." (TKr, col. 1234; Munich, Bayerische Staatsbibliothek, CLM 17703 [13th cent.], fols. 30r–30v)/"...qui supersunt dies menses inperfecti iam transacti de lunatione in qua sumus," agrees only superficially with the passage in the *Computus* of Campanus, "Quoniam autem perutile est scire annos et menses arabum predictos eo quod prima die cuiusque mensis ipsorum..." (Berlin, Staatsbibliothek, latin Q.455 [14th cent.], fol. 33r)/"...in secunda uero posui dies agregatos quibusque mensibus e directo respondentes et hee sunt tabule tres predicte." All the manuscripts listed in TKr, col. 1234, belong to the canon. Other manuscripts are: Bologna, Biblioteca Universitaria, 2408 (1225) (14th cent.), fol. 13v^a; Florence, Biblioteca Nazionale, II.II.67 (15th cent.), fol. 97r; Oxford, Bodleian Library, Ashmole 393 (15th cent.), fol. 60r^b; Oxford, Bodleian Library, Bodley 432 (15th cent.), fol. 33r^a; Paris, Bibliothèque Nationale, fonds latin, 7421 (14th cent.), fol. 209r. Additional manuscripts of the second canon are: Berlin, Staatsbibliothek, latin F.610 (14th cent.), 159r^a (incomplete); Berlin, Staatsbibliothek, latin Q.33 (14th cent.), fol. 33r^a; Florence, Biblioteca Nazionale, II.III.24 (14th cent.), fol. 243v; London, British Museum, Harley 13 (13th cent.), fol. 30v^{ab}; Oxford, Bodleian Library, Ashmole 393 (15th cent.), fol. 60v; Oxford, Bodleian Library, Bodley 432 (15th cent.), fol. 33v^a; Paris, Bibliothèque Nationale, fonds latin, 7416B (13th–14th cent.), fol. 67r; Paris, Bibliothèque Nationale, fonds latin, 7421 (14th cent.), fol. 118r; Rome, Biblioteca Apostolica Vaticana, latin 3133 (13th–14th cent.), fol. 28r^{ab}; Rome, Biblioteca Apostolica Vaticana, Palatinus 1414 (13th cent.), fol. 152v^b.

59. I have examined a microfilm of Vat. lat. 3118, and the title of the table on fol. 5v reads as follows: "[Heading] mccliiii 1254 [this indicates that 1254 is the first year of a nineteen-year lunar cycle; the next cycle begins on fol. 6r with 1273, and so on]. Hec tabula composita fuit a magistro Campana [*sic*] de Nauaria et cum [?]perpaulum completa fuerit a capite incipiat et dies incepit a meridie et inuenitur per eam quo die qua hora quo minuto sit coniunctio solis et lune secundum hebreos. [Heading of table] Almanach coniunctionum solis et lune mediarum

(8) An untitled astrological work (TKr, col. 1078) exists in three different manuscripts.[60] Only the Palatine manuscript ascribes this work to Campanus, and that in a marginal title and in a different hand: "Tractatus Campani." The Paris manuscript has no title; the Vienna copy reads merely, "Tractatus de astrologia iudicaria."[61] There is a curious similarity in its *incipit* to that of the often copied and printed familiar astrological work of Alchabitius (al-Qābīṣī), *Liber introductorius*,[62] but the resemblance must remain a curiosity for the time being.

(9) and (10) The next two works actually accompany two other treatises, the letter (TKr, col. 1682) prefixed to Simon of Genoa's *Synonyma medicinae*, which was dedicated to Campanus, and the *Epistola responsiua* (TKr, col. 843) attached to the treatise of Raner da Todi. Both works have been discussed elsewhere.[63]

(11), (12), and (13) The final three works that may have been written by Campanus are, like the first, commentaries on earlier works: the *De sphera* of Menelaus, the *De sphera* of Theodosius, and the *Planispherium* of Jordanus Nemorarius.[64] Björnbo and Vogl say that the commentary of Campanus on

---

ad primum circulum lune post meridiem nouarie." Below the table is another note in the same hand, and below that in a different hand: "incipiendo debent detrahi hore 5 minuta 47. Videtur quod hec tabula inceperit 1235." On fol. 1r occurs the first table entitled, "Tabule ad inueniendum annos mundi cum annis christi." This has dates: 76, 152, 228,..., 1520; beside 1292 there is a cross in the margin, which may signify that these tables belong to the *Almanach* of William of St. Cloud, for which 1292 is the beginning date. See Duhem, *Système*, vol. 4, pp. 10–19; Sarton, *Introduction to the History of Science*, vol. 2, p. 991. On the relationship of Campanus and William, see Duhem, *Système*, vol. 3, p. 326.

60. MS Paris, Bibliothèque Nationale, fonds latin, 7342 (14th cent.), fols. $1r^a$–$58r^b$: "Postulata a domino artis signorum ueritate..."; MS Rome, Biblioteca Apostolica Vaticana, Palatinus 1363 (15th cent.), fols. 66r–88r; MS Vienna, Nationalbibliothek, 5327 (15th cent.), fols. 14r–54v. The ?Campanus work has been edited by Catherine Sue Yielding as "An Edition of *De proprietatibus duodecim signorum*," M.A. thesis, Emory University, Atlanta, Georgia, 1967 (available in typescript in the Emory University Library).

61. Zinner, *Verzeichnis*, p. 114, no. 3342, cites only the Vienna copy.

62. "Postulata a domino prolixitate vite...." See F. J. Carmody, *Arabic Astronomical and Astrological Sciences in Latin Translation: A Critical Bibliography* (Berkeley, 1956), pp. 144–49.

63. See above, notes 23 and 25.

64. TKr lists no commentary for Menelaus, one anonymous commentary for Jordanus, and, in col. 515 (but not in the index), a commentary by Campanus from MS Venice, Biblioteca Marciana, VIII, 32 (14th cent.), fols. 1r–35r. An unexhaustive survey of the catalogues yields only the following: MS Cracow, Biblioteka Jagiellońska, 568 (DD.III.24) (1465), fol. 134, where, at the conclusion of Euclid, the scribe writes, "...de qua sequitur liber Theodosii Mylei [Menelai] uel Campani De speris..."; MS Rome, Biblioteca Apostolica Vaticana, lat. 3380, fol. $1r^a$, *rubric.*: "Theodosius de speris" (handwritten catalogue adds "cum commento Campani" to title).

the *Planispherium* was printed in 1536 by Jacob Ziegler.[65] I have examined this work and found no direct mention of Campanus; any indirect inferences might be highly misleading.

Ascription of the following works to Campanus is dubious although some of the geometrical and arithmetical works may be excerpts from the Euclid.

(14) Thorndike and Kibre list six different *incipits* for a work entitled *De figura sectore* (TKr, cols. 280, 630). All of them, either in the manuscripts or in the writings of later historians, have become attached to the name of Campanus,[66] but only one seems even possibly to be of Campanus's authorship.[67] This last version has been printed three times.[68]

(15) Another work attributed to Campanus is a *De proportione et proportionalitate* (TKr, col. 1139), which Carmody assigns to Ṯābit ibn Qurra.[69] It may possibly be an excerpt from Book V of the Euclid. It is not the *De proportione* printed in Schöner's edition of the *Algorithmus demonstratus* in 1534 as has been suggested.[70]

65. A. A. Björnbo and Sebastian Vogl, "Alkindi, Tideus und Pseudo-Euklid, Drei optische Werke," *Abhandlungen zur Geschichte der mathematischen Wissenschaften mit Einschluss ihrer Anwendungen*, vol. 26 (1912), p. 126, note 5: "*Sphaerae atque astrorum coelestium ratio, natura, & motus: ad totius mundi fabricationis cognitionem fundamenta*, Valderus (Basel?), 1536, 275–94 (mm 2r–oo 3v) *Iordanus de planisphaerii figuratione*."2d ed., Venice (1558), cf. A. A. Björnbo, "Studien über Menelaos' Sphärik," *Abhandlungen zur Geschichte der mathematischen Wissenschaften mit Einschluss ihrer Anwendungen*, vol. 14 (1902), pp. 149, 151–54.

66. "Continuet [Continet] deus conseruationem...," Ṯābit ibn Qurra (E. A. Moody and Marshall Clagett, eds., *The Medieval Science of Weights: Treatises Ascribed to Euclid, Archimedes, Thabit ibn Qurra, Jordanus de Nemore, and Blasius of Parma* [Madison, Wis., 1952], pp. 79ff.); "Cum cuiuslibet gradus...," az-Zarqāl, Canons on the Toledan tables (agrees with MS Madrid, Biblioteca del Escorial, O.II.10, fol. 99r, and on fol. 99v, "Kardaga est portio circuli..."); "Intellexi quod dixisti..." (listed twice in TKr, cols. 754–55, both times ascribed to Ṯābit; cf. Carmody, *Arabic Astronomical and Astrological Sciences*, pp. 122–24); "Quod de figura que nominatur...," Ṯābit (cf. Carmody, *Arabic Astronomical and Astrological Sciences*, pp. 122–23).

67. "Cum aliquis semicirculus diuiditur..." (TKr, col. 280). Some of the manuscripts ascribe this rendition to Campanus, but others to Tract 4 of the *Quadripartitum* of Richard of Wallingford.

68. Printed editions: (i) Venice: L. A. de Giunta, 1518. (ii) Venice: Octavianus Scotus, 1518. (iii) Rome: appendix 2, pp. 36–37 of "Intorno a Nasawi ed Abu Sahl el-Kuhi matematici arabi commentatori del Liber assumptorum attribuito ad Archimede, Lettera III di M. Steinschneider a D. B. Boncompagni," in *Intorno ad alcuni matematici del medio evo ed alle opere da essi composte, lettere di Maurizio Steinschneider a D. B. Boncompagni* (Rome, 1863; date corrected in pencil to 1867 in the copy at Widener Library, Harvard University): reprint of Scotus, 1518, edition.

69. Carmody, *Arabic Astronomical and Astrological Sciences*, pp. 127–28. He also admits its ascription in the manuscripts to Campanus and Alkindi, and points out the possible relationship of this work to a work with similar title but different *incipit* by Ahmad b. Yusuf (cf. Carmody, *Arabic Astronomical and Astrological Sciences*, pp. 130–31).

70. Pierre Duhem, "Sur l'*Algorithmus de-*

(16) The *Almagesti minoris libri VI* or *abbreuiati* or *parui* (TKr, col. 1006) exists in at least twenty manuscripts, one of which is ascribed to Campanus.[71] Other ascriptions are to Thomas Aquinas, Jābir (Geber), al-Battānī (Albategnius), Albertus Magnus, or anonymous. That it occurs in four manuscripts along with the *Theorica* is no assurance of Campanus's authorship.

(17) The *De solida sphera* (TKr, col. 1567) is ascribed to Campanus only in the printed edition.[72] In those manuscripts which date it, the year 1303 is usually given.[73] The ascription of authorship in the manuscripts is fairly evenly divided between Johannes de Harlebeke and Accursius of Parma.

(18) A slight work entitled *De dispositione aeris* (TKr, cols. 57, 293) is commonly assigned to Grosseteste or al-Farġānī (Alfraganus), but is ascribed to Campanus in at least two manuscripts.[74] Thorndike discusses the possibility that this work is the basis of the sixth chapter of the *Exafrenon pronosticorum temporis*, a work on a similar subject (astrological weather prediction) of uncertain date and authorship.[75] Both the *De dispositione aeris* and the sixth chapter of the *Exafrenon* contain calculations for the years 1249 and 1255, dates which are consistent with the authorship of Campanus.

(19) Daunou mentions a *Breviloquium duodecim signorum zodiaci cum peculiari modo erigendi thematis coelestis per divisionem verticalis primarii*, of presumed Campanus authorship.[76] He indicates that Fabricius, Weidler, and Marchand all attest to its being in a certain library.[77] Apparently it is the same

monstratus," *Bibliotheca mathematica*, ser. 3, vol. 6 (1905), p. 14. The *De proportione* of the *Algorithmus* begins (ed. Schöner [1534], sig. H.[iv]v), "De proportionibus appendix. Datis extremis duobus media inter eos...." Cf. TKr, col. 363, citing a fifteenth-century manuscript, Vienna, Nationalbibliothek, 5203, fols. 136r–37r.

71. MS Dresden, Sächsische Landesbibliothek, Db. 87 (15th cent.), fols. 104–62.

72. Cf. Daunou, "Campanus de Novarre," pp. 252–53; the printed edition of the *Tractatus* of Venice: L. A. de Giunta, 1531, where "*Campani compendium super tractatu de sphera*" is followed by "*eiusdem tractatulus de modo fabricandi spheram solidam.*" The latter has the usual *incipit* (TKr, col. 1576), "Totius astrologie speculationis radix...."

73. The two manuscripts dated 1300 probably represent a scribal omission of the digit 3. One manuscript, Berlin, Staatsbibliothek, latin Q.581 (15th cent.), which dates it 1301, also assigns it uniquely to Stephanus Arlandi.

74. Printed by Ludwig Baur among the works of Robert Grosseteste (*Die philosophischen Werke des Robert Grosseteste, Bischofs von Lincoln*, ed. Ludwig Baur [Münster, 1912], pp. 41–51); an expanded list of manuscripts of the *De dispositione aeris* is given in S. H. Thomson, *The Writings of Robert Grosseteste, Bishop of Lincoln 1235–1253* (Cambridge, 1940), pp. 103–4. For ascription to al-Farġānī, see Carmody, *Arabic Astronomical and Astrological Sciences*, p. 116. The manuscripts that assign the work to Campanus are London, British Museum, Sloane 332 (15th cent.), fols. 16r–v: "*Practica campani de dispositione aeris*"; Oxford, Bodleian Library, Ashmole 345 (later 14th cent.), fols. 23v–24: "*Pronosticacio Campani.*"

75. Thorndike, *History of Magic*, vol. 3, pp. 124–35.

76. Daunou, "Campanus de Novarre," p. 253.

77. Ibid.: "chez M. Drosser. Prosper Marchand (*Dictionaire historique, ou mémoi-*

work that Regiomontanus used in his *Tabulae directionum* (concerned with tables for calculating the limits of the houses and the rules for their use).[78] The *Breuiloquium* is also cited by the sixteenth-century Frenchman Orontius Finaeus and by Tycho Brahe.[79] It is undoubtedly either an adaptation of the chapter *De duodecim domibus que sunt in celo* from Campanus's *Tractatus de*

(Note 77 continued)
res critiques et littéraires [The Hague, 1758], p. 147) says of this *Breuiloquium*, "quem Gazulus, Raguzaeus Astrologus, qui anno 1438 floruit, secutus est." Fabricius, *Bibliotheca latina*, vol. 3, pp. 897–98. J. F. Weidler (*Historia astronomiae sive de ortu et progressu astronomiae* [Wittenberg, 1741], p. 274), who dates Campanus as of 1030, says, "Excogitavit etiam peculiarem modum erigendi thematis coelestis, per divisionem verticalis primarii, quem Gazulus secutus est," but mentions no library. Gesner (*Bibliotheca universalis*, augmented ed., p. 134, col. 2) writes, "Breviloquium 12 signorum Zodiaci, manuscriptum apud M. Dresserum," but that entry does not appear in Gesner's edition of 1545.

78. Ernst Zinner, *Leben und Wirken des Johannes Müller von Königsberg genannt Regiomontanus* (Munich, 1938), pp. 106–7. See Thorndike, *History of Magic*, vol. 6, pp. 103–4. J. B. J. Delambre (*Histoire de l'astronomie du Moyen Age* [Paris, 1819], pp. 288–90) says Regiomontanus distinguishes three different manners of dividing the sky into twelve houses, the second of which Campanus follows, the third Regiomontanus. Likewise, Regiomontanus listed in his announcement for publication a work entitled *De distinctione domiciliorum caeli contra Campanum et Joannem Gazulum Ragusinum: cuius et alia de horis temporalibus decreta ibidem retractantur*, which was never published according to Zinner (*Leben und Wirken*, pp. 106–7, 246); but the work was published as an addendum to Schöner's *Tabulae resolutae* in 1536 (Thorndike, *History of Magic*, vol. 5, p. 359 and note 127; see also pp. 336, 360). I have seen a microfilm of this addendum to Schöner's book, which is in the University of Michigan Library; beginning on sig. V.[1]r and running to the end of the book, it is entitled *De aequationibus duodecim domorum celi.* On sig. V.2r Schöner says that Orontius in his *Sphere* "hunc modum rationalem aequationis domorum coeli plane reijcit, conaturque Campani atque Gazuli modum in lucem rursus asserere, qui modus iam annos ante multos, a Regiomontano nostro iure meritissimo est improbatus." He then continues onto sig. V.2v with a eulogy of Regiomontanus. Sigs. V.2v through V.[4]r contain excerpts from Regiomontanus, chiefly stating the three theories of Ptolemy, Campanus and Gazulus, and Regiomontanus. On sig. V.[4]v is the title *Figuram coeli secundum hunc modum rationalem M. Ioannis de Monteregio, ad quodcunque tempus erigere*, followed by a series of tables concluding the work. I cannot say, without comparing this edition with the manuscript, if it still exists, whether the addendum represents the whole of Regiomontanus's work or just excerpts published by Schöner.

79. Orontii Finaei Delphinatis, regii mathematicarum Lutetiae professoris, *In eos quos de Mundi sphaera conscripsit libros, ac in Planetarum theoricas, Canonum Astronomicorum Libri II* (Paris: Michael Vascosanus, 1553), bk. 1, canon 11, fol. 17r: "Arcus igitur circuli verticalis, inter meridianum & datum positionis circulum comprehensus, iuxta rationalem domificandi modum, quem una cum Campano Novariensi, multis nominibus, vel argumentis, imitari compellimur (de quibus amplissimam conscripsimus digressionem) est graduum 30: cuius sinus rectus est partium itidem 30." Tycho Brahe (Cotta, *Museo Novarese*, p. 88) cites, "IX. *Breviloquium* duodecim signorum Zodiaci MS" (cf. Tycho Brahe, *Opera omnia*, ed. J. L. E. Dreyer [Copenhagen, 1913–29], vol. 8, p. 217), and incidentally, in his list of authorities, mentions a Dresser.

sphera (ed. Venice, 1518, chap. 29, fols. 155v[b]–56r[a]; see item (3), above), perhaps by Johannes Gazulus de Ragusa, as the manuscripts I have discovered indicate;[80] or else simply that chapter of the *Tractatus de sphera* itself (see also item (28), below).

80. MS Rome, Biblioteca Apostolica Vaticana, Palatinus 1375 (15th cent.), fols. 171v–76r[b] and 268v–70r: "[*Rubric.*] Tabula domorum secundum Campanum et Gasulum [halfway down right col., *rubric.*] Introytus in arietem et alia signa cardinalia anno domini 1468 ad meridiem Cracouiensem ubi est eleuatio poli 50 gradus) [fol. 172r] Tabula ad sciendum per lunam quanta est mora [fol. 172v] creature in utero matris et qua hora fuerit con [fol. 173r] cepta et qua exit de utero matris. [These tables end fol. 174v. Fol. 175r, star tables by constellations. Fol. 176r[b], *rubric.*] Deus laus et honor. Finis stellarum fixarum anno domini 1488 in studio Cracouiensi per me Baccalaureum Io. de Hasfurth. [Fol. 268v, *rubric.*] Modus equandi domos celi secundum opinionem Campani et Gazaly. [Bottom mg.] Hec figuratio ostendit equationem domorum quam posuit Campanus quem etiam sequutus est magister Joannes Gazulus Kausiensis in tractatu suo de directionibus et fit hec equatio tali modo inter circulos qui uadunt per uerticem capitis qui dicuntur azimut est unus medius qui extenditur per ortum et occasum equinoctialis hunc circulum subdiuidunt in 12 partes equales et per illas diuisiones ymaginantur extendi sex magnos circulos in ambabus intersectionibus meridiani cum orizonte concurentes[?] inter quos circulos meridianus et orizon sunt principales. Hy sex circuli semper sunt fixi et equalia spatia celi unaquaque domus continet. Sex etiam domus tote et integre sunt super terram et sex integre sub terra sunt quod non fit secundum primam equationem erroneam ubi domus 12 sunt diuise ab oriente sitque omnes[?] 12 secundum aliquam partem sunt super terram et econtrarium[?] secundum alteram sunt sub terra quod est contra omnium uetustissimorum astrologorum oppinionem. [Fol. 269r, *rubric.*] Modus equandi domos secundum modum rationabilem. [Fol. 269v] Incipit liber de spiritibus inclusis [ends fol. 270r; this may not be part of the work]." Also MS Paris, Bibliothèque Nationale, fonds latin, 10265 (15th cent.), fol. 262v: "[*Rubric.*] Tabula domorum secundum Campanum et Gazulum [first four cols.]. Tabula domorum regionalis secundum Johannem de Monte Regio [fifth and sixth cols.]." MS Karlsruhe, Badische Landesbibliothek, Rastatt 36 (15th–16th cent.), fols. 150v–51r: Two figures illustrating methods of division of the sphere into the twelve houses. Above the figure on fol. 151r: "Modus equandi domus caeli secundum Campani et Gasoli opinionem"; and below: "Hec figuratio ostendit equationem domorum quam posuit Campanus quam etiam secutus est magister iohannes Gasulus ragusiensis in tractatu suo de directionibus et fit hec equatio tali modo inter circulos qui uadunt per uerticem capitis qui azimut dicuntur est unus medius qui extenditur per ortum et occasum equinoxialis hunc circulum subdiuidunt in 12 partes equales et per illas diuisiones ymaginantur extendi magnos circulos in ambabus intersectionibus meridiei cum orizonte concurrentes: inter quos circulos meridianus et orizon sunt principales. Hii 6 circuli semper sunt fixi et equalia celi spacia unaquaque domus continet. Sex etiam domus tote et integre sunt supra terram et 6 integre sub terra quod non fit secundum primam equationem eroneam ubi domus 12 sunt diuise orizonte et sic quod omnes 12 secundum aliquam partem sunt supra terram et eedem secundum alteram sub terra quod est contra omnium uetustissimorum astrologorum opinionem etc." Fol. 151v has a figure entitled, "Pro dyametris solis et lune et umbre uisualibus." Fol. 152r, a figure entitled, "pro latitudine trium superiorum et cetera." Fol. 152v, top, a figure, "pro latitudine ueneris," and bottom, "pro latitudine mercurii." Fol. 153r, a figure with the inscription, "pro illo textu triplex est ratio cur luna post coniunc-

(20) Daunou conjectures that an article entitled *Nonnulla astrologica*, which he says that Fabricius, Weidler, and Marchand attribute to Campanus,[81] may be identical with that entitled *Astronomia cum quibusdam de Astrologia* in the 1512 Strasbourg edition of Gregor Reisch's *Margarita philosophica*. The words "Astronomia cum quibusdam de Astrologia" do indeed occur on the title page of that edition,[82] but they are merely a description of part of the text written by Reisch himself (works by others are in the appendix). Thus Daunou's conjecture is completely groundless.

(21) There exists one manuscript of a work entitled *De signis* (TKr, col. 130), which cannot be further identified.[83]

(22) A commentary on the *De longitudine et breuitate uite* from the *Parua naturalia* of Aristotle is ascribed to Campanus in a Munich manuscript.[84] Examination of a photostat of that manuscript reveals that it is the work of Averroës.[85] The false ascription arises from a notation in a later hand in the

---

(Note 80 continued)
tionem suam cum sole quandoque citius quandoque uero tardius appareat." Fol. 153v, a figure entitled, "Pro latitudine uisa et cetera lune etc." Fol. 154r, a figure; below: "Figuratio hec ostendit equationem domorum rationabilem que fit secundum magnos 6 circulos [?]et intersectiones communes orizontis cum meridiano transeuntes quorum duo cuilibet [*sic*] [?]sexaginta gradus equinoxialis concludunt." Fol. 154v, an incomplete figure; bottom: "Figura oualis quam describit centrum epicyli [*sic*] lune circa terre centrum. Inde concluditur lunam in quadraturis a sole terre esse propinquissimam in coniunctione uero et oppositione remotissimam."

81. Daunou, "Campanus de Novarre," p. 253; cf. Weidler, *Historia astronomiae*, p. 274, and Marchand, *Dictionaire historique*, p. 147. There appears to be no reference to *Nonnulla astrologica* in the article on Campanus in Fabricius, *Bibliotheca latina*, though Marchand, as well as Daunou ("Campanus de Novarre," p. 253), claims to have found one. However, it is mentioned by Gesner (*Bibliotheca universalis*, 1st ed., vol. 1, p. 161, and augmented ed., p. 134, col. 2). Gesner is presumably the ultimate source of all the above and of the reference in Cotta, *Museo Novarese*, p. 88.

82. See John Ferguson, "The *Margarita philosophica* of Gregorius Reisch: A Bibliography," *The Library*, ser. 4, vol. 10 (1930), p. 208.

83. MS Berncastel, Hospital zu Cues, 209 (14th cent.), fols. 64–67: "Campanus de signis." *Incipit*: "Aries est bestia [TKr, col. 130].../...Laudetur deus in operibus suis. Hec tibi dicta sunt ad bonum exitum." Cf. Zinner, *Verzeichnis*, p. 72, no. 1922.

84. MS Munich, Staatsbibliothek, CLM. 27256 (14th cent.; folio, 109 fols.), fols. 78–[80v]: "Campani Nov. de causis longitudinis et brevitatis vitae ex quibus erunt manifestae causae vitae et mortis." Described in C. Halm *et al.*, *Catalogus codicum latinorum Bibliothecae Regiae Monacensis* (Munich, 1868–81; 2 vols. in 7 pts.).

85. Fol. 78v[a]: "Intendimus hic dicere causas longitudinis et breuitatis uite..." (TKr, col. 757, citing "BN, fonds latin, 16222, f. 115" [should be fol. 44v] and "AL n.1, 224." AL is *Aristoteles Latinus*, codices descripsit Georgius Lacombe in societatem operis adsumptis A. Birkenmajer, M. Dulong, Aet. Franceschini, Pars prior [Corpus philosophorum Medii Aevi academiarum consociatarum auspiciis et consilio editum, Union académique internationale] [Bruges and Paris, 1957]). The *Aristoteles Latinus*, pp. 224–26, col. 2 only, gives an extensive excerpt from the *incipit* and *explicit* of the Paris manuscript, fols. 44v–45v, which confirms

bottom margin: "Incipit liber magistri Campani nouariensis." It would be interesting to know why a later scribe attributed it to Campanus.

(23–28) Two brief mathematical works, which may derive from one of his larger works, also exist in manuscript;[86] and in MS Cracow, Biblioteka Jagiellońska, 601 (DD.IV.5) (MS *w*), fols. 119r–20v, there are four slight *tractatus* all ascribed to Campanus. These four are excerpts from the *Tractatus de sphera* (item (3), above), corresponding to chapters 50, 52, 53, and 29, respectively, of the Venice, 1518 edition (fols. 158r^a–r^b, 158r^b–v^a, 158v^a–v^b, and 155v^b–56r^a). I checked specifically chapter 29 and discovered that the Cracow manuscript (MS *w*) agrees nearly exactly with the printed text except for the last section beginning "Idcirco autem de tot...." Instead of that, MS *w* reads, "Sciendum quod alzimuth sunt de maioribus transeuntes per polos orizontis qui sunt cenith et eius nadir Almucantarath uero sunt circuli breuiores equidistantes orizonti per alzimuth scitur distantia ortuum stellarum et per almucantarath scitur earum altitudo unde utrique ponuntur in astrolabio ut postea in suo tractatu patet." Except for the final phrase, "ut...patet," this sentence is a paraphrase of chapter 28 of the *Tractatus de sphera* (ed. Venice, 1518, fol. 155v^b). Clearly these four tracts should not be considered as separate works of Campanus.

(29) An otherwise unknown work by a Campanus (who may or may not be Campanus of Novara), on astrology in its application to medicine, is quoted by Nicholas of Lynn (late fourteenth century) in the canons to his calendar.[87]

---

the identification. The Munich manuscript ends on fol. 80v^b.

86. MS Dresden, Sächsische Landesbibliothek C.80 (15th cent.), fols. 366–67v: "Cautelae Magistri Campani ex libro de Algebra siue de Cossa et Censu"; MS Rome, Biblioteca Apostolica Vaticana, lat. 5335, fol. 1r^a: "Campanus. Numerus est principium agritionis omnis rei.../...[fol. 1r^b] et reducende sunt in idem genus."

87. On this calendar, see Thorndike, *History of Magic*, vol. 3, pp. 523–24. I have examined the quotations in three manuscripts of the work on astrological medicine: MS Oxford, Bodleian Library, Rawlinson C.895 (14th cent.), fols. 4r–4v; MS Munich, Staatsbibliothek, CLM.10661 (Pal. 661) (15th and 16th cents.), fols. 71(63)r^a–72(64)r^a; and MS London, British Museum, Arundel 347 (14th cent.), fol. 44r. I give the readings of the Bodleian manuscript, with variants from the Munich manuscript, with which the British Museum manuscript agrees closely: "[fol. 4r, Tit.:] Canon pro minutionibus et purgationibus. [*Incipit*:] Quia secundum sententiam Tholomei in suo centilogio et sui commentatoris hali propositione 56: in prima quadra lune et tertia humores corporum humanorum exeunt ab interioribus ad exteriora [TKr, col. 1231].../...[fol. 4v] Secundo CAMPANVS assignat causam dicens: Tangere cum ferro membrum illud uulnerando est causatiuum doloris et dolor causat fleuma [reuma: CLM] propter quod inquit in cirurgia cauendum est ab incisione in membro luna existente in signo significationem habente super illud membrum, aliam causam assignant alii astrologi dicentes humores confluere ad locum dolorosum ad confortandum naturam membri dolorem patientis qui congregati obtundunt calorem si non possunt exire et quia stant sine regimine nature putrescunt et inficiunt locum dolorosum. Si uero exeunt frequenter moritur patiens eo quod humores

If the ascription is correct, this is the only connection of Campanus with astrological medicine.

(30) Sarton suggests Campanus wrote a commentary on Ptolemy's treatise on music,[88] but no manuscript of this work exists to my knowledge.

The statement of Braunmühl that Campanus was also the author of a tangent table from 0°–45°, which would antedate the *tabula fecunda* of Regiomontanus by 200 years, seems without evidence.[89] Similarly there seems to be little substance in the belief of several authorities that Campanus was the first to propound or treat of several geometrical principles such as the stellated polygon, the trisection of an angle,[90] the study of continuous quantities, and the irrationality of the golden section. And finally Symon de Phares ascribes to Campanus an *Equatorium* with the *incipit* "Ut igitur habeas...," which cannot otherwise be identified.[91]

(Note 87 continued)
sunt subiectum caloris quibus expulsis expellitur calor sine quo membrum non potest in statum pristinum reuocari. Et hec est causa quare ex ictu lapidis uel baculi non frangentis cutem generat [generatur: CLM] tumor uidelicet propter fluxum humorum. Item narrat CAMPANVS se uidisse hominem imperitum in astris qui in periculo squinantie minuerat sibi de brachio luna existente in geminis quod signum dominatur super brachia et absque ulla manifesta egritudine excepta modica brachii inflatione die septimo mortuus est. Nouit etiam quemdam ut asserit patientem fistulam in capite membri uirilis et ipsum fuisse incisum luna existente in scorpione quod signum dominatur super partem illam corporis et eadem hora incisionis in manibus tenentium obiit nulla [alia *add.* CLM] causa concurrente."

88. Sarton, *Introduction to the History of Science*, vol. 2, p. 985.

89. A. von Braunmühl, *Vorlesungen über Geschichte der Trigonometrie*, pt. 1, p. 101, note 5. But in refutation of Braunmühl, see John David Bond, "The Development of Trigonometric Methods down to the Close of the Fifteenth Century," *Isis*, vol. 4 (1921–22), p. 316. Regiomontanus's "tabula fecunda" was published in his *Tabulae directionum*: see Zinner, *Leben und Wirken*, p. 107.

90. A solution to the problem of trisecting an angle is found in some manuscripts and the first edition of Campanus's version of Euclid as an addition to Book IV. It is printed, with translation and brief discussion, in Appendix VI of Clagett, *Archimedes*, vol. 1, pp. 678–81.

91. Wickersheimer, *Recueil... par Symon de Phares*, p. 168. There are some astronomical notes in MS Cracow, Biblioteka Jagiellońska, 1970 (BB.XXIII.13) (14th cent.), on pp. 104–7, which begin, "[V]t habeas uerum tempus mensis peragrationis," and which lie between the *Speculum astronomiae* of Albertus Magnus and the *Arismetrica* that begins "Numerus est duplex scilicet mathematicus" (TKr, col. 959) and is in the *explicit* here ascribed equally to Iohannes de Muris, Boethius, and Jordanus Nemorarius. However, the *incipit* preserved in Symon is far too slight to warrant any identification.

The *Theorica planetarum*

Title and Contents

It is not known what title, if any, Campanus himself gave to his treatise concerning the planets.[1] The titles given in the manuscripts vary enormously (for examples, see the critical apparatus accompanying the text and translation, section I, p. 129, and section III, p. 143, below). We have chosen *Theorica planetarum* because it is the title most commonly found. Furthermore, both the words in it were used by Campanus himself, whereas he never employed the word *equatorium*, a frequent component in other titles assigned to the work.

The main purpose of the work, as announced at p. 140, lines II.57–63, is to describe the construction of an instrument for finding the positions in longitude of the heavenly bodies (on this "equatorium," see p. 30, below). But in fact the scope is much wider. Campanus gives not only a description of the Ptolemaic solar, lunar, and planetary models on which the instrument is based, but also the dimensions of each model, with all its constituent parts, both relative to itself and absolutely (in terms of earth-radii and "miles").[2] He also provides information on the speeds of the various parts of the models and explains certain phenomena such as station and retrogradation in planets. For each body, too, he explains the way in which the tables commonly used in his day for computing celestial positions are laid out, the basis of their construction, and the method of operating with them. For the most part, however, the work is descriptive rather than explanatory. Except for the discussions of the instrument

1. In the only passage where he himself refers to it (*Tractatus de sphera*, chap. 13: fol. 154r[a] of the de Giunta, 1518, edition), he says, "in uno libro quem de modo equationis planetarum ad instantiam domini Vrbani pape quarti edidimus."

2. He is able to give absolute distances because of the world-picture of "nested spheres" which he adopts and describes: see "The Ptolemaic System," pp. 53–56, below. Campanus defines a "mile" as 4,000 cubits. This definition is taken from his source, al-Farġānī, and does not, to my knowledge, correspond to any "mile" actually in use in western Europe during any period. See the Commentary to section III, note 17, and, for the problem of the metrical values of the units involved, see the Nallino reference there.

and the absolute distances, it covers the same ground as books 3–5 and 9–12 of Ptolemy's *Almagest*, but it omits the underlying observation, argumentation, and calculation found in its source and merely extracts the resultant models and numerical data (with a few modifications, for which see p. 34, below).

The level of competence shown in the work is quite high. Campanus understands and transmits correctly most of what he has read (occasional mistakes are pointed out in the notes: see for example, the Commentary to section IV, notes 20 and 44, section V, note 95, and section VI, note 23). I have checked the numerous computations underlying the numerical results of the text and have found very few miscalculations indeed (such as I have found are recorded in the Commentary). On the other hand, there is very little original thought,[3] and in general Campanus does not give the impression of high intelligence (one derives the same estimate of his competence and intelligence from his commentary on Euclid). The solemn calculation of all the planetary distances to 660ths of a "mile" is only the most blatant example of a ludicrous pedantry (though it should in fairness be stated that the concept of "probable error" was completely unknown to the ancient and medieval worlds). When Campanus does produce an original argument, it can be a singularly poor one (see especially p. 334, lines VI.543–46, with the Commentary to section VI, note 88). But on the whole the work is an intelligible, if verbose, account of his subject.

A general outline of the contents of the *Theorica*, with references to the inclusive line numbers, follows.

Section I. Dedication: Praise of Pope Urban IV; justification of treatise

Section II. Prologue

- Lines 2–39: Place of astronomy among the sciences (2–26); subdivisions of astronomy (27–39)
- Lines 39–81: Purpose of equatorium, objectives of treatise, and general description of instrument

Section III. Theory of the Sun

- Lines 2–47: The solar model, its parts and motions
- Lines 48–71: "Absolute" sizes of the sun, the parts of the solar model, and the earth
- Lines 72–88: The earth's shadow
- Lines 89–156: Construction of the instrument for the sun
- Lines 157–202: Use of the instrument to find the sun's true position
- Lines 203–34: Use of tables for finding the sun's true position, clarified through instrument

3. There is one notable exception, the method of finding the period of a planet's retrogradation in days when the distance of the planet's station from apogee is known: see pp. 312–20, lines VI.234–340, and the Commentary to section VI, notes 29 and 33. The idea comes from Jābir, but the actual derivation of the method seems to be Campanus's own. I doubt that Campanus conceived of the planetary equatorium by himself: see pp. 32–33, below.

Section IV. Theory of the Moon

Lines 2–294: The lunar model, its parts and motions (2–166): diagram (167–273); summary (274–94)

Lines 295–377: The heavenly and elemental "spheres"

Lines 378–489: Sizes for lunar model: identifying diagram (378–98); relatively (403–13); in earth-radii (415–23); in miles (424–44); velocities of moon and its epicycle (444–51); relative volume of earth and moon (452–83); description of table of lunar magnitudes (484–89); table (pp. 356–57)

Lines 490–610: Construction of the instrument for the moon (methods for making moving parts, 546–82)

Lines 611–55: Use of the instrument to find the true position of the moon and to determine deviation from mean speed

Lines 656–749: Construction and use of the lunar tables

Section V. Theory of Mercury

Lines 2–52: Mercury's model

Lines 53–178: Motions of the parts of Mercury's model: peculiar motion of apogee (53–92); velocities (93–124); concomitant phenomena of orbital motion (124–78)

Lines 179–233: Forward motion, station, and retrogradation: variation in positions of stations (207–19) and periods of retrogradation and forward motion (219–33) as a function of epicycle position

Lines 234–329: Diagram of Mercury's model

Lines 330–33: Mercury's "sphere"

Lines 334–42: Summary of Mercury's circles and motions

Lines 343–479: Sizes for Mercury's model: identifying diagram (343–73); relatively (380–90); in earth-radii (391–426); in miles (427–35); Mercury's "sphere" (436–65); velocities of Mercury and its epicycle (465–73); relative volume of earth and Mercury (473–78); table of magnitudes ([478–79] p. 358)

Lines 480–596: Construction of the instrument for Mercury

Lines 597–650: Use of the instrument to find Mercury's true position (597–628) and to determine station, direct motion, or retrogradation (629–45)

Lines 651–1321: Construction and use of tables for Mercury, elucidated by instrument: mean and true centra, equation of center (663–766); mean and true anomalies (767–810); operations with tables to obtain foregoing (811–56); equation of anomaly (857–943); tables and "minutes of proportion" for equation of anomaly (944–1199); minutes of proportion for Mercury vs. those for the other planets (1200–1236) and for the moon (1237–62); summary of minutes of proportion (1263–75); remainder of equating procedure (1276–315)

Lines 1322–82: Use of the instrument to find tabular elements for all bodies

Section VI. Theory of Venus and the Three Superior Planets

Lines 2–42: The single model (2–12) for Venus, Mars, Jupiter, and Saturn (13–42)

Lines 43–194: Motions and velocities of the parts of the model (43–80); motions of outer planets related to solar motion bij a fixed law (81–128); all the bodies connected in some way with sun (129–41); diagram of motions (142–94)

Lines 195–379: Forward motion, station, and retrogradation (195–215); variation in positions of stations (216–33) and periods of half- and total retrogradation (234–60) as a function of epicycle position; determination of period of retrogradation (261–344); determination of greatest elongations of Venus and Mercury (345–79)

Lines 380–87: Summary of motions and circles of the four planets

Lines 388–467: Sizes for Venus's model: identifying diagram (399–407); relatively (408–17); in earth-radii (418–33); in miles (434–41); Venus's body and "sphere" (442–54); velocities of Venus and its epicycle (454–61); relative volume of earth and Venus (461–64); introduction to table of Venus's magnitudes (464–67); table ([467] p. 359)

Lines 468–535; Ptolemy vs. Campanus on calculating absolute distance of the sun (468–505) and parts of the solar model (506–35)

Lines 536–48: Proof of correctness of Ptolemy's and Campanus's planetary order

Lines 549–68: Sizes for solar model: in miles (549–53); the sun's body and "sphere" (553–60); velocity (560–64); relative volume of sun and earth (564–68); table of solar magnitudes ([568] p. 360)

Lines 569–621: Sizes for Mars's model: relatively (571–80); in earth-radii (581–93); in miles (594–602); Mars's body and "sphere" (602–9); velocities of Mars and its epicycle (609–17); relative volume of Mars and earth (617–20); table of magnitudes for Mars ([620–21] p. 361)

Lines 622–73: Sizes for Jupiter's model: relatively (622–31); in earth-radii (632–44); in miles (645–53); Jupiter's body and "sphere" (653–61); velocities of Jupiter and its epicycle (661–68); relative volume of Jupiter and earth (668–72); table of magnitudes for Jupiter ([672–73] p. 362)

Lines 674–720: Sizes for Saturn's model: relatively (674–79); in earth-radii (680–91); in miles (692–700); Saturn's body and "sphere" (700–709); velocities of Saturn and its epicycle (709–16); relative volume of Saturn and earth (716–19); table of magnitudes for Saturn ([719–20] p. 363)

Lines 721–34: Area of surface of sphere of the fixed stars (721–31), etc. (731–34)

Lines 735–41: Criticism of others' error in reckoning celestial magnitudes

Lines 742–46: Velocity of fixed stars in the ecliptic

Lines 747–866: Construction of the instruments for Venus, Mars, Jupiter, and Saturn (747–845); their use (846–66).

Authenticity of the Existing *Theorica* Text

It is hard to believe that the text of the *Theorica* represented by the consensus of our manuscripts faithfully reproduces that originally conceived by Campanus (I am speaking here not of detailed phraseology but of the general structure of the work).

Firstly, strong suspicion of interpolation attaches to all the passages concerning the *dieta* (the mean daily motion in "miles" of a heavenly body on its epicycle or eccentric). They contain a number of curious features. Contrary to the usual practice of Campanus, the noninteger part of the *dieta* is given as an *approximate sexagesimal* fraction, and not as an *exact common* fraction. The figures for the period of revolution of the planet on the epicycle and the epicycle on the deferent are computed from crude parameters for mean daily motion given elsewhere by Campanus: this procedure leads to significant errors in the periods of the return in longitude of Jupiter and Saturn (see the Commentary to section IV, note 73, section VI, notes 120, 135). Such errors are quite uncharacteristic of Campanus, who, as a table-maker, was well aware of the need to compute with more significant places than those he tabulates in order to achieve an accurate table. Furthermore, the curious and unique expression, "the moon's equant," occurs in a *dieta* passage (line IV.446 and the Commentary to section IV, note 70). None of these points by itself proves spuriousness, but their accumulation is suggestive, though we have not ventured to delete the passages.

Even more disturbing than these apparent accretions to the author's original text are the indications that there is at least one section missing from it. At lines V.15–17 Campanus promises to explain "in its proper place" how the epicycle of Mercury is inclined to the plane of the deferent (i.e., the mechanism of Mercury's motion in latitude). It might be claimed that this promise is fulfilled at lines V.338–42; however, all he says there is that Mercury's latitudinal motions are of great complexity and that discussion is pointless since they cannot be represented on his instrument. Regardless of one's interpretation of these passages, there is a clear promise at lines VI.386–87 (cf. lines VI.26–27) to explain the *models* for the latitudinal motions of the other four planets: "We will speak later of their motions in latitude and the circles necessary for them." No later passage can be said to fulfill this promise. One might argue that, because the latitudinal motions of none of the bodies can be represented on the instrument, they are not germane to the treatise. Nonetheless, he does describe the latitudinal motion of the moon (lines IV.6–46), though that motion too cannot be represented on the instrument; and we have the specific promise of lines VI.386–87.

Furthermore, the treatise as we have it ends extremely abruptly. It is true that the instrument and the way to use it have been fully described. But, remembering the pious invocation at lines V.647–50, celebrating the termination of what is only a subsection of the work, one might have expected a more graceful conclusion to the whole. This point by itself would be of little moment, but it does strengthen the probability that Campanus intended to add at least one more section, on the latitudinal motion of the planets.

Now Campanus's own words in the later *Tractatus de sphera* (quoted in note 1, p. 25, above) make it certain that he did publish ("edidimus") the *Theorica* as a completed work. There are then only three hypotheses which can explain its present apparent incompleteness:

(1) Campanus was so careless that he ignored his own earlier intentions;

(2) In order to get the work out as an offering to the pope, he "published" a

preliminary version, not complete in all respects; if there ever was a revised version, it never went into circulation;

(3) All known copies of the *Theorica* are descended from a single manuscript defective at the end.

Though none of these hypotheses is completely satisfactory, I am inclined to favor the third, since it could also explain why passages which are probably spurious occur in all extant manuscripts.

Campanus's Instrument

The instrument Campanus describes in the *Theorica* is what is known as a "planetary equatorium."[4] In fact it is a collection of seven separate instruments, combined in the following manner (see p. 140, lines II.74–80): there are three equal circular plates or boards, each bearing the representation of the systems of two of the heavenly bodies, one on each face; the moon is combined with Mercury, Venus with Mars, and Jupiter with Saturn. These three plates fit into a shallow circular container (the *mother*); on the back of the mother is the representation of the system of the sun. This arrangement is inspired by the familiar horizon plates and mother of the astrolabe, to which Campanus refers on p. 140, lines II.75–77.

Each representation is separate from and works independently of the others. Each is an exact scale reproduction of the Ptolemaic system for that body[5] with the addition of graduated circles along which mean motions are counted and true positions measured; threads attached to the point representing the observer and other necessary points facilitate reading off the graduated circles (see Figs. 10, 13, 16, and 22, on pp. 152, 196, 246, and 346, respectively). All the instruments except that of the sun have circular moving parts. The motion of these parts is effected by embedding individually carved circles or rings one into another (the process is described by Campanus at p. 200, lines IV.552–82).

To calculate a position in longitude of one of the heavenly bodies by means of the instrument, one requires a table of mean motions for that body. Extracting from that table the mean positions (in longitude, anomaly, and elongation as necessary) for the time in question, one sets up the instrument accordingly: it is then a reproduction of the configuration of the various parts of the Ptolemaic model for that body at the given moment. One finds the longitude by stretching the thread attached to the point representing the observer through

4. The term "equatorium" is not used by Campanus himself (he refers to it usually as "instrumentum nostrum"), but it is found in the headings of some of the later manuscripts. Its meaning is "that which is used to equate," i.e., to convert mean longitude to true longitude.

5. But restricted to a single plane, which is taken as the plane of the ecliptic: celestial latitude is ignored.

the point representing the body and reading off the position on the graduated circle representing the ecliptic.

Some predecessors and successors of Campanus's instrument

Campanus's instrument, as described above, is the simplest, most unsophisticated type of equatorium imaginable. Nonetheless, I think it improbable that Campanus hit upon the idea of such an instrument without inspiration from an outside source. Unfortunately our knowledge of the history of the equatorium in early medieval times is very inadequate. The best account is that of D. J. Price, to which, since I have nothing to add beyond what is found in Campanus's treatise below, I refer the reader for a history of the instrument (insofar as it may be sketchily traced at present) and merely extract the following tentative remarks.[6]

Campanus's *Theorica planetarum* is the first Western treatise on the equatorium known at present. It had a number of predecessors in Islamic countries. Of these, we may mention in particular the works of the eleventh-century Spanish writers ibn as-Samḥ and az-Zarqāl (Azarchel, Azarquiel).[7] The instrument described by ibn as-Samḥ is of a simple type, and the closest I know to that of Campanus. The instrument of az-Zarqāl is considerably more sophisticated. Both these treatises were translated into Castilian at the order of Alfonso X (the Wise) of Castile and Léon.[8] But the translations were made after

6. Price, *Equatorie of the Planetis*, pp. 119–33. Another good general account of the development of the equatorium is Emmanuel Poulle, *Astronomie théorique et astronomie pratique au Moyen Âge* (Paris, 1967). We are considerably better informed on the equatoria of the fifteenth and sixteenth centuries, thanks to the work of Poulle: see Poulle, *Un Constructeur d'instruments astronomiques au XVe siècle, Jean Fusoris* (Paris, 1963), pp. 41–64, 125–80; Poulle, "L'Équatoire de Guillaume Gilliszoon de Wissekerke," *Physis*, vol. 3 (1961), pp. 223–51; Poulle, "L'Équatoire de la Renaissance," in *Le Soleil à la Renaissance, sciences et mythes: Colloque international* (Brussels, 1965), pp. 129–48; Poulle, "Sur un Fragment d'instrument astronomique des Musées de Bruxelles," *Ciel et terre*, vol. 79 (1963), pp. 363–79; Poulle, "Théorie des planètes et trigonométrie au XVe siècle, d'après un équatoire inédit, le sexagenarium," *Journal des savants*, 1966, pp. 129–61; Poulle and Francis Maddison, "Un Équatoire de Franciscus Sarzosius," *Physis*, vol. 5 (1963), pp. 43–64. An interesting text describing the equatorium of the fifteenth-century Islamic scientist al-Kāshī has been edited and translated by E. S. Kennedy (*The Planetary Equatorium of Jamshīd Ghiyāth al-Dīn al-Kāshī* [Princeton, N. J., 1960]). But the instruments of the fifteenth century and thereafter are, for the most part, radically different from those of the earlier Middle Ages.

7. Price, *Equatorie of the Planetis*, pp. 120–23.

8. Printed in Manuel Rico y Sinobas, ed., *Libros del saber... del rey Alfonso X...* (Madrid, 1863–67), vol. 3, pp. 241–84. The Arabic text of az-Zarqāl's treatise has been published, with Spanish translation, in J. M. Millás Vallicrosa, *Estudios sobre Azarquiel* (Madrid and Granada, 1943–50), pp. 460–79.

Campanus had written his *Theorica*, and we know of no earlier translation of these or any similar Arabic works.[9]

From the late thirteenth century on, a number of works were written in the West on the construction of equatoria. The earliest known of these is the "semissa" of Petrus de Sancto Audomaro (perhaps to be identified with Petrus Philomena de Dacia). The treatise describing this was written in 1293 (see Olaf Pedersen, "The Life and Work of Peter Nightingale," *Vistas in Astronomy*, ed. Arthur Beer [Oxford, 1967], vol. 9, pp. 3–10). To the early fourteenth century belongs the equatorium of Johannes de Lineriis (John of Linières), which explicitly sets out to improve on Campanus's instrument.[10] Johannes de Lineriis obviously depends heavily on Campanus's treatise (see the Commentary to section VI, note 95), and he states specifically that Campanus was the first to construct an equatorium.

The instruments of Petrus de Sancto Audomaro and Johannes de Lineriis are considerably more compact and sophisticated than Campanus's. The same is true of the instruments of Richard of Wallingford and "Chaucer," and also of an equatorium which actually survives from medieval times, that of Merton College, Oxford. Price plausibly sees in these instruments, other than John's, the influence of az-Zarqāl's equatorium.[11]

In spite of the absence of any direct evidence, I am inclined to believe that Campanus derived the *idea* of constructing an equatorium from a Latin translation of an Arabic text. The plan of the instrument, however, may well have been his own: it is certainly crude compared with the earlier instrument of az-Zarqāl. As Price demonstrates from internal evidence, "Chaucer's" *Equatorie of the Planetis* is an adaptation of a text that was probably a Latin translation from the Arabic. Though this text has not been identified, it and other such texts may only await discovery.

Was Campanus's instrument ever constructed?

There is no doubt that it would be possible to construct Campanus's instrument. In practice, however, it would be difficult and costly to make and clumsy to use. Johannes de Lineriis's criticisms of it pinpoint its disadvantages very well: "Its construction is very wearisome, on account of the large number of plates [*tabularum*] contained in a single instrument, together with their various cavities, and also on account of the large size of the instrument; for it cannot

9. Alfonso reigned from 1252 to 1284. The publication date of the *Libros del saber*, of which the two treatises form a part, is 1276–77 (concerning these dates see Rico y Sinobas, *Libros del saber*, vol. 1, pp. 7–8, 153, and vol. 3, pp. 135, 287).

10. See p. 37, below, and Price, *Equatorie of the Planetis*, pp. 125–27, with a somewhat faulty text of Johannes de Lineriis's treatise on pp. 188–96.

11. Price, *Equatorie of the Planetis*, pp. 127–29.

easily be carried from place to place."[12] In my translation of Campanus's text I have rendered *tabulae* as "boards" or "disks." Campanus nowhere specifies the material of which they are to be made, but it seems probable from his instructions for "hollowing out" and "beveling" that he envisages them of wood. If that is so, a *tabula* accommodating one model on each face would have to be of considerable thickness, because each model requires three or even four layers of material for the various moving parts. Since the diameter of the *tabula* would have to be quite large (at least two feet) to accommodate the subdivisions specified by Campanus and to produce reasonably accurate results, the complete instrument, consisting of three thick boards stacked inside the mother, might well provide a transportation problem.

It would be possible to decrease the thickness by substituting metal or paper for wood, but to work metal in the way prescribed by Campanus would have been very difficult technologically in his time, and a paper model (which he clearly does not envisage) could at best illustrate the instrument:[13] it would be useless for accurate calculation.

I am inclined to think that Campanus had no actual experience in manufacturing his instrument,[14] and doubt very much whether anyone else ever manufactured it to his specifications in a really workable form. Both wooden and metal equatoria were constructed in later medieval times,[15] but, as far as we know, they were all of the sophisticated kind requiring only one plate.

Campanus's Sources

In this section I shall deal with the sources only for the technical parts of the *Theorica*. Literary echoes from the Bible, Boethius, etc., are pointed out in the Commentary (see especially the Commentary to lines I.4–91 for the influence of Petrus de Vineis on the "Dedication"). Discussions of certain philosophical and cosmological ideas occurring in the *Theorica* are likewise left to the Commentary for the passages concerned, where references to the appropriate authorities, such as Aristotle, Macrobius, Martianus Capella, and Chalcidius, are

12. Translated from ibid., p. 188.

13. Such paper or parchment models are found in MSS *M* and Γ, and probably in some other extant manuscripts of the *Theorica*. The best-known examples are to be found in the printed work *Astronomicum Caesareum* (Ingolstadt, 1540) of Petrus Apianus.

14. For inconsistencies in Campanus's account which have led to this opinion, see the Commentary to section V, note 161, and section VI, note 156.

15. The Merton equatorium mentioned above is metal: there is a photograph of it in Price, *Equatorie of the Planetis*, frontispiece. In the early fifteenth century, Merton College owned a wooden equatorium and a bronze one: see F. M. Powicke, *The Medieval Books of Merton College* (Oxford, 1931), p. 68, quoted by Price, *Equatorie of the Planetis*, p. 130, note 2.

given (see, for example, the Commentary to section III, note 14, and section IV, note 56).

The main source, as Campanus himself declares (p. 134, lines I.96–104), is Ptolemy's *Almagest*, probably in Gerard of Cremona's translation from the Arabic.[16] It provides him with the basic models of planetary motion (his sole qualitative modification is the assumption that the solar apogee is fixed siderеally, whereas Ptolemy had supposed it fixed tropically: see p. 42, note 2, below), as well as with all the basic parameters for the *relative* sizes of the parts of the model.[17] Even though all that information could have been derived from an intermediate source, it is clear from passages which closely imitate the wording used by Ptolemy (see, for example, p. 136, lines II.2–7, with the Commentary to section II, notes 1, 2, 3, 5) that Campanus had read the *Almagest* itself, evidently carefully enough to be able to reproduce its methods outside their immediate context (see the Commentary to section VI, note 33).

The most important secondary source is the astronomical work of the ninth-century Arabic author Aḥmad ibn Muḥammad ibn Kaṯīr al-Farġānī (better known in the West as Alfraganus).[18] The work of al-Farġānī is a short account of the elements of astronomy as they were understood in his time, and much of it is no more than a summary of the results of the *Almagest*. But it does also present the "system of spheres" taken from Ptolemy's *Planetary Hypotheses* (see pp. 53–56, below). There is no need to look beyond this work of al-Farġānī for Campanus's knowledge of that system (an assertion which still does not deny that he may have known it from other sources as well). From al-Farġānī Campanus also drew his figures for the sizes of the heavenly bodies in earth-radii, his parameter for the length of one degree of terrestrial latitude in "miles," and the metrical equivalence between 1 "mile" and 4,000 cubits. The passage concerning the length of one degree (p. 146, line III.60) incidentally proves that Campanus used Johannes Hispalensis's Latin translation of al-Farġānī, and not that of Gerard of Cremona: Campanus renders the name of the caliph al-Ma'mūn as "Almeon." In the relevant passage of al-Farġānī the manuscripts of Johannes Hispalensis read, according to Carmody, "in diebus Almeon (Almehon)," whereas Gerard's version, edited by Campani, says "in diebus Maimonis."[19] I have therefore referred to Carmody's edition of al-Farġānī throughout.

16. But other translations, both from Greek and from Arabic, were already in existence before his day: see C. H. Haskins, *Studies in the History of Medieval Science* (Cambridge, 1924), pp. 104–10, 157ff.

17. The one important discrepancy is in the size of Mars's eccentricity, which Campanus puts at 6;30, where Ptolemy gives 6. As explained in the Commentary to section VI, note 95, this is probably attributable to a mere error rather than to reference to a different source.

18. On al-Farġānī and the title of the work, see Suter, "Mathematiker und Astronomen," pp. 18–19, no. 39.

19. Alfraganus, *Differentie in quibusdam*

The other principal secondary source is the Toledan tables. The tables comprise a Latin translation of an Arabic compilation of astronomical tables; the translation was made probably by Gerard of Cremona.[20] It is likely that Campanus used them, not in the original version, but in the adaptation for the Christian era and the meridian of Novara, probably made by himself (see p. 15, above). In most instances where he draws on them in the *Theorica*, however, there is no difference between the two versions. The tables supplied him with the longitudes of the apogees of the sun and the five planets. These are *sidereal* longitudes in the Toledan tables, and must be so considered in the *Theorica* too: indeed, Campanus implies as much in his statements (p. 144, lines III.28–29, and p. 234, lines V.330–31) that the spheres of the sun and planets share the "motion of the fixed stars" (see also the explicit statement, p. 212, lines V.19–20). Whenever there are references to the layout of astronomical tables in the *Theorica*, they can be illustrated from the Toledan tables, as I have shown in the corresponding Commentary.

Campanus refers explicitly to "Geber" (Jābir ibn Aflaḥ) on p. 332, line VI.538.[21] He had certainly read Gerard of Cremona's translation of Jābir's work on astronomy, for he uses it not only in the above-mentioned passage but very probably in the procedure for finding the periods of planetary retrogradations from the lengths of the arcs of retrogradation, described on pp. 314–20, lines VI.261–340 (see the Commentary to section VI, note 30). I have noted a few other places where there may be verbal reminiscences of Jābir.

Campanus had read the work *De motu octaue spere*, the original of which is ascribed to the ninth-century Islamic astronomer and mathematician Ṭābit ibn Qurra.[22] His acquaintance with it is evident from the account he gives of Ṭābit's "trepidation" theory on p. 144, lines III.31–33 (see the Commentary to section III, note 11). I have noticed no other place where it influenced him.

We cannot doubt that Campanus was acquainted with most of the astronomical treatises published in the West between the time of the appearance of translations from the Arabic in the latter half of the twelfth century and his own time. Since many of these essentially duplicate each other in subject matter and in terminology, it is often difficult to state with confidence which one was *the* source. In the Commentary, I have pointed out parallels of the *Theorica* with the *De sphaera* of Robert Grosseteste and the *De spera* of John of Sacro-

---

collectis scientie astrorum, ed. F. J. Carmody (mimeo., Berkeley, 1943), pp. 13–14; Alfraganus, *Il 'Libro dell'aggregazione delle stelle'*, ed. Romeo Campani (Citta di Castello, 1910), p. 89.

20. On the Toledan tables in general, see G. J. Toomer, "A Survey of the Toledan Tables," *Osiris*, vol. 15 (1968), pp. 5–174.

21. On Jābir, see Suter, "Mathematiker und Astronomen," p. 119, no. 284. For the evidence concerning the approximate dates of his life, see the Commentary to section VI, note 84.

22. Concerning Ṭābit, see Suter, "Mathematiker und Astronomen," pp. 34–38, no. 66.

bosco, which were both written earlier in the thirteenth century.[23] There are also notable coincidences in terminology between Campanus's work and the *Theorica planetarum Gerardi* (see the index of technical terms for some examples).[24] The author of the latter work is certainly not the twelfth-century translator Gerard of Cremona. If he is indeed a Gerard, the most plausible identification is with Gerard of Sabbioneta (see the masterly article by C. A. Nallino).[25] A problem of priority then arises, since the chronological data on Gerard of Sabbioneta indicate that he was an exact contemporary of Campanus. But the development of Latin astronomical terminology during the twelfth and thirteenth centuries still waits to be unraveled. We hope that our index of technical terms will make a small contribution to the unraveling.

Influence of the *Theorica*

The continuing importance of the *Theorica planetarum* of Campanus of Novara is attested not only by the number of surviving manuscripts but by its subsequent use and citation, even into the sixteenth century. A definitive documentation of its utilization by others, however, would entail an endless investigation of large numbers of manuscripts, as well as printed sources. Therefore we have decided to limit our survey of Campanus's influence to a few authors who will be representative of the many in their times.

A contemporary or near-contemporary of Campanus was the Franciscan Bernard of Verdun, who wrote a work entitled *Tractatus super totam astrologiam*.[26] Overall, the work has little relation to that of Campanus in respect to organization, verbal parallels and echoes, and content. However, in one section of the work,[27] the measurements of the various distances of the planets are

23. Robert Grosseteste, *De sphaera*, in *Die philosophischen Werke des Robert Grosseteste, Bischofs von Lincoln*, pp. 10–32, ed. Ludwig Baur (Münster i.W., 1912); Thorndike, "*Sphere*" *of Sacrobosco*, pp. 76–117.

24. Gerardus, *Theorica planetarum*, ed. F. J. Carmody (Berkeley, 1942).

25. Nallino, "Il Gherardo Cremonese autore della *Theorica planetarum* deve ritenersi essere Gherardo Cremonese da Sabbioneta," *Rendiconti, Reale Accademia dei Lincei, Classe scienze morali, storiche e filologiche*, ser. 6, vol. 8 (1932), pp. 386–404.

26. Bernardus de Virduno, *Tractatus super totam astrologiam*, ed. P. P. Hartmann, O.F.M., *Franziskanische Forschungen*, vol. 15 (Werl, 1961). Hartmann edited this from two Paris manuscripts (Bibliothèque Nationale, fonds latin, 7333 and 7334), but there is a third of slightly later date at Erfurt, Wissenschaftliche Bibliothek der Stadt, F.386 (ca. 1359), fols. 1–25 (Zinner, *Verzeichnis*, p. 350, no. 11392). As Hartmann says in the introduction, Bernard is not listed in Lucas Wadding's *Scriptores Ordinis Minorum* (Rome, 1690) or its revisions, and there is no indication otherwise of his *floruit* except for the date of the manuscripts, the earliest of which is late thirteenth century.

27. Bernardus, *Tractatus* (tract 5, dist. 3, chap. 2), ed. Hartmann, pp. 111–14.

exactly those of Campanus, even to the form in which they are ordered. This use of Campanus's measurements can be found in the writings of later authors such as Andalò di Negro of Genoa (d. 1342), who wrote a work on the distances and magnitudes of the planets.[28]

Representative of the fourteenth century is the astronomer Johannes de Lineriis, whose extant works begin in the 1320s.[29] Seven of his works are concerned with an *instrumentum* or *equatorium*.[30] Although most of them do not show any close relationship to Campanus's *Theorica*, one in particular is usually entitled *Abbreviatio instrumenti Campani*.[31] Rather than an excerpt or partial transcription of Campanus's *Theorica*, the *Abbreviatio* seems to be an attempt on the part of Johannes de Lineriis to simplify and compress what he felt to be important but unserviceable in the original form. His attitude, then, is one of respect but not slavish acceptance.

The clearest example of dependence upon the work of Campanus can be found in a work of John of Gmunden,[32] who is noted principally for his teaching at the University of Vienna in the first half of the fifteenth century.[33]

28. On Andalò, see Duhem, *Système*, vol. 4, pp. 266–78 (citing the more extensive treatment in Cornelio de Simoni, "Intorno alla vita ed ai lavori di Andalò di Negro, matematico ed astronomo genovese del secolo decimoquarto, e d'altri matematici e cosmografi genovesi," *Bullettino di bibliografia e di storia delle scienze matematiche e fisiche*, vol. 7 [1874], pp. 313–38, and B. Boncompagni, "Catalogo de'Lavori di Andalò di Negro," ibid., pp. 339–76); Thorndike, *History of Magic*, vol. 3, pp. 191–204. This work occurs in MS Paris, Bibliothèque Nationale, fonds latin, 7272, fols. 85r–99v, which I checked from a photostat at Columbia University.

29. Duhem, *Système*, vol. 4, pp. 60–69; Thorndike, *History of Magic*, vol. 3, pp. 253–67.

30. TKr, cols. 402–3, 1106, 1224, and 1588. Zinner (*Verzeichnis*, p. 465, note to no. 6604) says that another work, opening, "Multiplicis philosophie variis radiis illustrato domino Roberto Lombardo..." (TKr, col. 889), with the title *Canones super tabulas magnas compilati ex tabulis Alfonsii*, deals with the instruments of Campanus, but no such material could be found in a microfilm of one of the Erfurt manuscripts, Wissenschaftliche Bibliothek der Stadt, Q.366, fols. 28–32v.

31. This has been edited by Price from a Brussels manuscript (*Equatorie of the Planetis*, pp. 188–96). Cf. Bodleian Digby 57, fol. 130r: "Quia nobilissima scientia.../...[fol. 132v] ex qua causantur longitudines longiores et longitudines propiores et his similia per datum instrumentum intelligenti theoricam satis patent. Et uoco instrumentum datum omnia instrumenta Campani simul iuncta uel equatorium magistri Iohannis de Lyneriis uel semissas prefacii iudei uel aliud equatorium de nouo compositum et pro parte abbreuiatum omnia predicta excellens in locorum certitudine et operis facilitate sed sciendum est quod iste canon precedens non tangit modum operandi cum albion sed cum instrumentis prius dictis...."

32. See John Mundy, "John of Gmunden," *Isis*, vol. 34 (1943), pp. 196–205; and Rudolf Klug, "Johannes von Gmunden, der Begründer der Himmelskunde auf deutschen Boden," *Sitzungsberichte der Akademie der Wissenschaften in Wien*, phil.-hist. ser., vol. 222 (1943), no. 4.

33. He was perhaps a pupil of Henry of Hesse or Langenstein, who moved to Vienna from Paris shortly after the founding of the university; such an association would explain the connection between the Paris school of astronomers of the fourteenth century and

John's work, *Compositio et usus instrumenti quod Magister Campanus in theorica sua docuit fabricare ad inueniendum uera loca planetarum*, is a direct excerpt of the portions of the *Theorica* that describe the instrument, coordinated with some linking material.[34]

Citations of the *Theorica* continue to appear, although with less and less frequency, into the seventeenth century.[35] Surprisingly, although more manuscripts exist for the *Theorica* than for any other of Campanus's works, with the exception of his edition of Euclid, it was never printed. This disregard for his work is similarly reflected in the disregard for the man.[36] By the late fifteenth century, the court astrologer, Symon de Phares, had confused him as two persons, and in the sixteenth century Jean Bodin, in his *Démonomanie*, classed Campanus as an Arabic writer along with al-Farġānī, Ṭābit, and al-Battānī.[37]

(Note 33 continued)

the German school of the fifteenth, leading to Peurbach, Regiomontanus, and ultimately Copernicus. On Henry, see Thorndike, *History of Magic*, vol. 3, pp. 472–510; Herbert Pruckner, *Studien zu den astrologischen Schriften des Heinrich von Langenstein* (Leipzig and Berlin, 1933); Joseph von Aschbach, *Geschichte der Wiener Universität* (Vienna, 1865–88), vol. 1, pp. 366–402; and Claudia Kren, "Homocentric Astronomy in the Latin West: The *De reprobatione ecentricorum et epiciclorum* of Henry of Hesse," *Isis*, vol. 59 (1968), pp. 269–81.

34. Edited from two Munich manuscripts in F. S. Benjamin, Jr., "John of Gmunden and Campanus of Novara," *Osiris*, vol. 11 (1954), pp. 221–46. In the list of manuscripts in Benjamin's article, to item no. 5 add 738 after 4°. See also p. 120, note 8.

35. Some of the most recent references are in works of Kepler: e.g., *Epitome astronomiae Copernicanae*, in *Gesammelte Werke*, ed. Walther von Dyck and Max Caspar (Munich, 1938——), vol. 7, p. 309. See also the index to that volume.

36. Thorndike, *History of Magic*, vol. 5, pp. 407–8, and vol. 6, p. 130.

37. On Symon de Phares, see pp. 9–11, above. On Bodin, see Thorndike, *History of Magic*, vol. 6, p. 43.

The Ptolemaic System

I give here an outline of the main features of the astronomical system developed by Ptolemy in the *Almagest* (μαθηματικὴ σύνταξις), which remained canonical until the sixteenth century. Like Campanus (p. 134, lines I.96–104), I must refer the reader to the *Almagest* itself for the basis of observation and calculation on which this system rests.[1]

In the Ptolemaic universe the earth is considered to be central and at rest. To explain the heavenly phenomena, one then assumes that the sphere of the fixed stars moves about the earth from east to west with uniform velocity (making one revolution in slightly less than twenty-four hours), carrying with it the sun, the moon, and the five planets known to antiquity and the Middle Ages; these seven bodies have an additional, slower motion in the opposite sense (all moving in or near the ecliptic, the apparent annual path of the sun through the fixed stars). This latter motion, however, is not uniform; rather, the apparent angular velocity of each body varies at different points on its orbit. The main problem in the Ptolemaic, as in any, astronomical system is to account for this variation in velocity (known as "anomaly" or "anomalistic motion"). For this purpose Ptolemy uses a general model which he inherited from his predecessors and which he modified to serve for all seven bodies. The general model can be represented in either of two forms, the "eccentric" and the "epicyclic."

The eccentric form is depicted in Figure 1, where O represents the observer (earth). The celestial body P moves on a circle (the "eccentric") with uniform angular velocity about its center M, which is distant from O by the amount e (the "eccentricity"). It is clear that, if the motion of P appears uniform from M, it will appear nonuniform ("anomalistic") from O, and P will seem to be moving most slowly at its greatest distance from O, in the "apogee" (ἀπόγειον, *aux*) A, and fastest at its least distance, in the perigee (περίγειον, *oppositio augis*), Π. The angle $\bar{\kappa}$ which P has traveled from A, with respect to the center of uniform

1. For a more detailed account of Ptolemy's work, see my forthcoming article "Ptolemy" in the *Dictionary of Scientific Biography.*

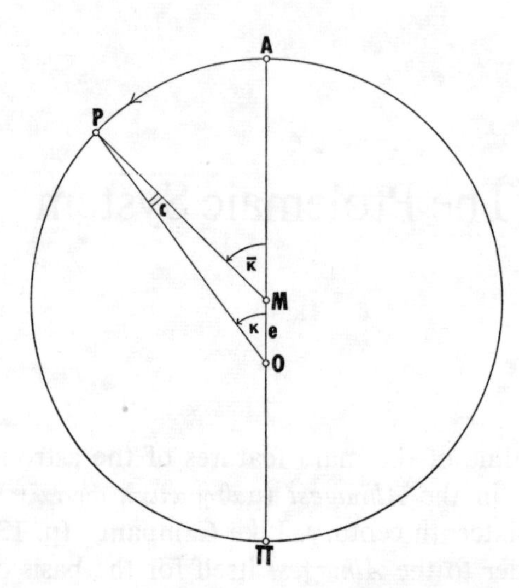

Fig. 1

Simple Ptolemaic eccentric model. The celestial body, P, moves with uniform angular velocity around a circle with center M. The observer (earth) is located at O, so that P, as seen from O, appears to move "anomalistically," being slowest at A, the apogee, and fastest at Π, the perigee. The distance OM is the eccentricity e. Angle AMP ($\bar{\kappa}$) is the "mean anomaly," and angle MPO (c) is the "equation." The angle at O (P's true angular distance from the apogee) is given by the equation $\kappa = \bar{\kappa} \mp c$.

motion M, is known as the "mean anomaly" (ἀνωμαλία μέση, *argumentum medium*). One can derive the angle κ (P's angular distance from the apogee as seen from O) by the formula $\kappa = \bar{\kappa} \mp c$, where c is known as the "equation" (προσθαφαίρεσις, *equatio*).

The epicyclic form of the general model is depicted by the solid lines in Figure 2. Here an epicycle, with center C, moves on a circle (known as the "deferent," *deferens*, because it "carries around" the epicycle), with uniform motion around that deferent's center O (the earth), while the body P moves uniformly around C. If the angular velocities of P around C and C around O are equal and in opposite senses, and if the radius of the epicycle is equal to the eccentricity (e) in the eccentric form, then the eccentric and epicyclic models are exactly equivalent. This is apparent in Figure 2, where the dashed lines represent the eccentric model, and $\bar{\alpha} = \bar{\kappa}$, and $r = e$.

Either model is satisfactory for Ptolemy's solar theory. However, it is also possible for the angular velocity of P around C to differ from that of C around O, as it does in Ptolemy's theory of the moon; or, again, P may rotate in the same sense as C, as in the theory of the planets. Thus the epicyclic form is adaptable to a variety of requirements.

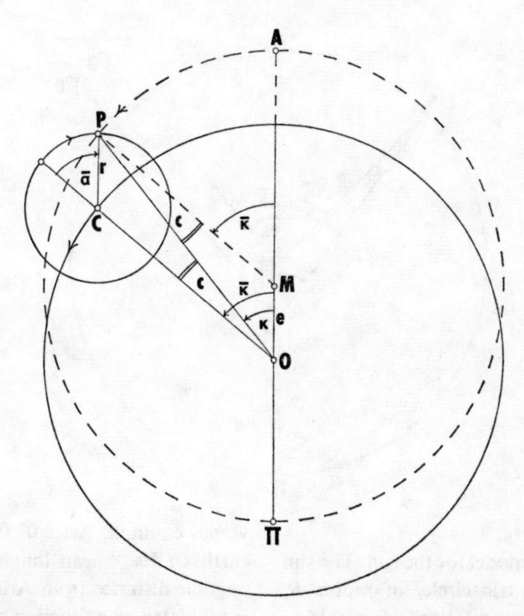

Fig. 2

Illustration of the equivalence between the eccentric and epicyclic models in the Ptolemaic theory. The eccentric model is represented by the broken lines, the epicyclic by the solid lines. In the latter, the epicycle's center *C* moves with uniform angular velocity on a circle (the deferent) with center *O*; the body *P* moves with uniform angular velocity on the epicycle around the center *C*. If (as depicted here) *P* and *C* move with equal angular velocities in the opposite senses (i.e., $\bar{\alpha} = \bar{\kappa}$, or *CP* is parallel to *OMA*) and if *r*, the radius of the epicycle, equals *e*, the eccentricity of the eccentric model, then *P* is brought to exactly the same position relative to *O* in either model; and therefore the two are completely equivalent. In both models $\kappa = \bar{\kappa} \mp c$.

The Sun

As explained above, Ptolemy represents the sun's motion by a simple eccentric (or the equivalent epicyclic model). The eccentric form is depicted in Figure 3, where the sun *S* moves with uniform motion about *M*, and the earth is located at *O*. The position of the sun (as of all the bodies) is given by its "longitude," i.e., its angular distance along the ecliptic from the vernal equinox (conventionally taken as the beginning of the ecliptic, Aries 0° [♈0]). Its "true longitude," λ, as measured at *O*, is related to its "mean longitude," $\bar{\lambda}$, measured at *M*, by the formula $\lambda = \bar{\lambda} \pm c$, where *c* is the "equation" mentioned above. Converting mean longitude to true longitude is known in medieval parlance as "equating" (*equare*) the body in question.

To compute the true longitude of the sun for a given instant from this model one needs to know (1) the ratio of the eccentricity $OM{:}MA$ ($e{:}R$); (2) the longitude of the apogee λ_A; (3) the rate at which $\bar{\lambda}$ increases (known as the "mean motion," ὁμάλη κίνησις, *medius motus*); and (4) the position of the sun

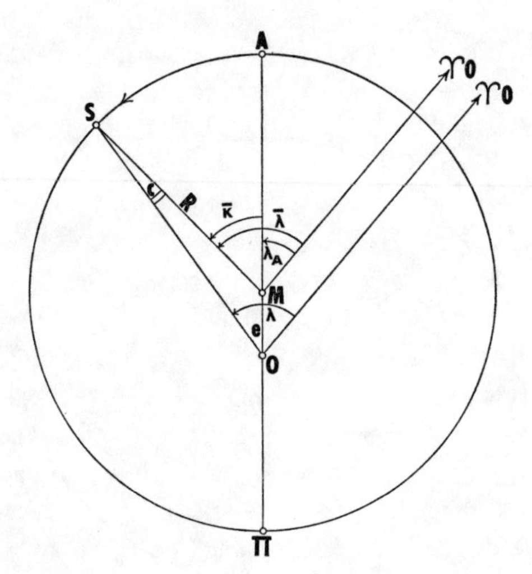

Fig. 3

Ptolemaic eccentric model for the sun. The sun S moves on an eccentric circle, of radius R, center M, and eccentricity relative to O of $OM = e$, with uniform angular velocity about the center M. The sun's position is given by its "true longitude," λ, which is its angular distance from the vernal equinox, Aries 0° (♈0), measured at the earth O. Its "mean longitude," $\bar{\lambda}$, which is its angular distance from Aries 0° measured at M, is tabulated as a function of time. The longitude of the apogee, λ_A, is taken as a constant. Then the mean anomaly, $\bar{\kappa}$, is given by $\bar{\kappa} = \bar{\lambda} - \lambda_A$, and the true longitude is found from $\lambda = \bar{\lambda} \pm c(\bar{\kappa})$.

for some fixed point in time ("epoch"). Having obtained these by observation and computation, Ptolemy computes (by trigonometry) the equation c as a function of $\bar{\kappa}$, the mean anomaly (which Campanus calls "argumentum solis"), and tabulates it. He tabulates $\bar{\lambda}$ from epoch as a function of time. He supposes the sun's apogee A to be tropically fixed.[2] Hence $\bar{\kappa}$ is derived from $\bar{\lambda}$ by $\bar{\kappa} = \bar{\lambda} - \lambda_A$.[3] Equation c being a function of $\bar{\kappa}$, the true longitude is found by $\lambda = \bar{\lambda} \pm c(\bar{\kappa})$.

The Moon

A simple eccentric like the one used for the sun could not adequately represent the lunar motion, since the period of the moon's return to the same

2. That is, that λ_A, the longitude of the sun's apogee from the vernal equinox, remains unchanged; in the case of the planets, however, he supposes that their apogees share the "motion of the fixed stars," which move slowly forward through the ecliptic from the vernal equinox (otherwise known as "the precession of the equinoxes"). Islamic and later medieval astronomers, including Campanus, suppose the sun's apogee, too, to share this motion.

3. Campanus (p. 154, lines III.158–59) refers to a mean motion table in which $\bar{\kappa}$ is tabulated directly: see Commentary to section III, note 33.

velocity (the "anomalistic month") is longer than the period of its return to the same fixed star (the "sidereal month"): in other words, the moon's apogee (where its apparent motion is slowest) moves rather quickly forward through the ecliptic. Such a motion could be represented by an eccentric in which the line of the apsides (corresponding to $A\Pi$ in Figure 1) rotates from west to east through the ecliptic around O. But to accommodate his further refinements, Ptolemy prefers to represent it by the geometrically equivalent model (Fig. 4) in which the moon P travels on the epicycle in the opposite sense and at a smaller angular velocity than that at which the epicycle (center C) travels about the earth O. Thus in Figure 4, if we imagine that at time t_0 the moon was in A, the apogee of the epicycle, when the longitude of C, the center of the epicycle, was ♈0 ($\bar{\lambda} = 0°$), then, at the instant depicted, $\bar{\alpha}$, which is the angular distance traveled by the moon on the epicycle since t_0, is less than $\bar{\lambda}$, which is the angular distance traveled by the epicycle center since t_0. To calculate the

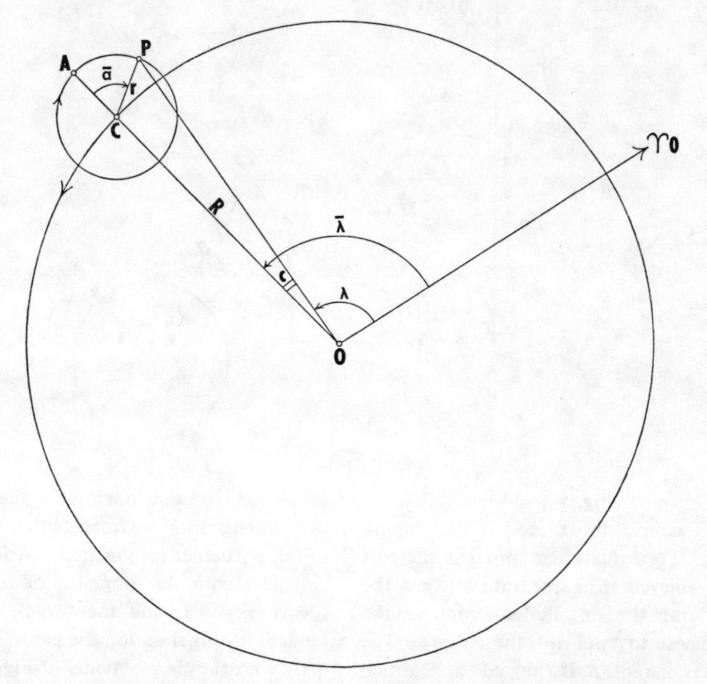

Fig. 4

Ptolemy's simple lunar model. The center C of the moon's epicycle moves on a deferent, with radius R, from west to east with uniform angular velocity around the deferent's center O (the earth). The moon's body, P, moves in the opposite sense on the epicycle, with radius r, with uniform angular velocity about C, but more slowly than C moves about O. Thus, over a given period of time, the increase in the mean anomaly $\bar{\alpha}$ is less than the increase in the mean longitude $\bar{\lambda}$ (the angular distance of the epicycle center from Aries 0°), as depicted here. For simplicity, we imagine that, when the moon P was last at apogee A, then C was at Aries 0° (i.e., both $\bar{\alpha}$ and $\bar{\lambda}$ were 0°); in the configuration shown, $\bar{\alpha} < \bar{\lambda}$. The equation c is a function of the mean anomaly $\bar{\alpha}$, and the true longitude, λ, is given by $\lambda = \bar{\lambda} \pm c(\bar{\alpha})$.

position of the moon for a given instant from this model, one needs to know the rates of increase of $\bar{\alpha}$ (the moon's mean motion in anomaly) and of $\bar{\lambda}$ (the moon's mean motion in longitude), as well as the ratio of the radius of the epicycle ($r = CP$) to the radius of the deferent ($R = OC$). From this ratio Ptolemy calculates and tabulates the equation c as a function of $\bar{\alpha}$. Then the true position of the moon is given by $\lambda = \bar{\lambda} \pm c(\bar{\alpha})$.

Ptolemy found that, although this model, with the parameters he had established, agreed well with observations at syzygies (at new and full moon, when the moon is in conjunction with or in opposition to the sun), at other configurations of moon and sun discrepancies occurred between observed and calculated positions, and these discrepancies reached a maximum at quadrature (when the moon is 90° away from the sun). To account for them it seemed necessary to increase the apparent size of the epicycle from syzygy to quadra-

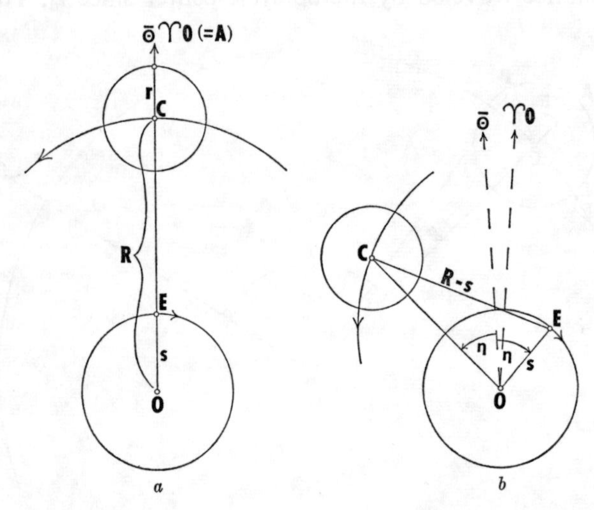

Fig. 5

Ptolemy's refined lunar model: the "crank mechanism." To increase the apparent effect of the moon's epicycle near quadrature (when the moon is 90° from the sun), Ptolemy contrives the following device to "pull in" the epicycle. The radius of the deferent, R, is "hinged" at E, which divides R into two sections, s and $R-s$. As C moves around the earth O from west to east, E moves in the opposite sense on the "small circle," with center O and radius s. E's motion is governed as follows: the angle η between E and the mean sun is always equal to the angle (the "mean elongation") between the mean sun and the epicycle center C. Figure 5a depicts the situation at mean conjunction: the epicycle center has the same longitude as the mean sun $\bar{\odot}$ (for simplification we imagine that the longitude of this particular conjunction is Aries 0°); E also coincides with the longitude of the mean sun (hence $\eta = 0°$); and the "crank" is fully extended. In this situation the model coincides exactly with the simple model of Figure 4. Figure 5b depicts the situation about four days later: both epicycle center C and mean sun have moved to the east, while E has moved in the opposite direction but over the same distance as has C from the mean sun, and the "crank mechanism" has brought C closer to O, so that $OC < R$. The length of OC is a function of the double elongation 2η; it is a minimum $(R-2s)$ when $2\eta = 180°$ and a maximum (R) when $2\eta = 0°$.

ture. Consequently Ptolemy incorporates a "crank" device which "pulls in" the epicycle ever closer as it moves from syzygy to quadrature. The device is as follows (Fig. 5): the epicycle center is considered to move, no longer on a circle of radius R about O, but on a circle of radius $R-s$ about a point E. E itself moves on a circle of radius s around O, in the opposite sense to the motion of the epicycle center. E's motion is governed as follows: if we give the name "mean sun" to an imaginary body which moves in the ecliptic at a constant rate equal to the mean motion of the sun and which coincides with the true sun at solar apogee, then E has the same longitude as the mean sun when the moon's epicycle center C has that same longitude (Fig. 5a), and thereafter moves away from the mean sun so that the angle η between E and the mean sun is always equal to the angle (known as the "mean elongation") between the mean sun and the moon's epicycle center C (Fig. 5b); in other words, η increases at a rate equal to the *difference* between the mean velocities of moon and sun. It is clear that, when η is 90° (at mean quadrature), the "crank" will pull the epicycle center in close so that its distance from O is a minimum ($R-2s$); but, when η is 0° or 180° (at mean conjunction or opposition), its distance from O will be R—i.e., the situation will be the same as that in the simple model of Figure 4. It must be emphasized that the uniform motion of the epicycle center in this refined model continues to take place about O, so that the refinement does not

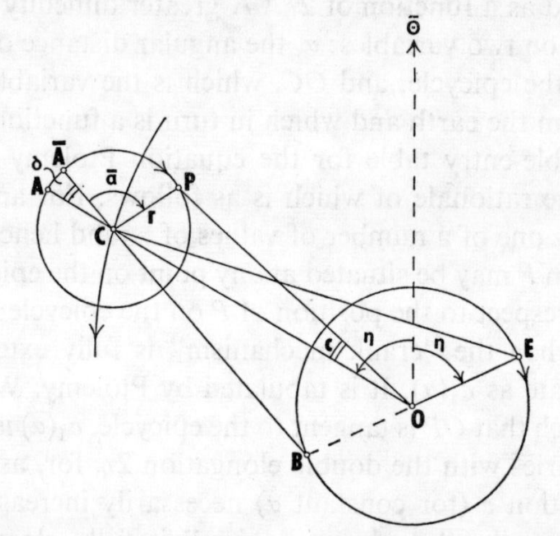

Fig. 6

The complete refined Ptolemaic lunar model. The epicycle's center C moves with uniform velocity about O. The point E moves in the opposite sense on the "small circle" about O, with uniform velocity with respect to the mean sun, bringing the epicycle alternately closer to and farther from O. The moon P moves uniformly about the epicycle center C: its mean anomaly $\bar{\alpha}$ is counted from the "mean apogee of the epicycle" $\bar{A}$, which is the point on the epicycle farthest from B, the point on the small circle diametrically opposite E: thus δ, the angle between $\bar{A}$ and A (the "true epicyclic apogee"), is a function of 2η. The equation c is a function of (1) the true anomaly, angle ACP ($\bar{\alpha} \pm \delta$), and (2) the varying distance OC, which is, in turn, a function of the position of E, i.e., of the angle 2η.

affect the *longitude* of the epicycle center, only its *distance* from O (and hence the size of the equation produced by the anomalistic motion).

One further refinement introduced by Ptolemy to achieve better agreement with observation is illustrated in Figure 6. The mean anomaly $\bar{\alpha}$ is no longer measured from A, on the extension of OC, but from a point $\bar{A}$ on the extension of BC, where B is the point diametrically opposite E on the small circle around O. $\bar{A}$ is known as the "mean (epicyclic) apogee" (μέσον ἀπόγειον, *aux media*), as opposed to the true apogee (*aux uera*) A.

A slight complication occurs because the moon's orbit lies, not in the plane of the ecliptic, but in a plane inclined at an angle of about 5° to the ecliptic and moving slowly westward through it. Thus the whole of the system of Figure 6 is put in such a plane; the two points where that plane intersects the ecliptic circle (the "nodes") are supposed to move slowly from east to west. But this is taken into account only in the computation of the moon's "latitude" (its vertical distance from the ecliptic); for longitude computations the system is treated as if it lay in the plane of the ecliptic.

To construct a table for the computation of the moon's position according to Ptolemy's refined model was no longer easy. First, the equation c is best taken as a function, not of the mean anomaly $\bar{\alpha}$, but of the true anomaly α (in Figure 6, $\alpha=\bar{\alpha}+\delta$). So δ, the difference between mean and true epicyclic apogees, is tabulated as a function of 2η.[4] A greater difficulty is that the equation c now depends on two variables: α, the angular distance of the moon from the true apogee of the epicycle, and OC, which is the variable distance of the moon's epicycle from the earth and which in turn is a function of 2η. To avoid a cumbersome double-entry table for the equation Ptolemy uses an interpolation technique, the rationale of which is as follows. For any given value of 2η there may be any one of a number of values of α (and hence of the equation c), because the moon P may be situated at any point on the epicycle; this simple variation of c with respect to the position of P on the epicycle (that is to say, as a function of α), when the "crank mechanism" is fully extended (i.e., when $2\eta=0°$), we designate as $c_1(\alpha)$. It is tabulated by Ptolemy. When P's position on the epicycle is such that OP is tangent to the epicycle, $c_1(\alpha)$ is at its maximum. However, c also varies with the double elongation 2η, for, as the distance OC decreases, the equation c (for constant α) necessarily increases. This increase will be greatest when the "crank mechanism" is fully closed (i.e., when $2\eta=180°$). At that point it is tabulated by Ptolemy. We designate that increase as $c_2(\alpha)$. Clearly the equation c at $2\eta=180°$ is the sum of $c_1(\alpha)$ and $c_2(\alpha)$. Furthermore, for every intermediate value of 2η between $2\eta=0°$ and $2\eta=180°$, there is a maximum equation c, which we will call $c_{max}(2\eta)$, when P is at the

4. In medieval astronomical tables this function, δ, is called "equatio centri," by analogy with the similar function for the planets (see p. 51).

aforementioned tangent point. The value of 2η which gives the largest possible value of c_{max} is $2\eta=180°$ (quadrature), for there OC is a minimum and $c_2(\alpha)$ is a maximum. Similarly the value of 2η which gives the least possible c_{max} is $2\eta=0°$ (syzygy), for there OC is a maximum and $c_2(\alpha)=0$. Once Ptolemy has computed $c_{max}(2\eta)$ for all values of 2η from 0° to 180°, he subtracts maximum $c_1(\alpha)$ from each to derive the increase in c resulting from the pulling-in effect of the "crank mechanism" at each value of 2η; let us call the consequent difference function φ. In using the maximum equations to compute φ, Ptolemy makes the assumption (not quite accurate but a reasonable approximation) that the ratio between two equations with the same α and different 2η is constant and equal to the ratio between the maximum equations, whatever value of 2η we take. That is, for $2\eta_1$ and $2\eta_2$,

$$c(\alpha_1,2\eta_1):c(\alpha_1,2\eta_2)=c(\alpha_2,2\eta_1):c(\alpha_2,2\eta_2)=c_{max}(2\eta_1):c_{max}(2\eta_2).$$

Now he is able to tabulate φ to be applied to any α when 2η is given. He norms the function, finally, so that its maximum at $2\eta=180°$ is 60 and its minimum at $2\eta=0°$ is 0; we designate the normed function as $c_3(2\eta)$. It is an interpolation coefficient and is known in medieval astronomy as "minuta proportionalia" or "minuta proportionis."

Then the true longitude of the moon is calculated from the tables as follows. For time t we find (from previously calculated mean motion tables) the mean motions $\bar{\lambda}(t)$, $\bar{\alpha}(t)$, and $2\eta(t)$. Entering the appropriate column of the table with argument 2η, we find δ and form $\alpha=\bar{\alpha}\pm\delta$. Then with argument α we find $c_1(\alpha)$ and $c_2(\alpha)$; with argument 2η we find $c_3(2\eta)$ and form the equation $c=c_1(\alpha)+c_2(\alpha)\cdot c_3(2\eta)$. Then the true longitude λ is given by $\lambda=\bar{\lambda}\pm c$. Finally, the latitude β is tabulated as a function of ω, the distance of the moon from the ascending node (in the *Almagest* the distance is counted from the north point of the lunar orbit).

The Planets

The most striking phenomenon in the apparent motion of the planets is that they become retrograde at certain points on their orbits. Their retrogradations are determined, not by their position in the ecliptic, but by their position relative to the sun: the outer planets are retrograde near opposition, Venus and Mercury near inferior conjunction (when they are in line with the earth and on the same side of the sun as the earth). This phenomenon is an obvious consequence of the earth's motion in a heliocentric universe. In his geocentric system, however, Ptolemy represents it by a model in which the planet moves on an epicycle in the same sense as that in which the epicycle moves around the earth. Then, for the outer planets, the radius-vector of the planet on the epicycle (CP in Figure 7) must always be parallel to the vector from the earth to the mean sun (line $O\overline{\odot}$ in Figure 7); for Venus and Mercury, the center of the epicycle must coincide with the mean sun. In both cases the planet will be retrograde when it is on the lower part of its epicycle. However, a simple epicyclic model, in which the earth is located at the center of the deferent, will not do: such a model would produce arcs of retrogradation of constant length and at regular intervals in time and space, whereas observation shows that neither of these conditions holds. In Ptolemaic terms, the epicycle center, too, has an anomalistic motion. To avoid a double epicycle in accounting for the

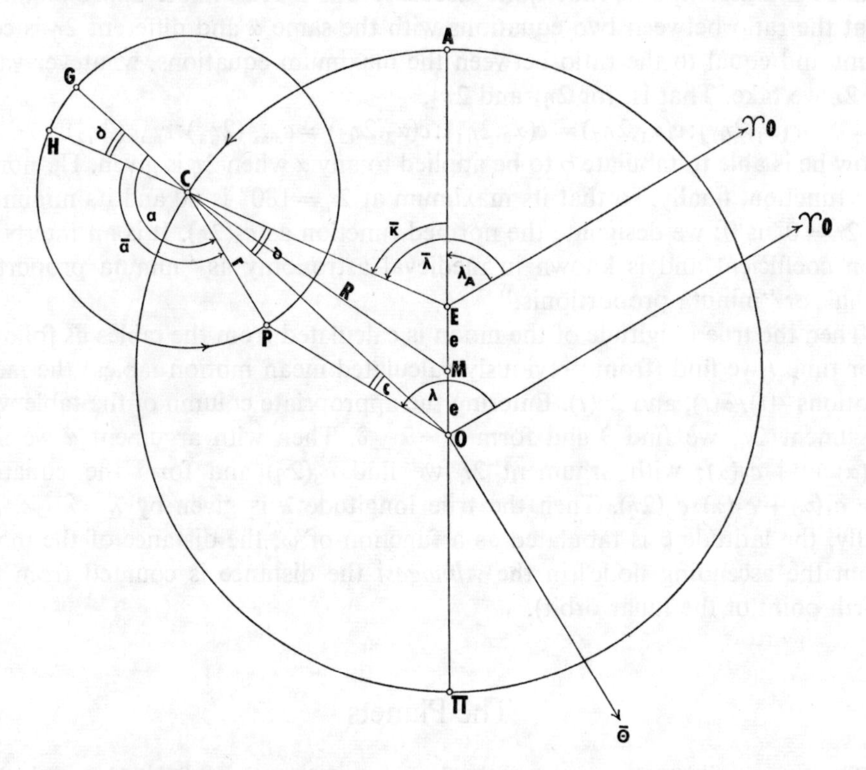

Fig. 7

Ptolemaic model for an outer planet. The epicycle's center C moves on the deferent, with radius R, center M, and eccentricity from earth O of $MO = e$. The uniform motion of the epicycle center takes place, not with respect to the center of its deferent, but with respect to the "equant" E, which lies on the line of the apsides $AMO\Pi$ at a distance from M equal to MO. The planet P moves on the epicycle, of radius r, with uniform motion about C in the same sense as that in which C moves about M, such that the radius vector CP is always parallel to the line of direction from the earth O to the mean sun. The mean anomaly $\bar{\alpha}$ is measured from the mean epicyclic apogee, H, opposite the equant E. The equation c, however, is a function of α (angle GCP), the "true anomaly" (G being the true epicyclic apogee, opposite O). Thus angle GCH (δ) must be applied to $\bar{\alpha}$ to find α; and this angle GCH is equal to angle ECO which must be applied to the "centrum," $\bar{\kappa}$, to find angle AOC (the distance of the epicycle center from the apogee as measured at O). The true longitude of the planet, λ, is given by $\lambda = \bar{\lambda} \mp \delta(\bar{\kappa}) \pm c(\alpha, \bar{\kappa})$.

anomaly of the epicycle center, Ptolemy uses an eccentric deferent. He then introduces an additional refinement to represent his observations better: he makes the uniform motion of the epicycle center take place, not about the center of the deferent (point *M* in Figure 7), but about another point on the line joining the center and the earth, on the opposite side of the center from the earth and equidistant from it (point *E* in Figure 7). This point, for which Ptolemy has no name,[5] is appropriately called "punctum equans" or "equans" in medieval astronomy, and hence "equant" in modern times. (This device of Ptolemy's leads to remarkably accurate results: it can be shown that the longitudes of a body moving according to Kepler's area law on a Kepler ellipse of eccentricity e are reproduced within the limits of ancient accuracy of observation by a Ptolemaic equant model of eccentricity $EM = MO = e$.[6])

The Ptolemaic model for an outer planet is depicted in Figure 7. The planet *P* moves on an epicycle with center *C*. *C* moves in the same sense on a deferent circle with center *M*. The earth is at *O*; the equant, or center of uniform motion ($\bar{\kappa}$), is at *E*; and $OM = ME$. The mean anomaly ($\bar{\alpha}$) of the planet *P* is counted, not from *G*, the point on the epicycle farthest from *O* (the "true apogee of the epicycle," *aux uera epicicli*), but from *H*, the point farthest from *E* (the "mean apogee of the epicycle," *aux media epicicli*). *CP* is always parallel to the line of direction from the earth to the mean sun (another way of saying this is that adding the motion of *C* on the deferent and *P* on the epicycle over a given interval of time gives the mean motion of the sun during the same time). The Ptolemaic model for Venus is identical with that of Figure 7, except that *EC*, not *CP*, is parallel to the line of direction from earth to mean sun.

On the other hand, the observed irregularities of Mercury's motions caused Ptolemy to introduce a device similar to that used for the moon. His observations led him to think that the epicycle appears greatest (i.e., comes closest to *O*) when it is about 120° away from the apogee *A* on either side. Therefore, to pull the epicycle closer to the observer at these points (Fig. 8), he made the center *M* of the deferent revolve about a circle of radius e at the same speed as, but in the opposite sense to, the revolution of the epicycle center about the equant *E*. The center *B* of this "small circle" is the same distance from *E* as is the earth *O*, but on the opposite side of *E*. With the parameters established by Ptolemy ($e = 3$ where $R = 60$), *C* does indeed approach closest to *O* when $\angle ABM$ is approximately 120°.[7] Thus, in one revolution of the epicycle about

5. Ptolemy tended to regard the equant point as the center of a notional circle (see the Commentary to section V, note 3), and hence we also find the medieval terms "circulus equans motum," "circulus equans," or plain "equans" used for the circle along which uniform mean motion is counted (though no body or point actually travels on such a circle).

6. See, for example, Max Caspar, trans., *Johannes Kepler*, *Neue Astronomie* (Munich, 1929), pp. 60*–62*.

7. Not exactly 120°: see the Commentary to section V, note 20.

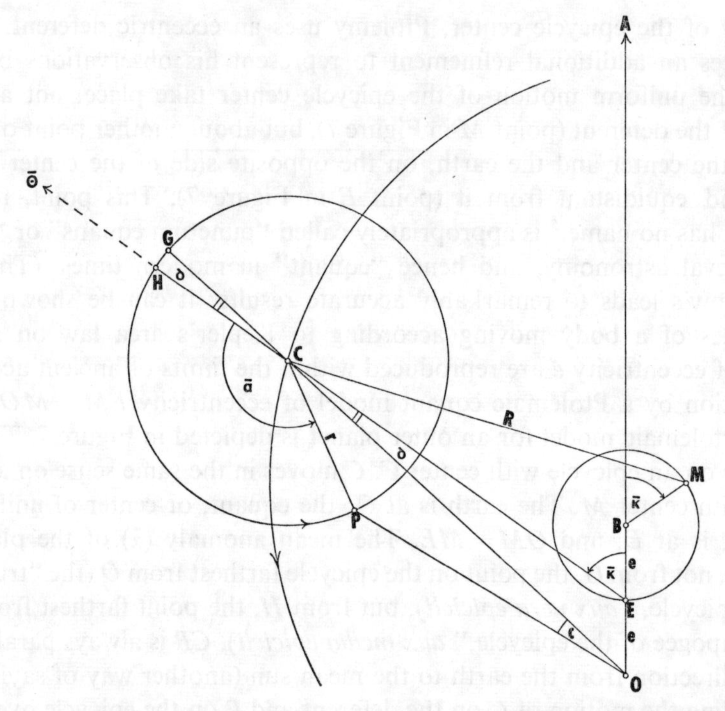

Fig. 8

Ptolemaic model for Mercury. The center C of the epicycle with radius r, moves on a deferent, with radius R and center M. The uniform motion of the epicycle center takes place about an equant point, E, distant from the earth O by the eccentricity e. M is not fixed with respect to O and E, but moves in the opposite sense to the motion of the epicycle center, although with the same angular velocity, around a small circle whose center B lies on the extension of OE at a distance e from E and whose radius is e. Thus, in each revolution of the epicycle center about E, the distance of M from O varies from a maximum of $3e$ ($\bar{\kappa} = 0°$) to a minimum of e (when $\bar{\kappa} = 180°$ and M coincides with E). The distance of the epicycle center OC also varies in each revolution of the epicycle, and its maximum, too, is at $\bar{\kappa} = 0°$, but its minimum occurs when angle ABM is about 120° ($\bar{\kappa} \approx 120°$ and 240°); consequently OC reaches a minimum (the epicycle is in the perigee) *twice* in each revolution. The planet P moves with uniform angular velocity about C in the same sense as that in which C moves about M. Its mean anomaly, $\bar{\alpha}$, is measured from the mean epicyclic apogee H, opposite E. As seen from E, the center of the epicycle C coincides with the direction of the mean sun.

O, C is once in the apogee (for $\bar{\kappa} = 0°$) but *twice* in the perigee (for $\bar{\kappa} \approx 120°$ and $\bar{\kappa} \approx 240°$).

In the construction of tables for the computation of the positions of the planets, Ptolemy was faced with the same problem as that which he encountered in the tables for the moon—namely, that the equation of anomaly c depends on two variables. He solves it in a similar fashion. For the planets, however, he computes the values of c not at *two* distances (maximum and minimum) but at *three*: maximum, minimum, and mean (when $OC = R$). He tabulates the values for mean distance: we call this function $c_1(\alpha)$; he then tabulates the *difference* functions, to be subtracted for maximum distance and to be added for minimum distance: we call these $c_2(\alpha)$ and $c_3(\alpha)$, respectively. He then computes an interpolation coefficient for the planets in exactly the same way as he did the coefficient for the moon, taking the ratios of the maximum equations of anomaly for a series of intermediate positions of the epicycle from apogee to perigee: clearly this coefficient is a function of $\bar{\varkappa}$ (called κέντρον, *centrum*: the "angle at the center").[8] It is normed so that it is 60 at maximum distance of the epicycle (at the apogee, where $\bar{\varkappa}=0°$), decreasing to 0 at mean distance of the epicycle and increasing to 60 again at minimum distance (at perigee, where, except for Mercury, $\bar{\varkappa}=180°$): i.e., it measures the decrease or increase, over the maximum equation of anomaly at mean distance, of the maximum equation of anomaly at greater or lesser distances of the epicycle.[9] We call this function $c_4(\bar{\varkappa})$. To find the true longitude of a planet, it is also necessary to know the angle δ, the difference between the angle $\bar{\varkappa}$ at E and the angle AOC at O (Fig. 7); δ is tabulated as a function of $\bar{\varkappa}$, the centrum, and hence is known as the "equation of center" (*equatio centri*).[10] Since $\angle GCH$, the difference between the mean epicyclic apogee H and the true epicyclic apogee G, is also equal to δ, δ is employed twice (with opposite signs) during the process of finding a planetary longitude from the tables.

The complete calculation for planetary longitudes is as follows (see. Fig. 7). For time t we find (from previously calculated tables) the mean motions $\bar{\lambda}(t)$ and $\bar{\alpha}(t)$. Knowing the longitude of the apogee λ_A, we form $\bar{\varkappa}=\bar{\lambda}-\lambda_A$. With $\bar{\varkappa}$ as argument we find δ and form $\alpha=\bar{\alpha}\pm\delta$. With α as argument we find $c_1(\alpha)$ and either $c_2(\alpha)$ or $c_3(\alpha)$, according to the position of the epicycle (the value of $\bar{\varkappa}$ tells us which to take). With $\bar{\varkappa}$ as argument we find $c_4(\bar{\varkappa})$ and form either $c=c_1-c_2\cdot c_4$ or $c=c_1+c_3\cdot c_4$. Then the true longitude λ is given by $\lambda=\bar{\lambda}\mp\delta\pm c$.

Planetary Latitudes

Since no extant section of the *Theorica* deals with planetary latitudes, a brief summary of Ptolemy's theory will suffice.

8. But see note 10, below.

9. For Mercury, which is twice in the perigee in one revolution of the epicycle, the function rises to 60 again at $\bar{\varkappa}\approx 120°$, then decreases once more to about 40 at $\bar{\varkappa}=180°$.

10. In the *Almagest* two functions have to be added to get the equation of center, but in all later ancient and medieval tables it is tabulated as a single function. The angle AOC ($\varkappa = \bar{\varkappa}\mp\delta$) is known as the "equated center" (*centrum equatum*). From Ptolemy's "Handy Tables" onward, merely for convenience in computation, functions which have to be used later in the procedure—the *minuta proportionalia* c_4 and the planetary stations (see the Commentary to section V, note 25)—are tabulated as functions of $\varkappa$ instead of $\bar{\varkappa}$.

All five planets move in orbits which are slightly inclined to the earth's orbit and intersect it on lines passing through the sun. The transposition of this situation to a geocentric universe is not simple, and Ptolemy's models do not solve the problem adequately. They are as follows:

For the outer planets the epicycle moves in a deferent plane which is inclined to the ecliptic at a fixed angle i_1 (Fig. 9), intersecting it in a fixed line passing

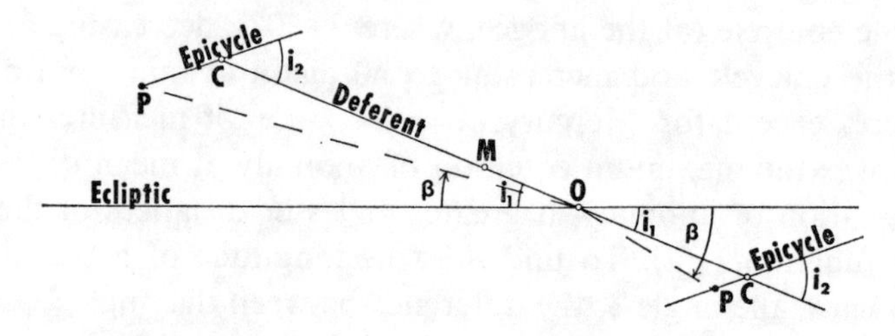

Fig. 9
Illustration of Ptolemaic latitude theory for an outer planet. The figure represents a vertical cross-section of the planet's orbit at right angles to the line of the nodes (the line of the intersection of the ecliptic and the planet's orbit). The epicycle is depicted in both the north point and the south point of the orbit. The orbit is inclined to the ecliptic at a fixed angle i_1. The epicycle, in the two positions depicted in the figure, is inclined to the orbit at an angle i_2, which is greater than i_1. When the epicycle is in the nodes, the inclination between epicycle and orbit becomes equal to i_1 (i.e., the epicycle lies in the plane of the ecliptic). Between the north or south point and the nodes (a distance of 90°), the inclination of epicycle to orbit varies sinusoidally between the maximum i_2 and the minimum i_1. The planet's latitude β is the angle between the ecliptic and the line of sight from the earth *O* to the planet *P*. Since the planet may be anywhere on the epicycle, β can vary considerably for any given position of the epicycle. In the upper part of Figure 9, β is a minimum for that position of the epicycle; in the lower part, β is intermediate, but nearer maximum for that position of the epicycle.

through the earth *O* (the line of the nodes). The plane of the epicycle in turn is inclined to the deferent plane so that the two planes intersect along a diameter of the epicycle parallel to the line of the nodes. If Ptolemy had made this inclination equal to i_1 (so that the epicycle was always parallel to the ecliptic),[11] he would have achieved a reasonable representation of the true situation. However, his observations led him to believe that the inclination of the epicycle is i_1 only when the epicycle is in one of the nodes, and that, when it is in the north or south point of the orbit (see Fig. 9), its inclination is greater (i_2), and that the inclination varies sinusoidally between i_1 and i_2 as the epicycle travels from the node to the north or south point of the orbit.

For the inner planets the situation in the *Almagest* model is even more complicated. As was true for the outer planets, the deferent plane intersects

11. This solution is in fact adopted in Ptolemy's later work, the *Planetary Hypotheses*, and also in his "Handy Tables."

the ecliptic in a nodal line passing through *O*.[12] But this plane is no longer at a fixed angle to the ecliptic. Instead its inclination varies with a period equal to that of the epicycle's passage around the deferent: when the epicycle center *C* is 90° from the nodal line, the inclination between deferent and ecliptic is at a maximum i_1; as the epicycle moves toward the node, the inclination between deferent and ecliptic decreases to zero, so that, when the epicycle is in the node, the deferent plane coincides with the ecliptic plane. The epicycle, too, is inclined to the deferent, so that the two intersect along a diameter of the epicycle which is at right angles to the line of the nodes. The angle of inclination between epicycle and deferent also varies sinusoidally from a maximum i_2 when the epicycle center is 90° from the node to a minimum i_3 when it is in the nodal line. This unnecessarily complicated model was simplified later by Ptolemy, but that is how it stands in the *Almagest*.

It is probable that very few in the Middle Ages thoroughly understood the workings of the above models for planetary latitude, or how the latitude tables were derived from them. An indication of the state of knowledge of planetary latitudes is the fact that, when the tables of al-Khwārizmī were adapted to a different base (60 instead of 150) in eleventh-century Spain, an error was made in the conversion of the planetary latitude tables such that the results derived from them would have been grossly wrong.[13] Yet these erroneous tables were copied and recopied, not only in Adelard of Bath's translation of al-Khwārizmī, but notably in the Toledan tables which had such a wide vogue in the West during the thirteenth century.

The System of Spheres

In the *Almagest* the dimensions of each planetary orbit are given, not in absolute distances, but only in terms of the radius of the planet's deferent, which is always taken as 60 units. The *relative* sizes of the epicycle, eccentricity, and deferent suffice for the prediction of planetary longitudes. Only for sun and moon are absolute distances (in earth-radii) established, from observations of the apparent diameters of the bodies and of the lunar parallax. The sole remark in the *Almagest* relevant to the subject of planetary distances is found in Ptolemy's brief discussion of the order of the planets (with regard to distance from the center) at Book 9, Chapter 1.[14] He there says that the order cannot be

12. For both Venus and Mercury this nodal line is at right angles to the apsidal line passing through *O* and the apogee of the deferent.

13. See G. J. Toomer, review of *The Astronomical Tables of al-Khwārizmī* by O. Neugebauer, *Centaurus*, vol. 10 (1964), pp. 205–6.

14. Ptolemy, *Handbuch der Astronomie*, trans. Karl Manitius (Leipzig, 1912–13), vol. 2, pp. 92–94 (cited hereafter as Ptolemy, Manitius).

decided by observation, since none of the five planets exhibits an observable parallax. He does indeed accept the (ascending) order moon, Mercury, Venus, sun, Mars, Jupiter, Saturn, but not on any cogent grounds.

However, we find a subsequent adaptation of the Ptolemaic system which allows absolute distances to be deduced for all parts of the system; it is widespread in medieval astronomy, both in Islam and later in the West. Its basic assumptions are that there is no space wasted in the universe and that the order of the heavenly bodies accepted by Ptolemy (see above) is correct. Then, adopting the relative dimensions of the planetary systems established in the *Almagest* (with occasional slight modifications) and taking just *one* absolute distance as given (usually the mean distance of the moon at syzygy as established in the *Almagest*, namely, 59 earth-radii), these *epigonoi* are able to calculate all the other absolute distances. Thus, the maximum distance of the moon in the Ptolemaic system (see Fig. 5) is $(R-s)+s+r$, where s and r are known in terms of $R-s$, and $R-s$ is known as an absolute distance. So the maximum distance of the moon can be calculated in earth-radii, and it is equal to the minimum distance of Mercury (because there is no space wasted). The minimum distance of Mercury is known in terms of the three parameters R, r, and e of Figure 8, and r and e are known in terms of R. Therefore R of Mercury can be calculated as an absolute distance, and hence the maximum distance of Mercury ($R_{☿}+3e_{☿}+r_{☿}$), which is equal to the minimum distance of Venus ($R_{♀}-e_{♀}-r_{♀}$) (cf. Fig. 7), and so forth right up to the sphere of the fixed stars, which is supposed to coincide with the maximum distance of Saturn.

Where the physical assumptions underlying this scheme are stated, they usually include the Aristotelian view that the universe is "full," i.e., that each planetary system is contained in a "sphere" (in fact the space enclosed by two concentric spheres) made up of the fifth essence (see Commentary to section IV, note 56). The universe as a whole consists of a series of contiguous hollow concentric spheres enclosing at their center the solid "sublunar" sphere comprised of the earth and its surrounding atmosphere (in the medieval view, itself arranged in the descending order of the "spheres" of fire, air, water, and earth).

Only recently has the conjecture been confirmed that this system was conceived by Ptolemy himself. An exposition of it is found in the last part of Book 1 of his *Planetary Hypotheses*, a part of the work that survived only in Arabic translation and was long overlooked (until the edition by Goldstein).[15] We can now see that nearly all the elements found in later expositions of the system had already been developed by Ptolemy.

15. B. R. Goldstein, "The Arabic Version of Ptolemy's *Planetary Hypotheses*," *Transactions of the American Philosophical Society*, n.s., vol. 57 (1967), pp. 3–55.

Indeed, Ptolemy, even more mindful of cosmic economy than his successors, did not consider it necessary to enclose each planetary system in a "sphere": he thought it sufficient to assign to each a "drum" cut out of such a "sphere," just wide enough to accommodate the planet's motions in latitude. Otherwise the order and arrangement of the systems are identical with those of the medieval picture. One feature that undoubtedly helped to convince Ptolemy of the truth of the system and to assure its acceptance by others was the remarkable fact that the greatest absolute distance of Venus, as derived from the system via the mean distance of the moon and the *Almagest*'s relative parameters for the moon, Mercury, and Venus, comes very close indeed to the minimum absolute distance of the sun derived in the *Almagest* (Book 5, Chapter 15) by a totally different method, which used the apparent diameter of the sun. This virtual agreement was the purest coincidence (Ptolemy's figure for the solar distance is too small by a factor of 20). But it assured the success of the system and also confirmed the belief that Mercury and Venus lay below the sun rather than above it, as some pre-Ptolemaic astronomers had assumed (cf. the Commentary to section VI, note 83).

In the *Planetary Hypotheses* Ptolemy derives not only the absolute distances but also the absolute sizes of the bodies of the planets. His determinations use the following "observations" of the apparent sizes of the planets' diameters as a fraction of the sun's apparent diameter: Mercury, $\frac{1}{15}$; Venus, $\frac{1}{10}$; Mars, $\frac{1}{20}$; Jupiter, $\frac{1}{12}$; Saturn, $\frac{1}{18}$; first magnitude stars, $\frac{1}{20}$. From his statement it appears that the "observation" of Venus, certainly, and possibly the "observations" of all were taken from Hipparchus. Accurate observations of the apparent diameters of these small objects were impossible before the invention of the telescope, and the above are at best estimations. But, taken as observations of the apparent size at mean distance, they yield absolute sizes of diameters, since the absolute mean distances have been established, and in the *Almagest* Ptolemy had already established the relative sizes of sun and earth diameters as $5\frac{1}{2}:1$. When the apparent diameters of the planets are given in later sources, they are always, as far as I know, mere repetitions of the above, though the absolute sizes derived from them often differ slightly from Ptolemy's results.

Other features introduced by Ptolemy which are frequently found in later sources are (1) a list of the volumes of the celestial bodies in terms of the earth's volume (immediately derivable from the absolute sizes of the diameters by cubing them) and (2) conversion of the absolute distances in earth-radii to absolute distances in metrical units. This entails an estimation of the size of the earth in metrical units: Ptolemy gives the circumference of the earth as 180,000 *stades* (a figure also known from his *Geography*; in modern terms, one stade is about an eighth of a mile). Later imitators employ different metrical units and different estimates of the earth's size, but the principle is the same.

Campanus adopts the system outlined above in every detail, with the omission only of the apparent diameters.[16] However, he takes from his sources (the *Almagest* and al-Farġānī's astronomical work: see p. 34, above) only the basic parameters, and himself calculates the absolute distances with great meticulousness. For instance, in calculating the absolute dimensions of a planet's sphere, he takes into account the fact that the planet is not a point but a body with some thickness; therefore the thickness of the sphere must be increased by one diameter of the planet's body. Campanus has an additional feature, which I have not found in any earlier source, that he probably introduced himself: he gives the circumference of the deferent and epicycle in metrical units (derived from their diameters), and thence calculates the absolute distances (in "miles") traveled by the epicycle and planet, respectively, in one day (the *dieta*).[17] By a similar calculation he finds the daily travel of the sphere of the fixed stars in "miles." In a final stroke of supererogation (p. 344, lines VI.729–31) he calculates the area of the sphere of the fixed stars in square "miles," but mercifully leaves us to find the areas of the other spheres for ourselves.[18]

16. He assumes directly the absolute sizes established by al-Farġānī, an unsystematic procedure since al-Farġānī's absolute distances differ from his own.

17. It appears probable, however, that the *dieta* passages are spurious additions by a later author. See pp. 28–29, above.

18. Another minor feature he gives, which I have not found earlier, is the size, in "miles," of the circumference (as well as the usual diameter) of the heavenly bodies.

Manuscripts and Sigla

The Manuscripts of the *Theorica*

Over sixty manuscripts containing the *Theorica planetarum* have been discovered, to my knowledge, and, although there may be others of which I am unaware, I have sought to survey and analyze all that were available. Of these known manuscripts, only two consist of the *Theorica* alone (MSS *H* and *N*).[1] In eleven others the *Theorica* is the opening work. In only one manuscript (MS *a*) is the *Theorica* alien from the remaining articles.[2] In all the rest of the manuscripts, the accompanying works are predominantly scientific and almost wholly confined to three of the four parts of the quadrivium, namely, to arithmetic, geometry, and astronomy.[3] Although scholars of the Middle Ages did not make the sharp distinction between astronomy and astrology that we do today, it is perhaps surprising to note the relative paucity of strictly astrological works accompanying the *Theorica*.[4] The most commonly associated astrological work is the *Introductorius* of Alchabitius (al-Qābīṣī) with ten copies, followed by six of the *Flores* of Albumasar (Abū Maʿšar), four of Ptolemy's *Centiloquium*, and three each of the *Centiloquium* of Hermes and the *De eclipsi lune* of Messehallah (Māšāllāh).

A cursory analysis, by title, of the frequency with which works from each of the aforementioned three parts of the quadrivium accompany the *Theorica* reveals an overwhelming preponderance of works on astronomy, with those on geometry next most frequent, and finally those on arithmetic lagging far

1. See the list of manuscripts that follows. MSS *G* and *g* seem to belong to this category since they are accompanied only by incomplete works; see their descriptions, pp. 70–71, below.

2. In MS *I* the *Theorica* is one of two items, the first of which is not scientific.

3. Only three manuscripts contain works concerning music (the second part of the quadrivium). They are MS *A*: Boethius, *De musica*; and MSS *b* and *M*: Johannes de Muris, *Musica*.

4. Although one or more astrological works may be found in over twenty of the manuscripts, only five (MSS *j*, *l*, *w*, *X*, and *Y*) include more than five treatises in astrology, and only one manuscript (*X*) seems predominantly astrological.

behind. General works on astronomy are by far the most prevalent, constituting perhaps half of that category; works on instruments comprise about one quarter and tables and works on the computus another quarter. This analysis is somewhat deceptive since many of the titles falling into the "general" astronomical group identify very short works, whereas the numerous copies of the Toledan and Alfonsine tables are of far greater substance. The most often-accompanying work is the *Theorica planetarum* ascribed to Gerardus, which appears in twenty different manuscripts, sometimes in more than one copy. The next most frequent is Messehallah's *Astrolabe* in thirteen copies. The Alfonsine tables are found in at least eleven copies, to which number might be added the apparent revisions of those tables by Johannes de Blanchiniis, Johannes de Lineriis, and Johannes de Muris. John of Sacrobosco's *Sphere* occurs in eleven copies, al-Farġānī's *Rudimenta* in nine, and az-Zarqāl's *Canons to the Toledan Tables*, Euclid's *Elements*, and Profatius Judeus's *Almanach* each in seven copies. There are also six copies of John of Sacrobosco's *Algorism* and one of his *Computus*. Of the other works of Campanus, six may be found, the most frequent being the *Computus*. The few medical works accompanying the *Theorica* may virtually all be classed as astrological medicine.[5] It would seem, then, that the manuscripts containing the *Theorica* were intended for university use in the standard curriculum for scholarly study of the later Middle Ages.[6] We must not, however, assume too much from the evidence in hand, because it is not always sufficient to indicate whether we are dealing with manuscripts in their original state upon which reliable conclusions could be based.

Within the larger manuscript categories we have established,[7] certain of the manuscripts are more intimately related than are others. On at least six occasions manuscripts *d*, *S*, and *t* (sometimes joined by *k*) have the same additions or marginalia. Manuscript *G* is similarly related to *w*, as are *F* to Σ, and *f* to *y* and Λ (manuscripts *y* and Λ also both contain a long addition at the end of the treatise describing an instrument for mechanically representing planetary motions with cogs). In some of these relationships one of the manuscripts could be a direct copy of another surviving manuscript of the *Theorica*, but in none can we prove it. Such dependence seems most likely for the fourteenth-century Bologna University manuscript (*z*) and the fifteenth-century Vatican Rossiano (*r*): the two manuscripts are of nearly the same length (142 and 137 folios, respectively) and contain the same works with hardly a variation (except in the order); one of these works is a calendar by Fucus de Ferrara which otherwise appears only in MS Paris, Bibliothèque Nationale, fonds latin, 13014. Yet even

5. Five copies of the slight work assigned to Hippocrates, three of the *Almansor*, and two of the *Urina non visa*, among others.

6. See Rashdall, *Universities of Europe*, vol. 1, pp. 247–49, 449, and corresponding notes.

7. α, β, δ, θ, and φ: see "Editorial Procedures," pp. xiv–xv, above.

here the text of the Campanus treatise in the later manuscript, *r*, though very similar, is not completely identical with that in the earlier, *z*, and direct copying is not absolutely proven.

The following descriptions of the manuscripts containing the *Theorica* depend upon the ready availability and reliability of the printed evidence, such as catalogues and articles; the information in those sources has been supplemented here only when examination of the manuscripts has revealed discrepancies. Those manuscripts for which printed material does not exist or is inadequate have been dealt with in greater detail within the limits of my resources. Since I have worked almost entirely through microfilm or photostats and have not always had the entire manuscript available even in this medium, my presentations must sometimes be limited accordingly. The listing of the manuscripts is ordered alphabetically in accordance with the sigla used for collating the text. Following that listing is a short list of manuscripts that could not be procured or are no longer extant.

For each of the manuscripts I give the following:

The *siglum* we have chosen to denote it, followed in parentheses by the Greek letter designating the group or subgroup of manuscripts to which we have assigned it (see pp. xiv–xv for the explanation of the groupings); the provenance and number of the manuscript; and a general physical description of the manuscript overall.

The approximate *date* of the manuscript.

The *text used*, that is, a description of the copy of the *Theorica* within that manuscript, including foliation, the first word of the *Theorica* in the particular manuscript ("Clementissimo" indicates that it begins at p. 128, line I.4, of our text; "Primus," at p. 136, line II.2, omitting the dedicatory letter; and "Sol habet," at p. 142, line III.2, omitting all introductory matter), and the condition (including presence or absence) of the figures and tables.

A *description* of all the works in the manuscript or a reference to a description printed elsewhere. If the description is primarily in my words, material quoted from the manuscript is set in quotation marks; however, if I have excerpted large amounts of material from the manuscript, I have omitted quotation marks and instead inserted my commentary within square brackets.

1. *A* (θ) = Oxford, Bodleian Library, Auct. F.3.13. Parchment, iii plus 229 fols., written in England (see Neil Ker, *Medieval Libraries of Great Britain: A List of Surviving Books*, 2d ed. [London, 1964], p. 105)

Date: 2d half of 13th century

Text used: fols. 115r^a–42r^a. Primus. Figures incomplete; tables

Description: For a description of this MS, see Falconer Madan, H. H. E. Craster, and others, *Summary Catalogue of Western Manuscripts in the Bodleian Library at Oxford* (Oxford, 1895–1953), vol. 2, pt. 1, pp. 245–46, no. 2177.

However, note that in the manuscript fol. [141b]r[a] is blank and fol. [141b]r[b] contains an incomplete table entitled "Tabula ad collocandas horas in chilindro secundum latitudinem londonensis et componitur hic gradus altitudinis solis cum fuerit in initiis signorum ad eandem ciuitatem."

2. *a* (α) = Maihingen, Schloss Harburg, Fürstlich Öttingische-Wallerstein'sche Bibliothek, II.1.F.10
   Date: 13th century
   Text used: fols. 184(170)r[a]–203(189)v[b] (incomplete). Primus. Figures incomplete; tables
   Description: The *Theorica* is the only scientific work in this MS, according to the librarian. Cf. fol. 204(190)r[a]: "Incipit opus super Gemmam regiminis."

3. *B* (β) = Basel, Öffentliche Bibliothek der Universität, F.II.33. Parchment, folio, 244 fols.
   Date: 14th century
   Text used: fols. 173r[a]–93r[a]. Primus. Figures mostly complete; tables
   Description: Although there is no adequate printed catalogue for this library, Björnbo and Vogl ("Alkindi, Tideus und Pseudo-Euklid," pp. 124–29) and Thorndike ("The Study of Mathematics and Astronomy in the Thirteenth and Fourteenth Centuries as Illustrated by Three Manuscripts," *Scripta mathematica*, vol. 23 [1957], pp. 72–76) have analyzed this particular manuscript. Björnbo and Vogl distinguish thirty-seven different works in nine different hands, of which *B* was written in hand 7. The only other work in hand 7 is that which immediately follows on fols. 193v–94r. *De motu octavae sphaerae* by Ṭābit. Clagett (*Archimedes*, vol. 1, pp. 238–54, 450–506) has edited fols. 116v–22r, *Verba filiorum* of the Banu Musa, and fols. 151r–53r, *De curvis superficiebus*.

4. *b* (δ) = Oxford, Bodleian Library, Bodley 300(500). Parchment, i plus 141 fols. (see Ker, *Medieval Libraries*, p. 25)
   Date: early 15th century
   Text used: fols. 19v[b]–40r[b]. Sol habet. Figures incomplete or absent and space left; tables
   Description: Described in Madan *et al.*, *Summary Catalogue*, vol. 2, pt. 1, pp. 386–88, no. 2474. The following is a correction of the foliation supplied in that catalogue and an addition of the *explicits* for those sections of the manuscript for which I have microfilm.

sec. 1: fols. 1r[a]–19v[a]
sec. 3: fols. 40v[a]–44r[b]. *Explicit:* "...et non corporaliter." Fol. 44v is blank.
sec. 4: fols. 45r[a]–53r[b]. *Explicit:* "...et ideo lune non retrograditur uide figuram."
sec. 5: fols. 53r[b]–64v[b]. *Explicit:* "...isti ymaginationi quare abiciantur etc."

sec. 6: fols. 64v[b]–81v[a]. *Explicit:* "...et hec est ratio et solertia in catis. Amen."
sec. 7: fols. 81v[a]–84r[b]. *Explicit:* "...at totam eius profunditatem."
sec. 8: fols. 84r[b]–90r[a]. *Explicit:* "...eius maiora erunt ut poterit aparere."
sec. 14: ends on fol. 132v[a]
sec. 15: fol. 132v[b]
sec. 16: fols. 133r[a]–v[b]
fols. 134r[a]–39v[b]. *Explicit:* "...illarum duarum horarum si deus uoluerit."
fol. 139v[b]: another, unnumbered work begins. "Incipit kata in numeris et primo coniuncta.../...[fol. 140r[b]] ut prius ex quo patet propositum. Amen."

5. *C* (α) = Brussels, Bibliothèque Royale, 1022–47, 252 fols.
Date: 1415 (see below)
Text used: fols. 53r[a]–82v[b]. Primus. Figures incomplete and complete; tables
Description: Described in R. Calcoen, *Inventaire des manuscrits scientifiques de la Bibliothèque Royale de Belgique*, vol. 1 (Brussels, 1965), pp. 34–37, but the most accurate printed description is in the old catalogue of 1842 (*Catalogue des manuscrits de la Bibliothèque Royale des Ducs de Bourgogne*, Brussels, 1839–42). What follows is from a microfilm of the complete manuscript, the bracketed numbers following the foliation being the separate articles, of which the *Theorica* is numbered as 1029.

fol. (a)v: In hoc libro continentur hec: Primo quatuor de eclipsi solis inuenienda per tabulas [*mg.*: "liber domus beate barbare in colonia ordinis carthusiensis"]; secundo tabule alphonsi correcte et bene distincte; tertio canones magistri iohannis [*mg.*: "danken"] de saxonia super easdem; quarto canones magistri iohannis de muris super tabulam tabularum [in a different hand: "anno 1321 item sermo eiusdem de regulis computistarum"]; quinto canones m. nicolai de aqueductu breues super tabulas alfontii; sexto theorica campani dupliciter uidelicet abbreuiata et ex integro [in a different hand: "uixit anno 1020"]; septimo speculum domini alberti de libris recipiendis et non rectis [in a different hand: "astronomorum"]; octauo opus petri de iuliaco facile de equatione planetarum; nono compotus campani optimus de festis kalendis etiam ipsius [?]indem [in a different hand: "idem calendarium eiusdem"]; decimo regule arismetrice de de minutiis etc.; undecimo tractatus de arte regulandi uasa per m. petrum de iuliaco; duodecimo expositio tabule tabularum m. pauli gerissem; tertiodecimo algorismus cum aliis diuersis de hac materia etc.; quartodecimo [in a different hand: "simonis tunscete si theologice electoris ars componendi instrumentum rectangulum"]; item eiusdem ars operandi cum illo; 15 tractatus arithmeticus ioannis de muris; 16 tractatus contra superstitiosos astrologos; 17 Tractatus de legibus et sectis contra superstitiosos astronomos petri de aliaco episcopi cameracensis et cardinalis.
fol. (b)r: [Modern hand] manuscrit complet et sans laceration en 252 feuillets. 1022–47 [library seal]. [Fol. (b)v is blank.]
fol. 1r[a]: [1022] uerus motus octaue spere [listed in TKr, col. 1691]. [Entries follow

for the years 60, 120, 180, 240,..., 3980 (ends fol. 1r[b]). Fols. 1v–12v are blank.]

fol. 13r: [1023] Per tabulam post hec descriptam solis equationem accessionis et recessionis.... [Fol. 13r[b]] Auges planetarum diuersis temporibus secundum Ptholomeum auges planetarum tempore nabugodonosor; tempore albategni secundum eundem; anno domini 1185; secundum alium, anno 1210.... [Bottom:] Magister rogerus infans composuit tabulas suas anno domini 1184 [cf. Haskins, *Studies in the History of Mediaeval Science*, pp. 124–25].... [Fol. 16v[b]] Ascensiones signorum ad latitudinem 51 gr 30 mi.

fol. 17r: [1024] [Top, in a different hand: "regis alphonsii decimi castelle et legionis."] Incipiunt tabule illustris regis Alfoncii et primo tabule differentiarum [listed in Thorndike and Kibre, *Catalogue*, 1st ed., col. 247; not in TKr] unius regni ad aliud/...[fol. 28v[b]] Tabula ad reducendum tempus christi ad 1ª 2ª 3ª 4ª...[for the years 28, 56, 84,..., 1232, 1260, 1288, 1316,..., 1680]. [Series similar to that in tables for Novara.]

fol. 29r[a]: [1025] Tempus est mensura motus ut uolt philosophus Aristoteles... [listed in TKr, col. 1561]/...[fol. 37r[b]] sicud in coniunctionibus planetarum dictum est et sic est finis canonum magistri iohannis de saxonia super tabulas illustris Alfoncii regis castelle [*mg.*: "Danken"].

fol. 37r[b]: Ad eclipsim solis inueniendum quere primo coniunctionem solis et lune mediam/...[fol. 39v[b]] figuram autem facies secundum doctrinam canonis magistri mei magistri iohannis de lyneriis. Expliciunt canones eclipsium solis et lune ordinati parisius per magistrum iohannem [*mg.*, in a different hand: "danken"] de saxonia anno domini 1330 scripti autem sunt isti anno domini 1410. [Paragraph] Longitudo augis a capite draconis sui. Saturni 0 S, 50 G; Iouis 1 S, 10 G; martis 2 S, 30 G. Lune uero et ueneris et mercurii continue uariatur propter motum proprium ecentricorum a motu octaue spere. Nota quod auges 3 superiorum semper sunt in latitudine equali septentrionali ab ecliptica quia aux et caput draconis in eis semper manent in eadem distantia abinuicem nec mouentur nisi ad motum 8ᵉ spere sed in aliis tribus scilicet uenere mercurio et luna non est ita.

fol. 40r: [1026] [Top, in a different hand: "sermo hic de regulis computistarum etiam infra habetur et adscribitur ioanni de muris."] [Fol. 40r[a]] De regulis computistarum quia cognite sunt a multis per quas se dicunt...[listed in TKr, col. 389; Zinner, *Verzeichnis*, p. 232, no. 7449]/...[fol. 40v[b]] Et hiis ebdomadas cum diebus de quibus est cura computistis inuenire facile est studenti. Explicit sermo de regulis computistarum [in a different hand: "io. de muris ut credo"]. [Followed by table of "aureus numerus"; then "Menses hebreorum" plus "Nota quod secundum hebreos october est prima mensis et illum uocant tysrim; secundus est nouember qui martesiban et sic de aliis per ordinem."]

fol. 41r: [1027] [Top, in a different hand: "sequentes canones ioannis de muris et iam infra [?]habentur."] [Fol. 41r[a]] Si quis per hanc tabulam tabularum proportionis...[listed in TKr, col. 1461]/...[fol. 43v[a]] que fuerint benedicta propter amorem scientie sollempniter exaltare et sic est finis de quo uirginis filius sit benedictus. Explicit canon tabule tabularum editus a magistro iohanne de muris anno domini 1321° mathematicorum excellentissimo.

fol. 43v[a]: [1028] [*Mg.*: "Nicolai."] [?*Rubric.*:] Canones tabularum illustris regis

Alfoncii. Quoniam canones tabularum regis Alfoncii per magistrum iohannem de lineriis... [listed in TKr, col. 1265] et iohannem de saxonia ordinati sunt plurimumque tediosi et longi et ambigui in operando idcirco ego Nicolaus de aqueductu/... [fol. 47v^b] Et est colonia orientalior unde motus tanti temporis in colonia debet subtrahi a radice et motu parisiensi unde *Addit in occasu regio sed subtrahit ortu* est sic est finis de quo primus motor infinitus et immobilis sit benedictus. Nota quod buth planete est idem quod uerus motus arcus planete in una die, scriptum anno 1410 die 16 octobris.

fol. 48r^a: [1029] [Top: "exerpta theorice campani."] Prologus. Primus philosophie magister philosophie negotium in tria... [listed in TKr, col. 1124]. [Fol. 50v^a] Nota hanc rationem que uidetur inter ceteras esse fortior pro reprobatione ecentricorum/... [fol. 50v^b] et sic semper mouetur circulus deferens titubando ab aquilone ad meridiem et super eandem omnino lineam econtrario a meridie ad aquilonem. [Followed by "Ex differentia 22^a Alfragani..."; then "Ex 5^a dictione almagesti Ptholomei.... Si numeri cubicentur erit corpus solis 6644^lum et sesquialterum fere corpori lune, corpus uero terre 170 fere et corpus terre corpori lune 36^lum sesquialterum, idest 36 P $\frac{1}{4}$."]

fol. 51r: [Top: "Expositio terminorum astronomicorum utilis."] [Fol. 51r^a] Ecentricus uel circulus egredientis centri siue egressi cuspidis... [listed in TKr, col. 539/... [fol. 52v^a] super quam est secundus motus latitudinis est in superficie deferentis etc. 1413°. [Bottom:] Nota auges 3 superiorum semper sunt in latitudine equali septentrionali ab ecliptica quia aux et caput draconis in eis semper manent in eadem distantia abinuicem nec mouentur nisi ad motum octaue spere sed in aliis tribus scilicet uenere mercurio et luna non est ita [cf. fol. 39v^b]. [Fol. 52v^b is blank.]

fol. 83r^a: [1030] Speculum alberti magni. Occasione quorundam librorum apud quos non est radix scientie... [listed in TKr, col. 975; 1842 catalogue has the *incipit* "Due magnae sapientiae sunt...," which is listed in the 1st edition of Thorndike and Kibre, *Catalogue*, but omitted in the revised edition]/... [fol. 89r^b] sed quia ambo inueniuntur ab eodem causata et sic est finis.

fol. 89v^a: [1031] Opus m. petri de iuliaco. [P]lurimi quidem scientie astronomie amatores tedio laboris... [listed in TKr, cols. 1055, 438] quapropter Campanus instrumentum materiale sensibile operi equationis planetarum satis conueniens/... [fol. 98v^b] nisi in gradibus parui circuli mercurii quorum computatio uadit uersus occidentem. [Fols. 99r–100v are blank.]

fol. 101r: [1032] [Top: "compotus campani"; in a modern hand: "uixit honeste circa annum 1020." In a different hand:] Rogauit me unus ex hiis quibus... [listed in TKr, col. 1365; Zinner, *Verzeichnis*, p. 72, no. 1924; 1842 catalogue gives a later *incipit*: "Computus est scientia mutationis...," which is listed in TKr, col. 243]/... [fol. 147v] pentheque iohan lau sumptio sancto horum uigilias ieiunes luceque marci. Explicit compotus bonus campani et sequitur kalendarius eiusdem.

fol. 148r: [1033] Prima dies iani timor est et septima uani [listed in TKr, col. 1089]. Ianuarius habet dies 31 luna eius 30 regularis solaris 3 lunaris 9 [listed in TKr, col. 653] et in kalendis eius renouetur litera dominicalis et aureus numerus et claues terminorum itemque annus cicli decemnouenalis et cicli lunaris quorum decemnouenalis incipit tribus annis ante lunarum et annus domini nostri ihesu christi secundum morem ytalie [this article ends with December on fol. 153v]. [Fol. 154r] Nota uersus de ciclo solari annorum numero tu iunge nouena minori que simul adiuncta partire per octo uiginti quod superest ciclum solarem significabit. Si nichil superest tunc octo uiginti nota illum et si partiri nequeunt dant hec tibi ciclum. Versus de ciclo lunari: annorum numero monadem tu iunge minori que simul adiuncta per denos diuide nonos quod superest ciclum lunarem significabit. Si nichil superest nota illum denaque nouem et si partiri nequeunt dat hic tibi ciclum.

fol. 154v: [1034] [Top, in a modern hand: "Canones ioannis de muris super tabulam tabularum—Nota eadem quae supra."] [S]i quis per hanc tabulam tabularum proportionis... [listed in TKr, col. 1461]/... [fol. 158v] que fuerint bene dicta propter amorem scientie solempniter exaltare et sic est finis de quo uirginis filius sit benedictus. Explicit canon tabule tabularum editus a magistro iohanne de muris anno domini 1321 mathematicorum excellentissimo.

fol. 159r: [1035] Regule ad soluendum diuersa enigmata per arismetriam. [*Mg.*: "1ª."] Ad multiplicandum numerum integrorum per qualescumque fractiones... [listed in TKr, col. 54]/... [fol. 161r] ut si 3º dabuntur una 3ª residui et 3 restabunt 5. [Bottom:] Quidam ad obuiantes sibi [?]aucas dixit saluate 100. [Fol. 161v is blank.]

fol. 162r: [1036] *Modum representationis minutiarum uulgarium et phisicarum proponere* [listed in TKr, col. 878]. Quia in fractionibus duo sunt numeri scilicet et numerus numerans... [listed in TKr, col. 1220]/... [fol. 168v] si potes uel ad duas uel ad quam pauciores poteris ut $\frac{2}{4}$ ad $\frac{1}{2}$ uel $\frac{3}{4}$ adduci dimidium et quartam.

fol. 169r: Nota ex algorismo de proportionibus proportionum aliquas regulas utiles et necessarias/... [fol. 172r] ut diameter inter costam et regulam coste etc.

fol. 172v: [1037] Incipit tractatus magistri petri de iuliacho de modo uirgulandi siue mensurandi uasa. Cum de quorumdam corporum regularium et uniformium adinuicem reductione/... [fol. 181v] quarta pars dyametri tunc ueniet pars 16ᵐᵃ ut hec faciliter uideri poterint in tabula ante pro diuisione uirge ordinata.

fol. 181v: Quoniam ars et uirtus sunt circa difficilia secundum philosophum in ethicis/... [fol. 184v] et sponse sue immaculate in regno patris sui cum quo uiuit et regnat in unitate spiritus sancti deus per infinita secula amen.

fol. 184v: [1038] Expositio practice tabule tabularum et propositionum Ptolomei pro compositione tabule sinuum et cordarum necessariarum facta pro simplicibus per me Paulum de gherisheym. Necessitatem et utilitatem tabule sinuum et cordarum astronomorum signifer Ptolomeus ostendit in prima dictione sui almagesti... [listed in TKr, col. 909]/... [fol. 189v] ut complete traderem utilitatem tabule tabularum que fere omnibus astronomie operationibus deseruit. [Paragraph] Declarato nunc usu tabule tabularum aggredier principale propositum uidelicet ad clare demonstrandum propositiones quibus Ptholomeus in suo almagesti ostendit compositionem tabule sinuum. Est itaque prima propositio talis. [In a large hand:] Data circuli diametro latera decagoni pentagoni exagoni tetragoni... [listed in TKr, col. 363, as Theorem i of *Parvum almagest*; cf. Carmody, *Arabic Astronomical and Astrological Sciences*, p. 164]/... [fol. 196r] et in hoc terminantur considerationes compositionis tabule sinuum et cordarum. Sequitur nunc ipsa tabula rectificata anno domini 1443. [Fols. 196v–97v] Tabula sinuum et cordarum rectificata. [Fols. 196v–97r, bottom:] cuiuslibet arcus propositi sinum rectum inuenire.... [Fol. 198r is blank.]

fol. 199r: [1039] [Top, in a modern hand: "simon tunscete author est."] Prologus in artem componendi instrumentum rectanguli. Rectangulum in remedium tediosi et difficilis operis...[listed in TKr, col. 1342]. Capitulum primum de arte componendi tibiam primam. Ad componendum igitur rectangulum preparemus laminam unam eris...[listed in TKr, col. 33]/...[fol. 200r] scribentur gradus trium signorum a paxillo uersus finem in inferiori spatio fiat econtra. Explicit ars componendi rectangulum.

fol. 200r: Incipit prefatio. Incipit prefatio in artem operandi cum illo [*mg.*, in a modern hand: "Simon tunscete author est"]. Rectangulus omnium circulorum magnorum in spera solis....Capitulum primum huius partis de nominibus partium rectanguli. Omnes partes rectanguli certis notabulis nuncupare...[listed in TKr, col. 987]/...[fol. 202v] que iam dicta sunt sufficienter docili discipulo patet uia ad omnia ea que per cetera instrumenta sciri possunt ponimus istic finem. Explicit tractatus operandi cum rectangulo compilatus per egregium uirum Symonem Tunscete sacre theologie doctorem qui et albionem construxit anno domini 1326.

fol. 203r: [1040] [Top, in a modern hand: "Ioannis de muris, NB est idem sermo qui supra."] De regulis computistarum quia cognite sunt...[listed in TKr, col. 389; Zinner, *Verzeichnis*, p. 232, no. 7450]/...[fol. 204v] de quibus est cura computistis inuenire facile est studenti. Explicit sermo de regulis computistarum. [Followed by a table of "aureus numerus," "Martius," and "Aprilis."]

fol. 205r: [1041] [Table of places, no title, with note. Fol. 206r contains a figure with a note. Below:] Si uolueris scire proportionem alicuius paralelli....[Fol. 206v contains a figure. Below:] *Et diuidatur etc* potest hoc fieri sic diuido primo lineam EF in 98 partes....[Fol. 207r is blank.]

fol. 207v: [1042] [Table with years 1440–70. Fol. 208r] Anno domini 1440 incepit hec tabula paschalis.... [Fols. 208v–9v are blank.]

fol. 210r: Omnia que a primeua rerum origine processerunt...[listed in TKr, col. 991]/...[fol. 216v] sub prima ultimi ternarii siue completi siue incompleti qui modus operandi idem est cum producto. Explicit algorismus de integris.

fol. 217r: [1043–44] [V]na medietas scribitur sic $\frac{1}{2}$ et una tertia...[identified in TKr, col. 1596, as a work of Nicholas Oresme]/...[fol. 222v] que est $\frac{1}{2}$ 2 P ueniet proportio AC ad AE et per nonam regulam patet quod hec proportio est $\frac{1}{2}$ 3 P. Sic igitur se habent aspectus signorum celi secundum istam considerationem ut patet in figura. [Fol. 223r is blank.]

fol. 223v: [1045] [Top: "Incipit tractatus arismetrice iohannis muris et subiectum eius sicut et totius arismetrice est numerus. Et diuiditur iste libellus in duas partes principales...."] [N]umerus est duplex scilicet mathematicus...[listed in TKr, col. 959]/...[fol. 226v] Octonarii ad senarium proportio sesquiterna et octonarii ad binarium proportio quadrupla ut patet in figura et sic patet [?]demonstrandum. [Fols. 227r–37v are blank.]

fol. 238r[a]: [1046] [M]ulti principes et magnates noxia curiositate soliciti...[listed in TKr, col. 887; edited in G. W. Coopland, *Nicole Oresme and the Astrologers*

(Cambridge, Mass., 1952), pp. 123–41]/...[fol. 241v[a]] quoniam in fabulis ueritas dicit rex insipiens perdit populum suum et principatus sensati stabilis erit etc. Explicit tractatus contra superstitiosos astrologos. [Fol. 241v[b] is blank.]

fol. 242r: [1047] [Top, in a modern hand: "sic lege nomen authoris huius Petrus de Aliaco episcopus cameracensis et cardinalis." Fol. 242r[a]] Incipit tractatus de legibus et sectis contra superstitiosos astronomos a reuerendissimo in christo patre et domino domino petro cardinali cameracensi. [C]ontemplatio conditoris in suis operibus et specialiter in celestibus corporibus... [listed in TKr, col. 259]/... [fol. 252r[a]] quod ipse nobis concedat qui in celis gloriose uiuit et regnat per infinita secula seculorum amen. Explicit tractatus de legibus et sectis contra superstitiosos astronomos a reuerendissimo in christo patre et domino domino petro cardinali cameracensi quondam per plures annos existente cancellario studii parisiensis et magister in theologia notabilissima compilatus anno christi 1417° die 24 mensis octobris et hoc anno christi 1418 completo die nona mensis martii in consilio constantie pronunciatus.

6. *c* (θ) = Cambridge University, Corpus Christi College, 37. Vellum, 99 plus 1 fols. (see Ker, *Medieval Libraries*, p. 11)

Date: 14th century

Text used: fols. 2r[a]–22v[a]. Primus. One partial figure; tables

Description: Described in M. R. James, *A Descriptive Catalogue of the Manuscripts in the Library of Corpus Christi College, Cambridge* (Cambridge, 1912), vol. 1, pp. 73–77.

7. *D* (δ) = Oxford, Bodleian Library, Digby 168. Parchment, in quarto minori, 231 fols.

Date: probably 14th century, possibly 13th century

Text used: fols. 40r[a]–61v[a]. Clementissimo. Figures mostly complete; tables

Description: Described in W. D. Macray, *Catalogi codicum manuscriptorum Bibliothecae Bodleianae*, vol. 9, *Codices a Kenelm Digby donatos complectens* (Oxford, 1883), cols. 172–77.

8. *d* (α) = Oxford, Bodleian Library, Digby 215. Paper, in folio, 98 fols., manu Italica

Date: 15th century

Text used: fols. 16v[a]–44v[a]. Primus. Figures complete and incomplete; tables

Description: For a description of the manuscript, see Macray, *Catalogi*, vol. 9, cols. 228–29. Two of the articles in the catalogue are inaccurately described. They should be amended and supplemented according to the following.

art. 4: [The *Theorica* ends on 44v[a], not 46[b] as stated by the catalogue. Immediately following the correct *explicit*, a new work begins, unlisted in the catalogue:] Scito quod astrolabium est nomen...[identified in TKr, col. 1409, as Messehallah's *Astrolabium*]/...[fol. 46r[a]] per quem uadit capud arietis et libre diuide circulum ABCD per 360 diuisiones equales sitque omnis quarta circuli. [See R. T. Gunther, *Chaucer and Messahalla on the Astrolabe* (Oxford, 1929), p. 202, sec. 7. Fol. 46r[b] is blank.]

art. 7: Incipit computus maior Campani Nauariensis. [Fol. 68v[a]] Rogauit me unus ex hiis...[listed in TKr, col. 1365]. [Fol. 68v[b]] Computus est scientia numerationis et diuisionis temporum...[cited in TKr, col. 243, as a work of Robert Grosseteste or Campanus]/...[fol. 92v[b]] Istorum uigilia ieiunes Luceque Marci. [Fol. 93r[a], *canon*:] Tabule quatuor que sequuntur non sunt de computo meo, sed feci eas extrauagantes.../...[fol. 93r[b]] in tabulis superioribus si uellis habere tempus coniunctionis ad meridiem rome. [Bottom half of fol. 93r:] Tabula. Almanach coniunctionum mediarum solis et lune ad ciclum 19[lem] post meridiem nouarie [added in right mg.: "in 1254"]. [Fols. 93v–94r: Tables of conjunctions for Novara continued (fol. 93v, left mg.: "1477"); bottom half ruled out but blank. Fol. 94v: calendar, with same title as that in MS *C*, for January–December, ending on fol. 96r[a]. Fol. 96r[b] is blank.]

9. *E* = Paris, Bibliothèque Nationale, fonds latin, 15122. Paper, petit format, 316 fols. (see below)

Date: end of 14th century

Text used: fols. 174r(172r)–75r(73r). Clementissimo (dedicatory letter only)

Description: First twenty-three folios are blank or contain French notes or mathematical rules. The following is from an examination of the manuscript and a microfilm of it.

fol. 24r: [John of Sacrobosco] Omnia que a primeua rerum...[listed in TKr, col. 991]. [Ends on fol. 31v; fols. 33–35 are blank.]

fol. 36r: [Bernard of Clairvaux] Incipit prologus in libro de precepto et dispensatione....[Ends on fol. 51v; fols. 52–53 are blank.]

fol. 54r: Sicut dicit philosophus omnis effectus habet causas sed.../...[fol. 86v] ad predicta caui duplex ex duo naui. Expliciunt flores quos dicimus artis odores. Frater Stephanus de uicima. [Fol. 87 is blank; fols. 88–151 are missing.]

fol. 152r: Canones in tabulas Alphonsi. Tempus est mensura motus ut uult...[listed in TKr, col. 1561]/...[fol. 170v] longiores quam estate et patet solis theoricam intuenti.

fol. 171r: Longitudines stellarum a terra secundum Thebith. Harum terre propinquior est luna...[cf. F. J. Carmody, *Astronomical Works of Thabit b. Qurra* (Berkeley, 1960), p. 137, section 43 of *De hiis*: "Longitudines uero earum a terra sunt iste: eorum terre propinquior est luna cuius longitudo..."]/...[fol. 171v] sed forte fuit error scriptoris hic uel ibi. Explicit tractatus thebith de longitudinibus stellarum a terra et de quantitate corporum planetarum. [Folio missing; fols. 176–227 also missing (presumably contained the remainder of the *Theorica*).]

fol. 228r: [Nicholas Eymeric] Incipit prologus in tractatum super demonum inuocatione.... [Ends on fol. 304r; fol. 305 is blank.]
fol. 306r: Decretum abreuiatum. [Ends on fol. 315v; fol. 316 is blank. Two blank sheets of parchment follow.]

10. *e* (θ) = Paris, Bibliothèque Nationale, fonds latin, 10263. Paper, moyen format, 172 fols.
Date: 15th century
Text used: fols. $150r^a$–$63r^b$. Primus. Figures incomplete; lacunae for missing tables
Description: For complete discussion of the manuscript, see Lynn Thorndike, "Notes on Some Astronomical, Astrological and Mathematical Manuscripts of the Bibliothèque Nationale, Paris," *Journal of the Warburg and Courtauld Institutes*, vol. 20 (1957), p. 146; Emmanuel Poulle, *La Bibliothèque scientifique d'un imprimeur humaniste au XVe siècle: Catalogue des manuscrits d'Arnaud de Bruxelles à la Bibliothèque nationale de Paris*, Travaux d'Humanisme et Renaissance, vol. 57 (Geneva, 1963), pp. 45–53.

11. *F* (θ) = Florence, Biblioteca Medicea Laurenziana, Plut. XXIX, cod. 46. Paper, folio, 44 fols.
Date: 14th century
Text used: fols. 2r–36r. Primus. Figures mostly incomplete; tables
Description: [Fol. $36r^a$] Incipit astrolabium spericum compositum anno domini 1303 dominus accursius de parma. Fuit principium huius operis prologus. [T]otius astrologice speculationis radix et fundamentum eius et prolixitatis immensitas... [listed in TKr, col. 1576]. [Fol. $36v^a$] Capitulum primum de formatione instrumenti. [C]um igitur fauente domino uolueris hoc instrumentum componere fac primo pilam de metallo uel ligno... [listed in TKr, col. 298]/...[fol. $44v^b$] Et quoniam de instrumentis tractare non est presentis intentionis hunc tractatum sub laude dei finiemus. Explicit tractatus de spera solida.

12. *f* (φ) = Frankfurt-am-Main, Stadtbibliothek, Barth. 134. Quarto, 163 fols.
Date: 14th century
Text used: fols. 46r–105v. Primus. Tables incomplete; figures generally complete
Description: The manuscript can be analyzed briefly from the following items in Zinner's *Verzeichnis*: p. 75, no. 2026; p. 72, no. 1901; p. 283, no. 9124; p. 213, no. 6797; p. 261, no. 8400; p. 115, no. 3359; p. 44, no. 923; p. 30, no. 460; p. 207, no. 6575; p. 208, no. 6615; p. 32, no. 547. However, more complete references, which I give below, have been taken from a microfilm of part of the manuscript.

fols. 1–44: [William of Conches, *Philosophia* II].
fol. 106r: [C]irculus ecentricus uel egresse cuspidis...[listed in TKr, col. 223]/...[fol. 115r] et non corporaliter. Explicit theorica planetarum.
fol. 115v: [Figure] Nota pro suprascripta figura equatio centri in zodiaco est arcus....
fol. 116r: Describemus circulum super centrum E et diuidamus ipsum...[possibly the work by Prophatius Judeus listed in TKr, col. 402]/...[fol. 116v] ut in astrolabio dictum est.
fol. 116v: Vtilitates noui quadrantis breuiter et lucide...[identified in TKr, col. 1627, as a work of Johannes Eligerus de Gondersleuen]/...[fol. 118r] quos ponere esset superfluum et onerosum et sic est finis. [An insert to the above follows and ends on the next folio, fol. 118v: "...nihilominus per digitos unius umbre scire digitos alterius diuidendo 144 qui est numerus quadratus 12 per digitos unius umbre et habebis digitos alterius."]
fol. 118v: [Left:] Tabula longitudinis et latitudinis stellarum fixarum; [right:] Hec tabula ostendit quoto gradu uel signo menses sua habeant initia secundum uerum motum ad meridianum Moguntie anno 1376 completo; [below:] Hec tabula certificata est anno domini 1376 completo ad meridianum Moguntie cuius longitudo scribitur esse 28 graduum et 24 minutorum.
fol. 119r: [Table with brief *canon*:] Cum per umbram hominis habentis in longitudine 6 pedes....
fol. 119v: Vt celum signis prefulgens est duodenis [in verse]...[TKr, col. 1615, identifies it as "Grand calendrier des bergiers"]/...[fol. 121r] addicias monadem dominum hore siue planetam. [Bottom:] Nota auctor in compendio theorice ueritatis libro 1 capitulum iii dicit lunam distare a terra xv miliaria dcxxv stadia cxxxvi; a luna usque ad mercurium sunt miliaria vii dccccxii et semisse....
fol. 121v: [Q]uoniam autem sapientes astronomi...[listed in TKr, col. 1264]/...[fol. 123r] saturnus in 12a et hec de domibus dicta sufficiant.
fol. 123v: Habent quoque planete in quolibet signo uirtutem....[Table] Nota pro tabula succedenti qui pertinet quadrantem....
fol. 124r: [S]apientes astronomi motus corporum celestium...[listed in TKr, col. 1377]/...[fol. 125r] puncta umbre recte opposite uero umbre uerse. [Rest of recto is blank: missing figure or end of treatise?]
fol. 125v: Cum gradum solis scire uolueris pone regulam...[listed in TKr, col. 356]/...[fol. 129r] cui a speculo et exibit altitudo quesita. [Fols. 129v–31v are blank.]
fol. 132r: Incipiunt tabule illustris regis Alphoncii. [Tables end on fol. 154v.]
fol. 155r: Sequntur canones in tabulas Alfoncii prescriptas. Prologus. Priores astrologi motus corporum supercelestium diligentissimis...[identified in TKr, col. 1127, as a work of Johannes de Lineriis; cf. Zinner, *Verzeichnis*, p. 207, no. 6575, where it is identified as a work of John of Saxony]. Capitulum primum. Numerum annorum incipientium ab aliqua erarum...[TKr, col. 958, gives, under Hermann Stilus de Norchem, anno 1355, "Numerum annorum mensium et dierum a principio alicuius ere..."; but cf. TKr, col. 959, "Numerus annorum mensium et dierum incipientium..."]/...[fol. 158r] capitulum 16m...sic tria signa complete uel per 12 gradus ante uel post die eclipsim lune possibilem in

illo mense. Expliciunt canones in tabulas illustris regis Alphoncii. [Followed by tables of latitude and longitude of places. In bottom mg.:] Nota diuersis inter meridianum Toleti et Moguntie....

fol. 158v: Incipit canon in tabulas Johannis de lineriis subsequentes. Volens inuenire medios motus planetarum.../...[fol. 159r] breuiorem in eis operandi modum componere satagebamus. Explicit canon Joh. de lineriis in tabulas subsequentes. [*Canon.*] Tabula subscripta dicitur tabula senarii ostendens quot senarii sint.../...istum modum facilius habebis in ultimo folio. [Table.]

fol. 159v: [Tables begin:] Motus solis in annis collectis [for 1335, 1363, 1391, 1419, etc.]. [Fol. 162v] Expliciunt tabule Johannis de lineriis complete una cum antecedentibus. Anno domini 1366 feria quarta ante Phil. et Jacobi [May 1].

fol. 163r: [Figure] Nota concurrens habet ortum....

fol. 163v: Cum uolueris ex prima linea cuiusque medii motus...[listed in TKr, col. 353]/...per unam aut duas figuras. Tabula senarii dicta.... [The full title and table are crossed out, but on the left is a small table:] Eclipses lune, Initium, Duratio [for the years 1377, 1377, 1378–82, 1384–86, 1384].

13. *G* (α) = London, British Museum, Additional 22772. Paper, folio, in double columns, 43 fols.

Date: 15th century

Text used: fols. 1r[a]–22r[a], 26r[a]–30r[a], 22r[a]–26r[a], 30r[a]–41v[b]. Clementissimo. Figures incomplete or blank spaces left; same for tables

Description: [Fol. 42r[a]] Imaginabor speram equatoris diei et circulos in ea ligatos...[identified in TKr, col. 661, as Thebit, *De motu octave spere*]/...[fol. 43r[a]] ut bona foret eius operatio et quod ipse aliquid profecisset [cf. "perfecisset," end of paragraph 22 in Carmody, *Astronomical Works of Thabit b. Qurra*, p. 104; occurs less than halfway through the treatise].

14. *g* (α) = London, British Museum, Additional 22773. Folio, 46 fols.

Date: July 25, 1477

Text used: fols. 2r–44r. Primus. Figures and tables missing or incomplete

Description: [Fol. 44v blank. Fol. 45r, top mg.:] Joh. Archangeli. [Col. 1:] Radices augium planetarum ad heram Christi; Aux solis: S 2, G 11, M 25, sec 23; Aux saturni: S 7, G 23, M 23, sec 42; Aux iouis: S 5, G 3, M 37, sec 0; Aux martis: S 3, G 15, M 12, sec 13; Aux ueneris: S 2, G 11, M 25, sec 23; Aux mercurii: S 6, G 10, M 39, sec 34. [Cols. 2 and 3:] Motus octaue spere [begins with the year 1000 and continues for every twenty years to the year 2020]. Motus octaue spere: 1000: G 14, M 17, sec 37, ter 45, quar 51; 1020: G 14, M 32, sec 34, ter 44, quar 24 [although there is a column for fifths, it contains no figures; after this date the seconds, thirds, and fourths are omitted]; 1040: G 14, M 47; 1060: G 15, M 2; 1080: G 15, M 16; 1100: G 15, M 35; [etc.]. [Fols. 45v–46r: four columns in two sections, including mean motions of sun, moon, Saturn, and Jupiter. Fol. 46r: Mars, Venus, Mercury, and caput et cauda draconis.]

15. *H* (δ) = Milan, Bibliotheca Ambrosiana, H.88 inf. Parchment, 32 fols.
Date: 14th century (catalogued as 13th century)
Text used: fols. 1v–32v. Clementissimo. Figures complete; tables
Description: Fly-leaf (?paper), fol (1)r: "Campani nouariensis mathematici theoricae liber 1^{s}. Codex characteris antiqui ann. H 88. [?]P^{a} inf. Felicibus auspiciis Illustrissimi Cardinalis Federici Borrhomaei Bibliothecae Ambrosianae fundatoris olgiatus uidit anno 1603."
Fol. 1r: "Campani nauariensis theorice olgiatus uidit anno 1603."
There is an alternate foliation, probably earlier, in the lower right corner of the rectos, but it has been trimmed off in some cases and is perhaps obscured in others by the thumb holding the manuscript for photographing in the microfilm I used. It is clear on the following rectos: 25(30), 26(31), 27(32), 28(33), 29(34), 32(37).

16. *h* (β) = Bamberg, Staatliche Bibliothek, Class. 84 (M.II.7). Paper, 298 fols.
Date: 14th century
Text used: fols. 223r^{a}–35r^{a}, 247r^{a}–59r^{a}, 235r^{a}–47r^{a}, 259r^{a}–r^{b}
Clementissimo. Tables; no figures
Description: Described in F. Leitschuh and H. Fischer, *Katalog der Handschriften der kgl. Bibliothek zu Bamberg* (Bamberg, 1887–1912), vol. 1, pt. 2, pp. 90–93.

17. *I* (θ) = Oxford, Bodleian Library, Canonici Latini, 192. Parchment, quarto, 123 fols., Auct. Classici
Date: 15th century
Text used: fols. 40r^{b} (ult.)–123v. Primus. Figures complete; tables
Description: Described in H. O. Coxe, *Catalogi codicum manuscriptorum Bibliothecae Bodleianae*, vol. 3, *Codices Graecos et Latinos canonicianos complectens* (Oxford, 1854).

18. *i* (α) = Oxford, Bodleian Library, Canonici Miscellanei, 501. Paper, in folio, 127 fols.
Date: 15th century
Text used: fols. 15r–36r, 43r–44r, 42r–43r, 41r–42r, 40r–41r, 39r–40r, 38r–39r, 37r–38r, 36r–37r, 44r–48r, 51v–53r (2d *explicit*), 48r–51v (1st *explicit*); also has summary preface. Sol unum. No figures; no tables
Description: Described in Coxe, *Catalogi*, vol. 3.

19. *J* (φ) = Rome, Biblioteca Apostolica Vaticana, Regina Sueviae 1924. 45 fols.

Date: 15th century

Text used: fols 6v–45r. Primus. No figures; only Table 3 and a table giving retrogradation times for Venus and the three outer planets

Description: [Fol. 1r] Circulus ecentricus dicitur uel egresse... [listed in TKr, col. 223] [second folio unnumbered; third folio numbered 2]/...[fol. 6r] et non corporaliter etc. Expliciunt theorice planetarum scripte anno gratie 1412mo.

20. *j* (α) = Rome, Biblioteca Apostolica Vaticana, Palatinus 1416. 251 fols.

Date: 15th century

Text used: fols. 198r–210v. Sol habet. Figures; no tables

Description: This manuscript is briefly described in Lynn Thorndike, "Some Little Known Astronomical and Mathematical Manuscripts," *Osiris*, vol. 8 (1949), pp. 45–47, and cited in Thorndike, "Notes on Some Astronomical and Mathematical Manuscripts," p. 143. The following is an incomplete analysis from the manuscript and a microfilm and the works of Thorndike mentioned above.

fol. 1r: [Top mg.: "Repertorium Astronomiae in 7 diuisum partes de naturis partium coeli stellarum et aeris mutatione tractantes."] Quia in antiquis uoluminibus sapientes antiqui...[identified in TKr, col. 1220, as a work of Firminus de Bellavalle]/...[fol. 48r] Item in sale et in [?]cardice et partibus corporis multis et sensibus presagia apparebunt infinita. Item peto ea exultantia hoc in eodem libro usque in finem. Laus deo omnipotenti hic finis est. Hic liber compilatus est anno 1318 sed scriptus est huius libri [?]presentia anno 1454 et completus 10 Ianuarii anni predicti per Io. de Blisia. [Fols. 48v–56v are blank; fols. 50r–54r are missing.]

fol. 57r: Alkindus de impressionibus aeris. Rogatus fui quod manifestare...[listed in TKr, col. 1364; cf. Carmody, *Arabic Astronomical and Astrological Sciences*, p. 80]/...[fol. 66v] similiter fac quando erit questio de pluuiis et hoc sufficit in hoc quod interrogasti. Explicit liber de impressionibus aeris.

fol. 66v: Apertio portarum dicitur cum coniungitur planeta inferior superiori... [listed in TKr, col. 112; cf. Carmody, *Arabic Astronomical and Astrological Sciences*, p. 81, who says the text may have been revised by John of Wasia; could this be the Io. de Blisia of fol. 48r?]/.... [fol. 67v] sicut predixi in libro mundi. Explicit Alkindus de impressionibus aeris. deo gratias [followed by a note].

fol. 68r: Nota aliqua de directione. Nota dirigere aliquem significatorem...[only 13 lines]. [Fol. 68v carries titles or notes.]

fol. 69r: Incipit epistola Messahala in rebus eclipsis lune...[listed in TKr, col. 729; cf. Carmody, *Arabic Astronomical and Astrological Sciences*, p. 32]/...[fol. 70v] Et quotienscumque cuncta fuerit fortuna malo apparebit natura [?]fortiorum etc.

fol. 70v: De debilitate planetarum et fortitudine. Quedam sunt debilitates et fortitudines...[listed in TKr, col. 1188; Thorndike, "Some Little Known Astronomical and Mathematical Manuscripts," p. 46; idem, "Notes on Some Astronomical and Mathematical Manuscripts," p. 143]/...[fol. 72r] et elongationem quia cum prope sunt tunc fortius cum elongatur operantur debilius.

fol. 72r: De temporis dispositione. Cum ergo dispositionem temporis ad quemlibet certum terminum prenosticare uolueris oportet...[listed in TKr, col. 293; Thorndike, "Some Little Known Astronomical and Mathematical Manuscripts," p. 46; idem, "Notes on Some Astronomical and Mathematical Manuscripts," p.143; cf. Carmody, *Arabic Astronomical and Astrological Sciences*, p. 116]/...[fol. 72v] aspexerint se et in signis aquosis etc.

fol. 72v: Item nota. Ad sciendum uerum et certum gradum alicuius necessitatis. Primo oportet uidere an illa necessitas est coniunctionalis uel preuentionalis.../...[fol. 73v] et tunc adde iterum partes horarum ut prius et cum hoc intra ut scis etc.

fol. 73v: De urina non uisa. [This work begins in German, then shifts to Latin.]/...[fol. 74r] cum natura loci demonstrat plene substantiam et colorem.

fol. 74r: De quatuor humoribus qui attribuuntur quatuor triplicitatibus. Ignea correspondent colore.../...[fol. 74v] applicabit se marti et ueneri.

fol. 75r: Nota de tempore medecinandi in omnibus. Quando uis conformare uim digestiuam sit luna in signo colerico....

fol. 75v: De triplicitatibus. Item triplicitas terrea habet agriculturam incisionis arborum et earum plantationes. Item aerea sunt uentis et humiditatibus et omnibus que per mare feruntur et ista figura solis hominibus proferuntur. Item aquea habent aquas et talia. Item ignea [followed by notes].

fol. 75v: [*Rubric.*:] Nota de etate lune seruiente ad egritudines. Si quis luna prima decubuerit et tertia die alleuiatus fuerit sanus erit....

fol. 76r: [German work crossed out with red hatching.]

fols. 76v–78v: [Astronomical and astrological data; on fol. 77v:] Item quidquid dant fortune dat in paucitate et tarditate et labore....

fol. 79r: [*Rubric.*:] Nota hic aliqua secreta astronomie in quibuslibet questionibus secundum horas planetarum et primo de Saturno. Si quis uenit ad te causa postulandi...[listed in TKr, col. 1462; Thorndike, "Some Little Known Astronomical and Mathematical Manuscripts," p. 46]. [Fol. 80v]...uni colori eorum dictorum. Nota de horis secundum aliquos. Hora Saturni...[ends on fol. 81r].

fol. 81r: [Top: "artofilax, arturus, ursa maior helix aliter, ursa breuior... fenix aliter, currus"—eight stars.] Nota quod una ursarum est maior et altera breuior/...[fol. 84r] et parum sibi ualebit ut lepores hoc est forma inter geminos et omus [followed by a diagrammed excerpt on signs from *Flores* of Albumazar].

fol. 84v: Nota de dominis nouene et duodene quomodo sciuntur per 12 signa sicut sciuntur domini terminorum et domini facierum ita hii domini sciuntur per 12 signa ut infra patebit/...[fol. 86r] et per planetam cui ipsum luminare applicatur et per dominum eius.

fol. 86v: De significationibus 12 domorum. Prima domus significat uitam nati...[listed in TKr, col. 1091; Thorndike, "Some Little Known Astronomical and Mathematical Manuscripts," p. 46]/...[ends on fol. 87v].

fol. 88r: Libra habet renes et umblicum [*sic*] signum nobilium.../...omnibus habet alas.

fol. 88v: [Two tables. Below: notes on triplicities. Fols. 89–100, inclusive, are missing.]

fol. 101r: Incipit liber Alfragani de 30 differentiis tractans. Differentia prima in annis arabum et latinorum...[listed in TKr, col. 429; cf. Carmody, *Arabic Astronomical and Astrological Sciences*, p. 114]/...[fol. 117r] Iam patefecimus de eclipsi solis et lune deo gratias. Explicit liber 30 differentiarum Alfragani deo gratias per manum iohannis magistri finitum et completum anno domini millesimo quadringentesimo xxxviii mensis may die xx[a] dies mercurii hora decima uel quasi ante prandium in domo henrici de zimhe protunc temporis magistri mei translatus a iohanne hispalensi [followed by a note].

fol. 117v: Ad inueniendum intentionem querentis per dominum hore...[10 lines].

fol. 118r: Incipit mathematica Alexandri Sidini astrologi. Luna frigide est nature et argentei coloris...[listed in TKr, cols. 835, 834; Carmody, *Arabic Astronomical and Astrological Sciences*, p. 76]/...[fol. 121v] primarum horarum dierum tochius ebdomidii tytulatur [followed by a table which concludes on fol. 122v]. explicit mathematica Alexandri ordinata anno 1040 [followed by notes].

fol. 122v: De caristia et bono foro rerum. Primo nota quod 3 sunt fortune...[listed in Thorndike, "Some Little Known Astronomical and Mathematical Manuscripts," p. 46]/...[fol. 124r] et cadentia cum hiis que hominibus non sunt necessaria [and short note].

fol. 124v: De eleuatione unius planete super alium. Quando planete eleuantur unus super alium...[listed in TKr, col. 1170; Thorndike, "Some Little Known Astronomical and Mathematical Manuscripts," p. 46]/...[fol. 138v] De pluuiis tonitruis et coruscationibus et uentis...sed si fuerint ponderosi non erunt [?]tanti precii.

fol. 139r: Dixit Perscrutator in anno christi 1325...[listed in TKr, col. 455]/...[fol. 155v] quarum intellectum ratione experimento ministrante deo dante denudaui Qui uiuit et regnat in secula seculorum amen. Explicit liber optimus de astronomie conclusionibus 8. [Followed by] Nota ex domino exaltationis et diuisore et domino radiorum et ex domino anni ex mutatione quoque planetarum in locis et eorum aspectibus sciuntur accidentia mundi in reuolutionibus annorum: hec hermes. [Fols. 156r–58v are blank.]

fol. 159r: Juliani Laodicensis mathematici ad dominum Marcum imperatorem de bello [listed in TKr, col. 1439]. [Ends on fol. 164v.]

fol. 165r: [*Rubric.*:] De aspectibus eorumdem 5 planetarum adinuicem. Saturnus in huiusmodi bellorum inceptionibus.... De ciuitate obsessa an capietur....

fol. 166r: Ad habendum eleuationem saturni ad meridiem cuiuslibet diei.../...[fol. 167v] et per inimicos et per medecinalia bona et mala excitatur ad multa. [Note: "Nota in Martianum: Astronomia est scientia mobilis magnitudinis.... Astrologia est scientia mobilis magnitudinis secundum situm terrarum...."]

fol. 168r: Compositio equatorii. Recipe tabulam ligneam pergameno...[listed in TKr, col. 1337]/...[fol. 169v] et abscindet uerum locum lune in circulo exteriori in nona spera.

fol. 170r: circuli equationis de quo prefertur et scilicet nona.... [Fol. 170v is blank.]

fol. 171r: [*Rubric.*:] De aliquibus gaudiis planetarum. [Has only a few lines.]

fol 171v: [*Rubric.*:] De aliquibus significationibus accidentium mundi causandis ex tonitruis fiendis in principio anni secundum quod luna fuerit in diuersis signis zodiaci. Aries. Cum luna fuerit in ariete et tonitruum auditur…[listed in TKr, col. 315; Thorndike, "Some Little Known Astronomical and Mathematical Manuscripts," p. 46]/…[fol. 172r] Nobiles et multum diuites pestilentia ledentur et pisces et animalia inter aquas. [Fols. 172v–[72a]v are blank.]

fol. 173r: Dixit Hermes quod sol et luna post deum…[listed in TKr, col. 453; cf. Carmody, *Arabic Astronomical and Astrological Sciences*, p. 54]/…[fol. 176v] et per planetam cui ipsius luminare coniungitur et dominum eius. Explicit liber Hermetis.

fol. 176v: [*Rubric.*:] Incipiunt propositiones Almansoris [also in black]. Incipiunt capitula stellarum oblata…[listed in TKr, col. 188; cf. Carmody, *Arabic Astronomical and Astrological Sciences*, p. 134]. Signorum dispositio…[listed in TKr, col. 1504]/…[fol. 181v] absque malo aspectu infortunarum dabit fortitudinem et regnum in qua nulla fiet iniustitia. Expliciunt propositiones Almansoris que sunt 150 numero. [Note:] Item recipe tartarum hoc est wulgariter…[*mg.*: "super laminam uitream"]. [Fols. 181bisar–81cterv are blank.]

fol. 182r: Zodiacus est quidam circulus latus uel zona…[listed in TKr, col. 1714; Thorndike, "Some Little Known Astronomical and Mathematical Manuscripts," p. 47]. [Fol. 182v] Duodecim signa sic uocantur scilicet aries…piscis. Et quia martius fuit primus mensis anni secundum Romulum qui primo ordinauit kalendarium…. [Fol. 183r] Luna existente in signo membri patientis illo membro nullum medicamen facias…[listed in TKr, col. 835]. [Fol. 184r] Memento [?]sy luna fuerit in ariete quod bonum est mutare de loco ad locum…[cf. TKr, col. 1454]. Aries habet caput et faciem et pupillam oculi [listed in TKr, col. 133] et intestina et quidquid accidit in eis etc in oculis atque naribus ex infirmitatibus. Cum luna fuerit in ariete signo orientali calido et sicco tropico igneo colerico et masculino…[listed in TKr, col. 315]. [Fol. 194r] in pisce prolongabitur et non exiet inde. [Paragraph] Huc usque de 12 partibus zodiaci compendiose uerba actorum [superscript: "auctorum"] in suis tractatibus inuenta ordinaui. Nunc autem de septem planetis et eorum naturis a qua scribo breuiter. Nunc itaque septem planete qui dicuntur errantia qui sunt omnium artium magistra sicuti et predicta 12 signa has planetas posuit deus opus sub celo sydereo….Primus harum septem planetarum inferius descendendo est saturnus qui est planeta frigidus et siccus atque maliuolus in rebus mundi malos habet affectus…[cf. TKr, col. 1384]/…[fol. 197r] dactiliorum et huiusmodi. Cf. same *explicit* in MS *r*, fol. 97r: "dactillorum et huiusmodi." Although a more detailed examination of this work did not prove an exact identification, TKr, col. 133, says this *incipit* belongs to the work which begins as does that in MS *r*.]

fol. 197v: [Two figures]

fol. 211r: [*Rubric.*:] Equationes 12 domorum. [Followed by table of ascensions of signs, fol 213r, and ending on fol. 213v.]

fol. 214r: Capitulum ad inueniendum gradum solis per diem mensis uel diem per gradum. Cum uolueris scire gradum solis pone regulam…[listed in TKr, col. 356; cf. Carmody, *Arabic Astronomical and Astrological Sciences*, p. 24]/…[fol. 221v] est enim omnibus hec utilis regula subtractione continentium facta [with a few lines of notes]. [Gunther, *Chaucer and Messahalla*, p. 231: II. 46.]

fol. 222r: [Top] Amice carissime nolite declinare a dextris uel sinistris quia maiora secreta sunt totius artis magice cum omnes fere philosophi deficiunt in eorum electionibus quia ista ignorant que subscribuntur. Et fuit uobis multum secreta. Ad habendum zenith solis... [listed in TKr, cols. 90, 43; Thorndike, "Some Little Known Astronomical and Mathematical Manuscripts," p. 47]. [Ends on fol. 222v.]

fol. 223r: Iudicium de infirmo quam infirmitatem patitur. Sic scies nota dominum hore; si hora saturni calores maximo patitur.... Item cum quis infirmatur et luna exeat de combustione tunc crescit infirmitas quousque uenerit ad oppositionis gradum et quando erit in coniunctione uide si fuerit causa malo aut in malo loco et si aspexerit domus mortis significat timorem. De luna in 12 signis. Cum infirmitas accidat luna in ariete... [identified in TKr, col. 309, as a work of Hippocrates]/... [fol. 227r] interiori fluminat si fortuna aspexerit eam uiuet si non morietur.

fol. 227v: [Top mg.: "Theorica planetarum."] Circulus ecentricus uel egresse cuspidis... [listed in TKr, col. 223]/... [fol. 234v] et non corporaliter. Explicit planetarum. Deo gratias. [Fols. 235r–37r are blank.]

fol. 237v: Incipit theorica regis Alfoncii. Circulus ecentricus uel egresse cuspidis... [listed in TKr, col. 223]. [Stops on fol. 239v, incomplete.] Equationes autem argumenti scripte in tabulis sunt equationes ac si habetur... [Alfraganus, Carmody, p. 22].

fol. 240r: [*Rubric.*:] Vt adimpleas id quod uis nota. Item pone dominum ascendentis....

fol. 240v: De planetis in domibus unius in alterum. Saturnus in domo propria facit scire edificare sedem.../... [fol. 243v] sed si est addita infortunie uel econtrario significat multas egritudines et malam spem.

fol. 244r: Saturnus in ascendente mortem causa debiti et terrarum... [cf. TKr, col. 1383]/... [fol. 247r] 12 impedimentum carcere si luna est significatrix in 12 inconstantiam negotiorum et ab inimicis omnino cauendum.

fols. 247v–48r: [Tables with notes]

fol. 248v: Saturnus significat saporem fetidum et aspidum.../...occidens diligit oriens.

fol. 249r: De stellis fixis in 12 signis. Aries est una in 21 uel....Piscis in 12 gradu alferam.

fol. 249v: [Astronomical notes]: De fleubothomia et medicinis laxatiuis.... Tabula de dignitatibus planetarum. Nota hic tabula eclysium solis et lune... [for the years 1447–62].

fol. 250r: [Nativities:] Natiuitas io. [?]magistri de blisia anno 1416 27 nouembris hora 6 minuta [?] in litera dominicali....

fol. 250v: [Paschal table with notes]

fol. 251r: [Table of contents: fols. 182–249.]

fol. 251v: Tabula principalium tractatuum huius libri [fols. 1–172].

21. *K* (α) = Berncastel, Hospital zu Cues, 214. Parchment, large quarto, 50 fols.

Date: 14th century

Text used: fols. 28r[a]–50v[b]. Primus. Figures incomplete or omitted generally; tables

Description: Described in J. Marx, *Verzeichnis der Handschriften-Sammlung des Hospitals zu Cues bei Bernkastel am Mosel* (Trier, 1905), p. 209. Examination of a microfilm of the manuscript reveals the following additional information.

fol. 10r: [Table] Medius motus solis in annis christi solaribus ad ciuitatem tholose.... [Fol. 19v blank.]

fol. 20r: [L]una frigide est nature et argentei coloris... [listed in TKr, cols. 835, 834; Carmody, *Arabic Astronomical and Astrological Sciences*, p. 76]. [Tables on fols. 24v–25r. Fol. 25v text resumes:] [H]is omnibus uidelicet horologiorum horoscopiorumque diuersitatibus.../... [fol. 27r] aliquando pro E ponitur alef semper sciafi suspicari [followed by a circular figure]. [Ends on fol. 27v.]

22. *k* (α) = Venice, Biblioteca Marciana, VIII, 69 (XI, 86). Paper, 246 fols.
Date: 15th century
Text used: fols. 67r–169r. Clementissimo. Figures mostly complete; tables complete except those for Jupiter and Saturn
Description: Described in J. Valentinelli, *Bibliotheca manuscripta ad S. Marci Venetiarum* (Venice, 1868–73), and in Lynn Thorndike, "Notes upon Some Medieval Astronomical, Astrological and Mathematical Manuscripts at Florence, Milan, Bologna and Venice," *Isis*, vol. 50 (1959), pp. 46–48.

23. *L* (α) = Rome, Biblioteca Apostolica Vaticana, latin 2225
Date: 15th century
Text used: fols. 71r[a]–83r[a] (incomplete). Clementissimo. Figures incomplete; no tables
Description: There is a partial analysis of this manuscript in Lynn Thorndike, "Some Medieval and Renaissance Manuscripts on Physics," *Proceedings of the American Philosophical Society*, vol. 104 (1960), pp. 195–200. The work of Nicholas Oresme (fols. 90r[a]–98v[b]) has now been edited: see Nicolas Oresme, *Quaestiones super geometriam Euclidis*, ed. H. L. L. Busard, Janus Supplements, vol. 3 (Leiden, 1961). A cursory survey of TKr suggests a total foliation of 198 for this manuscript (TKr, col. 1648: "Utrum febris sit propria passio...189r[a]–198r[b]").

24. *l* (α) = Rome, Biblioteca Apostolica Vaticana, Barberini 182. Paper and parchment, 138 fols.
Date: 15th century
Text used: fols. 103r[a]–19v[a]. Primus. Figures complete; tables
Description: Described by Theodore Silverstein, *Medieval Latin Scientific Writings in the Barberini Collection* (Chicago, 1957), pp. 57–60. One correc-

tion should be made. He foliates the Campanus *Theorica* as fols. 103–24, although it ends on fol. 119v[a] and is immediately followed by the *Theorica planetarum Gerardi* with the usual *incipit*: "Circulus ecentricus circulus egresse...." The *explicit* on fol. 124 is correct for the Gerardus work.

25. *M* (β) = Milan, Biblioteca Ambrosiana, C.241 inf. Parchment, except preliminary folios [a]–[c], a later addition, of paper; large folio, 192 fols., Parisius scriptus

Date: 1401

Text used: fols. 162v[a]–86v[b]. Primus. Tables; figures, beautifully drawn with cut-out volvelles, as follows:

[fol. 162r[a]] Figura de modo ymaginandi totam speram solis ex semicirculis circumductis etc.; [fol. 164r] Instrumentum equationis solis; [fol. 164v] Figura solis in suo ecentrico in qua declarantur tytuli tabularum equationis solis; [fol. 165v] Figura habitudinis motuum solis lune eius epycicli et eius augis necnon et capitis et caude [?]draconis; [fol. 167r] Figura quo ymaginanda sit speram lune est hec; [fol. 167v[a]] Figura magnitudinum et distantiarum in orbe lunare etc.; [fol. 169r] Forma instrumenti equationis lune est hoc [*mg.*: "Si libuerit uidere figuram ab autore primo et literaliter descriptam amoueas rotas et se ostendet"]; [fol. 170r] Figura pro declaratione terminorum in tabulas equationum lune; [fol. 172r[b]] Figura exemplaris motuum et orbium singulorum mercurii; [fol. 172v[b]] Figura prior repetita pro mercurio; [fol. 173r] Figura magnitudinum orbis mercurii; [fol. 174v] Forma instrumenti ad equandum mercurium [bottom mg.: "Si uidere uolueris huius instrumenti figuram literaliter descriptam remoueas tunc rotas et ostendet se quod petis etc."]; [fol. 181r] Figura motuum ueneris et trium superiorum est hec; [fol. 181v] Figura temporum retrogradationum ueneris et trium superiorum; [fol. 182v] Figura magnitudinum distantiarumque spere ueneris corporis et centrorum; [fol. 183v] Figura magnitudinum solis et eius orbis; [fol. 184r] Figura magnitudinis martis; [fol. 184v] Figura distantiarum magnitudinumque spere iouis corporis et centrorum; [fol. 185r] Figura distantiarum et magnitudinum spere saturni corporis et centrorum; [fol. 186r] Tabula motus centri epycicli, Forma instrumenti equationis ueneris necnon trium superiorum [in right mg.: "Si leuaueris rotam apparebit tibi ecentricus deferens literaliter descriptus per autorem"].

Description: The following analysis was made from an examination of the manuscript.

fols. [a]v–[b]r: [list of contents; fols. [b]v–[c]v blank]

fol. 1r[a]: Jordanus de Nemore de elementis arismetrice artis. [*Rubric.*:] Incipit liber primus Iordani de Nemore de elementis arismetrice artis. Vnitas est esse rei per se discretio...[listed in TKr, col. 1600]/...[fol. 27r[a]] tres medios assignare sit possibile. [*Rubric.*:] Explicit distinctio decima et per consequens totus liber de elementis arismetrice artis magistri Iordani de Nemore scripta Parisius anno domini millesimo quadringentesimo primo pro magistro Johanne Contareno de Venetiis protunc ibidem studio theologie insudante.

fol. 27r[a]: [John of Sacrobosco.] [*Rubric.*:] Algorismus de integris incipit. Omnia que a primeua rerum origine…[listed in TKr, col. 991]/…[fol. 30r[a]l quadratis quam in cubitis etc. [*Rubric.*:] Explicit algorismus de integris.

fol. 30r[b]: [Johannes de Lineriis.] [*Rubric.*:] Incipit algorismus de minutiis. Modum representationis minutiarum physicarum et uulgarium demonstrare…[listed in TKr, col. 878]/…[fol. 32r[b]] et sic radix 12arum erit 4a 9arum 3a 6arum 2a et 3arum integris etc. [*Rubric.*:] Explicit algorismus de minutiis. [Fol. 32v is blank.]

fol. 33r[a]: [*Rubric.*:] Incipit primus elementorum geometrie euclidis. Punctus est cuius pars non est…[listed in TKr, col. 1152]/…[fol. 124r[a]] quare assignato corpori constat nos speram quemadmodum propositum erat inscripsisse. [*Rubric.*:] Explicit quindecimus et ultimus liber elementorum Euclidis scriptus et completus Parisius anno 1401. [Considerable marginal notation referring to text as "pro additione Campani" ("correlarium Campani").]

fol. 124v[a]: [*Rubric.*:] Compendium proportionum ad musicam Muris que statim sequitur per maxime necessarium. Proportionum musicalium per uenerande memorie magistrum Iohannem de Muris…/…[fol. 125r[b]] Et 3 ad 11 se habent in proportione subtripla supertripartiente. Et hec sufficiant de generibus proportionum in generali. Et apparent etiam in sequenti arbore seu figura etc. [*Rubric.*:] Sequitur arbor proportionum patent et descriptiones et exempla necnon numerorum in omnium ipsorum uarietate denominationes liquent declarationes et exempla etc. [Fol. 125v contains diagram.]

fol. 126r[a]: [*Rubric.*:] Incipit Magistri Iohannis de Muris proportionum musicalium theorica. Etsi bestialium uoluptatum per…[listed in TKr, col. 527]/…[fol. 132v[b]] in hoc ordine consequentem. Et sic finitur secunda pars huius libri et per consequens totus. [*Rubric.*:] Iohannis de Muris musici precipui theorica proportionum musicalium explicit anno christi 1401 finita Parisius.

fol. 133r[a]: [*Rubric.*:] Theodosii de speris liber primus incipit. Spera est figura solida una tantum superficie… [listed in TKr, col. 1523]/…[fol. 150v[a]] Procedendum est autem in istius demonstratione antepremisse per quartam. [*Rubric.*:] Expliciunt libri tres Theodosii de speris anno domini 1401 Parisius scribendo perfecte. [On fol. 137v[a] (bk. 1, sec. 30), left mg.: "Correlarium [*abrupte*; ?read 'Campani']."]

fol. 151r[a]: [*Rubric.*:] Incipit tractatus de spera magistri Iohannis de Sacro Bosco. Tractatum de spera quatuor distinguimus…[listed in TKr, col. 1577]/…[fol. 157r[a]] machina dissoluetur. Natura enim compatitur suo creatori etc. [Figure of an eclipse.]

fol. 157v[a]: [*Rubric.*:] Incipit theorica planetarum communis. Circulus ecentricus uel egresse…[listed in TKr, col. 223]/…[fol. 162r[b]] et non corporaliter etc. [*Rubric.*:] Explicit theorica planetarum Johannis de Sacro Bosco scripta Parisius anno 1401. [Fols. 187r–91v plus one unnumbered folio are blank.]

26. *m* (β) = Bonn, Universitätsbibliothek, 497 (131, a). Folio, 109 fols.
Date: 14th century
Text used: fols. 86r[a]–104v[b]. Primus. Figures incomplete; tables

Description: Zinner, *Verzeichnis*, analyzes the manuscript briefly in the following sections, in the order indicated: p. 31, no. 485; p. 358, no. 11635; p. 205, no. 6515; p. 340, no. 11092; p. 343, no. 11185; p. 208, no. 6600; p. 208, no. 6620; p. 208, no. 6625; p. 254, no. 8166; p. 284, no. 9162; p. 296, no. 9544; p. 383, no. 12500. An examination of a microfilm of the manuscript supplies the following.

fol. 1r: Nota quod circulus interior est circulus solis siue zodiacus sed exterior est epyciclus lune.../...oportet ut prius narratum est. [In a different hand:] Liber monachorum sancte Marie in hymmenrode ordinis cisterciensis treuerensis dyocesis. [In a modern hand: "Table of contents."] fol. lv: [Figure. Below:] Si gradus solis plus distat ab A quam gradus argumenti tot hore sunt addende. Si uero gradus argumenti plus distat ab A quam gradus solis tot hore sunt subtrahende.

fol. 2r: Incipiunt tabule illustrissimi regis Alfonsii et primo tabula differentiarum unius regni ad aliud et nomina regum atque cuiuslibet ere coniunte. [Fol. 2v] Dixit Alkradius quod fuerunt inter annum coniunctionis in quo fuit significatio secte sarracenorum et inter annum yesdeiert qui fuit primus annus annorum arabum 61 solares et 57 dies. [Fol. 3v, bottom: "ad ciuitatem Magdeburg." Fol. 4r, bottom: "era Alphoncii anno 1437 completo fuit in annis 186 5 mensibus." Fol. 7r, bottom:] Nota super tabulam sequentem super loca stellarum suarum tabularum Ptholomeus in Almagesti addidit Alfoncius tempore suo 17 gradus et 8 minuta. Et anno domini 1350 addi debent 18 gradus 17 minuta.... [Fol. 8v, bottom: "Anno domini 1375 completo super Parysius." Fol. 25v, bottom, right:] Nota Calb Alezed tempore Alfoncii fuit 4 signo 19 gradus 38 minuta secundum Alfoncium sed secundum alios strictius obseruantes sunt uiginti octo minuta et sic anno domini 1350 erit in leone secundum Alfoncium 20 gradus 47 minuta sed secundum aliam diuisionem 20 gradus 37 minuta. [Fol. 38r, col. 4, lists "radices" for the years 1376 and 1375, with *canon*. Fol. 38v contains eclipse tables for Paris for the years 1321, 1345, 1369,..., 1609; second set with *canon*.] [Fol. 41r, bottom, *canon*:] Si uero latitudines trium superiorum inuenire desideras...[cf. TKr, col. 1453; possibly an excerpt from the *canones* to the Toledan tables] ueneris autem et mercurii cuius latitudinis sic habetur notitia intremus....[Fol. 43v, middle of left col., *canon*:] Scias quod introitum solis in primum minutum arietis equatum super Madeburgensem inueni de tempore anno 1339 incompleto die 12ª...; [right:] Scias quod hec tabula facta est ad meridianum parysiensem et est facta ad annum domini 1330.... [Fol. 44v, bottom:] Tabula ad sciendum moram nati in utero matris per horam natiuitatis. [Fol. 49v has two sets of eclipse tables for the years 1321, 1345,..., 1609; bottom, *canon*:] Si uis inuenire incensionem ianuarii que est radix aliarum, tunc intra cum annis incompletis christi tabulam annorum collectorum...[listed in TKr, col. 1471]. [Fol. 69v has notes at the top and tables of places in two parts. Pt. 1:] [P]aradisi medium, babilonia uetus, [?]geddo, Armenia, Baldach, Mecha nec corpus Me[?hometi], Babilonia noua in egipto, Alcufa, Anthiochia sedes Petri. [Pt. 2:] Iherusalem, Roma, Parisius, Toletum, Constantinopolis, Pysa, Bononia, Cremonia, Colonia, Maguntia, Herbipolis, Magdeburg, Brunswig,

Standalis, Mons Pessulanus, Erffordum, [?]Ludonia, Nouaria, Marsilia. [Tables followed by:] Nota coniunctio saturni et iouis fuit anno 1404 sit radix die 51 minuta 19 ad quam adde 19 52 annos 52 dies unum minutum diei 41 secunda et proueniet sequens uidelicet anno 1443 prima 44 20 minuta dierum. [Fol. 82r is blank. Fol. 83r contains tables of solar cycles, dominical letters, etc. To the right: "Tabula mensium"; below: "Signa phisica." Fol. 83v contains tables of solar cycles, dominical letters, etc.]

fol. 84r[a]: Cum uolueris scire uerum locum alicuius planete tunc quere prius medium motum illius quem scire desideras... [identified in TKr, col. 357, as a work of Johannes de Lineriis]/... [fol. 84v[a]] Nota medius motus lune et argumenti lune a coniunctione ad oppositionem et idem ab oppositione usque ad coniunctionem inuenies hic. Medius motus lune 6 S, 14 G, 33 M, 12 S, 6 T; Argumentum medium lune 6 S, 12 G, 54 M, 29 S, 28 T.

fol. 84v[b]: [Table:] Radices ad eram christi.

fol. 85r[a]: Aries est signum primum in ordine signorum et est calidum et siccum... [listed in TKr, col. 132]/... [fol. 85r[b]] et uulnus difficulter consolidatur hec de predictis dicta sufficiant. Nota quosdam uersus qualis se habere debet minutus post minutionem. Prima dies uene tibi sit moderatio cene [listed in TKr, cols. 1089–90]/ Altera leta dies sed tertia sit tota quies / Atque dies quartus eius fragiles facit actus / Sitque dies quintus sibi uires colligit intus / Balnea sexta petit septima uolt spatiari / Tale lote sta pranse uel frigeste minute. [Cf. *Collectio Salernitana*... (5 vols., Naples, 1852–59), vol. 1, p. 448: "Flos medicinae scholae Salerni. Pars prima," line 120; vol. 5, p. 6: idem, line 209.] [Fol. 85v is blank.]

fol. 105r[a]: Circulus ecentricus uel egresse cuspidis uel egredientis centri dicitur qui non habet... [listed in TKr, col. 223]/... [fol. 107v[b]] et habuit magnum astrolabium tricubitum per quod omnia ista uidit aut maioris quantitatis. Aspectus planetarum sic potest inuenire... [listed in TKr, col. 152]/... et non corporaliter. Explicit theorica. [Fol. 108r is blank.]

fol. 108v: Tabula more iuxta gradum ueri motus solis et intratur cum uero gradu solis. [*Canon*:] Pro inuentione uere coniunctionis solis et lune prescribatur media coniunctio ad lapidem uel ad asserem et postea quere argumentum lune... [listed in TKr, col. 1133]. [Fol. 109r] Tabula more iuxta argumentum et intratur cum gradu argumenti. [*Canon*, continued:] proueniunt subtrahe a tempore medie coniunctionis et remanebit coniunctio uera. Si autem minus fuerit in argumento quod 6 signa.../... per omnem modum sicud de uera coniunctione.

27. *N* (δ) = Oxford University, New College, 293. Parchment, in folio minori, 24 fols.
 Date: 14th century
 Text used: fols. 1r–24r[b]. Clementissimo. Figures and tables
 Description: The *Theorica* is the only work in this manuscript.

28. *n* (α) = Parma, Biblioteca Palatina, 984 (HH.3,17). Paper, folio, 145 fols.
 Date: 15th century

Text used: fols. 1r–45v (fol. 46r contains a figure; fol. 46v is blank). Primus. Figures and tables

Description: The following is an analysis of the whole manuscript from the handwritten catalogue and a microfilm.

fol. 47r: Incipit tractatus patris Asem Thebit filii Chore de accessione et recessione stellarum fixarum. Imaginabor speram equatoris diei... [listed in TKr, col. 661; cf. Carmody, *Arabic Astronomical and Astrological Sciences*, p. 121]/... [fol. 49r] cum quo intrasti in linea numeri. Sequitur tabula et figura. [Fol. 49v] Tabula et figura.

fol. 50r: Incipit liber quem edidit Thebit filius Chore de his que indigent expositione antequam legatur almagestum. Equator diei est... [listed in TKr, col. 502; cf. Carmody, *Arabic Astronomical and Astrological Sciences*, p. 121]/... [fol. 52r] erunt retrogradi. Finis.

fol. 52v: Incipit Thebit de imaginatione spere. [N]os iuxta rectam imaginationem... [listed in TKr, col. 924; cf. Carmody, *Arabic Astronomical and Astrological Sciences*, p. 121]/... [fol. 53v] primo imaginanda occurrunt. [This work is not listed in the handwritten catalogue.]

fol. 54r: Incipit tractatus Thebit de quantitatibus stellarum. [P]tholomeus et alii sapientes posuerunt... [listed in TKr, col. 1147; cf. Carmody, *Arabic Astronomical and Astrological Sciences*, p. 121]/... [fol. 55v] sicut potest uideri in spera. [These four works have been edited as manuscript *C* in Carmody, *Astronomical Works of Thabit b. Qurra*, pp. 82–148.]

fol. 55v: [D]ixit Thebit ben Chorat quod dixit Aristoteles qui legerit philosophiam... [listed in TKr, col. 458]/... [fol. 56r] antedicte. [All of this work is crossed out.]

fol. 56r: Incipiunt questiones super tractatum spere Johannis de Sacrobosco per Blasium de Parma Doctorem Excellentissimum, Mathematicum singularem. [C]irca tractatum de spera primo queritur utrum diffinitio spere... [listed in TKr, col. 222]/... [fol. 81v] ad rationem in oppositum patet solutio laus deo. Explete sunt questiones de spera secundum uenerabilem doctorem Magistrum Blasium de Parma Parisiensem.

fol. 82r: Questione [*sic*] de excentricis. [V]t ferrum ferro acuitur sic ignorantia... [listed in TKr, col. 1618]/... [fol. 85r] huius autem loca sunt centra. Et hoc de isto nunc. [Fols. 85v–86v are blank.]

fol. 87r: [Peter of Modena or Blasius of Parma.] [S]uper theoricas planetarum aliquas demonstrationes... [listed in TKr, col. 1545]. [T]res orbes mutuo ecentricos et deformes... [listed in TKr, col. 1584]/... [fol. 105v] centrum parui circuli et patet quomodo [?]rn ad demonstrationem contra istam etc.

fol. 106r: Quemadmodum Ptholomeus et ante eum nonnulli ueteris auctoritatis... [listed in TKr, col. 1190]/... [fol. 115r] et secundum axe proficietur tibi quod uolueris de scientia tabularum. Finis laus deo.

fol. 116r: [S]cito quod astrolabium est nomen grecum... [listed in TKr, col. 1409; cf. Carmody, *Arabic Astronomical and Astrological Sciences*, p. 25; Messehallah's *Astrolabium*]/... [fol. 126v] De utilitatibus astrolabii et primo epilogus. [N]omina instrumentorum sunt hec: primum est armilla... [listed in TKr, col.

916]/...[fol. 130v] Si uero est supra uersam et est minor longitudine. Expliciunt utilitates astrolabii. deo gratias. [Fols. 131r–34r are blank.] [Gunther, *Chaucer and Messahalla*, p. 231: II. 45.]

fol. 134v: [Left mg.:] Primum capitulum. Sic Leonardus Cremonensis prosequitur descriptionem cosmographie in plano. Terreni situs habitabilis partes describentium...[listed in TKr, col. 1566]/...[fol. 144r: end of decimum capitulum]. [Fols. 144v–45r contain notes; fol. 145v is blank.]

29. *O* (β) = Paris, Bibliothèque Nationale, fonds latin, 7295A. 193 fols.
Date: 15th century
Text used: fols. 1r–32v (fols. 33r–34v blank). Primus. Figures; tables
Description: The following is from an examination of the manuscript and a microfilm of it.

fol. 35r: De sole. Solis instrumentum sic faciemus uidebimus in tabula in quo signo...[identified in TKr, col. 1518, as a work of Prophatius Judeus]....[fol. 38r] et hec est tabula secundum opinionem Ptholomei et Alfragani.../...[fol. 40v] tantum filum equantis per successionem signorum. Et sic completum est opus nostrum ad preces uenerabilis magistri B[ernardi] de gordonio doctoris excellentissime in arte medicine in montepessulano. Explicit de armillis profacy.

fol. 41r: Notandum pro horalogys in trunco faciendis...[listed in TKr, col. 949; cf. Zinner, *Verzeichnis*, p. 302, nos. 9777–77a]/...[fol. 42r] uerificatorium ad omnes practicas horalogy trunci etc. et sic est finis trunci [with figure].

fol. 42v: [Top mg.:] Incipit arithmetica de rerum ac numerorum proprietationibus. Secuntur regule artificiales quibus fere totum factum...[listed in TKr, col. 1437]/...[fol. 45r] in principio tanto prescisius inuenies et magis quam etc. [Fols. 45v–48v are blank.]

fol. 49r: Si uis scire quantitatem cuiuslibet linee siue partes proportionales.../... aliarum quantitatum mensurabilium in longum latum et profundum.

fol. 49r: Incipit astrolabium messahale phylosophi. Scito quod astrolabium est nomen grecum...[listed in TKr, col. 1409; cf. Carmody, *Arabic Astronomical and Astrological Sciences*, p. 25; Emmanuel Poulle, "L'Astrolabe médiéval d'après les manuscrits de la Bibliothèque Nationale," *Bibliothèque de l'école des chartes*, vol. 112 (1954), p. 101]/...[fol. 62r] et super C pone regulam et ubi abscinderit dyametrum DB fac notam K et nota illa est polus zodiaci ut patet in hac figura etc. Explicit compositio astrolabii. [Gunther, *Chaucer and Messahalla*, p. 216: end of Pt. I.] [Fols. 62v–64r contain figures.]

fol. 64v: Hie wil ich anfachen bescheiden den sechs.../...[fol. 65r] das heisset trackensum. [Fols. 65v–73v contain tables in German.]

fol. 74r: Der fierd vettich als du gefragt hast uff welhe gegene.../...[fol. 74v] verstandest so besich die figur etc. [Fols. 75r–87v contain tables in German and Latin. Fol. 87v bears the date 1440.]

fol. 88r: Der erst vettich wiset dir die...[left mg.: "1462, 1473"]/...[fol. 88v] so gewinnest ein semlich zal etc. [with figures]. [Note:] Nun sol man mir diser zal gan in den andren vettich unn disen mittelgang warhafft machen als dich des canon ibidem leer. Nota anno 1440 habuerunt iudei 13 pro lunari numero christi anni 16. Nota anno 1461 habuerunt iudei pro aureo numero christiani 18 usque ad ianuarium et talis annus incepit in 76 post ianuarium et sic christiani habuerunt 18 sed post ianuarium 19 et ideo quasi numerus conclusionem intra 27...libet et in 15 aurei postea si inter septembrem introitus.

fol. 89r: Wer ander vettich ist dir [?]beturtem unn warhafft machen den mittell.../...[fol. 92r] die 24 mol 47 minutis die dosint etc. Explicit sex alarum magister. Ars esurit de certa tument lex lucre ministrat Moyses pontificat medicina thalamum subministrat.

fol. 92v: Notandum quod habendo aliquid de modo proferendi.../...[fol. 93r] secundus termynus secundi numeri ad suum tertium ut 8,4,2—4,2,1—et sic est finis de proportionibus. Finis deo gratias. [Fols. 93v–98v are blank.]

fol. 99r[a]: Ad intelligendum tabulas astronomie necessario opportet scire Quid sit radix planete...[listed in TKr, col. 48]/...[fol. 99v[b]] habitato quod distat gradus 17 a uero occidente. [Fols. 100r–v are blank except for brief note with date 1427 at top of fol. 100r.]

fol. 101r[a]: [Top: "io. de saxonia." Begins a new hand.] [*Rubric.*:] Canones super tabulas alfoncii olim regis castelle. Tempus est mensura motus ut uult Aristoteles...[listed in TKr, col. 1561]...[fol. 110v[a], left mg.: "Anno christi 1499"]/... [fol. 111v[a]] et non errabis deo gratias [concludes chap. 24]. [*Rubric.*:] Expliciunt canones super tabulas alfoncii [listed in Zinner, *Verzeichnis*, p. 80, no. 2158, as no. 4].

fol. 112r: [Tables; conclude on fol. 113v.]

fol. 114r[a]: Latitudo ciuitatis alicuius uel loci...Zenith...longitudo ciuitatis est distantia illius ab occidente Arym uero ciuitas Indie que recte sita est sub equinoctiali super quam Ptholomeus composuit almagestam suam. [Fols. 114r[b]–14v are blank.]

fol. 115r: [*Rubric.*:] Incipiunt tabule Alfoncii Olym Regis Castelle.... [Fol. 117v, bottom:] Radices huius motus Ad uiennam—Ad Parisius. [Fol. 127r is blank; fol. 127v] Tabula latitudinum quinque planetarum que est distantia eorum a uia solis. [Slip between fols. 131v and 132r and between fols. 135v and 136r. Fol. 136v] Tabula fortitudinum septem planetarum; [below:] Sensuient les figures des planettes. [Tables end on fol. 144r; fol. 144v is blank.]

fol. 145r: Prima tabula continens ueram latitudinem saturni ab orbe signorum pro omni loco et tempore in Oxonia constituta [ends on fol. 152r]. [Fols. 152v–54v are blank.]

fol. 155r: [*Rubric.*:] Incipiunt tabule magistri Johannis de Lineriis [end on fol. 171v]. [Fols. 172r–73v are blank.]

fol. 174r[a]: [Title in a different hand: "Canones de ligneriis quos habeo alibi melius."] Priores astrologi motus corporum celestium...[listed in TKr, col. 1127]/...[fol. 180v[b]] Gradum in ffinem 6 signorum et sic maior [different hand:] matutino. Si autem ultra 6 signa in 12 et sic in ortu uespertino et reliqua utraque in canone si compleatur. [Catchword:] matutino [bottom:] Hic deficiunt canones quinque scilicet 41us 42us 43us 44us et quadragesimus quintus.

fol. 181r[a]: [In a different hand:] Cum uolueris scire uarios colores eclipsis aspice longitudinem...[listed in (TKr, col. 357]/...[fol. 181r[a]] in lineis tabularum eclipsium super argumenta latitudinum inter se equipollentium uel latitudinis equipollentis etc. Expliciunt canones primi mobilis magistri iohannis de lineriis finiti Wienensi studio.

fol. 181r[a]: Nota secundum Alfraganum centrum ecentrici solis distat a centro terre duobus gradibus et dimidio.../...[fol. 181v] cum quo intrasti excedit breuiorem et numerus quotiens erit pars proportionalis. Nota denominatio proportionis constituitur...denominatoris diuidentis a denominatore diuise. Figura eclipsis solaris [with figure].

fol. 182r: [A]rabes maxime secundum motum lune tempora distinguentes annos et menses...[listed in TKr, col. 124]. [Fol. 185r[a]] [S]ecundum Alfraganum luna cum fuerit in superiori parte epycicli.... [Fol. 186r[a]] [N]ota in tabulis Tholetanis.../...[fol. 193v[b]] transitum lune ab initio eclipsis usque ad finem. Et hec omnia patent in figura sequenti. Figura eclipsis solaris [with figure]. [Last page is unnumbered and blank.]

30. *o* (φ) = Paris, Bibliothèque Nationale, fonds latin, 7401. 150 pp.

Date: 15th century

Text used: pp. 1–114. Primus (in top mg. in a different hand: "Joannes Deeus"). Figures partial; tables

Description: The following is from an examination of the manuscript and a microfilm of it.

p. 115: Si quis per hanc tabulam tabularum proportionis tabulam alio nomine nuncupatam...[listed in TKr, col. 1461]/...[p. 124] et que fuerint benedicta propter amorem scientie sollempniter exaltare. Explicit canon tabule tabularum edite a magistro Iohanne de Muris. Anno 1321.

p. 125: [Medical jottings.] luna noua iunii 14 hora 18 minuta 37. [P. 126 is blank.]

p. 127: [Incomplete table crossed out; p. 129 has incomplete table; pp. 128, 130–31 are blank.]

pp. 132–43: [Tables of proportions]

p. 144: [Incomplete]

pp. 145–46: Tabula more orientalis uel occidentalis.

p. 147: Tabula equationis casus spermatis.

pp. 148–50: [Unnumbered and blank.]

31. *P* (α) = Paris, Bibliothèque Nationale, fonds latin, 16198. Moyen format, 200 fols.

Date: 14th century

Text used: fols. 178r[a]–86v[b]. Primus. This manuscript omits the following lines of our text: III.72–156, III.160–99, IV.167–201, IV.210–71, IV.328–53, IV.383–99 ("premissas"), IV.490–655, V.234–333, V.350–75 ("quidem"), V.480–1382, VI.142–94, VI.285–336, VI.355–82, VI.747–866. Only two tables (Mercury and retrogradations of four superior planets); no figures

Description: Described by Thorndike, "Notes on Some Astronomical, Astrological and Mathematical Manuscripts," pp. 148–49.

32. *p* (α) = Paris, Bibliothèque Nationale, fonds latin, 7293A. 71 fols.
Date: 13th–14th century
Text used: fols. 26r–46v (fol. 47r contains a figure; fol. 47v is blank). Primus. Figures; tables
Description: The following is from an examination of the manuscript and a microfilm of it.

fol. 1r: [D]icit Johannes. Cum uolueris facere astrolabium accipe auricalcum optimum...[identified in TKr, col. 353, as a work of John of Seville; Carmody, *Arabic Astronomical and Astrological Sciences*, p. 169; Poulle, "L'Astrolabe médiéval," pp. 102–3]. [At fols. 4r, 7r, 9r, and 12r there are slips inserted containing figures; versos blank.] [Fol. 6r] [E]xpleto auxiliante deo almucantarach opere tractemus de azemuth que eas sequntur ordine.../...[fol. 25v] que est mensis decimus ex mensibus arabum intrabit annis 1036 10 die undecimi mensis A.
fol. 48r: In nomine domini Amen. In hoc tractatu breui et utili dicetur primo de concordia et adequatione annorum christi et ebreorum et arabum et aliorum... [in right mg., in a different hand: "Capitulum primum de annis christi uel latinorum"] [listed in TKr, col. 681] [in the tables on fols. 50r–v, those on fol. 50v begin, "1275, 1294, 1313, 1337, 1351, 1370," and in those on fol. 51v, they begin, "1260, 1288, 1316, 1344, 1372"]/...[fol. 61v] quamuis non uideamus in hoc ullam necessitatem. Explicit especulatio astrologie. [Fol. 61v] Postquam autem ostendimus dispositionem mundi et orbium celestium et motus. Nunc in tertia parte tractatus huius dicemus.../...[fol. 69r] que non sunt nisi puncta ymagi[n]abilia nichil habentia probabilitatis. Explicit suma astrologie edita per fratrem Johannem de ordine minorum. [Fols. 69v–70r contain tables; fol. 70v and an unnumbered folio are blank.]

33. *Q* (θ) = Paris, Bibliothèque Nationale, fonds latin, 7298. Parchment, 174 fols.
Date: 14th century
Text used: fols. 142r^b–74r^b. Primus. Figures and tables complete
Description: The following is from an examination of the manuscript and a microfilm of it.

fol. i verso: [Table of contents, dated 1549]
fol. 1r^a: In hoc primationum ciclo 4 linee descendentes...[identified in TKr, col. 680, as a work of Petrus Dacus] [in column b, the dates read, "1292, 1311, 1349,..., 1653"]/...[fol. 1r^b] primationes sicut prius [with table]. [Fol. 1v, *mg.*: "anno imperfecto 1294 in occasu solis." Fol. 3r, *mg.*: "anno 1323 ultima die aprilis." Fol. 8v] Tabula petri daci de loco lune.... [Fols. 9r–v are blank.]
fol. 9bis r: [In a late hand, *mg.*: "Compotus M. Badonini de Mardochio." Fol. 9bis r^a, *Rubric.*:] Incipit tractatus compoti manualis magistri Badonini de Mardochio continens tria capitula.... Ad habendum in manu prompte in quo die mensis... [listed in TKr, col. 42]. [Fols. 9bis r–v, bottom half, in a large hand:] Cisioianus. [Text continues.] [Fols. 12v^a–v^b have two tables. Fol. 13r^b] Modus componendi

tabulam Fungonis est iste rubrica. [Fol. $14r^a$ is the same, except the tables are for Bede. Tract concludes on fol. $15v^a$:]/...et literam etiam dominicalem ipsius.

fol. $15v^a$: Incipit algorismus rubrica. Omnia que a primeua rerum origine processerunt...[listed in TKr, col. 991]/...[fol. $20r^b$] et hec est de radicum extractione dicta sufficiant. Explicit algorismus.

fol. $20r^b$: [*Rubric.*:] Incipit tractatus de spera magistri Iohannis de Sacro Boscho. Tractatum de spera 4 capitulis distinguimus...[listed in TKr, col. 1577]/...[fol. $29r^b$] mundana machina cito dissoluetur. [*Rubric.*:] Explicit.

fol. $29v^a$: Incipit algorismus de minutiis phisicis. Cum multos de numeris tractatus uidisses hec in illo uel opus numeri...[listed in TKr, col. 320]/...[fol. $30v^b$] ac diuidendi exempla ne quid solum necessarium desit opera tum breuissimo. Explicit.

fol. $31r^a$: [*Rubric.*:] Incipit spera domini Roberti episcopi Lincolniensis. Intentio nostra in hoc tractatu...[listed in TKr, col. 763]/...[fol. $36v^a$] semidyametrorum solis et lune. [*Rubric.*:] Explicit spera episcopi linconiensis cuius compotus postea terminatur.

fol. $36v^a$: [*Rubric.*:] Incipit compotus magistri iohannis de sacro bosco. Compotus est scientia considerans...[listed in TKr, col. 243]/...[fol. $54v^a$] quod nos hinc fructificemus. Explicit.

fol. $54v^a$: Prohemium in compositionem et utilitatem quadrantis secundum modernos. Rubrica. Geometrice due sunt partes theorica...[listed in TKr, col. 585; edition by Tannery in Robertus Anglicus, "Le Traité du quadrant," pp. 33–72, was based on this manuscript (p. 16)]/...[fol. $59r^a$] productum dabit capacitatem. [*Rubric.*:] Explicit tractatus quadrantis. [Tables begin on fol. $59r^b$: fol. 60r, table for 1292 (*corr.* 1293); fol. 60v, table for 1294; fol. 61r, table for 1295; *mg.*: "Nota 1327 30 remoue."]

fol. $61v^a$: Prohemium in astrolabium messehallath Rubrica. Scito quod astrolabium est nomen grecum...[listed in TKr, col. 1409; Carmody, *Arabic Astronomical and Astrological Sciences*, p. 25; Poulle, "L'Astrolabe médiéval," p. 101]/...[fol. $70v^a$] polus zodiaci ut patet in hac figura. [Fol. $70v^b$] Nomina instrumentorum sunt hec: primum est armilla...[listed in TKr, col. 916]...[fol. $75v^a$] est comparatio stature tue ad planitiem. [*Rubric.*:] Explicit practica astrolabii.

fol. $75v^a$: [*Rubric.*:] Incipit theorica planetarum et primo de theorica motus solis et circuli eius. Circulus ecentricus uel egresse...[listed in TKr, col. 223]/...[fol. $81r^b$] et non corporaliter. [*Rubric.*:] Explicit theorica planetarum.

fol. $81r^b$: [*Rubric.*:] Incipit liber thebit bencorach de motu 8 spere. [Fol. $81v^a$] Ymaginabor speram equatoris...[listed in TKr, col. 661; Carmody, *Arabic Astronomical and Astrological Sciences*, pp. 117–19, 121]/...[fol. $84r^a$] in linea numeri. [Tables, fol. $83v^a$:] Motus accessionis et recessionis octaue spere ad annos arabum; [and fol. $83v^b$:] Tabula equationis octaue spere cum diuersitate capitis arietis et libre ab equatore.

fol. $84r^a$: [*Rubric.*:] Incipit liber thebith bechorath [*sic*] de his que indigent expositione antequam legatur Almagest. Equator diei est circulus maior...[listed in TKr, col. 502; Carmody, *Arabic Astronomical and Astrological Sciences*, pp. 117–19, 121]/...[fol. $86v^b$] oppositi erunt retrogradi. Expletus est liber thebith filii core de his que indigent expositione antequam legatur almagesti.

fol. 86v[b]: [*Rubric.*:] Liber thebith de ymaginatione spere et circulorum eius diuersorum. Nos iuxta ymaginationem... [listed in TKr, col. 924; Carmody, *Arabic Astronomical and Astrological Sciences*, pp. 117–19, 121]/... [fol. 88r[a]] ymaginanda occurrunt. Explicit thebit de ymaginatione spere.

fol. 88r[a]: [*Rubric.*:] Incipit liber thebit de quantitatibus stellarum et planetarum et primo terre. Ptholomeus et alii sapientes posuerunt... [listed in TKr, col. 1147; Carmody, *Arabic Astronomical and Astrological Sciences*, pp. 117–19, 121]/... [fol. 89v[a]] quia quidam modicum remansit. [*Rubric.*:] Explicit liber thebit de quantitatibus stellarum. [These four tracts of Ṯābit, fols. 81r[b]–89v[a], have been re-edited by Carmody, *Astronomical Works of Thabit b. Qurra*, pp. 82–148, using this manuscript.]

fol. 89v[a]: Capitulum primum de causa bisexti et de modis magis uerificandi calendarium... [listed in TKr, cols. 189, 243]. [List of chapter headings is then given.] [Fol. 89v[b]] Compotus est scientia numerationis et diuisionis temporum... [identified in TKr, col. 243, as a work of Robert Grosseteste or ascribed to Campanus, but comparison of the *explicit* indicates that it is by Campanus]. [Fol. 97r[b]] Hac premissa tabula quoniam ponit auctor ad inueniendum annos arabum per annos christi magis est ad inueniendum omnes annos christi per annos arabum. Quia uero perutile est scire annos... [listed in TKr, col. 1234; see p. 16, note 58, above]. [Fol. 97v has tables with dates, "1232, 1260, 1288,..., 1624." Fol. 102r[b], table:] Ciclus primationum secundum computationem calendarii nostri.... [Fol. 103v[b]]... per hos uersus: Aureus hac arte numerus formatur a parte... [listed in TKr, col. 166]/... [fol. 106r[b]] sumptio sancto istorum uigilia ieiunes luceque marci.

fol. 106r[b]: Composui tabulam hanc ad inueniendam diem et horam medie coniunctionis solis et lune et eorum oppositionis... [identified in TKr, col. 241, as Campanus, *Canon*, tables for Novara] et ad meridiem nouarie.../... [fol. 107r[b]] post medium diem ciuitatis parisius. Explicit compotus episcopi linconiensis [*sic*]. [Tables follow.]

fol. 107v[a]: Inuestigantibus autem astronomie homines primo ponendum est... [listed in TKr, col. 775, as possibly a work of Robert Grosseteste; Duhem, *Système*, vol. 3, p. 316; another copy with a different *explicit* exists in MS Rome, Biblioteca Apostolica Vaticana, latin 3133 (?15th century), fols. 20r[a]–27v[b]: "propter augmentationem et diminutionem equationum non augmentatur nec minuitur cursus planetarum in circulo signorum"; a third copy, in MS Berncastel, Hospital zu Cues, 212 (15th century), fols. 130r–36v (listed in Zinner, *Verzeichnis*, p. 72, no. 1917, incorrectly as fols. 131–41v), which cites the "instrumentum seu theorica Campani," seems to be a précis of this work since several of the folios either contain figures or are blank; finally, TKr, col. 775, lists with a slightly varied *incipit* two 13th century manuscripts, of which only one, MS Oxford, Bodleian Library, Bodley 625, fols. 86–120 (also note MS Paris, Bibliothèque Nationale, fonds latin, 16656, fols. 3r–35v), was available to me on microfilm, and a cursory examination reveals, beyond the similarity of the *incipits*, a much more extensive and elaborate treatment of similar material, the relationship of which to the shorter works only a transcription and close analysis would disclose. It might be hazarded that these last are the archetypes of the

former]/... [fol. 111v^{b}] in septentrionem uel meridiem per ea que hic dicta sunt. Explicit theorica planetarum.

fol. 111v^{b}: [*Rubric.*:] Incipit astrologia W. Marsiliensis. Quoniam astrologie speculatio prima figuram ipsius... [listed in TKr, col. 1261]/... [fol. 124v^{b}] que docentur in ipso auctore. Explicit astrologia massiliensis.

fol. 124v^{b}: [*Rubric.*:] Incipit liber de agregationibus scientie stellarum in principiis celestium motuum quem Ametus filius Ameti qui dictus est Alfraganus compilauit 30 continens capitula. [List of chapter headings follows and concludes on fol. 125r^{b}.] [Fol. 125r^{b}] Numerus mensium anni arabum... [listed in TKr, col. 960; Carmody, *Arabic Astronomical and Astrological Sciences*, p. 115]/... [fol. 142r^{b}] iam ergo declarauimus de eclipsibus solis et lune quod sufficit bene intelligenti. Expletus librum Amet Alfargani. Rubrica.

34. *q* (α) = Paris, Bibliothèque Nationale, fonds latin, nouv. acq. 176. Petit format, 90 fols.

Date: 14th century

Text used: fols. 1r^{a}–41v^{b}. Primus. Figures; tables

Description: The following is from an examination of the manuscript and a microfilm of it.

fol. 42r: incipit theorica planetarum magistri gerardi cremonensis. [*Rubric.*:] Incipit theorica magistri Gerardi Cremonensis. [Fol. 42r^{a}] Circulus ecentricus uel egresse cuspidis... [listed in TKr, col. 223]/... [fol. 47r^{b}] planete et non corporaliter.

fol. 47r^{b}: Imaginabor [*mg.*: "Tebith ben corath in motu accessionis et recessionis 8^{e} spera [*sic*]"] speram equatoris diei et 3 circulos... [listed in TKr, col. 661; Carmody, *Arabic Astronomical and Astrological Sciences*, p. 121]/... [fol. 48r^{b}] quod ipse aliquid profecisset. [Paragraph] Nos autem signabimus hunc motum et situs.../... [fol. 49r^{a}] sicut et arcuum P et LZ et illud est quod uolimus declarare. [Paragraph, beginning fol. 49r^{b}] Cum uolueris scire quantitatem longitudinis capitis arietis et libre ab equatore diei in omni tempore... [listed in TKr, col. 357] [on fol. 49v^{a}, *mg.*: "Istud capitulum debet esse in meo alio libro ubi est"]/... [fol. 49v^{b}] Hec est tabula motus accessus et recessus medii motus capitis arietis et libre in 2 circulis. [Tables follow, ending on fol. 50r; fol. 50v is blank.]

fol. 65r: [Different hand begins, with different numbering.] [*Rubric.*:] Incipit liber Arati de ymaginibus celi et ordine earum constellatione. [Fol. 65r^{a}] Quo sunt extremi uertices mundi quos appellant polos... [listed in TKr, cols. 1251, 473]/... [fol. 66r^{a}] ad ipsum usque decurrit accipiens. De ordine ac positione stellarum [?]uisi suis signis. Est quidem hic ordo et positio syderum que fixa.... [Fols. 66r^{b}–72r^{a} have space left with description of each star. Fol. 72v is blank.]

fol. 73r: [*Rubric.*:] Quedam dicta Ysidori de planetis. Cum sole et lune et septem astra numerantur que non sunt fixa in celo.... [Fol. 73v^{b}]... et mercurio demonstrata esse crescebuntur. [Blank space.] Lacteus circulus quem greci.... [Fol. 75v^{b}] Aratus patris [*Rubric.*: "Arati genus"] quidem est Athinodorii filius.... [Fol. 76r^{b}] [*Rubric.*: "De celi positione"] Caelum circulis quinque distinguitur.... [Fol. 76v^{b}] [*Rubric.*: "De stellis"] Stellarum alie cum celo feruntur.... [Fol.

77v[a]] [*Rubric.*: "De ordinatione stellarum"] Sic est stellarum ordo utrorumque circulorum.... [Fol. 781[b]] [*Rubric.*: "Liber Arati aliter excerptus"] Arati ea que uidentur ostensionem quoque.... [Fol. 82r[b]] [*Rubric.*: "Discretio duum semis eriorum est liber Arati cum dictis poetarum commixtia compilatus expositus"] Habet ante pondus totum medium terre terrenum.... [Next series of rubrics, beginning on fol. 82v[a], are names of stars or constellations.] [Fol. 88r[b]] [*Rubric.*: "De positione et cursu planetarum"] Inter celum et terram certis discreta spatiis vii sidera pende que ab [?]uicessu uocamus errantia...[listed in TKr, col. 765]. [Fol. 88v[b]] [*Rubric.*: "De interuallis planetarum"] Interualla earum a terra multi...[listed in Thorndike and Kibre, *Catalogue*, 1st ed., col. 772]. [Fol. 88v[b]] [*Rubric.*: "De absidibus planetarum"] Tres autem quos supra solem diximus sitas.... [Fol. 90r[b]] Quis enim cognouit sensum domini aut quis consiliarius eius fuit aut quis primo dedit illi et retribuetur ei. Quoniam ex ipso et per ipsum et in ipso sunt omnia ipsi honor et gloria in secula seculorum amen. [New line:] EGO BENEDICTVS DE NVRSIA

35. *R* (α) = Florence, Biblioteca Riccardiana, 885. Parchment, quarto, 380 fols.
Date: 14th century
Text used: fols. 58r–104r. Primus. Figures incomplete; tables
Description: Described by Thorndike, "Notes upon Some Medieval Astronomical, Astrological and Mathematical Manuscripts," pp. 38–40.

36. *r* (β) = Rome, Biblioteca Apostolica Vaticana, Rossiano 732 (X, 112). Paper, 335 mm. by 235 mm., 137 fols.
Date: 15th century
Text used: fols. 1r–31v. Clementissimo. Spaces left for figures and tables
Description: The following is from an examination of a microfilm of the manuscript.

fol. 32r: Incipit liber Alfragani etc. feliciter. [D]ifferentia prima in annis arabum et latinorum...[identified in TKr, col. 429, as a translation by John of Seville]. [Concludes on fol. 47r.]

fol. 47v: Incipit liber seu tractatus astrolabii Messehallam. Rubrica. [S]cito quod astrolabium sit nomen grecum...[listed in TKr, col. 1409]. [Fol. 62r]...Capitulum preambulum in usum astrolabii. [N]omina instrumentorum sunt hec. Prima est armilla...[listed in TKr, col. 916]/...[fol. 66v] ad totam planitiem etc. Explicit tractatus astrolabii.

fol. 67r: Incipit tractatus quadrantis. Rubrica. Geometrie due sunt species theorica et practica...[listed in TKr, col. 585]/...[fol. 72v] et productum dabit eius capacitatem. Explicit quadrans modernus.

fols. 73r–75r: Tabulae solis [five tables]. [Fol. 75v is blank.]

fol. 76r: [John of Sacrobosco.] Incipit algorismus. Omnia que a primeua...[listed in TKr, col. 991]/...[fol. 80r] tam in numeris quadratis quam in cubitis etc. deo gratias. Explicit algorismus [followed by a small table]. [Fol. 80v is blank.]

fol. 81r: Incipit liber de prenosticatione siue prescientia dispositionis temporum.

Ad prenotandam diuersam aeris dispositionem...[listed in TKr, col. 57]. [Section ends on fol. 82v; figure missing.] [Fol. 83r] Cum ergo dispositionem aeris ad aliquem certum terminum...[listed in TKr, col. 293]/...[fol. 83v] Gemini mercurio qui post uenerem fuerat deputati sunt. Explicit feliciter etc.

fol. 84r: Incipit liber pulcherrimus philosophie naturalis. Incipiam et dicam quod orbis prescitus...[listed in TKr, col. 722; cf. Carmody, *Arabic Astronomical and Astrological Sciences*, p. 33, where it is identified as Messehalla, *De orbe*]/...[fol. 93v] sapiens et sublimis. ffinitus est liber.

fol. 94r: Incipit liber cursus lune. Quicumque cursum lune recte scire uoluerit... [listed in TKr, col. 1236; cf. Carmody, *Arabic Astronomical and Astrological Sciences*, p. 66]/...[fol. 97r] dactillorum et huiusmodi. Et sic est finis huius libri etc.

fol. 97v: Nota quod 1080 puncta constituunt unam horam; 15 gradus constituunt unam horam. Item 60 secunda constituunt unum minutum et 60 minuta unum gradum. Nota quod sol currit inter diem et noctem 59 et 8 secunda. Item nota quod gradus in celo est 59 miliaria et due tertie unius miliaris in terra. Amen. [Remainder of folio is blank.]

fol. 98r: Incipit introductorium Alchabitii. Postulata a domino prolixitate uite... [listed in TKr, col. 1078]/...[fol. 114v] ad magisterium iudiciorum astrorum. Perfectus est introductorius Abdilaziz id est serui gloriosi scilicet dei qui dicitur Alzariti ad magisterium iudiciorum astrorum cum laude dei et eius adiutorio interpretatus a Iohanne Hispa. Explicit feliciter.

fol. 115r: Incipit theorica planetarum et primo de sole. Circulus ecentricus... [listed in TKr, col. 223/...[fol. 120v] tricubitum uidelicet ante maioris quantitatis. Explicit feliciter.

fol. 121r: Incipit theorica motuum latitudinis planetarum. Theoricam motuum latitudinis planetarum iam conueniens est perscrutari...[listed in TKr, col. 1571]/...[fol. 121v] Explicit capitulum de latitudinibus planetarum editum a magistro Petro de sancto Hodomaro secundum regulas albategni etc. Eplicit [*sic*] feliciter.

fol. 122r: De inuentione locorum planetarum in signis circuli zodiaci. Primo scire debes quod oportet te primo inuenire...[listed in TKr, col. 1113]/...in zodiaco ibi est luna uerissime. deo gratias. Explicit. [Fol. 122v is blank.]

fol. 123r: Incipit spera magistri magistri [*sic*] Iohannis de sacrobusco anglici. Tractatum de spera quatuor capitulis distinguemus...[listed in TKr, col. 1577]/...[fol. 130v] aut mundi machina dissoluetur. Sciendum uere est quod omnis planeta in quibusdam signis... cesserunt soli lune cancer et leo solo. Amen. ffinis spere magistri Iohannis de sacrobusco anglici ffeliciter.

fol. 131r: Calendarium. Prima dies iani etc....[listed in TKr, col. 1089]/...[fol. 136v] December. [Fol. 137r] Istud kalendarium factum est ad meridiem mediolani...[listed in TKr, col. 796]/...[fol. 137v] et operandum ut prius. Explicit [microfilm so blurred that the rest cannot be read; there is a small table on the quality of the signs].

37. *S* (α) = Florence, Biblioteca Medicea Laurenziana, Ashburnham 208 (134/140). Paper, 418 pp.

Date: 15th century

Text used: pp. 53a–175b. Primus. Tables

Description: Described in C. Paoli, *I codici Ashburnhamiani della R. Biblioteca Mediceo-Laurenziana di Firenze,* Ministero dell'Istruzione Pubblica, *Indici e Cataloghi,* vol. 8 (Rome, 1887–1917), pp. 221–23. It is the same as the manuscript listed in Jacob Tomasini, *Bibliothecae patavinae MSS publicae et privatae* (Udine, 1639), p. 211, col. 2 (top). Consult Antonio Favaro, "Intorno alla vita ed alle opere di Prosdocimo de'Beldomandi, matematico padovano del secolo XV," *Bullettino di bibliografia e di storia delle scienze matematiche e fisiche,* vol. 12 (1879), pp. 153–55, for its history.

38. *s* (θ) = Klagenfurt, Bischöfliche Bibliothek, XXX.b.7. 213 fols.

Date: 15th–16th century

Text used: fols. 34r–60r. Clementissimo. Some tables; no figures

Description: The following is from an examination of a microfilm of the manuscript.

fol. 1r: [Mathematical notes. Fol. 1v: musical notation]

fol. 2r: [A]in visier ruot mach also nym.../...[fol. 5r] als du hie hast un exempel also arbeit auch in den andern [with figure]. [Fol. 5v] Wild vinder ain centrum... [fol. 8v is blank]/...[fol. 9r] der selbig quadrat ist gleich dem circkel [with figure]. [Fol. 9v] Item das haist [?]multiplex.../... in zu proiche und gaussen. [Fols. 10r–v are blank.]

fol. 11r[a]: Incipiunt collectiones ad uirgas planam et scriptam pro capacitate uasorum inquirenda construendas et usu earum: propositio prima. Virgam uisoriam planam ad uasorum columpnarium capacitatem inueniendam construere; uirgam uisoriam dico quia ea uasi...[listed in TKr, col. 1699]/... [fol. 15v[a]] que sunt capacitas positionis uasis date quod erat propositum. Et sic explicit hoc opusculum in nomine indiuidue trinitatis patris filii spiritus quoque sancti amen. Secuntur tabule. [Fol. 15v[b] contains figure.]

fol. 16r: Verte folium et uide tabulas. [New line] Posses etiam etiam probare propositio infrascripta aliter quia noto circulo.... [Fol. 16v] Tabula ad mensurandum longitudinem spissitudinem et latitudinem datis correspondentem et per consequens uirgam uisoriam. [Tables end on fol. 20r.] [Fol. 20r] Explicit tractatus bonus de arte uisoria et de modo faciendi uirgam uisoriam et de usu eiusdem per geometricas et arismetricas doctrinas et de modo inuestigandi capacitates uasorum columpnarium pro quo indiuidua trinitas sit in seculorum secula benedicta amen.

fol. 20v: data area spere [crossed out]; datis area semydyametro quoque eius eius que dy [crossed out]; Datis spere dyametro et area ipsius capacitatem ipsius indagare esto spera proposita cuius dyameter 14 partium et area 154 etiam ducam dyametrum spere [?]unicum [illegible] in ipsius aream; Date spere capacitatem indagare esto spera proposita cuius area 154 cuius radix quadrata est 12 partes et $\frac{2}{5}$ fere. Ducam igitur 12 et $\frac{2}{5}$ que sunt radix quadrata in aream ipsius scilicet in 154 et prouenient 1905 et $\frac{3}{5}$ que sunt capacitas spere proposite; Spere proposite cubum equalem collocare sit spera proposita sicut prius cuius area radix quadrata 12 partes et $\frac{2}{5}$ fere sicut prius duco igitur 12 partes et $\frac{2}{5}$ in se cubos et proueniet cubus spere proposite equalis.

fol. 21r: [Top: cross, "I.N.R.I.," cross. Halfway down in large letters:] CAMPANVS. [Fols. 21v–22v are blank.]

fol. 23r: [A different hand begins.] [P]rimum capitulum continet epistolam ad dominum papam... [identified in TKr, col. 1115, as a work of Leo de Balneolis] [fols. 26v–27r have table of arcs and sines]/... [fol. 33v] usu predicti instrumenti superius. Explicit tractatus instrumenti astronomie magistri leonis de balniolis habitationis Aven. [Fols. 61v–62v appear from microfilm to be blank.]

fols. 63r–v: [?]Tabula canonum primi mobilis Joannis de Monteregio [followed by 61 chapter headings]. [Fol. 64r] Ad serenissimum principem ac cristianissimum Pannoniarum regem Mathiam Iohannis de regio monte in tabulas primi mobilis prefatio [listed in TKr, col. 63]. [A]udiui sepe numero... [listed in TKr, col. 164]. [Fol. 65r] Generalem tabulae usum in primis explanare. Omnes numeri in hac tabula positi representant.../... [fol. 100r] equationem octaue sphaere quesitum est 483. ffinis. [Fol. 100v is blank; fols. 101r–145v contain tables.]

fols. 146r–58v are missing from the microfilm.

fol. 159r: [Microfilm resumes:] equales sicut cancer cum gemini.../... [fol. 170v] ut dixit in libro fructuum Ptholomeus. Completa sunt nunc huius libri decem capitula et ideo illi diuino sit laux qui scientiam ampliauit. [T]erminatus est liber principium sapientie intitulatus. quem edidit abraam euenezre... petrus paduanus... 1293....

fol. 170v: [I]n nomine domini altissimi librum rationum componam uolo enim nunc ponere fundamentum libro de... [identified in TKr, col. 1710, as a work of Abraham ibn Ezra]/... [fol. 198v] ita enim dixerunt antiqui.

fol. 198v: [N]unc inchoabo librum de consuetudinibus in iudicys astrorum... [identified in TKr, col. 965, as Abraham Avenezra, *De consuetudinibus*, or Bethen, *Centiloquium*; cf. Carmody, *Arabic Astronomical and Astrological Sciences*, p. 74]/... [fol. 213r] iudicium planetarum et signorum. Amen.

39. *T* (θ) = Dublin, Trinity College Library, D.2.29 (403) (502). Parchment, quarto (M. R. James, *Lists of Manuscripts Formerly Owned by Dr. John Dee*, Supplement no. 1, *Transactions*, Bibliographical Society [Oxford, 1921], p. 31: Fr. 157)

Date: 16th century

Text used: fols. 107v–60r. Primus. Figures absent but spaces left; tables

Description: Contents listed in T. K. Abbott, *Catalogue of the Manuscripts in the Library of Trinity College, Dublin* (Dublin, 1900). Abbott gives only names and titles, without foliation. The following analysis is drawn from a microfilm of fols. 56v–160r.

fol. 56v: ...in eo enim quod dicit medicinam sauororum scientiam esse et egrorum et neutrorum signat. [Bottom:] quod et omnium.

fol. 57r: Incipit tractatus de composicione astrolabii secundum doctrinam Messahallah. [S]cito quod astrolabium sit nomen grecum cuius...[listed in TKr, col. 1409; cf. Carmody, *Arabic Astronomical and Astrological Sciences*, p. 24]/...[fol. 70r] et scindent diametrum AZ circuli qui queritur ut patet in sequenti figura. Explicit tractatus de compositione astrolabii secundum Messahalach. [Gunther, *Chaucer and Messahalla*, p. 213: I. 18.] [Fol. 70v] Tabula stellarum fixarum que ponuntur in astrolabio cum gradibus quibus celum mediant et cum distantia earum ab equinoctiali. [Fol. 71r] Secunda pars tabule stellarum fixarum que ponuntur in astrolabio cum gradibus quibus celum mediant et cum distantia earum ab equinoctiali. [Fol. 71v] Tabula stellarum fixarum uerificatarum per armillas parisius et est longitudo earum gradus circuli signorum circulum transeuntem per polos zodiaci et stellas latitudo uero earum est arcus eiusdem circuli cadens inter stellas et gradum longitudinis earum. [Fol. 72r] Incipit practica astrolabii siue rememoratio partium astrolabii. [N]omina instrumentorum astrolabii sunt hec. Primum est suspensoria armilla...[listed in TKr, col. 916; cf. Carmody, *Arabic Astronomical and Astrological Sciences*, p. 24]/...[fol. 78r] si uero cum punctis habueris fractiones uide quid debeatur sibi de gradibus ut supra determinatum est. Explicit tractatus de practica astrolabii. [Gunther, *Chaucer and Messahalla*, p. 230: II. 43.]

fol. 78v: Tractatus de compositione noui quadrantis Profacii iudei correctus a magistro Petro de Sancto Audomaro. [Q]uoniam conceditur opus huius instrumenti...[listed in TKr, col. 1267]. Incipit composicio. [S]i igitur quadrantem istam componere intendas...[listed in TKr, col. 1450]. [Fol. 88v] Sequitur secunda pars huius tractatus in qua ponuntur canones operationum huius instrumenti et primo de gradu solis inueniendo. [C]um scieueris mensem romanum et diem eius et uolueris scire in quo signo sit sol...[listed in TKr, col. 338]. [Fol. 95r] Secunda pars huius tractatus et primo de altitudine mensuranda. [C]onsequenter dicendum est de mensurationibus rerum inferiorum et primo de mensuratione rerum altitudinum. Si uis igitur [fol. 95v] scire altitudinem alicuius rei accessibilis...[listed in TKr, col. 250]/...[fol. 107r[a]] de 48 qui est totus numerus quartarum. [Fol. 107r[b] contains table: "umbre uerse."]

40. *t* (α) = Venice, Museo Civico Correr, Cicogna 2721 (3747). Paper, folio imperiale, two columns, 172 fols.

Date: 15th century

Text used: fols. 63r[a]–88v[b]. Primus. Some figures; tables

Description: The following is from an examination of a microfilm of the manuscript.

fol. 1r[a]: Incipit theorica planetarum magistri Gerardi Cremonensis. [C]irculus ecentricus...[listed in TKr, col. 223]. [Fol. 4r[b]]...tricubitum uidelicet aut maioris quantitatis. [L]inea egrediens a centro terre....[Fol. 4v[a]]...Quando uero est in longitudine longiori...et octaue spere collocatur et quali equaliter procedeant quia in tam paucis annis non est magna inequalitas anni. Et sic est finis tractatus theorice planetarum a magistro gerardo cremonensi uiri [*sic*] in hac scientia excellentissimi.

fol. 4v[b]: [Top mg.: "Astronomia est lex uel ratio figuras celestes ac eius motus in se et in suis efformibus ultima considerans."] [A]stronomie scientie doctrinalis due sunt partes una theorica alia practica inter quas est differentia.../...[fol. 36r^{b}] in comparatione ad alios planetas superiores mouetur in suo epiciclo ex dictis patet quod illi planete qui retrogradantur uelocius mouentur in suis epiciclis quam centrum epicicli eorum in suis deferentibus. [Fol. 36v is blank.]

fol. 37r^{a}: Dixit Y. qui fuit medicus et magister optimus. Cuiusmodi medicus est qui astronomiam ignorat. Nullus homo debet se intromittere in manus eius qui non est medicus perfectus...[listed in TKr, cols. 454, 277]/...[fol. 39r^{b}] et si fortuna eam aspexerit uiuet, sin autem morietur. Explicit.

fol. 39v^{a}: Hic est liber quem collegit Albumasar de floribus eorum que significant res superiores in rebus inferioribus...[listed in TKr, col. 616; cf. Carmody, *Arabic Astronomical and Astrological Sciences*, p. 94]/...[fol. 48r^{a}] et non uideatur aput te quod exposui tibi si deus uoluerit. Et ipse est auxiliator. Finitus est liber florum Albumasar sub laude dei et eius adiutorio. Deo gratias. amen. [Fol. 48r^{b} is blank.]

fol. 48v^{a}: Prologus Alchabicii. Postulata a deo prolixitate uite cephalica...[listed in TKr, col. 1078; cf. Carmody, *Arabic Astronomical and Astrological Sciences*, p. 148]/...et in libro meo quem edidi in confirmatione magisterii iudiciorum astrorum.... Incipit liber Alchabii Prima doctrina nistach id est circulus signorum diuiditur in tres partes...[listed in TKr, col. 913]/...[fol. 62v^{b}] Introductorium abdiladicir idest serui gloriosi scilicet dei qui dicitur algabitius ad magisterium iudiciorum astrorum cum laude dei et adiutorio interpretatus a Johanne Hispalensi deo gracias amen.

fol. 89r^{a}: Fiat primo circulus magnus cuiusuis quantitatis...[listed in TKr, col. 557]/...[fol. 90r^{a}] rota 63 argumentum mercurii. [Remainder of folio is blank; fol. 90v is blank.]

fol. 91r^{a}: Incipit compositio et operatio astrolabii secundum nouam compositionem. Cum plurimum ob nimiam quandoque accumulationem et magnam scriptorum sententiam canones astrolabii utilitates declarantes intelligere et memoria commendare non ualuerunt...[identified in TKr, col. 331, as a work of Prosdocimo de Beldomandi]/...[fol. 100r^{a}] et si contingat regulam cadere in partem puncti tunc operare ut superius ostensum est. Finis est.

fol. 100r^{b}: [Q]uamuis de astrolabii compositione tam modernorum quam ueterum...[identified in TKr, col. 1164, as a work of Prosdocimo de Beldomandi]/...[fol. 105v^{a}] Et alter clauus qui dicitur alforat id est equus restringens in foramen illud inmitti sicut hic patet. Et sic finitur operatio astrolabii. [Fol. 105v^{b} is blank.]

fol. 106r^{a}. De compositione et operacione chilindri. In compositione chilindri uocatur horologium uiatorum accipe lignum durum scilicet buxum...[listed in TKr, col. 667]/...[fol. 106v^{b}] que sunt post meridiem decrescit altitudo proportionaliter sicut ante meridiem creuit et sic est finis sit laus et gloria trinis.

fol. 107r^{a}: Magistri Joannis Symonis de Zelandia compositio specula planetarum incipit feliciter. In nomine dei misericordi Amen. Ad utilitatem comunem studentium in astronomia et specialiter medicorum...[listed in TKr, col. 64]/...[fol. 109v^{a}] De qua laudetur ueritas eterna regnans per infinita secula seculorum. amen. Et sic est finis sit laus et gloria trinis.

fol. 109v^{b}: Pro equatorio paralellario. Equatorium comune omnibus planetis componere recipe tabulam magnitudinis bipedalis.../...[fol. 110v[a]] per naturam lune in almuri positam ostendet eius uerum locum et sic est finis sit laus et gloria trinis.

fol. 110v[b]: Magistri Roberti Anglici quadrantis compositio ex qua geometrie exercitium habetur feliciter incipit. Geometrie due sunt partes theorica et practica...[listed in TKr, col. 585]/...[fol. 113v[b]] et productum dabit capacitatem. Finis.

fol. 114r: Tabule correcte ad annos christi M. CCCC LX a meridie uulgarium diei precedentis more astrologorum diem inchoando. Tabula solis prima in anno bisextilli ad inueniendam locum...[concludes on fol. 114v].

fol. 115r[a]: Incipit liber Messehalla de exceptione interpretatus a Johanne Hispalensi de arabico in latinum. Inuenit quidam uir ex sapientibus librum ex libris secretorum astrorum de illis quos thesaurizauerunt reges...[listed in TKr, col. 774; cf. Carmody, *Arabic Astronomical and Astrological Sciences*, p. 27]/...[fol. 122v[b]] Ideoque iungerebatur ad hoc nutu dei. Finitus est liber messehale de interrogationibus iudiciorum.

fol. 123r[a]: Incipit liber Thebit filii Choit de his que indigent expositione antequam legatur Almagesti. Equator diei est circulus maior qui describitur super duos polos...[listed in TKr, col. 502; cf. Carmody, *Arabic Astronomical and Astrological Sciences*, p. 121]/...[fol. 125r[a]] erunt retrogradi. Expletus est liber Thebit filii Chore de hiis que indigent expositione antequam legatur almagesti.

fol. 125r[b]: Incipit liber Messahallah de eclipsi lune. Incipit epistola messehallah in rebus eclipsis lune et in coniunctionibus planetarum...[listed in TKr, col. 729; cf. Carmody, *Arabic Astronomical and Astrological Sciences*, p. 32]/...[fol. 126v[b]] que protulimus in libro hoc et ex libro secretis scientie astrorum. Perfectus est liber Messahalla translatus a Johanne Hispalensi ex arabico in latinum in linea sub laude dei et eius auxilio. Laus Christo debet operis quod finis habetur.

fol. 127r[a]. Fiat columpna et locetur in congrua basi in capite columpne sit clauus concauus sustentans speram ligneam...[listed in TKr, col. 556]/...[fol. 127r[b]] sic habetur uerus motus planetarum preter duos circulos breues quoniam unus...equat cursum lune alter uno...equat cursum mercurii. Laus Christo debet operis quod finis habetur.

fol. 127v[a]: Incipit liber de quantitatibus stellarum et planetarum et primo terre. Ptolomeus et alii sapientes posuerunt centrum terre comunem mensuram sua metiebantur stellarum corpora...[listed in TKr, col. 1147; cf. Carmody, *Arabic Astronomical and Astrological Sciences*, p. 121]/...[fol. 128v[a]] sicut nauis in mari sicut potest uideri in spera. Laus Christo debet operis quod finis habetur.

fol. 128v[b]: Honestum quoque est huic salutem promittere alteri uero motum pronunciare et in breui et in prolixa egritudine est ut uult Ypocrates in fine pronosticorum...[listed in TKr, col. 638]/...[fol. 130r[b]] de quibus crisibus hic non intrando tractare breuitatis causa cum doctores sufficienter de talibus determinauerunt finis. [Fols. 130v–34v are blank.]

fol. 135r[a]: Inueni in pluribus libris practice arismetice algorismi nuncupatis modos circa numeros operandi satis uarios atque diuersos...[listed in TKr, col. 774]/...[fol. 145v[a]] uolentibus alium modum in hac arte operandi qui contineatur in Algorismo Johannis de Sacrobosco extrahere ad laudem omnipotentis Dei amen. Laus christo debet operis quod finis habetur. et hec de extractione radicum in numeris cubicis ac de totali tractatu per Prosdocimum de Beldemando de Padua compilata anno domini 1420 de mense iunii a compositione perfecta, sufficient uolentibus etc.

fol. 145v[b]: Quoniam operantibus circa artem calculatoriam ex maxime in ipsa arte non multum expertis siue expertis uel practicis sepe cum contingit...[listed in TKr, col. 1291]/...[fol. 147v[a]] in diuersis marginibus tot tabulas quot sibi placeret in sequendo ordinem predictum et sic sit finis huius canonis per Prosdocimum de Beldemando de Padua Padue compilati atque a compositione anno domini 1429 a compilatione perfecti ad laudem eius qui uiuit et regnat per infinita secula seculorum amen. Laus christo debet operis quod finis habetur. [Fol. 147v[b] is blank.]

fol. 148r[a]. Tractatum de sphera in quatuor capitula distinguimus dicentes... [identified in TKr, col. 1577, as a work of John of Sacrobosco]/...[fol. 155r[a]] machina istius mundi dissoluetur et sic est finis tractatus de spera. Laus christo debet operi quod finis habetur.

fol. 155r[b]: Si igitur quadrantem unum componere uolueris accipe tabulam eneam planam quam preparabis...[listed in TKr, col. 1450]/...[fol. 167r[b]] quarta pars numeri quadrati de 44 qui est totus numerus quartarum. Hiis autem completis auxiliante Deo huic tractatui finis est apponendo explicit tractatus quadrantis.

fol. 167r[b]: [Right mg.: "descriptio turketi et utilitatis eius incipit."] De omnibus partibus instrumenti quod turcetum dicitur...[listed in TKr, col. 384]/...[fol. 168v[a]] sagax lector per se facile intelligit ueniet ex predictis. deo gratias.

fol. 168v[a]: [Left mg.: "Constructio mirabilis horologii incipit feliciter."] [A]d faciendum horologium mirabile et secundum umbrae ueritas...[listed in TKr, col. 39]/...[fol. 168v[b]] residui altitudinis poli gradus. hec est figura.

fol. 169r[a]: Cum autem per instrumentum directionum uolueris dirigere significatores...[cf. TKr, col. 39]/...[fol. 169v[b]] debet dirigi ante et retro. Explicit compositio huius instrumenti per quod diriguntur significatores cum latitudinibus suis.

fol. 169v[b]: Accipe tabulam planam super cuius extremitatem fac circulum...[listed in TKr, col. 25]/...[fol. 170v[a]] latitudinis regionum hoc completo erit compositio instrumenti perfecta.

fol. 170v[b]: Pro compositione kilindri forma circulum...[cf. TKr, col. 1130]/...[fol. 171r[b]] hiis igitur perfectis poteris operari faciliter.

fol. 171r[b]: Girum lune componere super centro alicuius tabule.... Medium motum tunc per hoc instrumentum inuenire pone.../...[fol. 172r[b]] et alia multa huiusmodi que habito uno principio facile intelligantur.

41. *U* (α) = Vienna, Nationalbibliothek, 5273 (Philos. 61). Paper, folio, 355 fols.
Date: 16th century
Text used: fols. 1r–35r. Primus. Figures incomplete; tables
Description: Described in *Tabulae codicum manuscriptorum praeter graecos et orientales in Bibliotheca Palatina Vindobonensi asservatorum* (Vienna, 1864–1912), vol. 4.

42. *u* (θ) = Vienna, Nationalbibliothek, 5296 (Philos. 191). Paper, folio, 173 fols.
Date: 15th century
Text used: fols. 25r–71r. Primus. Figures incomplete; tables
Description: Described in *Tabulae codicum... Vindobonensi*, vol. 4.

43. *V* (δ, β) = Vienna, Nationalbibliothek, 5311 (Philos. 225). Paper, folio, 138 fols.
Date: late 14th–early 15th centuries
Text used: fols. 81r[a]–100r[b]. Clementissimo. Figures incomplete; spaces left for absent tables
Description: Described in *Tabulae codicum... Vindobonensi*, vol. 4.

44. *v* (α) = Vienna, Nationalbibliothek, 5412 (Rec. 1676). Paper, folio, 274 fols.
Date: 15th century
Text used: fols. 3r–69v. Primus. Figures; tables
Description: Described in *Tabulae codicum... Vindobonensi*, vol. 5. The catalogue gives fol. 71r as the final folio of Campanus's *Theorica*; actually it ends on fol. 69v, after which follows a series of partial excerpts from the description of the *instrumentum* in the various sections of the work (lines III. 157–60, III.193–202, IV.611–50, V.597–647, VI.844–66), ending with the usual *explicit* for the second time.

45. *W* (α) = Cracow, Biblioteka Jagiellońska, 589 (DD.IV.4). Paper, folio, 267 fols. (listed by W. Wislocki, *Catalogus codicum manuscriptorum Bibliothecae Universitatis Jagellonicae Cracoviensis* [Cracow, 1877–81], vol. 1, by pages; according to pagination, there are three preliminary pages and 538 text pages, of which the first four are blank. Therefore page 5 corresponds to fol. 1r. Pp. 419–44 [fols. 208r–20v] and pp. 497–504 [fols. 247r–50v] are also blank.)
Date: 1494–95
Text used: fols. 251r–66v. Clementissimo. No figures; no tables
Description: Described by Wislocki, *Catalogus codicum... Cracoviensis*, vol. 1. The following is from a microfilm of part of the manuscript.

fols. 1r–207v: Almagestum [see Wislocki, ibid.]

fol. 221r: [C]elestium siderum uarios multiplicesque decursus quos longis obseruationibus studiosi ingenioso intellectu uigentium professorum manifesti deprehendimus et quos... [listed in TKr, col. 198]/... [fol. 227v] apparet figuram diligenter inspicienti. [Fol. 228r is blank.] [Fol. 228v] [Q]uamuis ut iam antedictum est non multum intersit quantum ad operis huius propositum si augem ecentrici solis et eius centrum... /... [fol. 230v] quod facere uolebam et ad concipiendum leuius prescripta figura suffragatur. [Fol. 231r] Modum autem situationis instrumenti huius in latere secundo superioris casamenti quod inscriptum est latus ueneris et quomodo per rotam anni.../...[fol. 236r] et ad libitum possit inde leuari sequenti figura. [Fol. 236v, halfway down:] Expedita cum dei laude compositione instrumenti in quo omnes longitudinales motus mercury demonstrantur iam sequitur secundum ordinem ad compositionem instrumenti lune procedere in quo.../...[fol. 243v] computaueris sunt computatis magnitudinis et paruis numero quatuordecimi. [Fol. 244r] [R]estat nunc secundum propositum ordinem instrumenti iouis instructionem ostendere cuius quidem componere similem oportebat esse instrumento saturni quantitatibus.../...[fol. 245r] faciuntque in hoc opere rote inter paruas et maiores numero tantum [?]97 [remainder of folio is blank]. [Fol. 245v] [P]ost preterita reliquum est et ultimum compositionis instrumenti motuum martis.../...[fol. 246r] hys itaque pactis fabrica in tabulam consimili rotundam [remainder of recto is blank; verso begins abruptly:] et planam paulo latiorem uero epiciclo in superficie instrumenti descripto.../... quod proponere dederat dedit in effectu laudabiliter terminare. Telos. Astrary Johannis de dondis patauini libellus explicit scriptus per henricum Ragnetensem in studio uniuersali Cracouiensi. 1494.

fol. 267r: [Right mg.: "thebit benchorat fily chore de motu octaue spere"] [I]maginabor speram equatoris diei et circulos in eo ligatos qui sunt orizontis et circulus medie diei...[listed in TKr, col. 661]/...[fol. 267v] et illud est quod uoluimus declarare. Finis. 1494. Cracou.

46. *w* (α) = Cracow, Biblioteka Jagiellońska, 601 (DD.IV.5). Folio, 415 pp. (208 fols.) plus 9 fols.

Date: 15th century

Text used: fols. 96r–117r. Clementissimo. One figure (fol. 112v); no tables

Description: Wislocki, *Catalogus codicum...Cracoviensis*, vol. 1, only reproduces the index on fol. 1r of the manuscript. The following is from an examination of a microfilm of the manuscript.

fol. 10r: [T]ractatum de spera quatuor capitulis distinguimus...[identified in TKr, col. 1577, as a work of John of Sacrobosco; Zinner, *Verzeichnis*, p. 323, no. 10479]/...[marginal references to Campanus *passim*; fol. 16r] aut mundi machina dissoluitur etc.

fol. 16v: Nota quod sequentia septem climata sunt spatia partialia unius quarte ipsius terre quia solum una quarta est habitabilis...[cf. TKr, col. 940]. [Diagrams.] Nota quod ista septem climata non distinguntur abinuicem equalitate spatii.../...[fol. 17r] et principalior in affrica est cartago emula ratione in europa principalior est roma. [Fol. 17v is blank.]

fol. 18r: Circulus ecentricus uel egresse cuspidis uel egredientis centri dicitur qui non habet centrum cum mundo...[listed in TKr, col. 223]/...[marginal reference to Campanus; fol. 26v] et non corporaliter. [Fols. 27r–31v are blank. See Zinner, *Verzeichnis*, p. 285, no. 9199, for further details.]

fol. 32r: [D]ixit Ptholomeus iam scripsi tibi iesure libros de hoc quod operantur in quadripartito...[listed in TKr, cols. 456, 650]. Scientia stellarium ex te et illis...[listed in TKr, col. 1403]. [Fol. 37v] Sequitur expositio siue commentum scilicet: concordati sunt in hoc omnes philosophy ut in alio latere. [Fol. 38r] In hoc concordati sunt omnes philosophi quod more natorum uteris matrum sunt.../...[fol. 44r] in regibus et diuitibus apparebit et cetera. Deo laus. Explicit centiloquium ptholomei cum commento haly. [See Zinner, *Verzeichnis*, p. 272, no. 8758, for further details.]

fol. 44v: Incipit liber 157 uerborum almansoris regi magno saracenorum et dicuntur aphorismi Rasis. [S]ignorum dispositio est ut dicam unum scilicet est diurnum et alterum...[listed in TKr, col. 1504]/...[fol. 47r] et regnum in quo nulla fiet iniusticia. Completus est liber amphorismorum almansoris. [See Zinner, *Verzeichnis*, p. 33, no. 573, for further details.]

fol. 47r: Domino manfredo inclito regi sicilie stefanus de messana hos flores de secretis astrologie diui hermetis transtulit. In hoc titulo quelibet litera cuiuslibet dictionis ostendit principium propositionis. Centiloquium hermetis. [N]on diffinias neque eligas aliquid existente scorpione in ascendente neque cum anguli sunt obliqui aut mars existat in eis...[listed in Zinner, *Verzeichnis*, p. 141, no. 4239; Carmody, *Arabic Astronomical and Astrological Sciences*, p. 54]/...[fol. 48v] multi quidem interrogare nesciunt neque possunt exprimere quod intendunt.

fol. 48r: Tabula fortunarum et fortitudinum infortunarum et debilitatum planetarum. Fortuna solis. Si in aspectu fortune uel sibi iunctus.../...[fol. 49v] econtrario dic de eorum infortuniis et debilitatibus.

fol. 50r: Thebith benchorat de ymaginatione spere et eius circulorum liber incipit feliciter. [N]os iuxta imaginationem inchoantes astrologie [listed in TKr, col. 924; Zinner, *Verzeichnis*, p. 332, no. 10781; Carmody, *Arabic Astronomical and Astrological Sciences*, p. 119]/...[fol. 50v] sunt occurrunt etc. Finis.

fol. 51r: Compositio astrolabii. Quamuis de astrolabii compositione tam modernorum quam ueterum dicta pulcherrima habeantur tamen quia in eis...[identified in TKr, col. 1164, as a work of Prosdocimo de Beldomandi; Zinner, *Verzeichnis*, p. 45, no. 949]/...[fol. 53v] et ad principium capricorni applicare horum exempla patent in figura [tables of 12 signs]. Cum diuiseris circulum signorum certissime secundum aliquem modum ex predictis tunc...[listed in TKr, col. 294]/...[fol. 55r] et alter clauus qui dicitur alfarath idest equum restringens in foramen illud inmitti debet hec omnia in figura patent.

fol. 55r: [In a different hand, *mg.*: "Alter tractatus de compositione astrolabii."] Compositurus nouum quadrantem accipe tabulam planam ad similitudinem... [identified in TKr, col. 241, as a work of Prophatius Judeus]/...[fol. 57r] cuius foramina a predicta linea equedistent in latitudine. Et sic est finis.

fol. 57r: Ad intelligentiam circulorum astrolabii intelligatur stare homo sub equatore in uero centro terre.../... in quartis uero solsticialibus econtrario accidit.

fol. 57v: Prohemium messehallach in practicam astrolabii feliciter incipit. [I]nstrumentorum astrolabii plurima sunt nomina et ideo ut cognitionem eorum... [cf. TKr, col. 916; Zinner, *Verzeichnis*, p. 221, no. 7053]. De armilla. Primum igitur instrumentum astrolabii est armilla suspensoria.../...[fol. 60v] talis est comparatio stature tue ad planitiem. Finis.

fol. 60v: Messehallac de mensurationibus quadrati in astrolabio siue extra liber secundus feliciter incipit. [N]unc autem dicendum est de mensurationibus per quadratum in astrolabio uel extra.... Si uis scire alicuius rei accessibilis altitudinem per ambo foramina uno oculo...[identified in TKr, col. 1472, as a work of Robertus Anglicus]/...[fol. 61v] per eius altitudinem et productum dabit capacitatem. Finis.

fol. 62r: Arismetrica Algebre. In tota practica regularum algebre quatuor denominationes seu quatuor uocabula.... Volo multiplicare quatuor plus 2 res per 9 plus 3 pono figuras.../...[fol. 68v] et hec de radicum extractione sufficiant. Finis est. [Fols. 69r–71v are blank.]

fol. 72r: Predicere egritudinem urina non uisa. Sapientissimus omniumque medicorum peritissimus hypocras ait cuiusmodi est medicus que astronomiam ignorat...[listed in TKr, col. 1379]/...[fol. 74r] cum labore et difficultate nimia. Finitur deo gratia.

fol. 74v: Cuncta astronomie iudicia ad primum celum quod primum mobile nuncupatur referuntur in quo non sunt stelle...[cf. TKr, col. 359]/...[fol. 76v] sed sit iniuncta a malis uel nisi luna iungatur fortune ex trino uel sextili aspectu.

fol. 77r: Dolores autem oculorum et flegma aut albula et cetere eorum egritudines.../...quia talemin firmem sub prescripta figura in cura habui ego gaufredus de meldis [with figure].

fol. 77v: Nota quod in transmutationibus et negotiis cotidianis sunt magis fortes.../...[fol. 79r] in quo incepit egritudo ipsa luna. Iterum dies infelices grecorum in singulis mensibus quia quelibet res incepta in ipsis male terminatur et sunt hii infrascripti [with tables]. [Fols. 79v–82r are blank.]

fol. 82v: [Tables in two columns]

fol. 83r: Alkabitius. Introductorium ad iudicia astrorum alkabitii astrologi excellentissimi. [P]ostulata a deo prolixitate uite ceyfaddaula...[listed in TKr, col. 1078; Zinner, *Verzeichnis*, p. 18, no. 55; Carmody, *Arabic Astronomical and Astrological Sciences*, p. 147]/...[fol. 94r] astrorum ne quis erraret quasi in quam partem sumerentur. Perfectus est igitur introductorius liber abdilarich id est serui dei gloriosi qui dicitur alkabir id est magisterium iudiciorum astrorum sub laude et adiutorio dei.

fol. 94r: [Right mg.: "De orbe magno."] Ad cognoscendum orbes magnos qui sunt quedam reuolutiones 360 annorum et ad sciendum in quoto orbe mangno.../...in libris antiquorum de reuolutionibus annorum factis.

fol. 94v: Nota bene area esse planetarum a sole. Coniungitur soli saturnus et uniuersaliter quilibet aliorum planetarum cum fuerit in eodem minuto cum sole.... Item quod anno 1473° sol fuit dominus anni quia fuit in prima domo accedens ad literam ascendentem et non recedens.../...[fol. 95r] unde sol predicto anno fuit dominus anni et significator causarum in omni climate.... Venus autem tunc fuit inimica regis.... Fuit autem coniunctio precedens dictam reuolutionem die 26° februarii hora 13 minuto 19°.

fol. 117r: [Right mg.: "Hebith bencorat de motu octaue spere filii chore liber feliciter incipit."] Ymaginabor speram equatoris diei et circulos in ea ligatos...[listed in TKr, col. 661]/...[fol. 118r] et illud est quod uoluimus declarare. Finis est.

fol. 118r: Thebit bencorath de quantitate stellarum liber feliciter incipit. Et primo de corpore terre. [T]holomeus et alii sapientes posuerunt corpus terre comunem... [listed in TKr, col. 1147]/... [fol. 119r] sed iste non est in litera sed proximus ei maior scilicet 185 quia modicum quiddam remansit. Finis.

fol. 119r: Incipit tractatus Campani de retrogradatione 5 planetarum. In 5 planetis qui sunt Saturnus Iupiter Mars Venus et Mercurius [MS uses symbols for the planets] motus corporis planete in epiciclo est uelocior quam motus centri epicicli in suo deferente propter quod oportet quod in superiori parte epicicli in qua ambo predicti motus sunt... [Campanus, *De sphera*, chap. 50]/... [fol. 119v] potest non incongrue dici cursu media.

fol. 119v: Incipit tractatus Campani de coniunctione planetarum cum sole in suo motu. Mirabile illud omittere uolumus quod sol omnes alios planetas in motu eorum dirigit quasi omnes habeant respectum ad primum et secundum ipsius motum omnes alii sui motus fortunam... [Campanus, *De sphera*, chap. 52]/... bis describat deferentem in omni reditione lune ad solem.

fol. 119v: Incipit tractatus campani de eclypsibus lune primo secundo solis. Restat nobis ut dicamus causas eclypsium solis et lune pro quibus scientia est quod corpus... [Campanus, *De sphera*, chap. 53]/... [fol. 120r] deficiente omnis luminis uero fonte.

fol. 120r: Incipit tractatus Campani de 12 domibus. Intelligantur in spera alii quatuor uel 8 circuli per quos distinguntur secundum unumquemque situm... [Campanus, *De sphera*, chap. 29]/... [fol. 120v] scitur earum altitudo unde utrique ponuntur in astrolabio ut postea in suo tractatu patet [see p. 23, above].

fol. 120v: Tractatus de motu octaue spere. Octaue spere ad cuius motum orbes deferentes auges planetarum mouentur triplex inest motus unus.../... [fol. 121r] Cum autem fuerit maior semicirculo tunc erit substrahenda patent autem supradicta in figura posita in alio latere istius folii. Finis. [Fol. 121v contains a figure; fol. 122r, a table.]

fol. 122v: Abrahe Aben Esdre. Incipit introductorium Abrahe aben esdre in iudicia astrorum quod dicitur principium sapientie. Prohemium uniuersale eius. Cum initium sapientie dei timor existat hic utique... [listed in TKr, col. 309; Zinner, *Verzeichnis*, p. 19, no. 105]/... [fol. 148v] est in opositum signorum ut dixit Ptolomeus in libro fructuum. Finitus est liber intitulatus principium sapientie... quem quidem cum petrus de abano inuenisset in gallico ydeomate.... Fuit autem compilatus hic liber post creationem mundi seu ade [fol. 149r] 4908 annis. Nunc autem existentibus annis incarnationis domini 1293 sunt anni ade 5053 et 8 menses et circa 4 dies. Et sic est finis libri. Laus deo quoniam finis libri habetur in eternum memini. Tempus est mensura motus ut uult Aris. quarto phisicorum cum igitur motum scire desiderauimus. Nomina signorum sunt hec Aries Taurus Gemini Cancer Leo Virgo Libra Scorpion Sagittarius Capricornus Aquarius Pisces.

fol. 150r: Incipit liber florum Albumasar. [H]ic est liber quem collegit albumazar eorum que seruant res superiores in rebus... [listed in TKr, col. 616; Zinner, *Verzeichnis*, p. 23, no. 238]/... [fol. 157r] occultari ut non uideatur apud omnes secundum quod exposui si deus uoluerit.

fol. 157r: Incipit liber experimentorum eiusdem albumasar. [V]olens scire naturam aeris in singulis annis... [listed in TKr, cols. 1707, 1708, as anonymous; in TKr, col. 1706, as John of Seville]/...[fol. 157v] super tonitrua et choruscationes cum fuerit in domo sua uel in coniunctione cum ioue.

fol. 157v: Sequitur quot sunt cause debilitationis al. debilitatis planetarum. [D]ebilitas planetarum fit 10 modis...[listed in TKr, col. 396]/...ut planeta impediat semen ipsum id est ut sit in septima a domo sua.

fol. 157v[a]: Ab origine mundi usque ad christum....

fol. 157v[b]: Nota quod a coniunctione que significabat diluuium usque ad tempus presens....[Fols. 158r–61v are blank.]

fol. 162r: Canones super tabulas Alfonsy regis. Tempus est mensura motus ut uult Aris. quarto phisicorum cum igitur motum scire desiderauerimus...[listed in TKr, col. 1561; Zinner, *Verzeichnis*, p. 79, no. 2138]/...[fol. 168r] uel in uno minuto diei et tunc in omnibus idem est motus sicut dictum est in coniunctionibus planetarum. finis est. Sequitur quedam additio. Ad inueniendum partem proportionalem. Nota quod omnis arcus qui inuenitur ex tabulis mediantibus aliis arcubus.../...diuiseris per primum resultat pars proportionalis quesita. [Fol. 168v] Tabula stellarum fixarum que ponuntur in astrolabio uerificata per instrumentum armillarum parisius et est longitudo earum arcus orbis signorum...anno domini 1436 [followed by table and notes below].

fol. 169r: [Figure.] [Col. 1:] Hec est figura mansionum lune facta ad gradus zodiaci primi mobilis quod est decima spera secundum aliquos; [col. 2:] (Anno domini nostri iesu christi 1460); [below both columns:] Nota quod si uis uerificare mansiones lune quocumque tempore debes querere augem comunem....[Fol. 169v] Mansiones lune in 12 signis zodiaci primi mobilis sunt 28 quarum cuilibet de zodiaco cedunt gradus 12 minuta 51 et secunda 26 fere et fuerunt uerificate tempore alfonsii [followed by a list of the 28]. Nota quod in zodiaco sunt 346 stelle quarum 5 sunt de magnitudine prima et 9 de magnitudine secunda....

fol. 170r: Stelle uerificate per bartolomeum de manfredis ad meridianum mantue que sunt lucidiores in celo de prima scilicet secunda aut tertia magnitudine que sunt ponende in astrolabio et hoc anno domini nostri iesu christi 1440 de mense octobris Nomina stellarum cum nominibus ymaginum in quibus sunt situate... [concludes on fol. 171r with notes: "Mansiones correcte per M. Adrianum Zeeroliet anno domine 1473 completo..."]. [Fol. 171v is blank. Concerning Bartholomew, see TKr, col. 512.]

fol. 172r: Diuisio libri alarbamacalet id est quadripartiti ptolomei....[Fol. 172v] Capitulum primum in collectione intellectus scientie iudiciorum astrorum. Rerum iesure in quibus est pronosticabilis scientie...[listed in TKr, col. 1349; Zinner, *Verzeichnis*, p. 270, no. 8707]/...[fol. 176r] Idem quoque in omnibus aliis modis euenire secundum quantum comixtionem non dubitamus. Capitulum 12m in signis fixis et mobilibus et comunibus. [Fols. 176v–81v are blank.]

fol. 182r: [Top mg.:] particula octaua hali habenragel. [Fol. 182r[a]] [D]ixit halay filius abenragel laus deo qui est dominus subtilitatum nobilitatum mercedum pietatum creator omnium creaturarum cognitor occultorum rationum intellector et purificator recompensator dampnorum gubernator totius mundi....[Fol. 182r[b]] [S]cias quod reuolutiones annorum mundi accipiuntur...[listed in Zinner, *Verzeichnis*, p. 34, no. 605]. [Fol. 184v[a]]...et suorum lucrorum aspicitur in medium celi etc. [D]ixit Hermes quando uolueris cognoscere dominum anni aspice quem planetarum inueneris...[fol. 204r[b], at end:] particula octaua. [Fol. 204v[a], top mg.: "Particula quinta."] [D]ixit haly filius aberagel primitus uoluimus loqui in hoc capitulo.../...[fol. 211v[b]] et interficietur ab inimicis suis quando deus [*abrupte*].

fol. 212r: Introductorium in arismetrica ad calculum astrologie M. Mathei Moreti de brixia. Arithmetica est scientia de numero a quo etiam nomen accepit nam arithmion greci numerum appellant...[listed in TKr, col. 140]/...[fol. 213v] similiter si diuiseris quarta per secunda prouenient secunda quia si subtrahis 2 de 4 remanent 2.

fol. 213v: [Left mg.: "Canon primus."] Nunc accedam ad expositionem tabularum alphonsi pro quo est intelligendum quod dies apud Alphonsum incipit a meridie idcirco si habueris horas horologii.../...[fol. 216r] excederet proportionem que est secunde linee ad primam et tertie ad secundam et quarte ad tertiam et sic de aliis.

fol. 216v: Canones alii ad declarationem aliarum tabularum quas non composuit alphonsus quarum tamen noticia est necessaria ad calculum astronomie. Canon primus declarat tabulas de ascensionibus signorum. Cum igitur gradum ascendentem ac ceteras domos habere uelis reduc primo horas horologii ad horas... [cf. TKr, col. 300]/...[fol. 217r] habebis uerum locum solis tempore predicto et similiter facies de motu lune in hora.

fol. 217v: [Table]

47. *X* (δ) = Erfurt, Wissenschaftliche Bibliothek der Stadt, Q.361. Parchment, 156 fols., of English origin
 Date: 1st half of 14th century
 Text used: fols. 1r[a]–22r[b]. Clementissimo. Figures; tables
 Description: Described in Wilhelm Schum, *Beschreibendes Verzeichniss der Amplonianischen Handschriften-Sammlung zu Erfurt* (Berlin, 1887).

48. *x* (θ) = Erfurt, Wissenschaftliche Bibliothek der Stadt, Q.357. Parchment and paper, 133 fols., of southern European origin
 Date: late 13th-early 14th century
 Text used: fols. 33v–80r. Primus. Figures; tables
 Description: Described in Schum, *Beschreibendes Verzeichniss.*

49. *Y* (β) = Erfurt, Wissenschaftliche Bibliothek der Stadt, F.394. Parchment, 167 fols.
 Date: early 14th century
 Text used: fols. 92r[a]–110v[b]. Primus. Tables; incomplete figures
 Description: Described in Schum, *Beschreibendes Verzeichniss.*

50. *y* (φ) = Erfurt, Wissenschaftliche Bibliothek der Stadt, Q.356. Paper, 223 fols.
 Date: 1400

Text used: fols. 102r–63r. Primus. Tables; figures incomplete
Description: Described in Schum, *Beschreibendes Verzeichniss.*

51. *Z* (δ) = Cambridge, University Library, Mm.III.11 (2327). Parchment, folio, 199 fols.
Date: 15th century (it is possible that some of the items, including the Campanus *Theorica*, are 14th century)
Text used: fols. 109r–39v[b]. Clementissimo. Figures complete; tables
Description: The following is from an examination of a microfilm of the manuscript.

fol. 1r[a]: Incipiunt hic pauca notabilia excepta a libro Aristotelis de iuuentute et senectute.../...[fol. 1v[b]] Explicit.
fol. 1v[b]: Incipiunt pauca quedam excepta a libro de problematibus.../...[fol. 7r[b]] Expiciunt [*sic*] notabilia problematum Aristotelis.
fol. 7r[b]: Nunc autem incipiunt pauca notabilia excepta a libello Aristotelis de lineis [fuller title given in bottom mg.].../...[fol. 7r[b]] Explicit.
fol. 7r[b]: Incipiunt pauca notabilia Aristotelis de inundatione.../...[fol. 7v[a]] Explicit.
fol. 7v[a]: Avicenna in [?]methaphisica sua tractatu 10 capitulo primo.../...[ends on fol. 8r[a]].
fol. 8r[a]: Incipiunt notabilia excepta a libro auctoris uel ut plurimi opinantur Aristotelis de proprietatibus elementorum.../...[fol. 8v[a]] Explicit.
fol. 8v[a]: Incipiunt pauca excepta a libello de motu animalium.../...[fol. 10v[b]] Explicit.
fol. 11r: Incipit tractatus algorismi. Hec Algorismus ars presens dicitur in qua... [listed in TKr, col. 597] [fol. 11r[a], *mg.*, commentary: "Liber iste quem pre manibus habemus diuiditur in 2 partes, prohemium et tractatum, tractatus incipit, Addere si numerum...," which is listed in TKr, col. 791; commentary continues and finally concludes on fol. 15v[b]: "hec ratio patet per hos uersus, uel discontinua etc. Explicit comentum algorismi in integris. Explicit comentum algorismi in integris."]/...[fol. 18r] Productus numerus totam summam numerorum. Explicit algorismus integrorum, id est textus uersus sed comentum est ante terminatum.... [Fol. 18r[b]] Hic erit saltus de comento per medietatem folii ut habeatur melior corespondenda in scriptura ad textum uersuum et comentum.
fol. 16r[a]: [Commentary: "Incipit comentum minutiarum secundum philosophum primo elenchorum qui non sunt...," which is listed in TKr, col. 1421.] [Fol. 18v] Incipit algorismus fractionum siue minuciarum. [Fol. 18v[a]] Ista minutiarum presens ars dicitur in qua/Diffinire refert uoces quibus ars eget ista...[listed in TKr, col. 780]/...[fol. 24r[a]]——debes et ibi quot dimidium sunt. [Fol. 25r[b], commentary: "...et in altitudine stellarum de nocte." Fol. 25v is blank.]
fol. 26r[a]: [Top mg. in a different hand: "Ricardi Walingford siue Simonis Tunstede Canones de Instrumento Mathematico Albion dicto."] [A]lbion est geometricum instrumentum...[listed in TKr, col. 74]/...[fol. 37r[a]] a studio pietatis. Explicit albion. [Fols. 37r[b]–v are blank.]

fol. 38r[a]: In nomine domini nostri ihesu christi sciendum est quod quilibet planeta preter solem... [listed in TKr, col. 1395]/...quod multo plura contineat [followed by a brief note in a different hand]. [Fol. 39r[b] is blank.]

fol. 39v: Tabule mediorum motuum argumenti solis et planetarum [calculated to radix 1348]. [Fol. 41v has a list of cities, beginning, "Oxon., London., Colcestria,..."; list ends on fol. 42v.]

fol. 43r: [In different hand: "Vtrum ex piramide umbrae terrae luna possit eclipsari."] [Fol. 43r[a]] Vtrum extra piramidem umbre terre luna possit eclipsari... [listed in TKr, col. 1647] [lacuna on fol. 43r[b]]/...[concludes on fol. 43v[a]]. [Fol. 43v[b] is blank.]

fol. 44r[a]: Incipiunt canones tabularum illustrissimi principis et regis Alfonsi. Quia secundum philosophum 4[to] physicorum tempus et motus mutuo se mensurant... [listed in TKr, col. 1230]/...[fol. 51r[a]] secundum doctrinam magistri Johannis de Lineriis a quo habeo scientiam meam. Expliciunt canones tabularum regis Alfonsi. [Fol. 51r[b] is blank.]

fol. 51v: [Tables; the date 1351 is given in a footnote.] [Fol. 62r] Medii motus planetarum secundum triplicem opinionem in 36000 annorum secundum annos Christi [i.e., "Tholomeus," "Arzachel," "Alfonsus"]. [Fols. 62v–63v are blank.]

fol. 64r[a]: [*Tit.*, top mg.: "Incipiunt quedam notabilia super libro Aristotelis de coloribus."] [S]implices colorum sunt quicumque elementis consequntur... [listed in TKr, col. 1507]/...[fol. 64v[a]] Explicit.

fol. 64v[a]: [*Tit.*, top mg.: "Incipiunt quedam notabilia extracta de libro qui intitulatur de mundo Aristotelis."] [M]ultotiens michi diuina quedam ac mirabilius... [listed in TKr, col. 891]/...[fol. 65r[a]] Explicit. [In a different hand: "Anthonius de monte granario [?grauario] 1413 die prima mensis decembris regens in conuentu et postea [?]ladulo." Fol. 65r[b] is blank.]

fol. 65v: [Top mg., in a different hand: "Computus anni cum expositione (ut uidetur) Simonis Bredun."] Licet modo in fine temporum plures constat... [listed in TKr, col. 827]. Aureus in Jano numerus clauesque nouantur... [listed in TKr, col. 167]. [Fol. 65v[a], commentary: "Licet modo in fine temporum etc. presentis auctor negotii in compendioso prologo...," which is listed in TKr, col. 1085; commentary continues and concludes on fol. 72v[b]: "...eius locus in directorio principalis litere tabularis. Expleta est igitur expositio uersuum compoti ecclesiastici in nomine ihesu christi cui sit honor et gloria per infinita seculorum secula. Amen."] [Fol. 73r[b]]...qui sit natalis et ab aduentu titulabis [followed by note: "Quidam dicunt quod Dominus passus est 10 Kalend. April.../...a gregorio papa et a senatu romano fuit etiam consumatum"].

fol. 73v: Tabula bede concurrentes. Tabule fungonis.

fol. 74r: Cisioianus.

fol. 74v: Tabula dionisii, etc.

fol. 75r[a]: [Treatise on the sphere in three parts, incomplete.] Prima pars continet descriptionem et numerum... [listed in TKr, col. 1094]. [Second part]...in eadem superficie cum suo deferente. [Fol. 80r[b], right mg.: "tertia pars"] [L]ongitudo regionis est arcus circuli equinoctialis.../...[fol. 80v[b]] in orbe signorum inchoatur terminatur ad eundem centrum.

fol. 81r: [No tit., tables.] Anni arabum collecti etc. [Fol. 88v, bottom mg., in a different hand, a note on proportional minutes. Fol. 96v, bottom mg., a note. Fol. 98r, meridian of Toledo. Fol. 99r, Arzachel. Tables end on fol. 106r.]

fol. 106v: [Notes:] Nota quod quod operando in tabulis abreuiatis secundum canones Alfonsi...[four lines]. Ad habendum elongationem lune a sole... [thirteen lines]. [Fol. 107r has been erased.]

fol. 107v: [*Tit.*:] De annis planetarum. Et postquam Tholomeus memorauit in hoc libro suo multa de reuolutionibus planetarum...[twenty-three lines]. [Fols. 108r–v are blank.]

fol. 140r[a]: [*Tit.*, top mg. twice: "Iordanus de ponderibus."] Omnis ponderosi motum esse ad medium...[listed in TKr, col. 1000; edited, but not from this manuscript, in E. A. Moody and Marshall Clagett, *The Medieval Science of Weights: Treatises Ascribed to Euclid, Archimedes, Thabit ibn Qurra, Jordanus de Nemore, and Blasius of Parma* (Madison, Wis., 1952), pp. 173, 174–227]/... [fol. 145r[a]] toto conatu impulsum habebit trahere B. Explicit liber Io. de ponderibus.

fol. 145r[a]: [Left mg.: "Lincolniensis de spera."] Intentio nostra in hoc t[r]ac-[ta]tu...[listed in TKr, col. 763]/...[fol. 150r[b]] scilicet solis et lune. Explicit. [Fols. 150v–51v are blank.]

fol. 152r[a]: [O]mnis ponderosi motum esse ad medium...[listed in TKr, col. 1000; edited in Moody and Clagett, *Medieval Science of Weights*, pp. 128–42]. [Proposition:] [I]nter quelibet grauia est uelocitatis in descendendo.../...[fol. 153v[b], Proposition:] et medii alterius eadem sit distantia secundum hunc situm eque grauia fient. [Fol. 154r[a]]...dispositis et depensis ipsa erit eque grauia situ quare propositum. Expliciunt pondera iordani. [The ending of the proposition agrees with that of the first version edited by Moody and Clagett, but the ending of the text agrees with none; nor is this manuscript listed.]

fol. 154r[a]: [S]i fuerit canonium simetrum magnitudine et substantie...[listed in TKr, col. 1449; edited by Moody and Clagett, *Medieval Science of Weights*, pp. 62, 64–75]. [Proposition:] Esto canonium AB diuisum in duo equalia ad G.../... [fol. 155r[b], Proposition:] paralellum ebipedo orizontis longitudo uniuscuiusque portionis data erit. [Fol. 155v[a]]...et hic erit numerus minoris proportionis quod oportebat ostendere. Explicit liber de canonio. [This manuscript was listed but not used in the edition by Moody and Clagett.]

fol. 155v[a]: [T]ractatum de spera in quatuor capitula distinguimus...[listed in TKr, col. 1577]/...[fol. 161r[b]] aut machina mundana dissoluitur. Explicit iste tractatus.

fol. 161v[a]: [D]omino suo excellentissimo et in cultu religionis christiane serenissimo...[listed in TKr, col. 465]. Quantum luna ceteris stellis est lucidior tantum ingeniis uestri claritudo...[listed in TKr, col. 1176]. [Fol. 162r[b]] [D]eus omnipotens custodiat regem nostrum ad gloriam credentium...[listed in TKr, col. 410]/...[fol. 182v[a]] diuersa signa et aduersa declina semper ad meliorem et probationem partem. Completus est tractatus de signis et moribus naturalibus hominum ad regem magnificum Alexandrum qui dominatus fuit toti orbi dictus monarchia in septemtrione. [Note, in a different hand: "Sciendum est quod necesse est magnitudines equales...."]

fol. 182v[b]: [A]b oculo rectas ductas lineas...[listed in TKr, col. 7]/...[fol. 191r[a]] similiter enim demonstrabimus contingentia quemadmodum in circularibus. Explicit liber de uisu.

fol. 191r[b]: [V]isum rectum cuius media terminos recte continuant... [listed in TKr, col. 1704; cf. Carmody, *Arabic Astronomical and Astrological Sciences*, p. 21]/... [fol. 195r[a]] quare in eis stupa posita attenditur. explicit liber de speculis.

fol. 195r[a]: [P]relibandum est quoniam ysoperimetrorum ysopleurorum rectilineorum... [listed in TKr, col. 1083]/... [fol. 197r[a]] quare et solidum poliedrum minus spera.

fol. 197r[a]: Cuiuslibet rotunde piramidis curua superficies... [listed in TKr, col. 277; edited in Clagett, *Archimedes*, vol. 1, pp. 450–86]/... [fol. 199v[b]] circumuoluta ad circulum cuius semidiametrum est nb... [concludes in the middle of Proposition 7: p. 486, line 54, of the Clagett edition].

52. *z* (β) = Bologna, Biblioteca Universitaria, 132 (154). Parchment, 149 (142) fols.

Date: 14th century

Text used: fols. 41r–74r. Clementissimo. Figures incomplete; blank spaces left for tables

Description: The following description is based on Lodovico Frati, "Indice dei codici latini conservati nella R. Biblioteca Universitaria di Bologna," *Studi italiani di filologia classica*, vol. 16 (1908); Thorndike, "Notes upon Some Medieval Astronomical, Astrological and Mathematical Manuscripts," pp. 43–45; and primarily my own examination of a microfilm of the manuscript.

inside cover: Ms. 132

fly-leaf recto: [Blank]

fly-leaf verso: Contenta in hoc uolumine: Introductorium Alchabitii—fol. 1; Theorica planetarum—22; theorica motuum latitudinis planetarum—27; De inuentione locorum planetarum in figuris circuli zodiaci—28; De pronosticatione dispositionis temporum—28; Liber pulcherrimus philosophie naturalis—31; theorica planetarum Campani—41; Liber cursus lune—74; Liber Alphragani—81; Tractatus astrolabii Messehalle—94; Tractatus quadrantis moderni—120; Tabule solis—121; Algorismus—128; Tractatus de spera Jo. de sacro busco—133; Calendarium fratris fuci de ferraria—142. [Right mg.: "168."] Sec. xv. Ex Bibliotheca Ioannis Garzoni Bonon.

fol. 1r: [Top mg.: "Cod. num° 168. Aulo–11–A." Bottom mg.: "N° 132."] Incipit introductorium Alchabitii. Postulata a domino prolixitate uite... [listed in TKr, col. 1078; Pearl Kibre, "Further Addenda and Corrigenda," *Speculum*, vol. 43 [1968], p. 97; cf. Carmody, *Arabic Astronomical and Astrological Sciences*, p. 146]/... [fol. 20v] Introduximus quoque has partes nouissimas etsi est in eis narratio debilis ne dimitteremus aliquid quod possit esse introductorium ad magisterium iudiciorum astrorum. Perfectus est introductorius Abdilaziz idest serui gloriosi scilicet dei qui dicitur Alzabizi ad magisterium iudiciorum astrorum cum laude dei et eius adiutorio interpretatus a iohanne Hispa.

fol. 21r: Incipit theoricha planetarum et primo de sole. Circulus ecentricus uel egresse cuspidis uel egredientis centri dicitur quod non habet centrum cum mundo...[listed in TKr, col. 223; Kibre, "Further Addenda," p. 82]/...[fol. 27r] et habuit instrumentum magnum astrolabium tricubitum uidelicet autem maioris quantitatis. Explicit amen.

fol. 27r: Incipit theoricha motuum latitudinis planetarum. Theoricam motuum latitudinis planetarum iam conueniens est perscrutari...[listed in TKr, col. 1571/...[fol. 28r] Et hec de latitudinibus planetarum sufficiant. Explicit capitulum de latitudinibus planetarum editum a magistro Petro de Sancto Hodomaro secundum regulas Albatengni.

fol. 28r: De inuentione locorum planetarum in signis circuli zodiaci. Primo scire debes quod oportet te primo inuenire centrum medium...[listed in TKr, col. 1113]/...et ubi fiet contactus in zodiaco ibi est luna uerissime. Deo gratias.

fol. 28v: Incipit liber de prenosticatione siue prescientia dispositionis temporum. Ad prenotandam diuersam aeris dispositionem futuram...[listed in TKr, col. 57; cf. Carmody, *Arabic Astronomical and Astrological Sciences*, p. 116]/...[fol. 30r] in directo signi que reperiuntur; pono ergo pro domo 5 pro exaltatione 4 pro triplicitate 3 pro termino 2 pro facie 1. Et hec est figura. [Fol. 30v] Cum ergo dispositionem aeris ad aliquem certum terminum pronosticare uolueris...[listed in TKr, fol. 293]/...[fol. 31r] Taurus Veneri quam Mars sequebatur Gemini Mercurio qui post Venerem fuerat deputati sunt. Explicit.

fol. 31v: Incipit liber pulcerimus philosophie naturalis. Incipiam et dicam quod orbis prescitus spericus quem capiunt termini...[listed in TKr, col. 722; Carmody, *Arabic Astronomical and Astrological Sciences*, p. 33]/...[fol. 40v] sapiens et sublimis. Finitus est liber.

fol. 74v: Incipit liber cursus lune. Quicumque cursum lune recte scire uoluerit sciat primitus quod sol per signa uadit...[listed in TKr, col. 1236; cf. Carmody, *Arabic Astronomical and Astrological Sciences*, p. 66]/...[fol. 78r] qui sint strabi et in uno oculo lesi sunt humanarii aut uenditores mellis et olei ficus dactillorum et huiusmodi. Nota quod 1080 puncta constituunt horam et iii gradus constituunt unam horam. Item 60 secunda constituunt unum minutum et 60 minuta unum gradum. Nota quod sol currit inter diem et noctem 59 et 8 secunda. Item nota quod gradus in celo est 59 miliaria et due tertie unius miliaris in terra. amen. [Fol. 78v is blank; fols. 79–80 are missing.]

fol. 81r: Liber Alfragani. Differentia prima in annis arabum et latinorum et in omnibus mensium...[listed in TKr, col. 429; Kibre, "Further Addenda," p. 85, concerning fols. 81r–114; cf. Carmody, *Arabic Astronomical and Astrological Sciences*, p. 114]. Numerus mensium arabum et latinorum duodenus menses arabum incipiunt...[listed in TKr, col. 960]/...[fol. 95v: chap. 29] quod eclipsis solis fit diuersa ab hoc propter id quod accidit in aspectu de diuersitate locorum in quibus uidetur ex climatibus [end of chap. 29]. [Fols. 96–97 are missing.]

fol. 98r: [Acephalous; written in a different hand:] a cuspide usque in circulum sibi propinquiorem per 32 diuisiones et pones summitatem prime diuisionis ex parte cuspidis circuli signorum cuspidem et accipies ex hac diuisione...[Messehalla, Tractatus astrolabii; see Gunther, *Chaucer and Messahalla*, p. 197: I, chap. 2, line 32]. [Fol. 112r] Tabula stellarum fissarum que ponuntur in astrolabio cum gradibus quibus celum mediat et cum distantia earum ab equinoctiali linea. [Fol. 112v] Tabula stellarum fissarum uerificatarum per armillas parisius et est longitudo earum gradus circuli signorum per circulum transeuntem polos zodiaci et stellas latitudo uero earum est arcus eiusdem circuli cadens inter stellas et graduum longitudinis ipsarum. [Fol. 113r] Capitulum preambulum in usum astrolabii. Nomina instrumentorum sunt hec. Primum est armilla suspensoria ad aliquam altitudinem... [listed in TKr, col. 916; cf. Carmody, *Arabic Astronomical and Astrological Sciences,* p. 24]/... [fol. 119v] Post hec puncta umbre super que steterit regula ad 12 comparata et equalis fuerit compara et punctorum ad 12 talis est comparatio stature tue ad totam planitiem. Explicit tractatus astrolabii.

fol. 120r: Geometrie due sunt species... [listed in TKr, col. 585]/... [fol. 122r] et margarita perforata moueatur super filum de loco ad locum et erit completa compositio quadrantis. [Fol. 122v: mostly blank] Hec tabula docet in quo gradu signi sit sol initiis mensium [listed in TKr, col. 607; Kibre, "Further Addenda," p. 87]. [Fol. 123r] Sequitur de operationibus quadrantis et primo de altitudine solis. Scito de compositione quadrantis dicendum est de utilitatibus et operationibus que per ipsum exercentur... [listed in TKr, col. 1407]/... [fol. 125r] que area multipliciscatur per eius longitudinem siue altitudinem et productum dabit eius capacitatem. Explicit quadrans modernus.

fol. 125v: Tabula solis prima in anno bisextili ad inueniendum locum eius orbe decliui fixo. [Fol. 126r] Tabula solis secunda ad habendum locum eius in anno primo post bisextum. [Fol. 126v] Tabula solis tertia ad inueniendum locum eius in anno secundo post bisextum. [Fol. 127r] Tabula solis quarta ad habendum locum eius in anno tertio post bisextum. [Fol. 127v] Tabula solis quinta de eius declinatione.

fol. 128r: Incipit algorismus. Omnia que a primeua rerum origine processerant... [listed in TKr, col. 991; Kibre, "Further Addenda," p. 95]/... [fol. 133r] et hoc de radicum extractione sufficiant tam in numeris quadratis quam tubicis [*sic*]. Deo gratias amen. Explicit algorismus.

fol. 133r: Incipit tractus de spera. Tractatum de spera 4 capitulis distinguimus dicentes primo quid sit spera... [listed in TKr, col. 1577; Kibre, "Further Addenda," p. 106]/... [fol. 141v[a]] aut deus nature patitur aut cito mundana machina cito dissoluere. Deo gratias. Explicit [superscript: "spera"] magistri iohannis de sacro bosco. Amen. [Fol. 141v[b] has figures.]

fol. 142r: [Different hand begins. Calendar for Milan for 1311–86.] Prima dies iani timor est... [listed in TKr, col. 1089]. Ianuarius habet dies 31 et luna est 30 regularis solaris... [listed in TKr, col. 653]. [Calendar ends with December, on fol. 147v. Fol. 148r[a]] [I]stud Kalendarium factum est ad meridiem mediolani cuius longitudo est 31 graduum latitudo 45... [listed in TKr, col. 796]/... [fol. 148r[b]] addenda est unitas et operandum ut prius. Explicit kalendarium fratris Fuci de Ferraria. [Below this is a table of the signs, indicating whether they are good, bad, or indifferent for medicine and bleeding and what their qualities are—fiery, aerial, etc. Fol. 148v is blank.]

fol. 149r[a]: [Blank space.] Fuerit in medietate epicicli que respicit occidentem idest a dextris maior est medius motus quam uerus quare tunc equatio subtrahenda... [Gerardus, *Theorica,* p. 21 (chap. 2, line 19)]/... et recessum ab auge et ab opposito augis ut patet in figura sequenti. [Fol. 149r^b] Sequitur de capite draconis. Ecentricus lune declinat a uia solis in duas partes in septentrionem et meridiem et intersecat ecentricum solis in duas partem [*sic*]...remanet uerus locus eius computatus secundum successionem signorum ut patet in subiecta figura [followed by a blank space]. [S]equitur de tribus superioribus unde nota est quod quilibet trium superiorum habet 3 circulos ecentricos dispositos in eadem plana et superficie...[fol. 149v^b]. Quare autem diuersitates diametri ad longitudinem longiorem subtrahantur et ad longitudinem propiorem addantur equationi argumenti facile patebit animauertenti figuram. Quanto enim centrum epicicli plus appropinquat centro terre ut dictum est tanto plus maioratur equatio argumenti ut patet in figura posita [followed by a blank space]. [Gerardus, *Theorica*, p. 29 (end of chap. 4).]

53. Γ (β) = Cambridge University, Magdalene College, Pepys 2329 (C.M.A. 6767–6776, 6778, 6780–6784). Vellum, 12½ inches by 9¾ inches, 224 plus 4 fols., mostly in double columns of 31 lines (for further description, see M. R. James, *Bibliotheca Pepysiana: A Descriptive Catalogue of the Library of Samuel Pepys*, pt. 3, *Medieval Manuscripts* [London, 1923], pp. 86–89; Ker, *Medieval Libraries*, p. 27; James, *Lists of Manuscripts...Owned by...Dee*, p. 25: C. 91)

Date: Early 15th century (1407)

Text used: fols. 54r^b–93r^b. Primus. Figures with volvels; tables

Description: The following is from an examination of a microfilm of the manuscript. See also James, *Bibliotheca Pepysiana*, pp. 86–89.

list of contents: [In Marchall's hand]

fol. 1r^a: Incipit liber primus Iordani de nemore de elementis arismetrice artis. Vnitas est esse rei per se [listed in TKr, cols. 1601, 1600]. [The initial has a shield paly of six, argent and gules. The tract consists of text and comment, with marginal diagrams.] [Fol. 1r^b]...Sequitur propositio prima libri primi. *Omnis numerus minor maioris aut pars est aut partes*.../...[fol. 45r^a] assignare sit possibile etc. Expl. distinctio decima et per consequens totus liber de elementis arismetrice artis magistri Jordani de nemore scripta parisius per manus seruatii tomlinger de bauaria anno domini millesimo quadricentesimo septimo octaua die post festum pentheconstes finitus est iste liber.

fol. 45r^a: [Top mg.: "Algorismus magistri iohannis de sacro bosco."] Incipit algorismus de integris. Omnia a primeua rerum origine...[listed in TKr, col. 991]/...[fol. 50r^b] tam in numeris quadratis quam in cubicis etc. Explicit algorismus de integris.

fol. 50r^b: [Top mg.: "Algorismus in minuciis Johannis de Lineriis. Incipit algorismus de minuciis."] Modum representacionis minuciarum phisicarum et uulgarium demonstrare...[listed in TKr, col. 878] [Commentary: "Quia in fraccionibus duo sunt numeri," which is listed in TKr, col. 1220.]/...[fol. 54r^a] 4^a 9^{arum} 3^a 6^{arum} 2^a et 3^{arum} integrum etc. etc. etc. Explicit algorismus de minuciis etc.

fol. 93v^a: [Top mg.: "Tractatus magnus et utilis de proporcionibus proporcionum magistri nicholai horesme."] Omnis rationalis opinio de uelocitate motuum... [listed in TKr, col. 1002, as Peterhouse 277]/... [fol. 110v[a]] hic igitur 4[tum] capitulum fineatur tu autem domine miserere mei. Explicit 4[m] capitulum de proporcionibus huius tractatus editus a reuerendo magistro Nicolao horesme [*mg.*: "al. horem"] scriptis per me et uocatur tractatus de proporcionibus proportionum. [Two definitions follow:] Coniuncta proporcionalitas dicitur quotiens... [fol. 110v[b]] remoto utramque summorum similitudo proporcionum etc. Amen Amen dico uobis. [Fol. 111r is blank.]

fol. 111v[a]: [Begins imperfectly:] refert et etiam posterius uidebitur.../... [fol. 128r[a]] et ipse nescio quid super hoc iudex decreuit apollo etc. Explicit nobilis tractatus magistri Jordani de nemore de motibus celestibus etc. Si motus celestes sunt commensurabiles uel non. [This work is wrongly ascribed to Jordanus de Nemore. It is by Nicholas Oresme: see TKr, col. 1713; Thorndike, *History of Magic*, vol. 3, p. 405, note 18, which gives the identifying *explicit*.]

fol. 128r[b]: [Top mg.: "Algorismus proporcionum."] Algorismum proporcionum, reuerende presul Meldensis Philippe, quem pictagoram dicerem si fas esset credere finem ipsius de reditu animarum... [listed in TKr, col. 80]/... [fol. 130r[b]] de quibus determinatum est sufficienter etc. Explicit algorismus proporcionum per manus seruacii de monaco. [James, *Bibliotheca Pepysiana*, p. 88, says "Gervasius" is only the scribe although he states that both Dee and the *Cat. MSS Angl.* name him the author. TKr, col. 80, ascribes the work to Nicholas Oresme.]

fol. 130r[b]: [Written in a later hand. Top mg.: "Demonstrationes astrolabii."] Tres circulos in astrolapsu descriptos [right mg.: "Primum"]... [listed in TKr, col. 1583, as Peterhouse 277]/... [fol. 131v[a]] cetera in glosis presentis libri 2[e] propositionis inuenies etc. Expliciunt demonstrationes astrolabii.

fol. 131v[a]: [Same hand as that of the preceding work. Left mg.: "Tractatus de turketo."] Incipit turketus et in prima parte de tabula deseruiente equinoctiali. De omnibus partibus instrumenti quod turketum dicitur primo dicendum est... [listed in TKr, col. 384]/... [fol. 132v[b]] sagax lector facile per se inueniret. Explicit turketus deo gracias factus anno domini 1284 2[a] die Jullii in die dominica sole cancri 17, luna 20 aquarii, saturno 20 capricorni, ioue 21 sagitarii, marte 13 uirginis, uenus 16 leonis, mercurii 28 geminorum, capitis 12 capricorni ante meridiem. Deo gratias.

fol. 133r: [In a fine hand, written in red, blue, and black.] Incipiunt tabule Alfonsi olim regis Castelle illustris et primo tabula differentiarum unius regni ad aliud et nomina principum et ere cuiuslibet sub eis habentis exordium et sunt radices dierum cuiuslibet ere nominate in sequentibus tabulis posite. Tabula differentiarum unius regni ad aliud erarum sequentium. [Fol. 136r] meridiem Toletanam [ends on fol. 150r]. [Fol. 150v is blank.]

fol. 151r[a]: [Top mg.: "Canones tabularum Alfonsi Magistri Johannis de Saxonia."] Tempus est mensura motus ut uult aristotiles 4[0] phisicorum... [listed in TKr, col. 1561]/... [fol. 158v[b]] idem motus sicut in coniunccionibus planetarum dictum est et sic est finis etc. Expliciunt canones illustris regis Alfoncii quos magister Johannes Danckow de Saxonia compilauit.

fol. 159r[a]: [Top mg.: "Canones tabularum primi mobilis magistri Johannis de

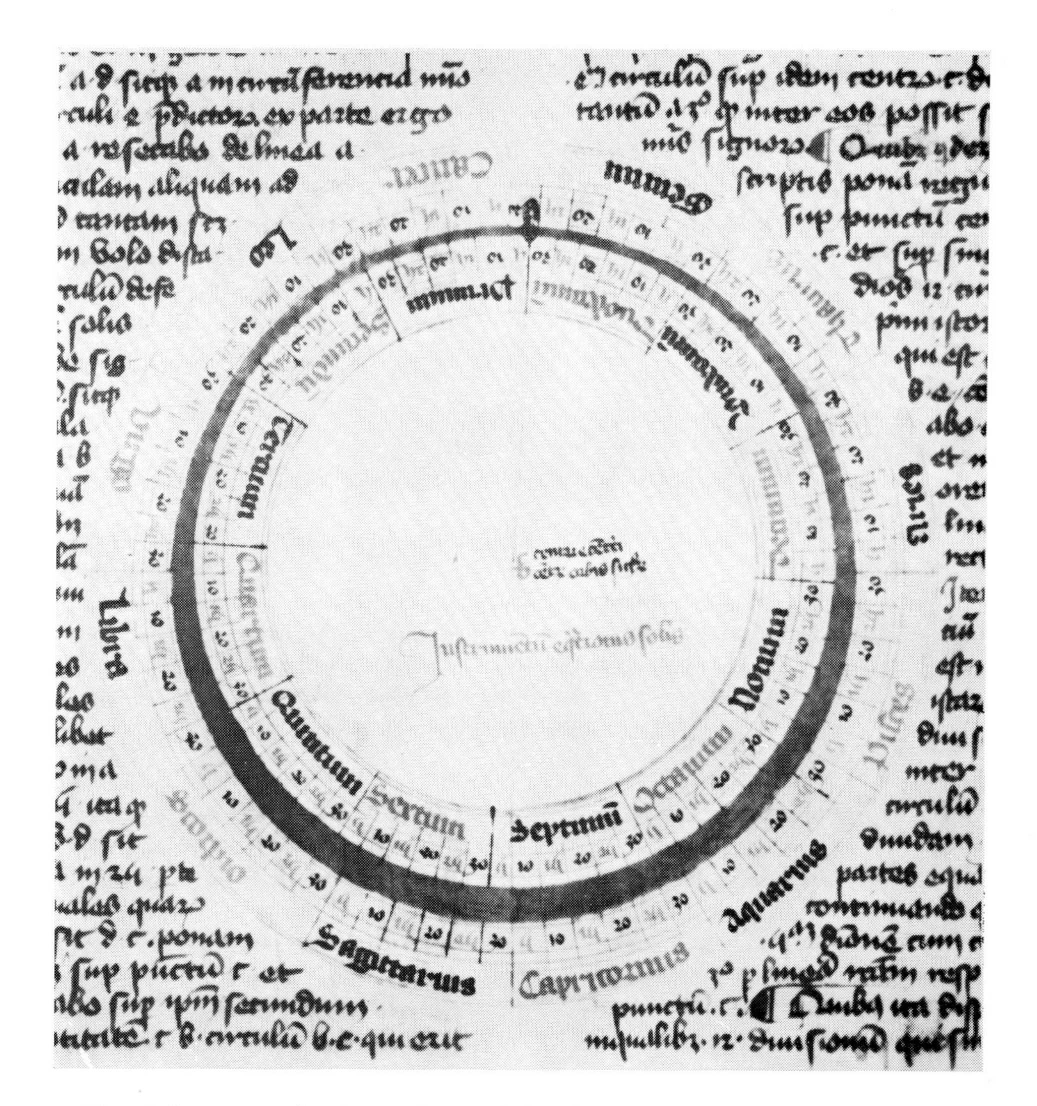

Plate 1. Instrument for the sun from MS Cambridge University, Magdalene College, Pepys 2329 (MS Γ, dated 1407), fol. 56v. Corresponds to Figure 10 (p. 152). *C* is labeled "centrum ecentrici" and *D* "centrum orbis signorum." Note that the apogee *A* is located, not at Gemini 17;50°, as Campanus directs, but at about Gemini 27°. This is a deliberate correction of Campanus's poor value (the apogee was in fact at the beginning of Cancer in his time). Compare the comments in the margins of MSS *Q* and *H* reported in the apparatus criticus, lines III.124–25.

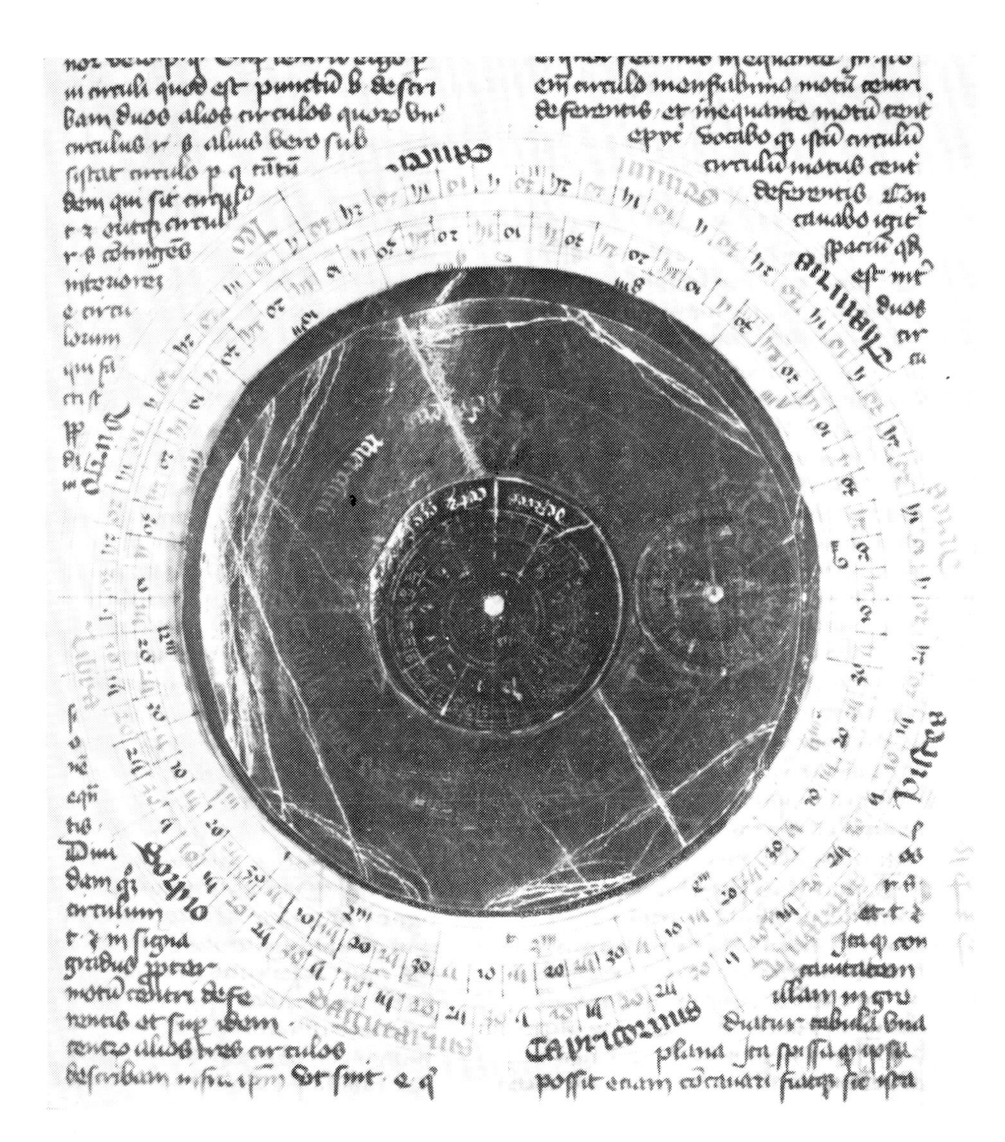

Plate 2. Instrument for Mercury from MS Γ, fol. 73r. Corresponds to Figure 16 (p. 246). The parts of the figure corresponding to the moving parts of the instrument, besides being drawn on the manuscript page, have also been inscribed on cut-out volvels, to produce a parchment model of the instrument. The volvels appear in the photograph as the dark inner portion, hiding the inner part of the figure on the manuscript page. The inscription of the volvels is indistinct, but one can distinguish the three separate moving parts: the "disk of the motion of the center of the eccentric," visible at the top and also just above the central graduated circles (where it is inscribed "deferens centrum ecentrici"); the "disk of the motion of the center of the epicycle," covering the whole of the rest of the dark area outside the central graduated circles and inscribed "deferens mercurii"; and the "disk of the motion of the epicycle," visible to the right of the dark area.

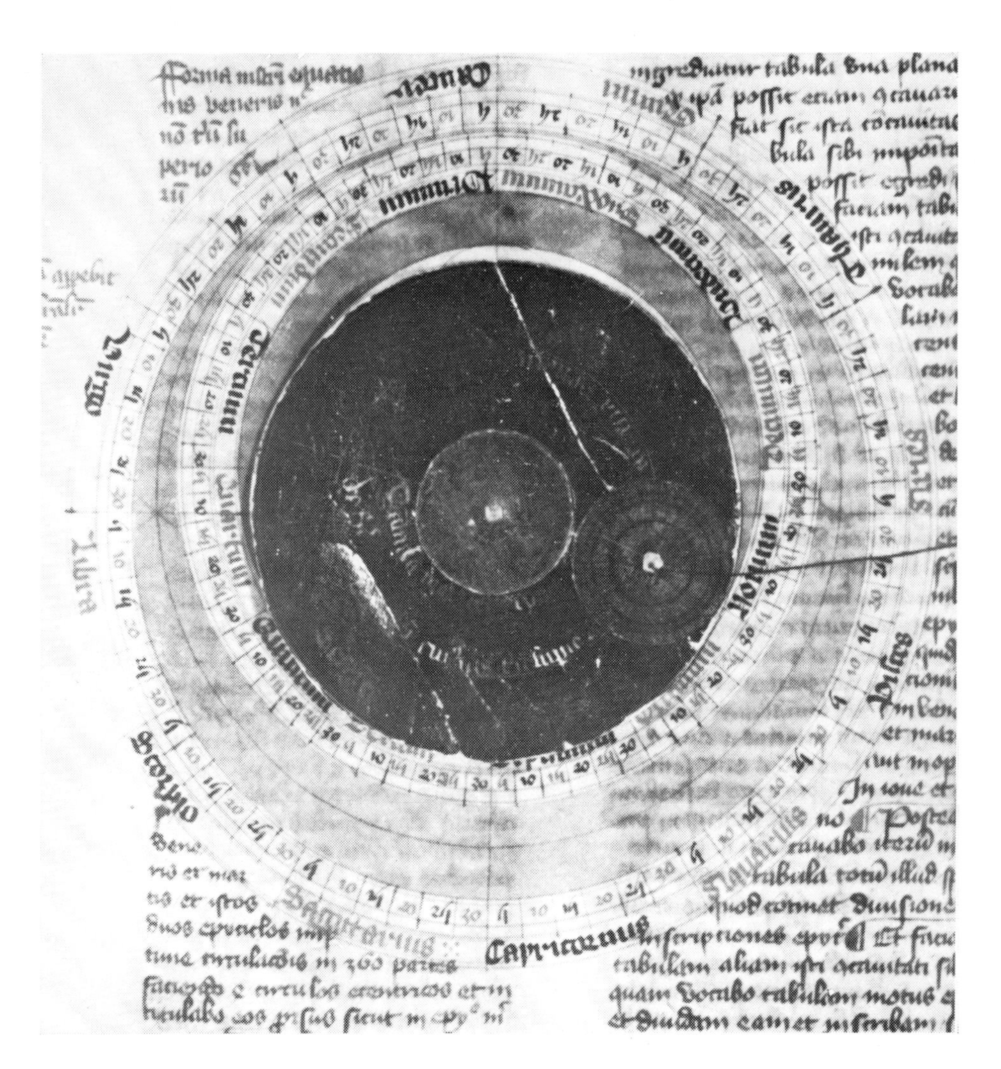

Plate 3. Instrument for Venus and the three outer planets from MS Γ, fol. 92v. Corresponds to Figure 22 (p. 346), which is drawn, however, for Saturn, whereas this diagram is for Venus. Hence the apogee here is in a different place, although not in Gemini 17;50°, as Campanus directs, but in Cancer 0°. This is a deliberate correction. Compare Plate 1 and the comments in MSS *G*, *W*, and *w* in the apparatus criticus for lines VI.38–39. The parts of the figure corresponding to the moving parts of the instrument are drawn on cut-out volvels as well as on the manuscript page. As in Plate 2, the inscription of the volvels is indistinct, but one can distinguish the two separate moving parts: the "disk of the motion of the center of the epicycle," covering the whole of the dark area except the central circle and duly inscribed (to the left of and below the central circle) "Tabula motus centri epicicli"; and the "disk of the motion of the epicycle," visible as a system of graduated circles at the right of the dark area. Attached to the center of this disk can be seen an example of the kind of thread Campanus prescribes for use as markers at the centers of various circles. Other threads attached to this model have probably been lost.

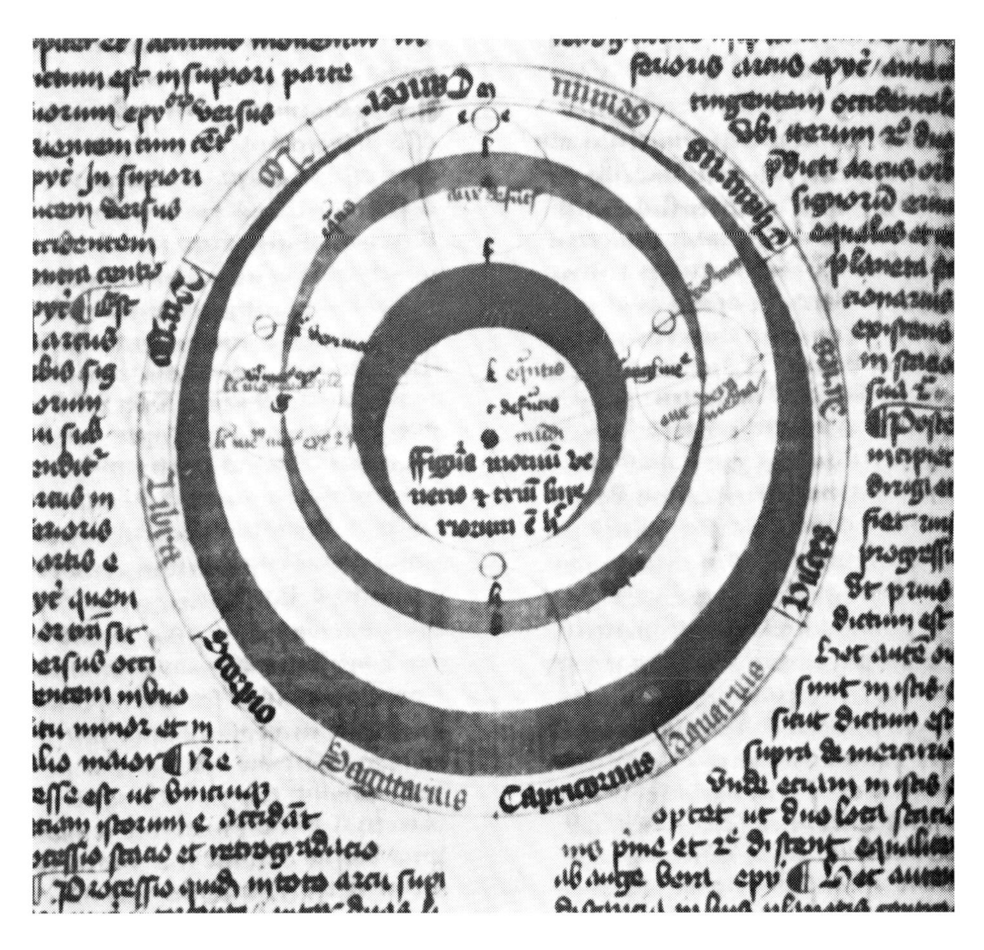

Plate 4. Illustration from MS Γ, fol. 84v, of the Ptolemaic model for Venus and the three outer planets (entitled "ffigura motuum ueneris et trium superiorum est hec"). Corresponds to Figure 18 (p. 308), but has been much elaborated. Thus the epicycle is drawn not only at the apogee of the deferent (point *A*, labeled "aux deferentis"), but also at the perigee (point *B*) and on either side of the deferent (at *G* and opposite it). Drawing it in these two side positions has enabled the scribe to distinguish the mean and true apogees of the epicycle (cf. Figure 26, p. 428). These are duly labeled "aux uera, media" on the right side of the illustration. Other elements from Figure 17 (p. 256) have also been incorporated; e.g., the "linea medii motus epycicli et planete" is drawn from *D* toward the left parallel to *KG*, the "linea medii motus planete," and *DG* is labeled "uerus motus epycicli." The planet is drawn not as a point but as a small circle (*EE* at the top) to represent its physical body.

Lineriis." An angel nicely drawn with the pen is beside the initial.] Cuiuslibet arcus propositi sinum rectum inuenire... [listed in TKr, col. 276]. [Fol. 167r[a]]... et uenias ceteras domus ut in 37 huius dictum et sic est finis canonum magistri iohannis lineriis super tabulas primi mobilis etc. [Top mg.: "Canones tabularum Alfonci per magistrum iohannem de lineriis."] Priores astrologi motus temporum celestium diligentissimis... [listed in TKr, col. 1127]/... [fol. 182v[b]] uero loco solis qui fuerit in radice et sic propositum etc. Expliciunt canones tabularum primi mobilis et equacionum simul et eclipsium ordinati per magistrum Johannem de Lineriis piccardum dyoc. Ambianensis etc.

fol. 183r[a]: [Top mg.: "Alkyndus de impressionibus aeris."] Ad prenotandam diuersam aeris disposicionem... [listed in TKr, cols. 57–58; cf. Carmody, *Arabic Astronomical and Astrological Sciences*, p. 116]/... [fol. 185r[b]] et in signis aquosis etc. Explicit tractatus Alchindi de disposicione aeris etc.

fol. 185v[a]: [Top mg.: "De utilitate arismetrice per Rogerum Baconensem."] De utilitate arismetrice potest sumi per infra scripta quomodo pones [?penes] res huius mundi [?]quodlibet ipsa utatur... [listed in TKr, col. 394, as Peterhouse 277]/... [fol. 188r[b]] 9° uenus 10° luna 11° mercurius etc. Et sic est finis huius operis magistri Rogerii Bakonis ut patet in sua summa ad Clementem. [From the *Opus majus*, bk. 4, chap. 1 (*The "Opus majus" of Roger Bacon*, ed. J. H. Bridges [Oxford, 1897–1900], vol. 1, pp. 224–36); see A. G. Little, ed., *Roger Bacon Essays*... (Oxford, 1914), p. 382.]

fol. 188v[a]: [Top mg.: "Compotus magistri Campani."] Rogauit me unus ex hiis quibus contradicere nequeo ut scienciam quam compotum appellamus... [listed in TKr, col. 1365, which incorrectly reads CUmc 2329 for CUma Pepys 2329]. [Fol. 188v[b]] Compotus est scientia numerationis uel diuisionis temporum secundum uulgares... [identified in TKr, col. 243, as a work of Robert Grosseteste, also ascribed to Campanus]/... [fol. 218v[b]] istorum uigilias ieiunes luceque Marci etc. Explicit compotus bonus Campani etc.

fol. 219r[a]: [Written in another hand. Top mg.: "Algorismus demonstratus per Jordanum ut creditur."] Figure numerorum sunt ix s. 1·2·3·4·5·6·7·8·9 et est prima unitatis... [listed in TKr, col. 558, as both Peterhouse 277 and CUma Pepys 2329]/... [fol. 224v[b]] que suffragantur ad hoc opus etc. [Fol. 225r is blank.]

The fly-leaf has a catchword, *textus*, but the tables of contents show no sign of mutilation. On the fly-leaf is another table of contents, not in Marchall's hand.

54. Λ (φ) = Memmingen, Stadtbibliothek, Folio 33, 257 fols.

Date: 1466

Text used: fols. 199r[a]–230r[b]. Primus. Figures; tables

Description: The following is from an examination of a microfilm of the manuscript.

fol. 1r: Quidam princeps nomine Albuguase quem scientiarum... [listed in TKr, col. 1245; Zinner, *Verzeichnis*, p. 268, no. 8631; cf. Carmody, *Arabic Astronomical and Astrological Sciences*, p. 15]/... [fol. 148v] in hac scientia preter quod inquiramus per ipsum prolongationem et abbreuiationem tunc iam sequitur et ostensum est ut ponamus hoc fine libri. Deo gratias. [Fols. 149r–50v are blank.]

fol. 151r: Omnium recte philosophantium non solum uisibilibus et credibilibus argumentis...[listed in TKr, col. 1006]/...[fol. 151v] [small hand:] cuius duo dyametri noti et 3 latera per 3 erit quartum notum scilicet DG ergo et corda residui arcus de semicirculo scilicet AG erit uero quod est propositum. [large hand:] Que linee inequales in circulo si protrahantur maiores ad minorem quam arcus longioris ad arcum breuioris minor est proportio. [About half this verso is blank.]

fol. 152r: Omnium recte philosophantium non solum uerisimilibus et credibilibus argumentis...[listed in TKr, col. 1006; Zinner, *Verzeichnis*, p. 73, no. 1935; cf. Carmody, *Arabic Astronomical and Astrological Sciences*, p. 164]/...[fol. 198v] super duo loca solis et lune nam uisus locus lune tunc ipse locus solis et inclinationes quidem tenebrarum sic se habent. Explicit liber sextus minoris Almageste.

fol. 230v^{a}: De introitu balnei. Melior pars sapientum astrologorum et dicit [?ostendit] quod electio in hoc est [?]uersa in una domorum...[not listed in TKr]. [Fol. 230v^{b}] Melior electio pro hoc est ut sit luna in libra...[listed in TKr, col. 864]/...[fol. 233r^{b}] tamen malys signis separando se ab infortunis et applicando fortune. [Fols. 233v–34v are blank.]

fol. 235r^{a}: [A]lbion est geometricum instrumentum Almanach autem arismetricum ...[listed in TKr, col. 74; Zinner, *Verzeichnis*, p. 356, no. 11583]. [Left mg.: "conclusio prima."] [D]istantiam centri deferentis solaris a centro terre una cum elongatione solis ab auge circuli deferentis experimentaliter demonstrare... [cf. TKr, col. 438]. [Fol. 239v^{b}] [A]lbion cum instar planisperii totius spere celestis concentricos et ecentricos et etiam epiciclos...[listed in TKr, col. 74]. [Fol. 244r^{b} is blank. Fol. 244v^{b}]...ita quod filum possit pertransire per medium eius eruntque facie instrumenti et completum est. [C]um ista perfecerimus aptemus armillam suam suspensoriam sine portatile et in circulo illo tertio quasi duabus [fol. 245r^{a}] marginibus circuli altitudinis et zodiaci primi limbi apud C faciam foramen et firmabimus ibi filum cum perpendiculo ad capiendum altitudines per utramque primulam perforatam ad perficiendum altera opera ad qua institutum est sicut loco suo dicemus et completum est hoc opus cum adiutorio dei anno domini 1321 inchoante. [Fol. 245r^{b} is blank. Fol. 245v^{a}] [A]lbion ad singulos motus nouem orbium se extendit eo quod stellas erraticas... [listed in TKr, col. 74]/...[fol. 250r^{b}] ipse ostendet eandem proportionem in omnibus fractionibus quam integris representat tabula etc. [Remainder of column is blank. Fols. 250v–57r contain tables. Fol. 257v is blank.]

55. Π (α) = Nuremberg, Stadtbibliothek, Cent. V.58. Paper, 165 fols.
Date: 15th century
Text used: fols. 1r^{a}–41v^{a}. Primus. Figures partly complete; tables
Description: Referred to in Zinner, *Leben und Wirken*, p. 222. The following is from an examination of a microfilm of the manuscript.

fol. [1a]r: [Top mg.: column of figures] Theorice campani et alia. [Text begins; acephalous:]...inquiens et immixtum est quoniam qualis motus principium intellectui [?]est sit enim utique solum mouebit....[Fol. (1a)v] [H]oc autem dupliciter aut non prope se ipsum mouens.../...unde Anaximander dixit recte intellectui impassibilem [*abrupte*; same hand as last folio, q.v.]

fols. 42r–43v: [Incomplete figures]

fols. 44r–61v: [Blank]

fol. $62r^a$: [A]lbyon est geometricum instrumentum almanach autem arismetricum utrumque tamen eisdem usibus...[listed in TKr, col. 74]. [Fol. $70r^a$] Incipit secunda pars huius. [A]lbyon est instrumentum inter cetera magis artificiosum ac pro uarys planetarum et stellarum fixarum motibus...[listed in TKr, col. 74]. [Fol. 77r] Tabula medii argumenti mercurii et medii motus lune ac argumenti medii lune [begins with March and ends on fol. 78r with February]. [Fol. 82v: table of stars, verified for 1430. Fol. $85v^a$: *explicit* of second part. Fol. $85v^b$ is blank. Fol. $94r^a$: *explicit* of third part. Fols. $94r^b$–94v are blank. Fol. $95r^a$] Quarta pars adiecte pro errore uitando in compositione albionis. [A]lbion pro magna sui parte non superficiem spere sed circulorum in spera descriptorum superficiem representat...[listed in TKr, col. 74]. [Fol. 95v] Tabula centri saturni. [Fol. 96r] Tabula centri iouis. [Fol. 96v] Tabula centri martis. [Fols. 97r–99v are blank. Fol. 100r] Tabula centri solis. [Fol. 100v] Tabula centri mercurii. [Fol. 101r] Tabula centri orbis lune. [Fol. 101v] Tabula ueri motus lune et equationis argumenti pro hora coniunctionis. [Fol. 102r] Tabula yomim siue circuli equantis motum diurnum. [Fol. 102v] Tabula latitudinis lune ab ecliptica; Tabula longitudinis cum 12^a; Tabula coniunctionum. [Fol. 103r] Tabula motus lune in una hora ad augem et longitudinem mediam et oppositum augis; Nomina stellarum fixarum. [Fol. 103v] Tabula medii motus mercurii. [Fol. 104r] Tabula medii motus lune. [Fol. 104v] Tabula medii motus argumenti lune. [Fol. 105r] Tabula ascensionis signorum in circulo directo. [Fol. 105v] Tabula ascensionis signorum in circulo directo. [Fol. 106r] Tabula ascensionis signorum in circulo obliquo. [Fols. 106v–9v are blank. Fol. $110r^a$, in a different hand:] [A]lbion ad singulos motus 9 orbium et cetera singulos circulos in prima facie instrumenti... [listed in TKr, col. 74, as encompassing fols. 110r–21v, but see what follows]. [Fol. $116v^b$] natum et pedem circini mobilem mutando circinum extende usque ad finem arcus quem immediate produxeras et circina iterum unam quartam arcus et sic usquoque semper placet.

fol. $116v^b$: *Partes instrumenti circulosque et lineas eius prosequentibus facilius intelligendis nominatim discutere*. [Right mg.: "1^a."] Hoc instrumentum quod extractum est ex albione habet duas partes...[identified in TKr, col. 1027, as a work of Johannes Schindel]/...[fol. $121v^a$] Deinde pone filum super lineam minutorum casus et id quod est inter centrum instrumenti et primum almuri est semidyameter solis et id quod est inter primum almuri et secundum est semidyameter lune. [Fol. $121v^b$ is blank.]

fol. 122r: [In a different hand] *Cenith solis ad quamcumque horam inuenire*. Cenith dicitur arcus orizontis inter sectionem eius...[sixteen lines]/...Hec Albategni capitulo 56 [followed by table: "Pro latitudine regionis 49 graduum 28 minutorum"]. [Fol. 122v has another table, "Ad latitudinem regionis 49 gradus 28 minuta."]

fol. 123r: [A]d faciendum horalogium horarum inequalium quod est de intentione Albategni capitulo 56. Apta tabulam planam ualde que quanto amplior tanto certius horam ostendet...[cf. TKr, col. 39]/...[fol. 123v] Est autem rotundius secundum umbras rectas. Hore quas uocant inequales per arcus circulorum in quadrantibus et semissis et dorsis astrolabiorum designate ueritatis certitudine carent; grossa tamen est acceptio aliquando modicum a rectitudine declinans. [Fol. 124r contains a figure; fol. 124v, a table: "Ad latitudinem 41 graduum et 50 minutorum ut Roma"; fol. 125r, a table: "Ad latitudinem 45 graduum ut Venetie"; and fol. 125v, a figure.]

fol. 126r: [I]nstrumenta [?]amussis que quadrantis partes submultiplices sunt in altitudinibus solis et stellarum accipiendis utilia dinoscuntur....[Fol. 126r] Ad faciendum igitur unum ex illis describatur circulus super centro D sit AECB duabus dyametris orthogonaliter sese secantibus quadratus.../...[fol. 126v] aput pinnulam per quam radius tunc incedit signatos et habebis altitudinem quesitam.

fol. 127r: [M]onocordi diuisio secundum modernorum musicam talis est.../...et suis semitonys respectu illorum. [Fol. 127v contains a table; fols. 128r–32v are blank.]

fol. 133r: Incipiunt canones tabularum iohannis de blanchinis super primo mobili. Capitulum primum. [N]on ueni soluere legem sed illuminare his qui in tenebris sedent...[listed in TKr, col. 923]. [Fol. 134r[a]] [O] sapientissime Ptholomee uir optime amator scientie illuminator celestis scientie astronomie quantum tulisti laborem.../...[fol. 154r[b]] multum confert pro inueniendo precise tempus accidentium respectu latitudinum planetarum.

fol. 154r[b]: [In a different hand] [A]d inueniendum gradum cum quo stella quelibet celum mediat in omni regione. Primo quere sinum latitudinis stelle ab ecliptica....[A]d inueniendum gradum cum quo stella oritur in quacumque regione uolueris.../...et proueniunt ascensus stelle in regione tua quibus correspondet gradus ecliptice cum qua oritur.

fol. 154v: [In a different hand] Anno domini 1415º 6ª dies iunii hora 18ª minuto 58 medium eclipsis solis in 24[to] gradu geminorum duratio 1 hora 58 minuta. Puncta 12.46. Latitudo lune uisa in principio 3 minuta septentrionalis in medio. In fine 3 minuta meridionalis. Semidiameter solis 16. Lune autem 18 minuta. [Eclipses to February 17, 1440, are listed.]

fol. 155r: [Tables:] Ad clima septimum. [Fols. 155v–62r are blank.]

fol. 162v: [Partial table for a musical text. Fol. 163r is the same. Fols. 164r–v are blank.]

fol. 165r: [In a different hand; see fol. (1a)r, above; acephalous:]...quia reflexum necesse est stare non solum in recto etsi secundum circulum feratur. Non enim idem est circulo fieri et secundum circulum....[Fol. 165v] [Q]ue autem loci mutatio prima sit nunc mirandum est....[Q]uod autem [?]continuit quemdam infinitum unum existentem et continuum motum et hic circularis est nunc dicemus... maxima autem manifestum est quod impossibilem continuum esse qui est in rectitudine motum [*abrupte*].

56. Σ (θ) = San Juan Capistrano, California, Library of Robert B. Honeyman, Jr., 23 (Astron. 12). Vellum, 45 fols., 22 cm. by 16 cm., in Italian hand in two parts originally separate (formerly Munich, J. Rosenthal, Katalog 90, 127)

Date: second half of 14th century

Text used: fols. 11r–45v. Primus. Figures, some incomplete; tables

Description: The following is from an examination of a microfilm of the manuscript.

fol. 1r: [*Tit.*:] Tabula prima saturni. Anno domini 1272 fuit saturnus in 13ª linea annorum sue tabule. [*Canon*:] Cum uolueris certificare per has tabulas de annis christi perfectis et hec nota [?]diligenter ne [?]decipiaris minue 20 uel adde eis 39 et quod residuum fuerit post augumentum uel diminutionem diuide per 59 et cum hoc quod remanserit intra tabulas saturni annorum mensium et dierum et in directo signi et mensis et diei inuenies saturnum in gradibus signis et minutis et habet tres tabulas huic addentur 4ᵒʳ gradus pro motu spere 9ᵉ extractis minutis que alio modo inueniuntur.

fol. 1v: Tabula saturni secunda.

fol. 2r: Tabula saturni tertia.

fol. 2v: Tabula iouis prima. Anno domini 1272 perfecto fuit iupiter in 77 linea annorum sue tabule.

fol. 3r: Tabula iouis secunda. [*Canon*:] Nota ad iouem inueniendum de annis christi perfectis minue 33 uel adde 50 et diuide per 33 et quod post diuisionem remanserit cum eo intra tabulas iouis annorum mensium et dierum et in directo signis mensis et diei iouem inuenies in signis gradibus et minutis et habet 4ᵒʳ tabulas et adduntur sibi pro motu 9 spere.

fol. 3v: Tabula iouis tertia.

fol. 4r: Tabula iouis quarta.

fol. 4v: Tabula martis prima. Anno domini 1272 perfecto fuit mars in 25 linea annorum sue tabule.

fol. 5r: Tabula martis secunda.

fol. 5v: Tabula martis tertia.

fol. 6r: Tabula martis quarta.

fol. 6v: Tabula martis quinta.

fol. 7r: Tabula martis sexta.

fol. 7v: Tabula martis septima.

fol. 8r: Tabula martis octaua.

fol. 8v: Tabula ueneris. Anno domini 1272 completo fuit uenus in 4ª linea sue tabule.

fol. 9r: Tabula mercurii prima. Anno domini perfecto 1272 fuit mercurius in 6ª linea annorum sue tabule.

fols. 9v–10v: [Tables left blank.]

57. Ψ (α) = Paris, Bibliothèque Nationale, fonds latin, 7280. 126 fols.

Date: 14th–15th century

Text used: fols. 15rª–43vª. Primus. Blank spaces left for figures and tables

Description: The following is from an examination of a microfilm of the manuscript.

fol. 1r: De la mare 427; Reg. 3022 [and scattered figures]. [First table on left numbered 1–12, of which 1–10 are complete:] Ferordimeh, Ardahasmeh, Bediadmes, Limeh, Marduneh, Marame, Obenmeh, Adrameh, Dimeh, Halimameh. [Below:] Menses arabum—His respondebant anno 622 quo coepit mahometus; Hi menses nostri. 1. Almuharan–maius; 2. Saphar–iunius; 3. Rabe primus–iulius; 4. Rabe secundus–augustus; 5. Iumedi primus–september; 6. Iumedi secundus–october; 7. Raieb–nouember; 8. Sahabem–december; 9. Ramadam–Ianuarius; 10. Scauueh–februarius; 11. Dulcheda–martius; 12. Dulhega–aprilis. Mense Romadam Mahometes fugit ex urbe mecha mense Dulhega celebratum fuit pascha anno nostro 623 die decima mensis aprilis omnino. [Table on right side:] Cum suscepit regnum Adhesit hoc anno numeramus. Ab orbe condito 4696; A diluuio noe 2940; Post abrahamum 2648; Ab exitu mosis 2143; Ab excidio Troiano 1817; A Dauidis regno 1705; Anno Olympico 1409 [at the right: "cccliii anno primo"]; Ab urbe Roma 1385; Anno Nabonassarii 1380; Anno Nabugodrosoris 1258; Anno Captiuitatis 1240 [?1246]; Monarchia Cyri 1169; Ab exactis Tarquinii [*sic*] 1141; A morte Alexandri 659 [*sic*]; Anno Seleuci familiae 943; Anno Imperatorum 680; Anno conspirationis Iulii 677; Aera Caesaris Hisp. 670; Anno Actiano 662; Anno appellationis Augustie 659; A nato Christo 637; Anno latini calculi 632; Anno Gratae Supputationis 614°; Anno Hegira Mahom. 10; Anno Heraclii Imp. 24; Indictione Quinta exeunte anno persico primo ineunte Anno Diocletiani 348; Aurei circuli numero sexto [then followed by "7280"]. [Fol. 1v is blank.]

fol. 1r[a] [renumbered 2r]: Liber de agregationibus scientie stellarum et principiis celestium motuum quem ametus filius ameti qui dictus est Alfranus compilauit 30 continens capitula. [C]apitulum primum de annis arabum et aliorum omnium et nominibus mensium ipsorum et dierum eorum et diuersitate inter eos adinuicem.... [Fol. 1r[b]] [N]umerus mensium anni arabum et aliorum omnium est duodecim menses [in the right mg., in a different hand: "Menses arabum"]... [listed in TKr, col. 960; Carmody, *Arabic Astronomical and Astrological Sciences*, p. 115]/... [fol. 14v[b]] Iam igitur declarauimus de eclipsibus solis et lune quod suficit intelligentibus.

fols. 44r–v: [Blank]

fol. 45r: [In a different hand] Tractum de spera etc. corporum ysoperimetrorum idest circulatorum seu circularium et dicitur ab ysoper; abydos quod est forma et pari quod est centrum et metros quod est mensura... [see Thorndike, "*Sphere*" *of Sacrobosco*, p. 153 (the commentary of Robertus Anglicus)] et hic est Virgillius fabullosus intuens infernum esse in parte opposita...in isto uersu Vergilius in georicis loquitur in quo homo debet seminare fabes milium... [ibid., p. 173]/... [fol. 45v] quod solum supple non premeretur ab illa regione idest ab alica parte signiferi poli idest zodiaci nisi ultima ungula tauri curuati procederet idest excederet ultra tropicum cancri scilicet et hoc dico poplite lapsso [ibid., p. 185]. Deo gratias. amen [about half of this verso is blank].

fol. 46r[a]: [In a different hand] Vna scientia est nobilior altera duobus modis... [edited in Thorndike, "*Sphere*" *of Sacrobosco*, pp. 143–98, and designated therein as MS *R*; see ibid., p. 73, for a description of the peculiarities of the scribe]/... [fol. 57v] pro quo fuit terre motus et miraculosa eclipsis. amen. [Fols. 58r–60v are blank.]

fol. 61r: [In a hand similar to that of the Alfraganus and Campanus works.] [Top mg.: "In nomine omnipotentis dei."] Consideratis omnibus regulis michi notis a sapientibus repertis... [listed in TKr, col. 255; Thorndike, *History of Magic*, vol. 4, p. 100]/... [fol. 72v] quem sibi accidat falsitates falsitatibus comulare.

fol. 73r: [In a hand similar to that of the Robertus Anglicus work.] [Top mg.: "Prohemium in astrolabium messehallach."] Scito quod astrolabium est nomen grecum eius interpretatio est... [listed in TKr, col. 1409]/... [fol. 81r[b]] est comparatio stature tue ad planitiem. Amen. Explicit astrolabium Messehala. Deo dicamus gratias. alleluya.

fol. 81v: [T]abula stellarum fixarum que ponuntur in astrolabio cum gradibus quibus celum mediant et cum distantia earum ab equinoctiali linea. [Fol. 82r] [T]abula stellarum fixarum... [same title as that on fol. 81v; followed by:] [T]abula stellarum fixarum uerificatarum per armillas parisius et est longitudo earum gradus circuli signorum per circulum transeuntem polos zodiaci et stellas. Latitudo uero earum est arcus eiusdem circuli cadens inter stellas et gradus longitudinis earum.

fol. 82v: Loca stellarum fixarum uerificata anno christi 1364. Aries... Pisces. [Bottom mg.:] Nota quod motus 8^{ue} spere a tempore Alfonsi usque ad 1406 fuit Gr. 1, Mi 44, Se 30.

fol. 83r: [Possibly in the hand of the Messehallah work] Incipit liber stelarum de locis stelarum fixarum cum immaginibus suis uerificatis ab Ebennesophin [= aṣ-Ṣūfī: cf. A. Hauber, "Zur Verbreitung des Astronomen Sūfī," *Der Islam*, vol. 8 (1918), pp. 50–51] philosopho anni arabum 325 qui fuit Christi 937. [Begins:] Vrsa minor.... [On the right a column has been added in a different hand, entitled "Ista posterior uerificatio pertinet ad annum 1406."] [The stars are numbered to 1023 and the list ends on fol. 100v. Thereafter are added notes:] Anno domino nostri 1406 incompleto quia die prima martii ipsius anni. Verificatio harum stellarum dicitur fuisse anno arabum 325 et sic hera incarnationis tenebat annos 936 dies 310 tunc temporis unde tempus elapsum a dicta uerificatione.../... hac ergo sic intelligantur.

fol. 101r: [Tables of stars, beginning with Ursa Minor and numbering 1023, as before. Tables end on fol. 111r. Fols. 111v–20v are blank.]

fol. 121r: [Top mg.:] Ptolomeus Halihabenrudi Sumista Angli[?cana] Habraam aven [remainder of folio is blank]. [Fol. 122r] Ptolomeus Haly Anglicanus. [Tables begin with "Aries" and continue through "Pissces" on fol. 123v.]

fol. 124r: [Lists of stars, probably belonging to the preceding. "Ptolomeus" appears on top mg. of all folios, recto and verso.] Vrsa maior... Corona meridionalis [fol. 126r].

The following manuscripts of the *Theorica* were not available to me:

1. Cambrai, Bibliothèque Communale, 1330 (1180) (Cathédrale, ancien 251). Parchment, 75 plus 12 fols.
   Date: 14th century
   Text of *Theorica*: fols. 1–36. Primus

Description: For a description of this manuscript, see *Catalogue général des manuscrits des bibliothèques publiques de France: Départements* (Paris, 1886–1911), vol. 17, p. 487. It is in such a bad state that it was impossible to procure microfilm or photostats of it.

2. Mostyn Hall, Flintshire, no. 82 (old catalogue no. 69). Old vellum, octavo (see *Reports of the Royal Commission on Historical Manuscripts* [London, 1870——], vol. 4, p. 350).

   Description: Begins with "Euclidis elementorum libri xv." At the end of this is a treatise on geometry in Latin dedicated "Clementissimo patri et piissimo dno Urbano" by Campanus. See Ker, *Medieval Libraries*, p. 24. The manuscript was sold July 13, 1920, lot 38, to Maggs, who sold it to S. Babra, bookseller, Barcelona, in 1923. Recent correspondence with Maggs and a former employee of theirs who knew Babra elicited no further information.

3. Pisa, Seminario Arcivescovile, Biblioteca Cateriniana, 69 (Mazzatinti 65). Parchment, 129 fols. (1b–10b, 342–409, 410–15, 117–64)

   Date: 14th century

   Text of *Theorica*: fols. 342r–76v. Primus

   Description: Described in G. Mazzatinti, *Inventari dei manoscritti delle biblioteche d'Italia*, fasc. 24, p. 78. It has long been lost, according to a letter from the librarian.

4. Salamanca, Biblioteca Universitaria, 2353 (San Bartolome 80; Palacio VII D4, 2 e 4 et 439). Parchment, 94 fols.

   Date: 14th century

   Text of *Theorica*: fols. 15v–34v. Primus

   Description: Described in Guy Beaujouan, *Manuscrits scientifiques medievaux de l'Université de Salamanque et de ses "Colegios mayores,"* Bibliothèque de l'école des hautes études hispaniques, fasc. 32 (Bordeaux, 1962), pp. 146–48.

Related Manuscripts

Among the manuscripts that I have omitted in establishing the text of the *Theorica* is a small group of mostly late manuscripts which include only the sections on the instrument or equatorium for the planets.[8] In addition, a large

8. MS Berncastel, Hospital zu Cues, 212 (15th century), fols. 131–41v, followed by measurements and notes to fol. 144r; MS Cracow, Biblioteka Jagiellońska, 575 (CC.I.

number of anonymous brief treatises describing instruments and equatoria may be found in the catalogues, but they are only remotely related to portions of Campanus's *Theorica*. To illustrate this statement, I shall give analyses, based on my own examination of microfilm and photostats, of two manuscripts from the Staatsbibliothek in Munich. The first, CLM.11067, is a fifteenth-century manuscript, in which four brief treatises are found together in seven folios.

fol. 180v^{a}: Canones de compositione equatorii planetarum. Cum uisus fueris tibi formare equatorium pro equandis planetis notandum est quod ad eius practicum due requiruntur leues tabule prima magna... [listed in TKr, col. 352, with *incipit*, "Cum visum," along with two other MSS, Vienna 5296, fols. 152r–54v, and Catania, Biblioteca Universitaria, 85, fols. 300r–302v]. [Two words, "asser" and "lamen," which do not appear in Campanus's *Theorica*, are used.] [Fol. 181r^{a}] De compositione epycicli.... [Fol. 181v^{a}] De formatione centrorum que sunt in tabula magna.../... [fol. 181v^{b}] item non oportet te facere specialem epiciclum lune si scieris secundum doctrinam in practica lune dictum est. Et sic finitur compositio equatorii pro equationibus planetarum etc. [Fol. 182r contains a table; at the bottom:] Sciendum quod centrum deferentium saturni iouis martis et ueneris inuenies in medio inter centrum terre et centrum equantis etc. [Fol. 182v] Exemplare equatorii.

fol. 183r^{a}: [Second of the treatises] Canones pro practica equatorii. Cum per instrumentum equatorii uera loca planetarum uolueris inuenire primo scias unum

---

30) (15th century), fols. 222r^{a}–27r^{a}; MS Cracow, Biblioteka Jagiellońska, 613 (DD.III.48) (14th–15th century), fols. 63r–71r; MS Leipzig, Universitätsbibliothek, 1475 (15th century), fols. 245r^{a}–49r^{b}, followed by excerpts of measurements and some notes to fol. 250r^{a}, the notes citing the *Albion* of Richard of Wallingford, the *Almagest*, and the Alphonsine tables; MS Oxford, Bodleian Library, Bodley 464 (Madan *et al.*, *Summary Catalogue*, vol. 2, pt. 1, pp. 375–76, no. 2458) (ca. 1318), fols. 85r–92v (listed in Kibre, "Further Addenda," p. 79 as BL 464); MS Munich, Universitätsbibliothek, Q.738 (1442), fols. 151r–59r (I wrongly identified this work as the excerpt of John of Gmunden from the *Theorica* of Campanus [Benjamin, "John of Gmunden and Campanus of Novara," p. 222] and even inadvertently omitted the shelf mark. Since then I have examined a microfilm of the manuscript and report herewith: "Excerptum ex theoricis Campani. Primeui temporis philosophi diligenti intuitu astra speculationis laboriosis conatibus... [seems to be a poor reading of the introduction of the *Theorica*]/... [fol. 159r] Directionem uero stationem et retrogradationem istorum quatuor planetarum per istud instrumentum faciliter inuenies quemadmodum prius dictum est de mercurio. Sed iste modus debet fieri in epiciclo uere circulationis et non in epiciclo oportune circulationis ut supra etc. et sic est finis 1442 in die s. petri." [See Zinner, *Verzeichnis*, p. 72, no. 1912, who describes this as Gmunden on Campanus.] To be sure, this manuscript has numerous works of Gmunden); MS Salamanca, Biblioteca Universitaria, 2621 (15th century), fols. 1r–6v; MS Tabley House, Cheshire, no number; a Latin treatise called *Aequatorium Campani* (6 leaves; see *Reports of the Royal Commission on Historical Manuscripts*, vol. 1, no. 46); MS Wolfenbüttel, Landesbibliothek, 81.26. Aug. fol. (2816) (15th century), fols. 145v–52v.

esse modum equandi saturnum iouem martem et uenerem sed aliorum scilicet solem mercurium et lunam quemlibet modum equandi habere proprium... [listed in TKr, col. 329]. [Fol. 183r^{b}]...Solem...; [fol. 183v^{a}]...Mercurium...; [fol. 183v^{b}]...Lunam.../...[fol. 184r^{b}] et ubi hec cordula [a word not used by Campanus] in orbe signorum ceciderit ostendit uerum locum lune etc. Et sic finitur practica equatorii cum compositione eiusdem deus sit benedictus.

fol. 184r^{b}: [Third of the treatises] Sequitur instrumentum Campani de equationibus planetarum. Ad faciendum equatorium planetarum de quo primo presumitur quod media loca planetarum ex tabulis sciantur. Accipe asserem concauatam ad modum astrolabii qui interius sit bene pergamenis bitumnatus quod sit planum et album et ille asser habeat lymbum distinctum per 360 gradus... [listed in TKr, col. 38]. [Fol. 184v^{a}]...Isto modo potes inuenire centra equantium saturni iouis martis et ueneris. Accipe in linea numeri communis in tabulis alphoncii tria signa prescise.... Post hoc extende circinum in tali amplificatione.... [Fol. 184v^{b}]...Deferentem uel ecentricum solis uel lune sic facies protende filum prescise a centro terre.... [Fol. 185r^{a}]...Deferentem et equantem mercurii de quo est specialis modus eam faciendo trahe lineam.... [Fol. 185r^{b}]...cum cultello uel plumbeta.... Deferentem lune scias quod ipse debet diuidi ex zodiaco super centro zodiaci uel terre.... Epyciclum lune impone pone centrum epycicli ad augem sui deferentis.... [Fol. 185v^{a}]...Epyciclum saturni ueneris iouis et martis sic impone. Accipe unum asserem rotundum.... Epiciclum mercurii sic impone.../...[fol. 185v^{b}] secabit uel tanget et secundum illam extensionem describe circulum qui erit epyciclus mercurii etc. Practica huius equatorii. Ad habendum uerum locum solis oportet te primo scire augem sub illa auge pone lineam.... Verum locum lune inuenire oportet te primo scire.... [Fol. 186r^{a}]... Verum locum saturni iouis martis ueneris et mercurii sic inuenies.... Medium motum cuiuslibet planete et argumentum ueneris et mercurii inuenies in tabula mediorum motuum in astrolabio; auges quere in tabulis Alphoncii.../...tot etiam gradus transiuit planeta ab auge sua media epycicli ut dicit canon in tabulis Alphoncii etc. Et sic est finis deus sit benedictus.

fol. 186r^{b}: [Fourth of the treatises] Incipit canon de compositione equatorii pro coniunctione et oppositione solis et lune. Ad faciendum figuram ad inueniendum coniunctionem ueram et oppositionem solis et lune describe primo circulum perfectum in superficie aliqua quam latius poteris super centrum E quem quadrabis duabus dyametris... [not listed in TKr]. [Fol. 186v^{b}]...tangent...ex utraque parte equedistantem linee superioris OP et sic perficitur illa figura cum adiutorio dei etc. Vide figuram ex opposito pro exemplari etc. [Fol. 187r contains a figure; at the bottom:] Nota si gradus solis plus distat a puncto A quam gradus argumenti adde horas equatas medie coniunctioni. Si autem gradus argumenti plus distat tunc subtrahe horas equatas a media coniunctione et habentur uera etc. [Fol. 187v^{a}] Practica equandi cum instrumento coniunctionem ueram et oppositionem solis et lune. Si per instrumentum presens et figuram ad hoc factam scire uolueris ueram coniunctionem et oppositionem solis et lune. Quere primo ex tabulis ad hoc factis.../...[fol. 187v^{b}] Notandum quod si gradus solis plus distat a puncto A quam gradus argumenti adde horas curtas medie coniunctioni; si uero gradus argumenti plus distat tunc subtrahe horas a media coniunctione etc. Explicit equatorium coniunctionis et oppositionis solis et lune anno domini millesimo cccc° xlvii uicesima quinta die mensis octobris etc.

In addition to the words pointed out in the foregoing, the words "amplificatio," "tango," and "curtus," which appear in CLM.11067, do not occur in Campanus's *Theorica*. The only seeming reflection of the use of the Campanus work is the ordering of the planets: Venus and the three superior planets, sun, moon, and Mercury. The *Theorica planetarum Gerardi* usually separates Venus from the three superior planets and associates it with Mercury. But, despite the similarity in the planetary groupings, the CLM.11067 manuscript usually lists Saturn first whereas Campanus places Venus first.

The other manuscript is CLM.19689 of the sixteenth century, and the relevant article is on fols. 148r–50r, accompanied by tables on fols. 151r–52v for the mean motions of Saturn, Jupiter, Mars, and Venus, respectively. The article begins as follows.

fol. 148r: Compositio theorice ueri motus saturni ex Campano abreuiata. Primo fiat circulus signorum super centro D cum suis interstitiis et diuisionibus quemadmodum factum est in theoricis luminarium uel quemadmodum in astrolabio dorso fieri solet. Secundo quadretur prefatus signorum circulus duabus dyametris orthogonaliter... [not listed in TKr; compare this with the opening passage of the work beginning on fol. 186r[b] of the CLM.11067 manuscript, in particular the last phrase].

Campanus is cited twice thereafter (fols. 148v and 149v], and certain words used do suggest knowledge of the *Theorica*: "uera et oportuna circulatio," "seorsum," "filum." The work ends:

fol. 150r: ...et nota intersectionem istius fili in epiciclo uere circulationis super quam ultimo pone filum D et tactus eiusdem fili in orbe signorum ostendit uera loca eorumdem. Correlarium: Peurbachi subtracto medio motu planete de medio motu solis remanet argumentum medium; aliud correlarium subtracta auge planete a medio motu eiusdem planete remanebit centrum medium.

From these examples, then, one might conclude that even the citation of Campanus in the title or in the text does not affirm a close relationship with the *Theorica*. However, one of the anonymous and brief treatises on the instrument claims in its introduction to depend on Campanus's description in the *Theorica*, and, indeed, the claim appears valid.[9] Although much of the

9. MS Paris, Bibliothèque Nationale, fonds latin, 7443 (15th century), fols. 243v–46r, opening, "Equatorium planetarum facilis compositionis paruarum expensarum non minoris utilitatis septem instrumentorum Campani per quod ad omne tempus datum certa loca planetarum...." Listed in TKr, col. 502; Thorndike and Kibre, *Catalogue*, 1st ed., col. 237, extend the foliation to fol. 247v, but on fol. 246r this treatise seems to end as follows: "...ne uoluatur et sic erit compositio nostri instrumenti. completa." Then it begins immediately, "Vera loca omnium planetarum eleuationem [supralinear:

vocabulary of that manuscript is not found in the *Theorica* ("competens," "clauiculus," "bytumen assere," "euacuo," "rotula," etc.), many technical terms are, and several passages reflect the phraseology of Campanus.[10]

Three manuscripts known to me have extracted from the *Theorica* the tables containing the sizes and distances of the various parts of the planetary systems.[11] Another manuscript has excerpted those parts of the text giving the sizes and distances and periodic times of the planets.[12]

In yet another manuscript a very brief excerpt from the *Theorica*, which has not been collated for the text herein, warrants description because of the inadequacy of the description in the catalogue.[13] The *incipit* (fol. 169r), "Dicit Campanus in theorica sua quod..." (listed in TKr, col. 417), is followed immediately by the opening phrase ("Primus...") of the introduction (section II) and then by a faithful transcription of lines II.2 through II.44.[14] At this point, a commentary begins with specific citations of the following: Nicholas Oresme, Aristotle, Cicero *De natura deorum*, Psalms, Isidore of Seville, Ptolemy ("sapiens sapientie 13° almagestum"), Dionisius, and Euclid. Then in the middle of

---

(Note 9 continued)
'in longitudine'] elongationem..." (identified in TKr, col. 1686, as the canons for the tables of Oxford, 1348), which is not completed and ends abruptly on fol. 247v, a new treatise in a different hand beginning on fol. 248r.

10. The date 1360 appears in the margin. For other manuscripts with the same *incipit*, see MS Rome, Vaticana Biblioteca Apostolica, lat. 3127 (14th century), fol. 34r: "[E]quatorium planetarum facilis compositionis.../...[fol. 37r] quemadmodum in argumento solis non oporteret te nisi semel intrare cum diebus et tunc esset opus facilimum et sic est finis"; also fol. 66r: "Equatorium planetarum facilis compositionis.../...[fol. 69r] quod pone rotulam in cuius dyametro stabit corpus mercurii in rota magnitudinis sui epicicli per quam tale filum a centro terre ad zodiacum et abscindet uerum locum mercurii in nona spera et in hoc fuerit practica equatorii planetarum deo gratias"; MS St. Gallen, Stadtbibliothek (Vadiana), 412 (15th century), fol. 59r: "Equatorium planetarum facilis compositionis...[listed in Zinner, *Verzeichnis*, p. 105, no. 3045, with incorrect foliation which TKr, col. 502, copies]. Auges autem planetarum anno christi 1360 sunt hee posite in sequenti iam tabula/...[fol. 62r] [the same as Vat. lat. 3127, fol. 69r:] in nona spera. Sequitur igitur descriptio figure huius instrumenti. [Fols. 62v–67r are blank.]"

11. MS Berncastel, Hospital zu Cues 212, fols. 142r–43v (cf. note 8, above); MS Bergamo, Biblioteca Civica Σ.2.2 (15th century), fol. 133v (only the distances in miles); MS Paris, Bibliothèque Nationale, fonds latin, 7281, fols. 249r–50v.

12. MS Brussels, Bibliothèque Royale, 10117–26 (14th–15th centuries), fols. 79v, 110v, 112v–18v (with other material interspersed). Cf. also the précis of the *Theorica* in MS *C* (Brussels, Bibliothèque Royale, 1022–47), fols. 48r[a]–50v[a], described on p. 63, above.

13. Henry Stevenson and J. B. de Rossi, *Codices Palatini Latini Bibliothecae Vaticanae...*, vol. 1 (Rome, 1886), pp. 141–42: MS Rome, Biblioteca Apostolica Vaticana, Palatinus 446 (paper, in quarto, 15th century), fols. 169r–[92a]v (of 254 fols.).

14. A hasty transcription and comparison has shown that it belongs to the β group of manuscripts, but has the marginal notations concerning the four parts of the quadrivium as in MS *G*.

fol. 171r, blocked in, written in the largest of the three hands found in the manuscript, and interspersed with extensive interlinear additions, begins the familiar *Theorica planetarum Gerardi*, "[E]centricus circulus dicitur uel egresse cuspidis uel egredientis centri...." The aforementioned commentary, in the middle-sized hand, continues without interruption in the top and bottom margins and in the right margin of the rectos and the left margin of the versos. In addition, marginalia begin, in the smallest hand, in the left margin of the rectos and the right margin of the versos. (Campanus is cited on fols. 171r–v and 183v.) Several of the folios that follow, numbered consecutively, are of smaller size and do not contain either the Campanus or the Gerardus text or the commentary of the smaller hand but only the commentary of the larger hand or figures with notes. The Gerardus text continues on the normal-sized folios, where figures also occur. The commentary in the smaller hand tapers off after fol. 177v. The Gerardus text ends on fol. 189v with the usual *explicit*: "...et non corporaliter. Et sic est finis theorice communis." There is no commentary in the smaller hand on that folio; the last author cited by the commentary in the larger hand is Albumazar,[15] and then it, too, ends on fol. 189v: "demonstrabit mallui enim stari prolixitate gaudere quam temerari aut inuidie denigrari reprehensiones. Et in hoc terminatur theorica de motibus miri artificii primi motoris scilicet in honore ipsius cuncta [right mg.: 'cuncta'] gubernantis tui cunti seruiunt motores et subsunt mouentes orbes absque fatigatione et pena; immo toto tuo optimam uitam cum primo ducente cui laus optans et gratiarum actio per omnium seculorum secula. Amen." Figures follow on fols. 190–[92a]v. Fols. [192a]r–v are blank. Fol. 193r: "[Q][?Ve] ego in doctorum libris inueni 100 capitula pro posse compilabo de aspectibus lune aliorumque planetarum cum significationibus eorumdem. Cum unus planeta est inter duos planetas bonos assiliantur [assimilantur] uiro in bono statu.../...[fol. 194r] et non combustus. Caue etiam ne intres primam horam iouis seu horam martis quia malum est." Fol. 194v is blank. On fol. 195r the familiar work of Alchabitius opens: "Postulata a domino prolixitate uite..."; but in this manuscript the work is entitled, in the top margin, "Alchabitius in theorica planetarum." I have not attempted a detailed analysis of this somewhat peculiar use of an excerpt of the introduction of Campanus's *Theorica* to precede a more extensive commentary of the *Theorica planetarum Gerardi*.

15. "...De hiis dubiis require defferentiam tertiam tractatus septimi introductorii Albumezar quia presentis negotii non interest...."

Theorica planetarum

I

[Incipit theorica planetarum magistri Campani quam misit ad dominum papam Vrbanvm etc.]

Clementissimo patri ac piissimo domino unico mundane pressure solacio
5 domino Vrbano Qvarto electione diuina sancte romane ecclesie summo pontifici Campanvs Novariensis sue dignationis seruus inutilis beatorum pedum osculum cum qua potest reuerentia. In uobis, beatissime presul, sub cuius felici ducatu tota feliciter militat sancta mater ecclesia, quadam prerogatiua mirabili sicut monarchicum decet principem caritas intus ar-
10 det; pietas foris nitet; scientia uero radiat utrobique. Quis enim est affectu tam purus? Quis beneficio tam fecundus? Quis consilii claritate tam certus? Ecce, pater, noctes insompnes ducitis; ecce, dies in libramine consiliorum expenditis; ecce, thesauros uestros quasi cinerem spargitis; ecce, uos ipsum quasi uictimam continuis laboribus immolatis ut uestre clementie
15 subditis paretis in inuisa turbatione quietem, et Petri nauiculam inter tirannice tempestatis procellas ad portum tranquilla nauigatione ducatis. Ad istius sancte nauicule regimen solius diuine dispositionis salutifero beneplacito de milibus estis electi omni prorsus humani consilii spe succisa, nec uestram reuerendam electionem aliqua labes humani faminis fermen-
20 tauit. Sic eligi decuit ruenti mundo sub multis erroribus se basem solidam prebiturum, ut diuino potius quam humano consilio cunctos errores elimi-

1–110 *Codd. DEGHhkLNrsVWwXZz*
5 electione: dictione *G* dispositione *k*
7 oscula *V*
10 enim *om. DEhNVXZ*/effectu *Dh*
12 insompnes: uos *add. DENVXZ*
18 electi: editi *kWw*
19 aliqua labes: aliquis labor *DEhNVXZ*/ humani: humida *Lrz*/faminis: flaminis *h* fauoris *k*
20 ruenti: tuenti *NV*
21 preserituram *V*
21–22 eliminet: illuminet *DGksWwX* illuminat *h*

I

[Theory of the Planets by CAMPANUS Dedicated to Pope URBAN]

To the most merciful father and most pious master, the only comfort against the burden of the world, Lord URBAN THE FOURTH,[1] by divine election highest priest of the holy Roman church: CAMPANUS OF NOVARA, a useless slave of his worthiness, with all possible reverence [bestows] a kiss on his blessed feet.[2] Most blessed chief, under whose auspicious leadership the entire holy mother church soldiers on with good success: thanks to a certain marvelous privilege, as is fitting for a prince who is sole ruler, charity burns within you, piety shines abroad from you, while knowledge casts its rays both in you and from you. For who is so pure as you in affection? Who so bountiful in good works? Who so unerring in the brilliance of his planning? Behold, father, you pass your nights without sleep;[3] behold, you spend your days weighing counsels against each other; behold, you scatter your treasures broadcast like ashes; behold, you offer yourself up like a sacrificial victim to ceaseless labors, [all] in order to bring tranquility in the dreadful turbulence to those who are placed in subjection to your clemency, and to guide the ship of Peter,[4] amid the gusts of the tyrant storm, on a calm course into port. You were elected out of thousands only by the saving decree of divine providence to direct the course of that holy ship when all hope that rests on human counsel had been utterly destroyed; and no stain of human speech spoiled[5] your venerable election.[6] It was fitting that such should be the manner of election of the one who was to prove himself to be a solid foundation for a world collapsing under the weight of many errors,[7] so as to remove all the errors and shore up the rickety structure on all

---

1–3 *rubric.* Teoricae CAMPANI mathematici liber secundus INCIPIT *H*; *tit.* Equatorium planetarum CAMPANI *k*; CAMPANI NOVARIENSIS Theorica motuum planetarum et instrumenta eorum *L et mg.* theorica motuum planetarum et instrumenta eorum magistri CAMPANI NOVARENSIS et quantitas orbium planetarum; *tit.* Incipit theorica planetarum magistri CAMPANI (in *add. z*) quam misit ad dominum papam VRBANVM etc. *rz*; [p]rimum capitulum continet epistolam ad dominum papam predictum etc. *s*; *tit.* Theorica CAMPANI incipit *V*; *tit.* Theorica CAMPANI w_2; *tit.* theorica campani de motibus planetarum *Z*; *caret tit. DEGhNWX*

net et undique fulciat ruitura. Quia uero dispositionem diuini beneplaciti nullius creature potest operatio uacuare, futurum est ut per uos, diuine rector, mundus pacis recipiat libertatem et ecclesia purgatis erroribus sit 25 secura Petrique nauicula per strati blandicias equoris ad tutum portum applicet sine fluctu.

Exultet igitur totus mundus; psallat sancta mater ecclesia; iubilet felix Petri nauicula; quoniam mundus principem, pastorem ecclesia, nautam nauicula gubernantis dei munifica prouisione recepit qui cuiusque pressuras 30 alleuiet, opituletur defectibus et languoribus adhibeat medicinam. Talia certe meretur roridus uestre pietatis affectus ut sub uestri regiminis umbra sitientes aquas hauriant cum securitate plenaria salutares, et defessi quilibet membra sua super stratum uirentis graminis dent sopori, et ut cesset galea cesset ensis omnisque militaris strepitus conquiescat. Affectus huius 35 gloriosissimi magnitudinem sic predicent qui sentiunt per effectum. Puto uociferantium sonos in omnem mundi distantiam intonare: "Is quippe non est inuentus despicere carnem suam sed cunctis egenis subuenit; uagos recipit; nudos operit; famelicos satiat, et languenti compatitur et potum tribuit sitienti."

40 O dulcissima mirande misericordie necessaria uena, que tante profunditatis fontem suauissimum parturit a cuius potu nullus excluditur, sed eo quisque iuxta sue capacitatis modulum deebriatur. Ad hunc fontem purissimum unda continue scaturiginis effluentem defessi labore studii concurrunt undique litterati fecundos ex ipso calices bibituri. De puluere, pater, 45 Philosophiam erigitis que lugere solet in sue mendicitatis inopia nostrorum presulum auxiliis destituta. Nunc autem ad uestre serenitatis aspectum facie reuelata consurgit quam hactenus obduxerat uerecundie pallio rei familiaris angustia macerata. Latere mallebat tenuis et pudica quam aulicorum impudice se largis dapibus immiscere. Quippe semper est in uere 50 domesticis arbitrata ridiculum, ut in risum histrionum more uocari soleat que mores instruere debet et uitam hominum mensurare. Ad uos autem qui non solum intellectu uigetis, polletis ingenio et scientia radiatis sed etiam affectu multiplici soli sue uidemini pulcritudinis amatores tam secura

24 mundi *L*
27 felix: fidelis *Lrz*
29 cuiusque: cuiuscumque *GkWwX* eiusque *Lrz*
31 roribus *DENVX*
34 Affectus: Effectus *Lrz*
35 magnitudinem *om. Lrsz*
35–36 Puto...intonare *om. V*
36–37 non est *om. DEhNV₁XZ supra lineam V₂*
40 miranda *DHX*/necessaria: uestre *LrsZ* uestra *H* uera *k*
41 parturit: percurrit *Lrz*
43 undam *DhLrWz*
43–44 concurrunt: confluunt *GHLrsWwz*
46 ad...aspectum: a nostre (uestre *w*) seruitutis aspectu *kw*
48–49 aulicorum: anglicorum *hWw*
50 arbitratum *V*/rudiculum *DEVXZ*

sides with the aid of divine rather than human counsel. Moreover, since what is ordered by divine decree cannot be made void by the operation of any created being, it will come to pass that through you, God-appointed guide, the world will receive freedom to live in peace, and the church, its errors cleansed away, will be secure, and the ship of Peter, [sailing] through the gentle caresses of a smooth sea, will reach safe anchorage without [encountering] great waves.[8]

Therefore let the whole world be jubilant; let the holy mother church sing joyful psalms; let the ship of Peter rejoice in its good fortune; seeing that through the bountiful provision of God, who guides us, the world has gained a chief, the church a shepherd, and the ship a sailor, to lighten each one's burdens, bring aid to its deficiencies, and administer healing medicine for its sicknesses. Such at least is the virtue of your piety's dewy compassion that, beneath the shade of your rule, the thirsty may drink deep of the waters of salvation in complete security, and all those who are weary may stretch out their limbs in sleep on the smooth expanse of green grass;[9] and that helmet and sword may be put aside,[10] and all the din of war may die away. Let the greatness of this most glorious compassion be thus proclaimed by those who feel it in its working. It seems to me that the noise of their cry thunders out to every distant part of the world: "For he has not been found to despise his own flesh, but he comes to the aid of all who are needy; he takes in the homeless; he covers the naked; he fills the hungry; and he feels compassion for the sick and gives drink to the thirsty."[11]

O most sweet and necessary[12] vein of admirable pity, which gives forth a most delightful spring of such depth, from drinking at which no one is barred: rather each is regaled with it according to the measure of his capacity. To this purest of springs, pouring forth in a never-ceasing welling gush, men of letters, weary with the toil of study, hasten from all quarters to drink abundant draughts from it. Father, you raise up Philosophy[13] from the dust [where] she is accustomed to lie, sorrowing in the helplessness of her poverty, destitute of all aid from our princes. Now, however, she rises and reveals her face to the gaze of your serenity, though until now she had kept it covered with the cloak of shame, distressed as she was by the poverty of her circumstances. She preferred to lie hidden, poor but modest,[14] rather than to thrust herself shamelessly into the sumptuous banquets of princes' courts. For she has always been thought a joke amongst those who are truly servile,[15] with the result that she whose proper function it is to give moral instruction and be a yardstick for the life of man is usually summoned, as are actors, to be laughed at. To you, however, who not only are strong in understanding, of powerful talent, and brilliant in learning, but also, by reason of your many-sided affection [for her], seem the only true lover of her beauty, she comes both confident and joyful, for she sees that she

uenit quam leta cum non ad peregrina sed ad propria uideat se uocari.
55 Sumptis namque fecundis dapibus placet ut illud uenerabile capellanorum uestrorum collegium quos sibi uestra coadesse clementia uoluit uos sequatur. Quibus ad uestre sanctitatis pedes sedentibus iocundum sapientie certamen indicitis in quo militaribus armis accincti militaliter dimicant partes agrediens et agressa. Hec quidem instat ualide iaculis rationum;
60 illa uero responsionum clipeis strenue se defendit. In hoc uestro Philosophia camerali gignasio iocundatur, ubi sicut et uos estis ipsi domestici sic eidem domestica problemata disquirenda proponitis, eaque rationum collectione pensatis; postremo iubetis quid in hiis tenendum Philosophia censeat diffiniri. Habent itaque Philosophiam professi de uestre mense bene-
65 dictione quo uentrem reficiant et quo mentem. Ista uero sunt illa saturnalia festa quorum sollempniis prothophilosophos legimus uacauisse. Iste uero sunt epule quas reuerendus SOCRATES discipulis suis legitur ministrasse et quas sibi uice mutua ministrari postulat ab eisdem. Ad has tam sanctas tam uenerandas epulas, clementissime domine, licet tantis indig-
70 num muneribus pietate propria me uocastis, et huius duplicis sancte mense participem me fecistis ut me nobilitaretis titulis uestre dignitatis amictum, qui tenuitate proprie scientie plebescebam. Propter quod possum uere dicere, gratia domini mei VRBANI sum id quod sum.

Sed ne gratia tanti patris in me uacua remaneret, a recepte beneficientie
75 tempore iugiter mente discussi sollicita si quid saltem uel minimum inuenirem quod uestre maiestatis honori possem in signum purissime deuotionis offerre. Cumque mihi sedule perquirenti nichil inueniretur in mee paupertatis armario quod auderem tante celsitudini presentare, tandem diuina largitas que datorum nichil inproperat et dat omnibus habundanter
80 mihi quiddam aperuit quod oblatione desiderata diutius arbitror non indignum. Hoc enim sicut est in nostris inuentionibus nouum, sic sue iocunditatis utilitate uestre magnitudini censeo placiturum. Licet equidem si forsan ad uestre plenitudinis habundantiam referatur entitatis speciem uix defendat, collatumque splendori uestre scientie tenebrescat; tamen in
85 uestre potentie subditis quorum uobis cura peruigil puto satis inueniet

55 capellanorum: capitulorum *DhX* capellorum *Ww*
58 militariter *HhkrsWwz*
61 iocundant *r*
62–63 collectione: collatione *Vz*
63–64 censeat: sentiat *DENVZ* sensiat *X* signat *h*
66 prothophilosophos: Ptholomeus *Lrz* / uocauisse *hks* uocasse *Ww*
69 tam: quam *E*
70 muneribus: mercedibus *Lrz* / et: in *DEhNVXZ*
71 participem: principem *DEhNsXZ*
71–72 amictum: amicitia *DEhNVXZ*
77 offerre: conferre *Lz* conferri *r*
80 mihi *om. DEhNVXZ*
82 utilitati *DEhLNXZ* / censeo: sentio *DEhNVXZ*

is summoned, not to something alien, but to what is her own. For it is your pleasure that, when your bountiful dinner has been consumed, that venerable company of your chaplains whom your clemency has decreed to be your companions[16] should follow you out. They sit at the feet of your holiness, and you give the signal to engage in a merry battle of learning, in which the attacking and defending parties, accoutered with military weapons, wage war in military fashion. The former presses the assault strongly with the javelins of reasoning; the latter defends itself stoutly with the shields of rebuttal. Philosophy is delighted in this domestic gymnasium of yours, where you, who are also a familiar of hers, accordingly propose to her familiar problems for disputation;[17] you weigh these problems up by comparing the arguments; finally you bid [the participants] to determine what Philosophy decrees should be the received opinion on these matters. Thus those who follow the profession of Philosophy have the means to replenish their bellies and their minds from the blessings bestowed by your table. These feasts of yours are [like] those saturnalian banquets in celebrating which, we read, the foremost philosophers spent their leisure.[18] Such are the feasts which the venerable SOCRATES is reported to have served to his pupils, and which he demands that they in turn serve him.[19] To these feasts, so sacred, so awe-inspiring, you have summoned me, most merciful lord, though I do not deserve such great gifts through my own piety, and you have made me a member of this doubly sacred[20] table, to ennoble me by clothing me in the titles of your dignity, when I was a commoner in virtue of the poverty of my own knowledge. For this reason I may truly say, "I am what I am by the favor of my lord URBAN."

But in order that the favor of so great a father toward me might not remain unrequited, from the moment I received your benefaction I cast about ceaselessly with anxious mind in the hope of finding something, even of very small consequence, which I could offer to honor your greatness as a sign of the purest devotion. And when, in spite of my diligent search, nothing in the store of my poverty could be found that I dared present to such loftiness, at length the generosity of God, which rejects no gift and gives to all in abundance,[21] revealed to me a thing which I think not unworthy of the offering I have desired so long to make. For this thing is a novelty among my discoveries, and furthermore I think it will please your greatness by the usefulness of its amusing approach.[22] And though indeed, if it were by any chance to be compared with the wealth of your vast resources, it would hardly make good a claim to be considered as existing at all and would grow dim when contrasted with the splendor of your learning, nevertheless amongst those who are subjected to your power, for whom you exercise sleepless care, it will, I think, find enough to supplement

---

61 gignasio: aliter gimnasio *add. s*

quid repleat, quid illustret. Audacter igitur uestre glorie, qui etiam parua contempnere non soletis sed offerentis affectum metiri potius in eisdem, offero presens opus firmiter arbitratus quod dignissimus Illius uicarius Qui pauperculam mulierem in oblatione dragmule non despexit pietate
90 solita non dedignabitur agregare sanctitatis sue referto gazophilatio dragmam meam, quin immo paruitatem eius laudibus sue curialitatis attollet.

Sed suppliciter deprecor omnes in quorum manus propositum opus deuenerit ne mee paupertatis munus, si decenter compositum esse cognouerint, inuidie dente dilacerent aut nouacula liuoris extenuent, sed si
95 fructus ex eo collegerint quibus philosophie dulcedo sit insita fraterna caritate comedant et commendent. Quin imo quicumque propter subscriptas meas nouas ymaginationes alicubi uoluerint me mordere, primum in suis demonstrationibus mordeant PTHOLOMEVM; nam ymaginationes mee quas hic inuenient eius demonstrationibus sunt cognate quasi con-
100 clusiones proprie sint earum. Vere quidem nichil hic quisquam ymaginatum inueniet quod ibi non reperiat demonstratum. Hec igitur adeo sunt super irrefragabiles demonstrationes mirifici PTHOLOMEI fundamentaliter solidata quod mordacis dentis impressionem uel minimam quamuis adamantinam duriciem habeat non pauescant. Potius enim in mordendo dens
105 ipse lesus retunditur quam inferat lesionem. Coerceat ergo morsum saltem ob inpossibilitatem effectus qui cohercere non solet eundem pro ueritatis affectu, et studeat ipse magis ut eo maiora uobis offerenda, pater, inueniat quo in philosophie thesauris se nouerit ditiorem.

Deinceps uero quale sit istud munus qualemque cernentibus utilitatem
110 parturiat est dicendum.

89 mulierem: minorem *DEhNVXZ*
90 sanctitati *DEGHkLNrsWwXZz*
91 immo: modo *DEhNsXZ*
91–92 quin...deprecor *om.* V
92 suppliciter: simpliciter *hNw*
95 philosophi *DEhNVXZ*/fraterna: superna *GkLrWwz*
96 Quin...propter: quod minimo quisque pre *Lrz*
97–98 primum...nam: almagesti Ptolomei respiciant et uidebunt quod *V*
98 ymagines *L*
99 cognate: concordes cum *V*
101 reperiatur *HLrs* reperiet *w* inueniatur *X*
102 mirifici: musici *DhNXZ* munifici *HsWw* uiuifici *Lrz*
103 uel minimam *om.* *DEhNVZ* uel numquam *kw* numquam *W*
104 habeat: habenat *DEhNw* habeant *HXZ* *om.* *Lz*/pauescat *Lrz* pauescent *DNVX* pauescunt *E*
105 retenditur *E* reconditur *h*
106 effectus: affectus *DEhNVXZ*
107 uobis: nobis *DELNrVXz*
108 nouerit: inuenerit *Lrz*/dictiorem *GLrz* doctiorem *W* doctorem *w*
109 cernentibus: comedentibus *Lrz*

and illuminate [in their knowledge]. So I boldly offer the present work to your glory, since it is your custom not to despise even small things, but rather to measure the sentiments of the giver by them; I am firmly convinced that the most worthy vicar of Him Who did not despise the poor woman when she offered her small mite will, with his usual piety, not disdain to add my mite to his saintliness's full treasury;[23] no, he will even exalt its smallness by the praises of his courteousness.

But I humbly beg all those into whose hands the work I now lay before them shall fall not to tear this gift of my poverty apart with the tooth of envy or pare it with the razor of malice, if once they have recognized that it is properly written; but rather, if they should pluck from it fruits in which the sweetness of philosophy is to be found, to eat them and commend them with brotherly charity. Furthermore, if any should wish to attack me on any point because of the novel models which I shall describe, let them first attack PTOLEMY for his proofs. For my models which they will find in this work are related to his proofs as closely as if they were actual conclusions from them. In truth no one will find anything depicted in a model here which he would not find proven there.[24] These models, then, are so solidly based on the irrefragable proofs of the marvelous PTOLEMY that they need not fear the attacker's tooth will make the least impression, no, not though its hardness should be that of adamant: instead of inflicting a wound, the tooth itself is damaged and blunted[25] by the bite. Therefore let him who does not restrain his bite for love of truth at least restrain it by reason of the impossibility of its having an effect, and let him rather study to make discoveries to offer to you, father, which shall be greater than mine in proportion to the greater riches which he knows himself to possess in the treasures of philosophy.

But next I must say what this gift is, and what the advantage it produces for those who sift it.[26]

II

[Prohemium; incipit tractatus]

Primus philosophie magister ipsius negotium in tria prima genera dispartitur quorum primum diuinum nominat, secundum mathematicum et tertium naturale. Fitque medium quodammodo naturam participans extre-
5 morum eo quod intentio mathematica communiter reperitur in naturalibus et diuinis, et sicut subiecti nobilitate primo subsidet sic tertium antecedit, licet utrumque doctrinalis modi certitudine sibi cedat, propter quod et doctrinale genus anthonomasice nuncupatur, eo quod docendi modum habeat cui nequit discipulus contraire. Inchoat enim a conceptibilibus in-
10 tellectu que uidelicet sunt omnibus per se nota, et ex hiis prima demonstrabilia post media post ultima certissime sillogizat a primis ad ultima per ordinata media gradiendo.

Negotiatur autem hoc totum genus de propriis passionibus quantitatis, et quia quatuor habet partes, quadriuium nominatur; suntque de discrete
15 quantitatis genere due prime, cetere uero due sunt de continua quantitate. Prima itaque quam arismetricam dicimus quantitatem discretam considerat absolute et perquirit que genera que species numerorum. Secunda uero que musica dicitur quantitatem discretam sonis applicat, querens que con-

1–81 Codd. *AaBCcDdeFfGgHhIJKkLlMm NnOoPpQqRrSsTtUuVvWwXxYyZz*ΓΛ ΠΣΨ
2 magister: primum *add.* β
4–5 Fitque...quod: sed ut patet medium namque naturam participans est extremorum sic quoque *f*
5 mathematicorum α/communiter: quodammodo β similiter *f* quodammodo *add.* *Fp*Σ
7 certitudinem δ/cedat: uendicet *V*
9 conceptibilibus: conceptionibus βφ in *add.* *AaOpqR*Ψ
9–10 intellectu...se: qui intellectui sunt primo *f*
11 certissime *om.* *aBdghkORrStz*Γ
14 nominatur: nuncupatur *Bo* uocatur *J*
17 que sunt genera et que species δ

II

[Prologue]

The foremost master of philosophy divides the province of that [subject] into three primary genera; the first of these he names theological, the second mathematical, and the third natural.[1] And the middle term becomes in a way a partaker in the nature of the two extreme terms,[2] because mathematical principles are found in the realms of nature and theology alike, and because it ranks below the first and above the third in nobility of subject matter, although both of them yield place to it with respect to certainty of the method of teaching;[3] this is the reason, moreover, why it is called, by a transfer of epithet,[4] "the teaching genus," on the grounds that it possesses a method of teaching which the student cannot contradict.[5] For it begins with things which are grasped by the intellect, namely, things self-evident to all men, and from these it deduces, by an infallible process, the first demonstrables, then the middle ones, then the last, proceeding from first to last through the middle ones in their due order.

This whole genus deals with the phenomena peculiar to quantity, and, because it has four parts, it is called the "quadrivium":[6] the first two parts are concerned with the genus of discrete quantity, the second two with continuous quantity. The first, which we call arithmetic, considers discrete quantity in an absolute sense and inquires what are the genera and species of numbers. The second, which is called music, applies discrete quantity to sounds, inquiring

---

1 de motibus planetarum *mg. a*; Theorica CAMPANI planetarum *mg. B*; Theorica CAMPANI ex integris *mg. C*; *tit.* Incipit theorica planetarum edita a magistro CAMPANO de ciuitate nouarie *d*; *tit.* Incipit theorica CAMPANI *I*; teorica CAMPANI *mg. K*; *tit.* Primum capitulum *k*; *tit.* Theorica magistri CAMPANI nouariensis *l*; *tit.* Incipit Theorica CAMPANI *n*; *tit.* CAMPANI theorica planetarum *o*; *tit.* Incipit opus CAMPANI de modo adequandi planetas siue de quantitatibus motuum celestium orbiumque proportionibus centrorumque distanciis ipsorumque corporum magnitudinis *rubric. Q*; de astrologia—id est theorica *mg. q*; *tit.* Incipit theorica CAMPANI Nouariensis et Compoxitio instrumenti equatorii planetarum *R*; *tit.* Incipit tractatus *r*; *tit.* Incipit theorica planetarum edita a magistro CAMPANO de ciuitate NOVARIE *St*; *tit.* capitulum declarans partes philosophie uidelicet quadruuium *s*; *tit.* Liber CAMPANI de aequationibus planetarum per instrumenta U_2; *tit.* Incipit theorica planetarum Magistri CAMPANI *Y*; *tit.* Theorica CAMPANI *y*; *tit.* Prohemium; incipit tractatus *z*; *tit.* Incipit theorica planetarum CAMPANI Γ; *tit.* Incipit equatio planetarum magistri CAMPANI Σ

2 ARIST. sexto methaphysica *mg. G*

sonantiarum species et que genera sunt melorum. Tertia geometria uide-
20 licet in continue quantitatis cognitione secundum quod est immobilis conuersatur. Sed quarta continue mobilis notitiam profitetur quam astronomiam idcirco nominant ab antiquo quia sperarum celestium orbium et stellarum motus et proportiones motuum inuestigat: cuius quanta sit nobilitas et quanta sit pulcritudo, ne fiat sermo concepta mensuratione pro-
25 lixior, omittamus. Hoc enim intuebitur facile qui considerabit ad ipsam quadriuiales ceteras ordinari tamquam ipsa sit finis et terminus earundem.

Hanc autem eximie nobilitatis scientiam antiqui professores ipsius in duo capita diuiserunt. Nam celestes motus et in se considerare possumus et ad inferiora, prout in ipsis dum irradiant influunt, retorquere, eritque
30 prior consideratio scientie demonstrantis sed alia iudicantis.

Ea quoque pars que demonstrationibus nititur rursus in suam theoricam suamque practicam est diuisa: et est sua theorica que singulorum motuum celestium quantitates orbiumque proportiones centrorumque distantias necnon et corporum magnitudines ceteraque talia per certissimas consi-
35 derationes tamquam per prima principia geometrie sillogizat. Sua uero practica est que prefatas conclusiones per conuenientes figuras geometricas demonstratas applicat operi eas propriis numeris arismetrice uestiendo, propter quod nullus potest aptus esse discipulus in hac arte nisi geometricis et arismetricis theoreumatibus primo fuerit informatus.

40 Hec autem practica quia est ualde perutilis, cum sit finis proximus demonstrantis et antecedens necessarium iudicanti, ad tabularum promptitudinem quo facilior inquirenti fieret a philosophis est redacta per quas etiam modicum litterati dummodo prompti sint in opere numerorum inuenire possunt de facili omnium planetarum ad omnia data tempora certa loca.
45 Verum quoniam in hoc opere oportet numeros plures adinuicem agregare aliosque ab aliis maioribus remouere, sepe etiam multiplicare sepeque

19 sunt: sint *fHlPsz*ΛΠ / melodorum (melodiorum, melodiarum) β melodie *kWw*
20 cognitione: argumentatione *f*
21 notitia perficietur β
21–22 astrologiam *BO*
22 sperarum: spatium *BChmy*Γ / celestium: et *add.* δ
24 quanta sit *om.* φ / concepta: de certa β deincepta φ*Rx* decepta Π
28 possimus *DGNZ*
30 iudicantis: mediantis φ
31 nititur: innititur βφ utitur *JPQq*
36 conclusiones: demonstrationes *f* accomodationes *p* considerationes *y*Λ
37 demonstratas: detractas *KV* / numeris: uel nominibus *K*
38 nisi: qui *add.* δ
38–39 geometricis: stigmatibus *add. f* stigmatibus id est demonstrationibus *add.* *y*Λ
39 primo *om.* α / informatus: eruditus φ
41 antecedat β antecedere *Lrz* antecedit *Oq* genus *o* / necessarium: notitiam β negotium *f* / iudicandi α iudicantis Λ indicandi *f* iudicatur *H*
42 inquirenti: intuenti β / redacta: ibi data *f*
43 inuenire: numerare α
45 plures: plurimos φ

what are the species of harmonies, and what the genera of tunes.[7] The third, namely, geometry, is concerned with the knowledge of continuous quantity insofar as it is without motion; whereas the fourth claims the knowledge of continuous quantity in motion; this they have from ancient times called astronomy because it investigates the motions of the celestial spheres and circles and of the stars, and also the proportions among these motions. Lest our account exceed a concise[8] length, let us omit to say how great is the nobility and beauty of this subject. For that will be easily perceived by anyone who considers that the other divisions of the quadrivium are ordered in relation to it as if it were their goal and end.

This science of great nobility was divided by those who professed it in ancient times into two headings:[9] for we may both consider the celestial motions by themselves and also relate them to earthly things, according to their influence on the latter as they cast their rays[10] on them; the first study belongs to a science of proof, but the second to a science of judgment.[11]

Again, that part which relies on proofs is divided in its turn into theoretical and practical parts:[12] its theoretical part is the one which deduces the quantities of the individual celestial motions and the relative sizes of the [heavenly] circles and the distances of their centers, as well as the sizes of the bodies and other such things, by infallible methods, for instance, by the elements of geometry.[13] Its practical part is that which applies the above conclusions, which have been demonstrated by suitable geometrical figures, to use by clothing them in the numbers which are peculiar to arithmetic; for that reason no one can be an apt student of this art unless he has first been instructed in the theorems of geometry and arithmetic.

However, since this practical part is very useful, being the immediate end of the science of proof and the necessary antecedent of the science of judgment, it has been reduced by philosophers to a handy form of tables to make it easier for the inquirer. By means of these tables even those with only a modest education, provided that they are handy in dealing with numbers, can easily find the exact positions of all the planets for any given time. But since it is necessary in this operation to add several numbers consecutively, and to subtract some numbers from other greater numbers, and also often to multiply and often to

diuidere, sed et in tabulas plures, nunc cum centris, nunc cum argumentis intrare eorumque equationes suscipere et quandoque per duplicem introitum adequare, rursusque per longitudinem longiorem et propiorem aut
50 circulum breuem et minuta proportionalia corrigere, fit opus istud non solum inexpertis difficile, sed etiam exercitatis et sapientibus tediosum. Distrahuntur igitur ab hoc opere plurimi, licet sint huius scientie nobilis amatores, quoniam uel indebiles uel aliis occupati tantorum anfractuum uarietatibus nequeunt se donare, sed annuales equationes quas almanach
55 uocant ab aliis emendicant, defectum sue uel occupationis uel ignorantie consolantes.

Vt igitur habeant omnes qui uel negotiorum occupatione uel experientie paruitate uel intellectus debilitate predictarum difficultatum examinis sunt expertes, quo circumscriptis prefatis scrupulosis uarietatibus numerorum
60 certa loca semper inueniant planetarum, et ea possint per instrumentum sensibile rotationi celesti similem motum efficiens uisibiliter intueri, studui tale materiale instrumentum operi prenominato conuenientissimum fabricare, quod, ni fallor, qui uiderit sueque modum operationis cognouerit et in eius pulcritudine pascet aspectum et in eius utilitate reficiet intellec-
65 tum. Volo igitur in hoc opusculo declarare qualiter instrumentum hoc componi debeat qualiterque per ipsum loca planetarum conueniat inuenire. Narrabo autem in quolibet planeta modum motuum et orbium et spere ipsius planete, secundum quod conueniet dictis PTHOLOMEI, et magnitudinem ipsius et distantiam eius a terra et proportiones orbium distan-
70 tiasque centrorum. Et explanabo etiam opera tabularum et docebo componere materiale instrumentum conueniens modis motuum eius et magnitudinibus orbium distantiisque centrorum. Deinde subiciam modum equandi planetam per ipsum instrumentum.

Erunt uero in ipso instrumento tres tabule rotunde, equales habentes
75 ambas superficies planas quantum possibile erit ad modum tabularum in quibus ponuntur regiones in astrolabio; et locabuntur omnes ab eadem

47 sed et: sed α / tabulis δ*A*Ψ / plures: pluries *ACdeFfHLlnOoPQrstuVz*ΠΣΨ plurimos *G*
48 equationem β
50 proportionis δ proportionum *P* proportionabilia *AI* / fit: sit δ sic *M*
53 indebiles: in ea (in eo *ew*Λ) debiles αφ*HZ* / aliis: negotiis *add. f*
55 ignorantie: ignauie δ
56 consonantes δ
57–59 uel negotiorum...numerorum: tantis difficilibus nequeant uacare *f*
59 expertes: inexperti *M*Γ
61 rotationi: rationi β ?rationem *p* locationi *fI*
61–62 studui: docebo *f*
62 prenotato βδ
63 ni fallor: infallor δ*qRu*₁*z*Γ ni fallar *V* / operationis: compositionis *V*
67 et[1] *om. Af*
68 planete: plane δ*A*
73 equandi: eundem *add.* α*H*
75 superficies: superiores facies *C*
76–77 ab...matre: alie eiusdem nature *f*

divide, and furthermore to enter several tables, at one time with the "centrum," at another with the "anomaly", and to take the equations corresponding to them and sometimes to interpolate by double entry,[14] and, again, to correct [the equation] by the "greatest distance" and the "least distance" or the "small circle" combined with the "minutes of proportion"[15]—this operation proves not only difficult for the untrained, but wearisome even for the trained and wise. Therefore very many are deterred from this operation, even though they may be lovers of this noble science, because either through weakness[16] or through occupation with other things they cannot give themselves over to the complications of such involved procedures; instead they beg from others the true places computed for a year, which they call an "almanac,"[17] and thus make up for the deficiency arising from their preoccupation or ignorance.

Therefore, in order that all who, either through occupation with affairs or lack of training or weakness of understanding, are unable to deal with the above-mentioned difficulties may have the means to find out the exact positions of the planets at all times, while avoiding the detailed numerical complications which I mentioned, and that they may be able to see [those positions] with their eyes by an instrument which is perceptible to the senses and which brings about a motion similar to the rotation of the heavenly bodies, I have striven to manufacture such a material instrument[18] most suitable for the above-mentioned purpose, and, if I am not mistaken, anyone who sees it and learns the mode of its operation will both feast his gaze on its beauty and delight his intellect with its utility. It is then my wish in this little work to make clear how this instrument should be constructed and how the positions of the planets may conveniently be found with it. I shall describe for each planet the particulars of the motions, of the circles, and of the sphere of that planet, following what PTOLEMY says, and its size and its distance from the earth and the proportions of its circles and the distances of their centers; I shall also explain how to use the tables, and I shall teach how to construct a material instrument appropriate to the particular nature of the [given planet's] motions, the sizes of its circles, and the distances of their centers; then I shall append the way to equate the planet by means of the instrument.

In the instrument there will be three circular boards[19] with both plane surfaces as level as possible, similar to the boards in which the horizons are placed in the astrolabe; all will be located in the same "mother," as also happens in

matre, sicut etiam in astrolabio contingit. In dorso autem matris ponam figuram solis, in una uero tabularum trium ponam figuram lune ex una parte et figuram mercurii ex alia; in secunda uero ponam figuram ueneris 80 et martis; in tertia autem figuram iouis et saturni. A solis ergo tractatu sumamus initium.

III

[De sole]

Sol habet unum circulum super cuius circumferentiam corpus eius semper equaliter mouetur omni die naturali 59 minuta et 8 secunda; unde linea recta exiens a centro eius uadens ad centrum corporis ipsius, cum 5 ymaginati fuerimus quod ipsa deferat uniformiter corpus solis per ipsius circumferentiam, describet super centrum ipsius equales angulos in equalibus temporibus et in eius circumferentia arcus equales. Dicitur autem circulus iste ecentricus eo quod centrum eius non est centrum orbis signorum sed distat ab eo uersus finem geminorum duabus partibus et medie- 10 tate partis fere de partibus illis de quibus semidiameter ipsius ecentrici habet 60 partes; unde linea que inter centrum orbis signorum et centrum circuli solis est si sumatur 24 uicibus erit equalis semidiametro circuli ecentrici. In quo ecentrico si protrahatur diameter que transeat per centrum orbis signorum erit punctus in ecentrico solis qui terminat hanc 15 diametrum sub fine geminorum maxime remotus a centro orbis signorum et ipse dicitur aux solis quod sonat solis eleuatio. Punctus uero qui terminat eandem diametrum in parte opposita, scilicet sub fine sagittarii, erit inter omnia puncta ecentrici solis maxime propinquus centro orbis sig-

77 contingit: conuenit *Ff*ΠΣ

1 *hic incipiunt codd. bij*
3 secunda: pertransiens *add.* δ
4 cum: si *F*Σ
7 equales: resecat *add.* φ
8 iste: ille solis φ *om.* α
17 eandem: hanc β/in: sub β

the astrolabe. On the back of the mother I shall place the figure of the sun; on one of the three boards I shall place the figure of the moon on one side and the figure of Mercury on the other; on the second I shall place the figure of Venus and the figure of Mars; on the third the figure of Jupiter and the figure of Saturn. Let us then begin with our treatment of the sun.

III

[Theory of the Sun]

The sun has a single circle on the circumference of which its body moves with a perpetually uniform motion of 59 minutes and 8 seconds[1] every natural day;[2] hence, if we picture that the straight line which proceeds from the center of that circle to the center of the sun's body carries the sun's body around its circumference with a uniform motion, it [the line] will describe in equal times equal angles with respect to its center and equal arcs on its circumference. This circle is called the "eccentric" because its center is not coincident with the center of the ecliptic but is removed from it in the direction of the end of Gemini[3] by almost two and a half of those parts of which the radius of the eccentric contains 60;[4] hence, if you take 24 times the line between the center of the ecliptic and the center of the sun's circle, it will be equal to the radius of the eccentric. If in this eccentric a diameter is drawn to pass through the center of the ecliptic, the point on the sun's eccentric which terminates that diameter toward the end of Gemini will be the point farthest from the center of the ecliptic, and this point is called the sun's "apogee," which means[5] the elevation of the sun.[6] The point which terminates the same diameter on the opposite side, i.e., toward the end of Sagittarius, will be the nearest of all points on the sun's eccentric to the

---

1 *tit.* De circulis et motibus solis Rubricha: capitulum secundum *d*; Hic incipiunt equatoria de ueris motibus septem planetarum ipsius CAMPANI et etiam aliqua communia extracta de theorica planetarum secundum ipsum super meridiem NOVARIENSEM *mg. j*; *tit.* Capitulum secundum de sole *k*; *tit.* theorica solis *o*; *tit.* De circulis et motibus solis capitulum *R*; *tit.* De circulis et motibus solis *S*; *tit.* Sequitur compositio spere solis *s*; *tit.* De circulis et motibus solis Rubrica *t*; *tit.* Theorica solis *Y*; *tit.* De motibus solis *Z*; *tit.* de sole *z*

norum et ipse dicitur oppositio augis solis quod sonat solis depressio.
20 Si itaque super centrum orbis signorum lineaueris duos circulos quorum unus supergrediatur punctum augis solis secundum quantitatem semidiametri corporis solis et alius subsistat puncto opposito augi secundum quantitatem eandem et tu ymaginatus fueris medietatem utriusque istorum duorum circulorum, manente fixa sua diametro, moueri a quouis situ 25 quousque redeat ad eundem, erit corpus quo spera descripta a maiori circulo excedit speram descriptam a minori spera solis. Et hoc est spatium inter uestigia duorum motuum predictorum que sunt due superficies sperice interceptum. Ista quoque spera, sicut et omnes alie, mouetur super polos orbis signorum secundum motum stellarum fixarum trahens secum 30 augem et oppositum augis qui est secundum PTHOLOMEVM omnibus 100 annis uno gradu ad successionem signorum, licet THEBITH posuerit eis motum accessionis et recessionis per motum in circulis paruis descriptis super capita arietis et libre.

Ecentricus autem iste est in eadem superficie cum orbe signorum diui- 35 dens latitudinem zodiaci in duo equalia. Inde fit ut sol semper sit in medio zodiaci equaliter distans ab extremitate septentrionali et meridionali; sed omnes alii planete quandoque declinant ab orbe signorum in partem septentrionis, quandoque in partem meridiei, quandoque etiam sunt in superficie orbis signorum, quoniam eorum ecentrici deferentes secant orbem 40 signorum super eius centrum in duobus locis qui dicuntur nodi: unus, nodus capitis et alius, nodus caude. Nodus autem capitis dicitur illa sectio in qua planeta incipit moueri uersus septentrionem ab orbe signorum; reliqua uero in qua incipit moueri uersus meridiem dicitur nodus caude. Declinat igitur una medietas omnium istorum ecentricorum uersus sep- 45 tentrionem ab orbe signorum et alia uersus meridiem. Et est ista declinatio in quibusdam maior, in quibusdam minor; sed in nullo est maior medietate latitudinis zodiaci que est 6 graduum.

Distantia autem centri solis a centro terre continet semidiametrum terre 1210 uicibus. Diameter uero corporis ipsius solis continet quinquies dia-

21–22 semidiametri: diametri *AQ*
24 circulorum *om.* δ
25 quousque redeat: usque δ
27 inter predicta uestigia β/predictorum *om.* β
30 augis: eius *B*/omnibus: in omnibus β
35 in duo: per duo δ
40 nodi: duo nodi δ*A*
41 capitis$_1$: draconis *add. jVy*
48 centri *om.* α

center of the ecliptic: it is called the sun's "perigee,"[7] which means the depression of the sun.

Thus, if you draw two circles on the center of the ecliptic, one of them passing outside the point of the sun's apogee by an amount equal to the radius of the sun's body, the other passing inside the point of its perigee by the same amount, and if you imagine that a half of each circle moves from any given position, its diameter remaining fixed, until it returns to the same position, [then] the body by which the sphere described by the [rotation of] the greater circle exceeds the sphere described by the [rotation of] the lesser circle will be the sphere of the sun.[8] This is the space intercepted between the traces[9] [i.e., loci] of the two motions I have described, which are two spherical surfaces. Moreover, this sphere, just like all other [spheres], rotates about [an axis whose ends are] the poles of the ecliptic, carrying with it the apogee[10] and perigee, with a motion [equal to that] of the fixed stars, which is, according to PTOLEMY, one degree in the direction of the succession of the signs every hundred years, though ṬĀBIT attributed to them a motion of progression and regression by means of a movement on small circles drawn around the initial points of Aries and Libra.[11]

This eccentric [of the sun] lies in the same plane as the ecliptic and divides the zodiac latitudinally into two equal halves.[12] Thus the sun is always in the middle of the zodiac, equally distant from its northern and southern limits; whereas all the other planets sometimes decline from the ecliptic toward the north, sometimes [decline] toward the south, and sometimes lie in the plane of the ecliptic, since their eccentric deferents intersect the ecliptic [along a line passing] through its center in two points which are called "nodes": one, the "head node," and the other, the "tail node."[13] The "head node" is the name of that intersection in which the planet begins to move from the ecliptic toward the north; the other [intersection], in which it begins to move toward the south, is called the "tail node." Therefore one half of each of those [planetary] eccentrics declines from the ecliptic toward the north and the other half [declines] toward the south. The [amount of the] declination is greater in some, less in others; but in none is it greater than half the width of the zodiac, which is 6 degrees.[14]

The distance between the center of the sun and the center of the earth is 1,210 times[15] the earth's radius. The diameter of the actual body of the sun is

---

19 depressio: et ita uulgat quod spera solis est concentrica mundo siue octaue spere sed orbis eius est ecentricus infra eam *add. f*

38 meridiei: et dicitur motus sectionis. istud autem commune est quia appellatur caput et cauda (draconis *L*) (illa sectio qua planeta incipit moueri uersus septentrionem ab orbe signorum et *rz*) *add. Lrz*

44–45 Dat caput ad boriam Draconis; caudam dat ad austrum *mg. D*

50 metrum terre et medietatem eius fere et corpus eius continet quantitatem corporis terre 166 uicibus. Quia uero spatium quod in superficie terre supponitur uni gradui celi continet 56 miliaria et 2 tertias unius miliaris, secundum quod miliare constat ex 4000 cubitis, erunt in circuitu terre 20400 miliaria. Huius enim probatio est satis euidens: sumpta nempe in 55 quauis regione altitudine poli, si sub orbe meridiei ambulaueris directe uersus septentrionem aut uersus meridiem quousque altitudo poli augeatur aut minuatur ab altitudine priori quantitate unius gradus, erit spatium inter duo loca sexagesima trecentesima pars circumferentie terreni ambitus, eo quod spera terre concentrica est spere celi, ipsumque repertum fuit 60 in tempore ALMEON a multis sapientibus qui ad huius rei probationem conuenerunt in quantitate predicta. Si itaque huius numeri sumamus secundam uigesimam partem et eam remoueamus ab ipso et de residuo sumamus tertiam, ipsa erit longitudo diametri terre in miliaribus, et ipsa est 6490 et 10 undecime unius miliaris. Erunt igitur a superficie terre usque ad 65 centrum corporis solis miliaria 3923754 et 6 undecime unius miliaris, et erit diameter corporis solis 35700 miliaria, ambitus uero rotunditatis eius erit miliaria 112200; et si subtraxerimus 17850, que est medietas diametri corporis solis in miliaribus, de 3923754 et 6 undecimis, que est distantia centri corporis solis a superficie terre in miliaribus, remanebit distantia 70 superficiei corporis solis a superficie terre et ipsa est miliaria 3905904 et 6 undecime unius miliaris.

Quia uero sol est maior terra quantitate predicta, oportet ut terra, illustrata a sole in parte una, in partem oppositam extendat umbram piramidalem, ita quod forma umbre terre sit forma piramidis rotunde cuius 75 basis est circulus descriptus in superficie terre diuidens partem terre illus-

51 uicibus: et 3 octauas eius *add. DJK*
59 concentrica sit αφ*H*
61 in quantitate predicta: quantitas predicta α predicte quantitatis φ scilicet quantitas predicta *H*Λ
63 longitudo: magnitudo β
67 subtraxeris αφ
69 superficie: centro αφ
71 miliaris: precise *add.* δ
72–73 illustrata: illuminata β
75–76 illustratam: illuminatam β

about five and a half times the diameter of the earth, and its volume is 166 times the amount of the earth's volume.[16] Because the distance which is subtended on the earth's surface by one degree in the heavens is $56\frac{2}{3}$ miles, where one mile is defined as containing 4,000 cubits,[17] the circumference of the earth will be 20,400 miles. The test of this [assertion] is obvious enough: namely, if you take the altitude of the pole at any spot on earth and walk along the meridian straight toward the north or south until the altitude of the pole is increased or decreased from its previous altitude by the amount of one degree, the space between the two places will be $\frac{1}{360}$ of the circumference of the earth's girth,[18] because the earth's sphere is concentric with the heavenly sphere; and the distance was found to be the above amount in the time of AL-MA'MŪN[19] by many wise men who assembled to examine this matter. Thus, if we take the twenty-second part of that number [20,400], and subtract that from the whole, and then take a third of what remains,[20] the result will be the length of the diameter of the earth in miles, and that is $6{,}490\frac{10}{11}$ miles.[21] Therefore from the earth's surface to the center of the sun's body will be $3{,}923{,}754\frac{6}{11}$ miles, and the diameter of the sun's body will be 35,700 miles, while the circumference of its rounded surface will be 112,200 miles.[22] And if we subtract 17,850, which is the radius of the sun's body in miles, from $3{,}923{,}754\frac{6}{11}$, which is the distance in miles of the center of the sun's body from the surface of the earth, the remainder will be the distance of the surface of the sun's body from the surface of the earth, and that is $3{,}905{,}904\frac{6}{11}$ miles.

Because the sun is greater than the earth by the aforesaid amount, it follows that the earth, which is illuminated by the sun on one side, must cast a cone-shaped[23] shadow on the opposite side, in such a way that the shape of the earth's shadow is that of a circular cone whose base is the circle described on the earth's surface dividing the illuminated part of the earth from the non-

---

51 uicibus: sicut (quod *o*) probatur 18 propositione quinti libri almagesti determinati (demonstratione *o*) *add. fOo* sicut potest 28ª propositione quinti libri almagesti demonstrari *add.* Λ

51–54 Ambitus terre 252000 stadia quia 700 stadia corespondet uno gradui in firmamento et continet 21000 miliares de miliari 12 stadiorum. Aliter circuitus terre est xx mille iiii centum miliaria profunditas terre vi mille v centum miliaria et in medio terre est infernus et usque ad medium inferni sunt iii mille cl miliaria anglicanica *mg. A* Quot miliaria terrae correspondant uni gradui in coelo ALBVMASAR in differentia prima tractatus quarti sui magni introductorii: dicit 87 miliaria cum dimidio. Nota quod cubitum constat ex quatuor palmis; palmus ex quatuor polis; polus uero ex sex granis ordei secundum latitudinem *mg. G* Alfraganus differentia prima tractatus quarti magni sui introductorii dicit 87 miliaria cum secundis et hoc est uerum prout miliare constat ex 8 stadiis sed Campanus intendit prout miliare constat ex 12 stadiis *mg. w*

53 cubitis: et est uerior lectura (litera *BCJ*Γ) sicut patet per (in *J*) ALFRAGANVM differentia octaua secundum (secundum *om. C*) quod miliare constat ex 4000 (40 *B* 20400 *J*) cubitis *add. BCJM*Γ

tratam a parte eius non illustrata. Conus uero piramidis istius est punctus in etherea regione in quo concurrunt radii circumferentiales a corpore solis egressi terram circumferentialiter contingentes, id est punctus in quo concurrunt radii corpus solis et corpus terre communiter contingentes. Sicut 80 autem a terra que corpus est spericum illustrata a sole fit umbra piramidalis, sic a quolibet 6 planetarum reliquorum illustrato a sole, cum quilibet eorum sit corpus spericum, fit umbra piramidalis. Plura autem corpora non sunt citra speram stellarum fixarum que possunt sistere solarem radium et umbram efficere. Quare sequitur ut quidquid sit extra istas 7 85 piramides umbrosas erit in luce solis. Conus autem umbre terre distat a superficie ipsius per quantitatem semidiametri terre 267 uicibus sumptam. Erunt itaque a superficie terre usque ad extremitatem umbre ipsius que est conus piramidis umbre 866536 miliaria et 4 undecime unius miliaris.

Ad hoc autem ut in instrumento nostro sensibili ponantur ea que ad 90 solis equationem sunt necessaria describam primo orbem signorum in dorso matris super centrum *D* et diuidam ipsum in signa 12 que sunt aries taurus etc., et unumquodque signum in 30 gradus, ita scilicet quod distinguam eos in uno circulo singulos et singulos et in alio colligam eos quinos et quinos quod faciam hoc modo. Describam primo super centrum *D* 95 circulum unum contingentem fere extremitatem matris. Postea restringam circinum et super idem centrum describam alium circulum tantum distantem a primo quod inter ipsos possint comode nomina signorum scribi. Itemque restringam circinum minus tamen quam primo et super illud idem centrum lineabo circulum alium tantum solummodo distantem a secundo 100 quod inter ipsos possint cadere singulares distinctiones graduum. Rursus iterum tertio restringam circinum et faciam super idem centrum circulum quartum tantum distantem a tertio quantum distat secundus a primo ut inter eos possit scribi numerus graduum distinctorum per 5 et 5. Postea diuidam circulum exteriorem in 12 partes equales et ponam regulam super 105 centrum *D* et super quamlibet illarum diuisionum et continuabo circulum exteriorem istorum 4 cum interiore eorum per lineam rectam que non transeat interiorem. Deinde inter duos circulos exteriores scribam nomina signorum incipiendo ab ariete quem scribam in prima predictarum 12 diuisionum que erit in dextra parte et procedam scribendo uersus sinistram 110 ponendo scilicet taurum in secunda et geminos in tertia et sic de ceteris. Post ista diuidam spatium quod est inter secundum circulum et tertium

76 illustrata: illuminata β
77 corpus solis βδ corporis solis θφ
85 erit: sit β
89 in hoc instrumento αφ
96 describam: protraham δ scribam *f*
105 illarum: 12 *add.* α 22 *add.* *Q*
107 scribam: describam δ
108 incipiendo: scilicet *add.* θ*H*Γ
110 de ceteris: deinceps β

illuminated part. The apex of that cone is the point in the upper air at which the rays proceeding from the circumference of the sun's body and tangent to the circumference of the earth meet each other, namely, the point at which the rays that are common tangents to the body of the sun and the body of the earth all meet. And just as the earth, which is a spherical body, forms a cone-shaped shadow as it is illuminated by the sun, so each of the other six planets illuminated by the sun, since each is a spherical body, forms a cone-shaped shadow. However, there are no more bodies on this side of the sphere of the fixed stars which could intercept the solar rays and form a shadow. Hence it follows that whatever is outside those seven shadowy cones will be in the light of the sun. The apex of the earth's shadow is distant from the earth's surface by an amount equivalent to 267 times[24] the earth's radius. Thus from the earth's surface to the farthest point of its shadow, which is the apex of the cone of the shadow, will be 866,536$\frac{4}{11}$ miles.[25]

Now in order to put on our material[26] instrument those things necessary for equating the sun [see Fig. 10],[27] I first, with center *D*, draw the ecliptic on the back of the "mother" and divide it into the 12 signs, namely, Aries, Taurus, etc., and [divide] each sign into 30 degrees, in such a manner that on one circle I mark every degree, while on another I mark them in groups of five. I do this as follows: first I draw, on center *D*, a circle almost touching the edge of the mother. Then I pull in one leg of my compasses and on the same center draw another circle far enough from the first that the names of the signs can conveniently be written between them. I pull in the leg of my compasses again, but less than the first time, and on that same center draw another circle, just enough distant from the second for the marks of the individual degrees to be put between them. Then again for the third time I pull in the leg of my compasses and make on the same center a fourth circle as far distant from the third as the second is from the first, so that the numbers of the degrees which are marked off in groups of five can be written between them. Next I divide the outer circle into 12 equal parts, and, placing my ruler on center *D* and each of those divisions [in turn], I draw a straight line joining the outer of those four circles to the inner, but not passing beyond the inner one. Then I write the names of the signs between the two outer circles, beginning with Aries, which I write in the first of the aforementioned 12 divisions on the right-hand side, and working toward the left, that is, putting Taurus in the second division, Gemini in the third, and so forth for the others. After that I divide the space under each sign

---

89 *tit.* Hic docet facere instrumentum *A* ad componendum figuram solis *o* De compositione instrumenti *R*/Nota compositionem instrumenti equandi solis *mg. D*/ descriptio *mg. T*

sub quolibet signo per 30 partes equales que erunt gradus quod faciam hoc modo. Primo enim diuidam spatium illud in 6 partes equales et ponam regulam super quamlibet illarum et super centrum *D* et protraham lineam
115 rectam a secundo circulo usque ad quartum et ita diuidam spatium quod unicuique signo supponitur in tribus circulis interioribus in 6 partes. Spatium uero quod est inter quamlibet illarum 6 partium inter duos circulos medios diuidam in 5 partes equales eritque spatium inter istos duos circulos medios in circuitu diuisum in 360 partes. Post hoc scribam in 6 di-
120 mensionibus que sunt inter duos circulos intimos scilicet tertium et quartum sub quolibet signo: in prima quidem illarum 6 dimensionum 5; et in secunda 10; et in tertia 15; et in quarta 20; et in quinta 25; et in sexta 30. Et hec omnia potes manifeste uidere in 4 circulis exterioribus sequentis figure. Postea continuabo punctum ubi terminantur 17 gradus et 50 minu-
125 ta geminorum cum centro *D* per lineam *AD* sitque *A* in circumferentia interioris circuli 4 predictorum. Ex parte ergo puncti *A* resecabo de linea *AD* portiunculam aliquam ad libitum tantam scilicet quantum uolo distare circulum deferentem solem ab orbe signorum sitque portiuncula illa *AB*. Residuam autem uidelicet lineam *DB* diuidam in 5 partes equales et quam-
130 libet illarum in alias 5, ita quod linea *DB* sit diuisa in 25 partes equales quarum una sit *DC*. Ponam itaque centrum super punctum *C* et lineabo super ipsum secundum spatium *CB* circulum *BE* qui erit circulus solis ecentricus quem diuidam incipiendo a puncto *B* in 12 partes equales, et restringam circinum aliquantulum et lineabo iterum alium circulum super
135 centrum *C* minorem primo, ita quod inter eos ponatur diuisio singulorum graduum. Rursus restringam circinum parum plus quam prius et faciam alium circulum super idem centrum tantum distantem a secundo quod inter ipsos possint poni dimensiones graduum per 5 et 5 cum inscriptione suorum numerorum quemadmodum factum est prius in diuisione orbis
140 signorum. Iterum quoque restringam circinum et faciam quartum circulum super idem centrum *C* tantum distantem a tertio quod inter eos possit scribi numerus signorum. Circulis hiis descriptis ponam regulam super centrum *C* et super singulas 12 diuisiones circuli primi istorum 4 qui est

112 per: in β
116 partes: equales *add.* *BFWw*Σ
117 quamlibet: quaslibet *H*
117–18 inter[2]...medios *om.* φ
119 partes: equales *add.* β
119–20 dimensionibus: diuisionibus αφ*H*
120 scilicet: inter *add.* δ
124 terminatur δ/17: 27 *fkMmSVX*Γ
125 lineam: obscuram *add.* Π
127 portiunculam: particulam β partiuncula *F* partiunculam Σ
128 solis αφ
129 lineam: particulam β portiunculam *adip*Ψ
131 centrum: et circinum *add.* *o* circinum *mg.* *G*
135 diuisio: circulorum *add.* δ
136 Rursus: iterum *add.* δ

between the second and third circles into 30 equal parts, which will be degrees. This I do in the following manner: first I divide that space into six equal parts and, placing my ruler on center *D* and each [division in turn], I draw a straight line from the second circle to the fourth and thus divide the space under each sign in the three inner circles into six parts. I divide the space that is under each[28] of these six parts [and] between the two middle circles into five equal parts, and [thus] the space between those two middle circles will be divided into 360 parts around the whole circumference. After that I write in the six divisions which are under each sign between the two inner circles, namely, the third and fourth: in the first of those six divisions [I write] "5," in the second "10," in the third "15," in the fourth "20," in the fifth "25," and in the sixth "30." You may see all this clearly in the four outer circles of the subjoined figure [Fig. 10, p. 152]. Then I join the point marking 17 degrees and 50 minutes of Gemini[29] to center *D* by the line *AD*, where *A* is on the circumference of the innermost of the four above-mentioned circles. From point *A* I mark off a small section of the line *AD* to suit myself, as much, that is, as I want the distance of the circle carrying the sun to be from the ecliptic; let that small section be *AB*. The remainder, i.e., the line *DB*, I divide into five equal parts, and each of these again into five, so that the line *DB* is divided into 25 equal parts; let one of these be *DC*. So I put my center on point *C* and with radius *CB* draw around it a circle *BE* which will be the sun's eccentric; this I divide into 12 equal parts, beginning from point *B*; then I pull in the leg of my compasses a little and again draw another circle on center *C*, smaller than the first, so that the division into single degrees may be put between them. Once again I pull in the leg of my compasses, a little more than before, and make another circle on the same center, far enough distant from the second that the five-degree divisions, with the numbers written by them, can be put between them, as was previously done in the dividing of the ecliptic. Yet again I pull in the leg of my compasses and draw a fourth circle on the same center *C*, far enough distant from the third that the numbers of the signs[30] may be written between them. Having drawn these circles, I place my ruler on center *C* and on each [in turn] of the 12 divi-

---

116 signo: Hic nichil deficit sed spera uel instrumentum solis habet huius locum sed quia in papiro non poterat stare propterea posui in franceno *add. j*
124 Hic debet poni figura *mg. A*
124–25 17 gradus et 50 minuta geminorum: in alia super 29 graduum *mg. Q* Nota locum augis solis tempore nouariense *mg. V* errat ut credo quia tempore CAMPANI non erat ibi aux scilicet anno 1262 forsan uoluit dicere 27 gradus *mg.* H_2

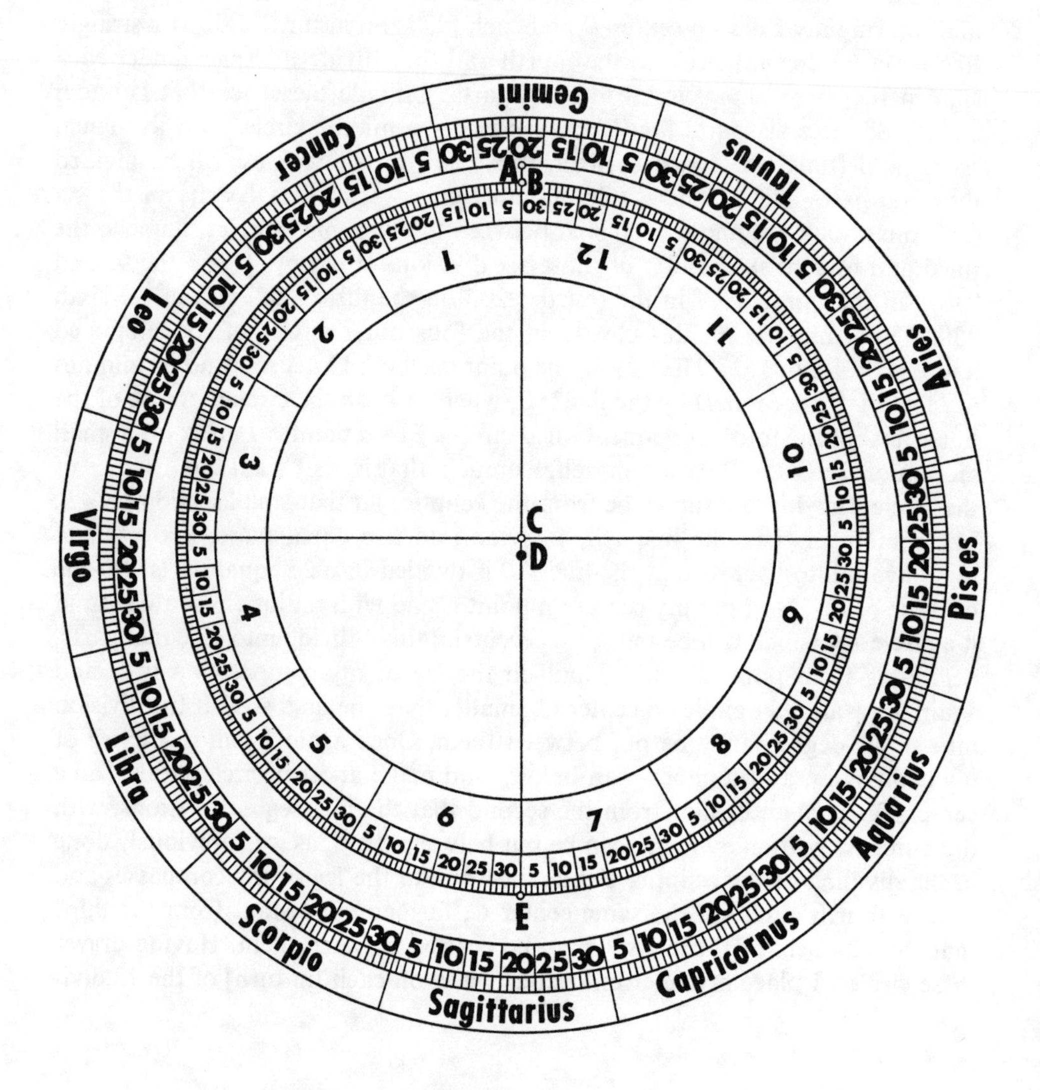

INSTRVMENTVM EQVATIONIS SOLIS

Fig. 10

Legend for Fig. 10

Instrument for the sun (after Campanus). The outer graduated system of circles represents the ecliptic (with center *D*, the earth). Point *A* is the sun's apogee, in Gemini 17;50°. The inner graduated system of circles, with center *C*, represents the sun's eccentric. Both systems are divided into twelve signs, which are subdivided successively into groups of five degrees and single degrees. To find the sun's true position, a thread, attached by one end to point *D*, is stretched through the point on circle *BE* marking the sun's anomaly (expressed in signs, degrees, and minutes); the point on the ecliptic through which the thread passes is the sun's true longitude. Compare Figure 3 (p. 42) and Plate 1.

circulus *BE* et continuabo inter ipsum et interiorem per lineam rectam.
145 Item spatium quod est in qualibet istarum 12 diuisionum inter primum circulum et secundum diuidam in 30 partes equales continuando quamlibet quintam diuisionem cum circulo tertio per lineam rectam respicientem punctum *C*. Quibus ita dispositis in qualibet 12 diuisionum que sunt inter duos circulos intimos scribam proprium numerum incipiendo scilicet
150 ab ea que sequitur punctum *B* uersus sinistram et in ea scribam 1. Deinde in alia que hanc sequitur uersus sinistram scribam 2, in alia quoque 3; et ita de ceteris usque ad ultimam in qua scribam 12 que omnia clare uidere potes in interiori parte proposite figure. Post hec figam clauum unum habentem hastam tenuissimam perpendiculariter super punctum *D* quod
155 positum est esse centrum orbis signorum et ligabo ad pedem eius filum unum de seta tenuissimum et equale in omni sui parte.

Cum igitur uoluero equare solem per illud instrumentum intrabo cum tempore ad quod uolo ipsum equare quod uoco tempus datum in tabulam argumenti solis. Est autem argumentum solis distantia eius ab auge sui
160 deferentis; et queram ipsum in tabula annorum collectorum; et si ipsum inuenero ibi, sumam quod in directo eius inuenero de signis gradibus atque minutis et scribam illud seorsum eo ordine quo ponitur in tabula et ipsum est argumentum solis. Si autem non inuenero ipsum, sumam minorem eo propinquiorem tamen et accipiam quod in directo eius inuenero de
165 signis gradibus atque minutis et scribam illud seorsum eo ordine quo ibi ponitur. Postea intrabo cum annis residuis in tabulam annorum expansorum, et quod in directo eorum fuerit de signis gradibus atque minutis sumam, et sub prioribus scribam, signa uidelicet sub signis, gradus sub gradibus, et minuta sub minutis. Intrabo quoque postea cum mensibus in
170 tabulam mensium si in tempore dato fuerint menses; et cum diebus in tabulam dierum si fuerint in eo dies; cum horis quoque in tabulam horarum si fuerint in eo hore; et sumam similiter quod in eorum directo inuenero et scribam extra sub aliis unumquodque scilicet sub suo genere. Deinde colligam omnia simul incipiens a minutis ex quorum omnium agregatione,
175 si resultet numerus minor 60, illum numerum scribam super omnia minuta agregata, si autem 60, scribam 0 super omnia minuta, et pro illis 60 minutis addam unitatem gradibus antecedentibus. Si autem maior 60, quod erit

144 inter *om.* βφ
145 in qualibet: inter quamlibet θ
152 de ceteris: deinceps β
156 seta: sera *Bi* serico *AahJMwy*ΓΛ serica *j*
158 tabula βδ
166 tabula δ
170 tabula δ
174–91 ex...mutant: etc. *B*
175 resultat *adfjKkpRt*Λ
176 60[1] *om. AeFfglOopqx*/si autem...minuta, et: et si resultent 60 δ et *Hz*/illis *om.* αφ
176–77 si...antecedentibus *om. CM* Si autem sint 60 scribam cifrum super omnia aggregata et pro 60 illis minutis addam unum gradum sequentibus *J*

sions of the first of the four circles, namely, circle *BE*, and join it to the inner circle by a straight line. Again, I divide the space in each of those 12 divisions between the first circle and the second into 30 equal parts, producing every fifth division in a straight line toward *C* as far as the third circle. These dispositions completed, I write the appropriate number in each of the 12 divisions between the two innermost circles, beginning with that division which extends from point *B* toward the left: in that I write "1." Then in the [division] next to this toward the left I write "2," in the next "3," and so forth for the rest up to the last, in which I write "12." All this may be clearly seen in the inner portion of the figure I have set out. After that I affix a nail with a very thin spike perpendicularly upon point *D*, which was defined as the center of the ecliptic; at the base of the nail I attach a thread made of silk,[31] very fine and of equal thickness in all its length.

Therefore, when I want to equate the sun by this instrument,[32] I enter the table of the sun's anomaly with the time at which I want to equate it, which I call the "datum time."[33] The sun's anomaly is its distance[34] from the apogee of its eccentric.[35] I look for it in the table of "collected years," and, if I find the actual [number of the datum year] there, I take the amount in signs, degrees, and minutes which I find opposite that number and write it down separately in the same order in which it is set down in the table; and that is the sun's anomaly. But if I cannot find the actual number, I take the number which is less than, but nearest to it, and take the amount in signs, degrees, and minutes which I find opposite it and write that down separately in the order in which it is set down there. Then I enter the table of "individual years" with the [number of] remaining years[36] and take the amount I find opposite it in signs, degrees, and minutes and write it down under the first amount, that is, putting the signs under the signs, the degrees under the degrees, and the minutes under the minutes. Then I enter the table of months with the [number of] months, if there are any months in the datum time, and the table of days with the days, if there are any days in it, and also the table of hours with the hours, if there are any hours. And in the same way [as before] I take the amounts I find opposite them and write them down separately underneath the other amounts, that is, putting each under its own kind [of unit]. Then I add them all up, beginning with the minutes: if the result of adding up all the minutes is a number less than 60, I write down that number above[37] all the minutes I have added; if, however, [the result is exactly] 60, I write 0 above all the minutes, and add one degree for those 60 minutes to the degrees to the left of it; but if the result is greater than 60, I write the number by which it exceeds 60 above

---

156 in omni sui parte: dictorum (punctorum *B*) autem omnium figura que sequitur si iuste descripta fuerit (et perfecte diuisa *add. m*) erit (est *MΓ*) lucidissima demonstratrix (declaratrix *MΓ*) β

ultra 60 scribam super omnia minuta et pro 60 addam unitatem gradibus antecedentibus ut prius. Et ut breuius dicam in agregatione minutorum
180 pro quibuslibet 60 addenda est unitas gradibus, quod autem ultra 60 relinquitur est supra minuta scribendum. Postea uero agregabo gradus adinuicem superadditis eis tot unitatibus quotiens in agregatione minutorum superfuerint 60, si aliquotiens superfuerint, et quod ex eorum agregatione minus 30 uel ultra 30 prouenerit, super omnes gradus scribam, uel 0 si 30
185 uel eius multiplex prouenerit; pro quibuslibet uero 30 unam unitatem signis apponam. Colligam quoque postea signa in unum superadditis eis tot unitatibus quot ex tricenariis graduum prouenerint et quod ex ista agregatione prouenerit minus 12 aut plus 12 uel plus quolibet eius multiplice scribam supra omnia signa. 12 aut quodlibet eius multiplex quotiens mihi
190 euenerit abiciam. Nam ipsum numerat reuolutiones integras que situm solis uel alterius cuiusuis planete in orbe suo non mutant.

Cum igitur omnia hec fecero, signa gradus et minuta ex prefata agregatione prouenientia erunt argumentum solis ad propositum tempus. Queram itaque hoc argumentum in circulo solis ecentrico; et ubi numerus signo-
195 rum et graduum et minutorum terminabitur, faciam notam uel materialiter uel solum in mente; et tunc accipiam filum quod est ligatum ad punctum *D* et faciam ipsum transire per notam predictam et considerabo ubi terminetur exterius in circulo signorum. Nam in gradu illo et minuto illius gradus erit sol secundum eius uerum motum ad horam propositam.
200 Verum autem motum in omni planeta generaliter uoco punctum celi ubi terminatur linea recta egressa a centro terre transiens per centrum corporis ipsius planete.

Per hoc autem instrumentum patet ratio operis tabularum in sole. Sumunt tabularii primo medium motum eius. Medium autem motum desig-
205 nat in sole linea ducta a centro orbis signorum usque ad firmamentum equedistanter linee ducte a centro sui ecentrici per centrum corporis eius. Hec enim linea semper mouetur equaliter in orbe signorum quemadmodum ducta a centro ecentrici mouetur equaliter in ecentrico. Nam cum ipse sint equedistantes, necesse est ut super centra a quibus egrediuntur descri-

179 Et ut breuius dicam: Breuiter (similiter *f*) dico quod αφ*H*
182 tot unitatibus: totiens unum Λ
184–85 super...prouenerit: aut plus quolibet eius multiplice super omnes gradus scribam. Si autem 30 scribam 0 *F*
185 quibuslibet uero 30: quolibet uno δ quolibet uero *Z*
189 mihi: inde δ
190 abiciatur δ
191 mutant: immutant δ monstrant *f* immutauit *el* minutant *Lz*
194 ecentrico: secundo assignato in dicto instrumento *add.* β
198 exterius *om.* β
199 motum: cursum αφ
203 operationis β
204 tabularii: enim δ enim tabule *AfJ* enim tabularum Ψ/motum²: eius *add.* αφ

all the minutes and I add one degree for the 60 minutes to the degrees to the left as before. To put the matter briefly, in summing the minutes, one must add one to the degrees for every 60 minutes and write down the surplus beyond 60 above the minutes. Next I sum up the degrees in their turn, adding to them, besides, as many single degrees as there were 60s when the minutes were added up, that is, if there were any 60s at all; when those degrees have been added up, above all the degrees I write the result if it is less than 30, or the amount by which it exceeds 30 [if it is more than 30], or 0 if it is [exactly] 30 or a multiple of 30; for every 30 I add one to the signs. Then I sum all the signs, too, adding, besides, as many single signs as there were sets of 30 degrees. I write above all the signs the result of that addition if it is less than 12, or the number by which it exceeds 12 or any multiple of 12 [if it is more than 12]. The 12, or any multiple of it which results, I neglect, for it tells the number of complete revolutions, which do not make any difference to the position of the sun or any other planet in relation to its own orbit.

Therefore, when I have done all this, the signs, degrees, and minutes which result from the above summation will be the anomaly of the sun at the given time. So I look for this anomaly on the eccentric circle of the sun, and at the point to which the number of signs, degrees, and minutes brings me, I make a mark, either on the actual instrument or merely in my memory; then I take the thread attached to point *D* and make it pass through the above mark and note where it cuts the ecliptic on the outer part [of the instrument]. For in that degree and minute will be the place of the sun at the given time according to its true motion. I define "true motion" for every planet in general as that point of the heavens where the straight line proceeding from the center of the earth and passing through the center of the planet meets [the ecliptic].

Now through this instrument the reasoning behind the mode of operation with tables for the sun becomes clear.[38] First the table-makers take its mean motion. The mean motion of the sun is determined by the line drawn from the center of the ecliptic to the firmament[39] parallel to the line drawn from the center of the sun's eccentric through the sun's body. For that line always moves with uniform motion in the ecliptic, just as [the line] drawn from the center of the eccentric moves with uniform motion in the eccentric. For, since [the two lines] are parallel, they must necessarily describe equal angles in equal times about the centers from which they are drawn, and therefore [they must describe

193 erunt: medius motus solis a quo subtraham augem solis que est 2 signa 18 gradus 50 minuta et residuum erit *add. T*

210 bant equales angulos in temporibus equalibus, ideoque in circumferentiis circulorum quorum sunt illa centra arcus equales. Quia ergo linea que egreditur a centro ecentrici describit in equalibus temporibus super centrum eius equales angulos et in circumferentia eius arcus equales, necesse est quoque ut linea ducta a centro orbis signorum equedistanter ei descri-
215 bat super eius centrum in temporibus equalibus equales angulos et in circumferentia ipsius arcus equales. Distantia igitur huius linee ab ariete dicitur medius motus solis. Ab ista distantia subtrahunt augem solis que dicitur distantia augis ab ariete et relinquitur distantia solis ab auge sua que uocatur argumentum ut dictum est prius. Vbicumque igitur fuerit sol
220 in ecentrico ducatur etiam linea a centro terre per centrum corporis eius indicabitque hec linea uerum motum solis. Ideoque distantia eius ab ariete dicitur uerus motus. Distantia autem linee medii motus a linea ueri motus dicitur equatio solis que nulla esset si eadem esset linea medii motus solis et ueri quemadmodum accidit sole existente in auge uel in eius oppositione.
225 Vbique autem alibi quia sunt diuerse, aliqua est solis equatio uariaturque secundum solis distantiam ab auge ideoque cum ea inuenitur. Quia uero in medietate ecentrici que est ab auge usque ad eius oppositionem linea medii motus precedit lineam ueri, in alia uero medietate sequitur eam, ideo in medietate priori medius motus est maior uero, et in secunda minor
230 secundum quantitatem equationis. Quapropter si argumentum solis est minus 6 signis quod est in medietate prima, subtrahimus equationem de medio motu; si idem argumentum est maius 6 signis quod est in medietate secunda, equationem inuentam addimus medio motui solis et habetur uerus.

211 centra: resecant *add. f*Λ describunt *add. N*   219 que: quod 80

in equal times] equal arcs on the circumferences of the circles of which those are the centers. Then, since the line which proceeds from the center of the eccentric describes in equal times equal angles with respect to its center and equal arcs on its circumference, the line drawn parallel to it from the center of the ecliptic must also necessarily describe in equal times equal angles with respect to its center and equal arcs on its circumference. Therefore, the distance of this line from [the beginning of] Aries is called the mean motion of the sun. From that distance they [the table-makers] subtract the sun's "apogee," which is the name given to the distance of the apogee from [the beginning of] Aries; the result is the distance of the sun from its apogee, which is called the anomaly, as stated before. Thus, wherever the sun may be in its eccentric, let a line be drawn from the center of the earth through the center of its [the sun's] body: this line will show the true motion of the sun; and so the distance of the line from [the beginning of] Aries is called the true motion. The distance between the line of mean motion and the line of true motion is called the "equation" of the sun. This equation would be zero if the line of mean motion of the sun and the line of its true motion were one and the same, as occurs when the sun is in its apogee or perigee. But for every other position of the sun, since the lines are different from each other, the equation of the sun has some quantity; it varies according to the distance of the sun from its apogee and, therefore, is found from that distance. Now because in that half of the [sun's] eccentric which extends from the apogee to the perigee the line of mean motion is ahead of the line of true motion, while in the other half the former is behind the latter, it follows that in the first half the mean motion is greater than the true, and in the second half less than the true, by the amount of the equation. Therefore, if the anomaly of the sun is less than six signs, as it is in the first half, we subtract the equation from the mean motion; if that anomaly is greater than six signs, as it is in the second half, we add the equation, when we have found it, to the mean motion of the sun: thus we get the true [motion].

---

234 uerus: Et nota quod operatione tabularum fit hec additio uel subtractio equationis solis (solis *om. o*) propter hoc quod ibi accipitur argumentum in zodiaco et certum est quod non semper est sol *secundum uerum eius motum* (ubi terminatur medius motus [eius *o*] *Jo*Λ) in zodiaco sed in operatione per illud instrumentum non fit hec additio uel subtractio equationis solis quia hic accipitur argumentum *solis in circulo* (in solis *J* in circulo solis Λ) ecentrico ut (ut *om. o*) *in figura precedente* (*om. o* prius *J* in pagina precedenti Λ) certum autem est quod ubi terminatur (ubique *f*) uerum (uerum *om. Jo*Λ) argumentum sumptum in ecentrico (quod *o*) ibi etiam (etiam *om. o*Λ) est sol (sol *om.* Λ) secundum uerum eius motum *fJo*Λ

IV

[Incipit tractatus de luna]

Luna quoad completam suorum motuum cognitionem 5 habet circulos. Habet enim epiciclum in cuius circumferentia mouetur corpus lune pertransiens de eo omni die naturali 13 gradus et 4 minuta fere et mouetur in
5 superiori quidem parte uersus occidentem et in inferiori uersus orientem. Et habet circulum ecentricum deferentem qui declinat a superficie orbis signorum in una sui medietate uersus septentrionem et in alia uersus meridiem. Secat enim iste circulus deferens circulum signorum et secatur ab eodem in duobus locis qui dicuntur capud draconis et cauda. Dicitur au-
10 tem capud draconis locus illius sectionis quam sequitur maxima declinatio septentrionalis; cauda uero dicitur locus illius sectionis quam sequitur maxima declinatio meridiana; differentia uero communis istius sectionis est linea recta transiens per centrum orbis signorum et per utramque duarum sectionum predictarum que dicuntur duo nodi. Fit enim huius
15 sectio super diametrum orbis signorum. Hec autem declinatio fixa est quantum ad sue magnitudinis figuram, quantum uero ad partes deferentis et quantum ad partes orbis signorum continue uariatur. Partes nempe deferentis prout inuicem sibi succedunt uersus orientem, sic fiunt successiue in utrolibet duorum nodorum et in utralibet duarum maximarum
20 declinationum. Vterque quoque duorum nodorum et utraque duarum

2 quoad: uero ad β
4 omni: in omni θ*Hz* in δ
9 draconis *om.* β
10 quam: quem *BcfgHhMWw*
15–17 Hec...signorum *om.* Π/autem...signorum *om.* *fu*Σ
16–17 quantum uero...signorum *om.* *oy*Λ
20–21 Vterque...declinationum *om.* β

IV

[Theory of the Moon]

The moon has five circles[1] [which one has to take into account] for complete knowledge of its movements. It has an epicycle on the circumference of which the body of the moon moves, traversing about 13 degrees and 4 minutes of that circumference every natural day;[2] it moves toward the west in the upper part [of the epicycle] and toward the east in the lower part. And it has an eccentric deferent circle which is inclined to the plane of the ecliptic toward the north in one half and toward the south in the other [half]. For that deferent circle intersects the circle of the ecliptic and is intersected by it in two places which are called the "head of the dragon" and the "tail [of the dragon]."[3] The name "head of the dragon" is given to that place on the intersection which precedes the greatest northern declination, while the name "tail [of the dragon]" is given to that place on the intersection which precedes the greatest southern declination.[4] The common section of these intersecting [circles] is the straight line passing through the center of the ecliptic and through both of the aforesaid two points of intersection, which are called the two nodes. For the intersection of this [deferent circle] takes place along the diameter of the ecliptic. Now this declination is invariable with respect to its amount, but varies continuously with respect to the parts of the deferent and the parts of the ecliptic [in which it is found]. For as one part of the deferent succeeds another coming from the east,[5] each point in turn becomes each of the two nodes and each of the two maximum declinations. Moreover, each of the two nodes and each of the two

---

1 Capitulum 2 de motu lune et figura eius *mg. A*; *tit.* Incipit tractatus de luna *aRStz*; *tit.* Incipit tractatus de luna et centera: capitulum sextum *d*; *tit.* Incipit tractatus de luna et primo de quinque circulis ex quibus tota sua theoricha componi intelligitur *g*; *tit.* De equatorio et theorica lune *j*; *tit.* Tractatus secundus de luna capitulum primum *k*; *tit.* De theorica lune et eius instrumenti compositione *l*; *tit.* sequitur secunda pars principalis de theorica lune etc. *mg. M*; *tit.* Incipit theorica lune *m*; *tit.* De theorica lune *o*; *tit.* De luna *r*

18 orientem: occidentem hoc donec (?dicit) quantum ad partes deferentis nam deferens mouetur ad occidentem. Ergo quando pars deferentis que facit sectionem transiuit locum sectionis uersus occidentem alia pars occidentalis succedit locum eius in ipsa sectione; pars alia deferentis que prius erat uersus orientem et cum hec transiuit locum sectionis ad occidentem succedit ad eandem partem alia pars orientalis illi continua et sic deinceps *fJ et sim. y*Λ

maximarum declinationum successiue mutant partes orbis signorum sibi inuicem succedentes uersus occidentem quemadmodum manifeste apparebit ex hiis que de motu capitis dicentur. Quantitas itaque maxime declinationis predicte septentrionalis aut meridiane est semper 5 graduum et ipsa
25 est in puncto qui diuidit per equalia utrumque duorum arcuum quorum unus est inter nodum capitis et nodum caude et est totus uersus septentrionem ab orbe signorum, alius uero est inter nodum caude et nodum capitis et est totus uersus meridiem. Suntque inter utramlibet istarum duarum maximarum declinationum et utrumlibet nodorum predictorum
30 tria signa de partibus orbis signorum. Vocamus autem istam declinationem latitudinem lune et determinat eam circulus transiens per polos orbis signorum et per quemuis punctum ecentrici deferentis. Quantus enim fuerit arcus huius circuli interceptus inter orbem signorum et deferentem, tantam dicemus esse latitudinem ipsius puncti, septentrionalem quidem cum
35 fuerit punctus ille in arcu deferentis septentrionali, meridianam uero cum fuerit in arcu meridiano.

Superficies autem epicicli est in superficie istius ecentrici numquam declinans ab ea in aliqua parte sui. Ideoque luna nullam habet latitudinem qua declinet ab orbe signorum nisi propter latitudinem deferentis. Dicitur
40 autem iste circulus deferens eo quod deferat centrum epicicli. In ipsius enim circumferentia mouetur centrum epicicli uersus orientem et ad successionem signorum. Quare manifestum est quod motus centri epicicli non fit super polos orbis signorum sed super polos alios distantes ab illis quantitate 5 graduum qui sunt poli sui deferentis, et describunt circa polos
45 orbis signorum duos circulos paruos quorum semidiametri sunt secundum quantitatem distantie illorum ab illis.

Mouetur autem centrum epicicli in circumferentia istius circuli deferentis omni die naturali de partibus circuli in ipsius superficie descripti cuius centrum sit centrum orbis signorum 24 gradus et 23 minuta fere. Si enim
50 a centro orbis signorum ducantur due linee recte, una ad punctum orbis lune ecentrici in quo erat centrum epicicli ipsius in principio alicuius diei naturalis, et alia ad punctum eiusdem orbis ecentrici ad quem peruenit in fine ipsius diei naturalis, angulus ab istis duabus lineis contentus erit 24 gradus et 23 minuta fere. Verbi gratia, si hodie ad meridiem ciuitatis

22 occidentem: orientem β
28 meridiem: ab orbe signorum *add.* β
29 nodorum: duorum nodorum δ
35 ille *om.* β / punctus ille *om. f*
39 qua declinet: quia non declinat Γ a uia solis et per consequens *add.* β
40 iste *om. AFfoPuy*ΛΠΣ
42 signorum: mouetur enim in ipsius (eius *v*) circumferentia *add. aBJKpqRUv*
48 de partibus circuli: et ille circulus dicitur equans lune *add.* φ
52 perueniat *bNPV* proueniant *D*
53 naturalis: et hec omnia (que etiam Λ) patent in figura lune instrumento (instrumenti lune Λ) posita modicum post illud (idem Λ) *add. fy*Λ

maximum declinations occupy each point of the ecliptic in succession as they [the points of the ecliptic] succeed one another in turn coming from the west,[6] as will be quite clear from what we shall say about the motion of the node.[7] The amount of the aforementioned maximum northern or southern declination is, then, always 5 degrees;[8] it is found at the point of bisection of both of the two arcs, one of which lies between the head [ascending] node and the tail [descending] node and is entirely to the north of the ecliptic, while the other lies between the descending node and the ascending node and is entirely to the south [of the ecliptic]. The distance between either of those two maximum declinations and either of the above-mentioned nodes is three signs of the ecliptic. Now we call that declination the "lunar latitude"; it is determined by the circle passing through the poles of the ecliptic and through any point on the eccentric deferent [of which we want to know the latitude]. For we say that the latitude of that point is as great as the size of the arc of the above circle which is cut off between the ecliptic and the deferent: the latitude will be northern when that point is in the northern arc of the deferent, but southern when it is in the southern arc.

The plane of the epicycle lies in the plane of that eccentric and is never inclined to it at any point on its surface. Therefore the moon has no latitude to make it decline from the ecliptic except what is due to the latitude [i.e., inclination] of the deferent. That circle is called the deferent [i.e., carrying circle] because it carries the center of the epicycle along. For the center of the epicycle moves on its circumference toward the east in the succession of the signs. Hence it is clear that the motion of the center of the epicycle takes place not about [an axis passing through] the poles of the ecliptic but about other poles 5 degrees away from them; these are the poles of its deferent, and they describe around the poles of the ecliptic two small circles whose radii are the size of the distance between those [poles of the deferent] and those [poles of the ecliptic].

Now the center of the epicycle in its motion on the circumference of that deferent circle traverses in every natural day about 24 degrees and 23 minutes[9] of a circle described about the center of the ecliptic in the plane of the deferent. For if two straight lines are drawn from the center of the ecliptic, one to the point on the eccentric circle of the moon in which the center of its epicycle was at the beginning of any natural day and the other to the point on the same eccentric circle which the epicycle has reached at the end of that day, the angle contained by these two lines will be about 24 degrees and 23 minutes. For example, if today at noon[10] at the city of NOVARA the center of the moon's

NOVARIE sit centrum epicicli lune in auge deferentis et aux deferentis sit in primo minuto arietis, cras ad meridiem NOVARIE distabit centrum epicicli ab auge deferentis 24 gradibus et 23 minutis fere de partibus circuli descripti super centrum orbis signorum et in superficie ipsius deferentis. Nam centrum epicicli lune mouetur proprio motu super circumferentiam sui deferentis uersus orientem et ad successionem signorum de partibus circuli descripti super centrum orbis signorum et in superficie ipsius deferentis 13 gradibus et 10 minutis et 34 secundis, et aux deferentis lune mouetur de partibus eiusdem circuli uersus occidentem et contra successionem signorum 11 gradibus et 12 minutis et 18 secundis. Et si coniunxeris istos duos motus in unum inuenies 24 gradus et 22 minuta et 52 secunda.

Scire autem debes quod centrum epicicli lune cum uersus orientem mouetur super circumferentiam sui ecentrici deferentis semper respicit centrum orbis signorum, et respectu eius uniformiter mouetur, unde super ipsum describit in equalibus temporibus equales angulos, et de circumferentia cuiuslibet circuli descripti super ipsum et in eadem superficie cum suo ecentrico deferente equales arcus. Centrum quoque deferentis mouetur uersus occidentem in circumferentia cuiusdam parui circuli descripti in superficie ipsius deferentis cuius centrum est centrum orbis signorum, et semidiameter eius est distantia centri deferentis lune a centro orbis signorum, fitque motus iste etiam respectu centri orbis signorum. Describit enim super ipsum in equalibus temporibus equales angulos et de circumferentia cuiuslibet circuli super idem centrum et in eadem superficie cum deferente descripti equales arcus. Mouetur itaque centrum deferentis lune in circumferentia prefati circuli parui uersus occidentem equaliter et mouet secum longitudinem longiorem ipsius deferentis et totum circulum deferentem secundum situm suum quantitate 11 graduum 9 minutorum et 7 secundorum.

Circulus autem alius descriptus super centrum orbis signorum et in ipsius superficie qui necessario secat ecentricum lune predictum in eisdem duobus locis in quibus ipsum secat orbis signorum mouetur super centrum orbis signorum uersus occidentem et mouet secum utraque loca abscisionis que dicuntur capud et cauda, et cum hoc etiam longitudinem longiorem ecentrici et etiam totum ecentricum secundum situm suum super polos orbis signorum quantitate trium minutorum et 11 secundo-

55 et aux deferentis: que *DZ*
68 respectu: cum centrum δ/unde: ad successionem signorum *DZ*
73 cuius: scilicet circuli parui *add.* β
76 enim: etiam *add.* α*H*
76–77 circumferentia: secundum hoc *add.* β
79 parui *om.* β
84 qui: que *ABCMRT*ΓΨ

epicycle is in the apogee of the deferent and the apogee of the deferent is in the first point of Aries, tomorrow at noon at NOVARA the center of the epicycle will be distant from the apogee of the deferent by about 24 degrees and 23 minutes of a circle described about the center of the ecliptic in the plane of the deferent. For the center of the moon's epicycle moves, by its proper motion on the circumference of the deferent toward the east in the succession of the signs, an amount of 13 degrees 10 minutes 34 seconds[11] [a day] of a circle described about the center of the ecliptic in the plane of the deferent,[12] while the apogee of the deferent of the moon moves, toward the west counter to the succession of the signs, an amount of 11 degrees 12 minutes 18 seconds[13] of the same circle. If you add these two motions together, you get 24 degrees 22 minutes 52 seconds.

Now you ought to know that the center of the moon's epicycle, in its motion toward the east on the circumference of its eccentric deferent, always has regard to[14] the center of the ecliptic, in respect to which its motion is uniform; hence it describes in equal times equal angles about it [the center of the ecliptic] and equal arcs on the circumference of any circle described about it [the center of the ecliptic] in the same plane as its [the epicycle's] eccentric deferent. Moreover, the center of the deferent moves toward the west on the circumference of a certain small circle described in the plane of the deferent: the center of this circle is the center of the ecliptic, and its radius is the distance between the center of the moon's deferent and the center of the ecliptic; this motion, too, takes place [uniformly] with respect to the center of the ecliptic. For it [the center of the deferent] describes in equal times equal angles about the center of the ecliptic and equal arcs on the circumference of any circle with the same center and in the same plane as the deferent. Therefore the center of the deferent of the moon moves with uniform speed on the circumference of the above-mentioned small circle toward the west and carries with it the "greatest distance"[15] of the deferent and the whole deferent circle, with respect to its position, an amount of 11 degrees 9 minutes 7 seconds [a day].[16]

Another circle described on the center of the ecliptic and in its plane, which must of necessity cut the aforementioned eccentric of the moon in the same two places [as those] in which it [the eccentric] is cut by the ecliptic, moves about the center of the ecliptic toward the west and carries along in its motion both the points of intersection, which are called ascending and descending nodes; it also carries the "greatest distance" of the eccentric and the whole eccentric, with respect to its position, an amount of 3 minutes and 11 seconds[17]

---

68 signorum: ideo in equatione lune non indigemus equante sed eius loco utimur orbe signorum, puta (scilicet $y\Lambda$) zodiaco, qui est in margine matris *add.* φ

71 arcus: sed in deferente ecentrico non describit centrum epycicli in temporibus equalibus equales arcus *add.* φ

90 rum, et hic est motus qui dicitur motus capitis draconis. Componitur igitur motus augis deferentis lune uersus occidentem ex motu centri deferentis in circumferentia parui circuli descripti in superficie ecentrici deferentis super centrum orbis signorum et ex motu capitis draconis predicto. Si itaque istos duos motus agregaueris adinuicem inuenies motum augis
95 deferentis lune a quouis puncto orbis signorum in omni die naturali uersus occidentem 11 gradus 12 minuta et 18 secunda. Et inuenies motum eius ab alterutro duorum nodorum capitis scilicet et caude uersus occidentem in omni die naturali 11 gradus 9 minuta 7 secunda. Motum quoque centri epicicli lune ab alterutro eorundem duorum nodorum uersus orientem
100 inuenies 13 gradus 13 minuta et 45 secunda, et motus iste dicitur motus latitudinis, et est compositus ex motu centri epicicli uersus orientem et motu capitis uersus occidentem.

Ipse autem motus centri epicicli uersus orientem factus super centrum orbis signorum equaliter dicitur medius motus lune. Motus uero lune in
105 suo epiciclo cuius principium est ab auge epicicli que respicit punctum qui in circumferentia parui circuli descripti a centro deferentis diametraliter opponitur ipsi centro deferentis dicitur medium argumentum lune, et ista aux dicitur aux media. Nam aux uera epicicli lune est punctus respiciens centrum orbis signorum maxime eleuatus ab eo et per compa-
110 rationem corporis lune ad hanc augem sumitur argumentum uerum, et est manifestum quod iste due auges epicicli fiunt una et eadem cum centrum epicicli fuerit in auge deferentis uel in eius opposito. Cum autem fuerit in illa medietate circuli deferentis que est ab auge sua usque ad eius oppositionem, tunc aux uera epicicli precedit augem mediam uersus orientem.
115 In altera uero medietate ipsius aux media precedit ueram, unde in istis duabus medietatibus deferentis iste due auges se habent modo contrario quantum ad precessionem et successionem. Eritque inter eas maior distan-

90 draconis: et caude *add.* β
94 agregaueris: congregaueris *AFfIx*ΛΣ
99 orientem: occidentem *f*
108 media: lune ut patet in figura (lune in medietate *add.* Λ) sequenti *add. JM*ΓΛ
109 eo: quem punctum determinat linea deducta usque ad ipsum per centrum epicicli ueniens ab orbe signorum *add. gSt*
114 precedet δθ
115 precedet δ procedit Γ
117 successionem: secutionem *AbcGHInorst uwxz*ΛΠ seq(u)utionem δ subsecutionem β/et successionem *om.* φ/eas: eos βδ

[a day] about the poles of the ecliptic: this is the motion which is called the motion of the node. Therefore the motion of the apogee of the moon's deferent toward the west is composed of the motion of the center of the deferent on the circumference of the small circle which is described on the center of the ecliptic in the plane of the eccentric deferent, and of the above-mentioned motion of the node. If, then, you add those two motions together, you will find the motion of the apogee of the moon's deferent from any point on the ecliptic toward the west in any natural day to be 11 degrees 12 minutes 18 seconds, and you will find its motion from either of the two nodes, that is, the ascending node or the descending one, toward the west in any natural day to be 11 degrees 9 minutes 7 seconds. Moreover, you will find the motion of the center of the moon's epicycle from either of the same two nodes toward the east to be 13 degrees 13 minutes 45 seconds[18] [a day]: that motion is called the "motion of [the argument of] latitude"; it is composed of the motion of the center of the epicycle toward the east and the motion of the node toward the west.

The uniform motion of the center of the epicycle toward the east about the center of the ecliptic is called the "mean motion" of the moon. The motion of the moon on its epicycle, which is measured from that apogee of the epicycle which is [an apogee] with respect to[19] the point that is diametrically opposite to the center of the deferent on the circumference of the small circle, which is the locus of the center of the deferent, is called the "mean anomaly" of the moon, and that apogee is called the "mean [epicyclic] apogee." For the true epicyclic apogee of the moon is the point [on the epicycle] which has regard to the center of the ecliptic and is farthest removed from it: it is by comparison of [the position of] the body of the moon with this apogee that the true anomaly is found. It is obvious that these two epicyclic apogees become one and the same when the center of the epicycle is in the apogee or perigee of the deferent; but when it is on the half of the deferent circle that lies between its apogee and its perigee, then the true epicyclic apogee precedes the mean apogee, [i.e., is] to the east of it; when, however, it is on the other half of the deferent, the mean apogee precedes the true. Hence, the relative position of these two [epicyclic] apogees on one half of the deferent, with regard to which precedes or follows the other, is the contrary of what it is on the other half. The maximum

99 orientem: ex quo apparet (patet *M*ΓΛ) quod motus capitis qui est uersus occidentem non facit retrocedere epyciclum cuius centrum mouetur uersus orientem alias (aliter *M*ΓΛ) motus latitudinis non esset compositus ex motu centri epycicli uersus orientem et motu capitis uersus occidentem *add. fM*ΓΛ (*post* occidentem *linea 102 M, post* lune *linea 104* Γ), ut hic dicitur. Ex hoc patet falsum (factum *f*) esse quod dicitur in composito 19ne (illius *f*) propositionis quarti libri parui (parui *om.* Λ) almagesti, *ut etiam ibi notaui* (ubi et in naturam *f*) et ad hoc uidelicet ymaginatio motus epycicli quam notaui (naturam *f*) *add. f*Λ

tia que esse potest circa utramlibet duarum longitudinum mediarum, quando scilicet centrum epicicli erit quasi equaliter distans ab auge defe-
120 rentis et eius oppositione, quod quidem erit parum ultra medium medietatis prime deferentis in qua centrum epicicli descendit ab auge sui deferentis ad eius oppositionem, et parum citra medium medietatis secunde in qua ascendit ab oppositione augis ad augem.

Fuit autem ex precepto Dei gloriosi et sublimis ut in prima coniunctione
125 solis et lune secundum medios eorum motus esset centrum epicicli lune in auge sui deferentis. Quid autem sit medius motus solis, quid etiam medius motus lune, dictum est prius. Igitur sol secundum ipsius motum medium, centrum quoque epicicli lune, aux etiam deferentis, fuerunt tunc in eodem puncto orbis signorum. Ponamus ergo hoc fuisse in principio arietis et in
130 instanti quod erat meridies NOVARIE, et adiciamus ibidem fuisse nodum capitis. In crastino igitur in instanti meridiei eiusdem loci fuit centrum epicicli lune in ariete 13 gradus 10 minuta et 34 secunda et aux deferentis in piscibus 18 gradus 47 minuta et 42 secunda et nodus capitis similiter in piscibus 29 gradus 56 minuta et 49 secunda, distantia quoque centri epi-
135 cicli ab auge sui deferentis de partibus orbis signorum 24 gradus 22 minuta et 52 secunda. Quia uero sol mouetur omni die naturali secundum medium eius motum de partibus orbis signorum 59 minuta et 8 secunda, fuit ipse sol secundum medium eius motum in prefato crastino et in instanti prefato meridiei eiusdem loci in ariete 59 minutis et 8 secundis. Distabat ergo
140 tunc sol secundum medium eius motum a centro epicicli lune 12 gradibus 11 minutis et 26 secundis. Ab auge quoque deferentis lune distabat tantumdem. Cum igitur prefati tres motus sint in perpetuum uniformes, sequitur ut sol secundum medium ipsius motum semper sit medius inter centrum epicicli lune et augem sui deferentis et equaliter distans ab utro-
145 que eorum. In omni enim tempore motus centri epicicli lune uersus orientem excedit motum augis sui deferentis uersus occidentem in duplo motus solis medii in eodem tempore. Quia uero motus solis est in eandem partem cum motu centri epicicli et in partem contrariam motui augis

118 circa: citra *H*
126 quid ctiam: et quid *ΛΜ*ΓΛΣΨ et Π
130 quod: quo β
131 capitis: draconis *add.* β
138 crastino: die *add.* β/prefato[2]: prefati δ
139 secundis: et 13 tertiis *add.* *C*/distabit βφ
141 distabit βφ
147 motus[2]: medius motus β

possible interval between them will occur at about the two mean distances, that is, when the center of the epicycle is approximately equidistant from apogee and perigee of the deferent: this [situation] will occur [when the epicycle center is] a little beyond halfway around the first half of the deferent, in which the center of the epicycle descends from the apogee of its deferent to the perigee, and a little before halfway around the second half, in which it ascends from perigee to apogee.[20]

Now it so happened by the ordainment of glorious and high God that, when the sun and moon were at the first conjunction in their mean motions, the center of the moon's epicycle was in the apogee of the deferent.[21] We have already explained[22] what the mean motion of the sun is and also what the mean motion of the moon is. Therefore the sun in its mean motion and also the center of the moon's epicycle and the apogee of the deferent were at that time in the same point of the ecliptic. Let us then suppose that this took place in the first point of Aries and at the instant of time that was midday[23] at NOVARA; let us add the further supposition that the ascending node, too, was in the same point. Then on the following day at the instant of midday at the same place the center of the moon's epicycle was in 13 degrees 10 minutes 34 seconds of Aries[24] and the apogee of the deferent was in 18 degrees 47 minutes 42 seconds of Pisces,[25] and the ascending node was also in Pisces, in 29 degrees 56 minutes 49 seconds.[26] The distance of the center of the epicycle from the apogee of the deferent was 24 degrees 22 minutes 52 seconds of the ecliptic. Moreover, because the sun in mean motion moves 59 minutes and 8 seconds of the ecliptic every natural day, on the above-mentioned following day at the above-mentioned instant of midday at the same place the sun was in 59 minutes and 8 seconds of Aries according to its mean motion. Therefore the mean sun was at that moment 12 degrees 11 minutes 26 seconds away from the center of the moon's epicycle and the same distance away from the apogee of the moon's deferent also. Thus, since the above-mentioned three motions[27] are always uniform, it follows that the mean sun is always halfway between the center of the moon's epicycle and the apogee of the moon's deferent and equidistant from both. For the motion of the center of the moon's epicycle toward the east over any period of time exceeds the motion of the apogee of its deferent toward the west by double the mean motion of the sun during the same time. But because the motion of the sun is in the same sense as the motion of the center of the epicycle, and in the opposite sense to the motion of the apogee

---

134 49 secunda: sed a nodo capitis plus distat quam ab ariete (ipsum centrum epicicli lune *add. J*) quia nodus interim motus uersus occidentem est 3 minutis et 11 secundis ut dictum est supra *add. JΛ*

147 tempore: id est motus solis excedit motum augis deferentis in duplo quantum mouetur sol in eodem tempore *add. T*

deferentis, minuetur distantia centri epicicli lune a sole quantitate motus
150 solis et augebitur distantia solis ab auge deferentis lune quantitate eiusdem motus. Itaque sol semper erit in medio. Erit igitur in omni coniunctione et in omni oppositione mediis solis et lune centrum epicicli lune in auge sui deferentis et in omni quadratione lune, cum uidelicet centrum epicicli lune distabit a sole secundum medium eius motum quarta parte orbis
155 signorum, siue hoc fuerit in tempore quo luna augetur in lumine, siue in tempore quo minuitur in lumine, erit centrum epicicli in opposito augis sui deferentis. Ex premissis autem manifestum est quod centrum epicicli lune bis percurrit in omni mense lunari circumferentiam sui deferentis; bis quoque est in auge, scilicet in coniunctione sua media cum sole et in
160 oppositione sua media; bis etiam est in opposito augis, scilicet in quadratura crementi sui luminis et in quadratura diminutionis eiusdem. Et est totum tempus quo semel describit suum circulum deferentem 14 dies 18 hore et 22 minuta, et ipsum est tempus medietatis unius lunationis equalis. Tempus autem quo bis ipsum describit est 29 dies 12 hore et 44 minuta, et
165 ipsum est tempus unius integre lunationis equalis que per medios motus solis et lune accipitur.

Vt autem ea que de motibus lune diximus figurali exemplo comprehendantur, describam circulum in superficie orbis ecentrici lune et supra centrum orbis signorum, qui circulus sit *ABG* cuius centrum *D*, et protraham
170 diametrum *AD* ponamque ut punctum *A* sit principium arietis et quod aux deferentis lune, centrum quoque epicicli ipsius corpusque lune et nodus capitis itemque sol secundum medium motum fuerit in puncto *A* heri ad meridiem NOVARIE. Linea igitur *AD* designabat locum augis et centri epicicli et corporis lune et nodi capitis et medii motus solis heri in
175 meridie predicti loci, et tunc inceperunt hec omnia moueri a puncto *A*: aux quidem corpusque lune, si erat in auge epicicli, et nodus capitis uersus

152 mediis: medii motus β
159 sua *om.* δ
162 semel *om.* αφ
164 describit: pertransit β
169 qui circulus sit: circulum αφ*H*
170 AD: ADI δ ABD φ
172 fuerint βδ
173 NOVARIE: MOLINIS *O*
174 centrum α

of the deferent, the distance of the center of the moon's epicycle from the sun will be diminished [from the above motion of the center of the epicycle] by the amount of the sun's motion, and the distance of the sun from the apogee of the moon's deferent will be increased [over the above motion of the apogee] by the amount of that same motion [of the sun].[28] So the sun will always be halfway [between the two]. Therefore at every mean conjunction and mean opposition of sun and moon the center of the moon's epicycle will be in the apogee of its deferent, and at every quadrature of the moon, i.e., [at that point] where the center of the moon's epicycle is separated from the mean sun by a quarter of the ecliptic, whether it be at the time when the moon is waxing or at the time when it is waning,[29] the center of the epicycle will be in the perigee of the deferent. From the foregoing it is obvious that the center of the moon's epicycle traverses the circumference of its deferent twice in every lunar month, and that it is also in the apogee twice, namely, at its mean conjunction with the sun and at its mean opposition, and in the perigee twice, namely, at the quadrature in which it is waxing and at the quadrature in which it is waning. The total time it takes to traverse its deferent once is 14 days 18 hours 22 minutes: that is the period of half of one mean lunation.[30] The time in which it traverses it [the deferent] twice is 29 days 12 hours 44 minutes:[31] that is the time of one whole mean lunation, which is derived from the mean motions of sun and moon.

But in order that what we have said about the motions of the moon may be understood by an illustrating diagram [see Fig. 11, p. 172], I draw [a representation of] the circle in the plane of the moon's eccentric and about the center of the ecliptic: let this be circle *ABG*, with center *D*; I also draw the diameter [through] *AD* and suppose that point *A* is the beginning of Aries, and that the apogee of the moon's deferent, as well as the center of its epicycle and the body of the moon and the ascending node and the mean sun also, was at point *A*[32] yesterday noon at NOVARA. Therefore the line *AD* marked the position of the apogee, of the center of the epicycle, of the body of the moon, of the ascending node, and of the mean sun yesterday at noon in the aforementioned place, and at that moment all these [points] began to move away from point *A*: the apogee and the body of the moon, if it was in the apogee of the epicycle, and the

---

151 medio: hocque patet quia sol semper est in medio inter centrum epycicli et augem deferentis lune. Quando (unde *f*) autem centrum epycicli est in auge deferentis ad hoc quod sol equaliter distet (distat *f*) ab eis, id est, a centro epycicli et ab auge deferentis, oportet quod uel (uel *om. f*) simul sit cum eis et sic est in auge deferentis in coniunctione (e contrario *f*) cum luna (est *add. f*) secundum medium motum uel (tunc *f*) oportet quod sit in opposito augis deferentis et sic est in opposito a luna *add. fy*Λ

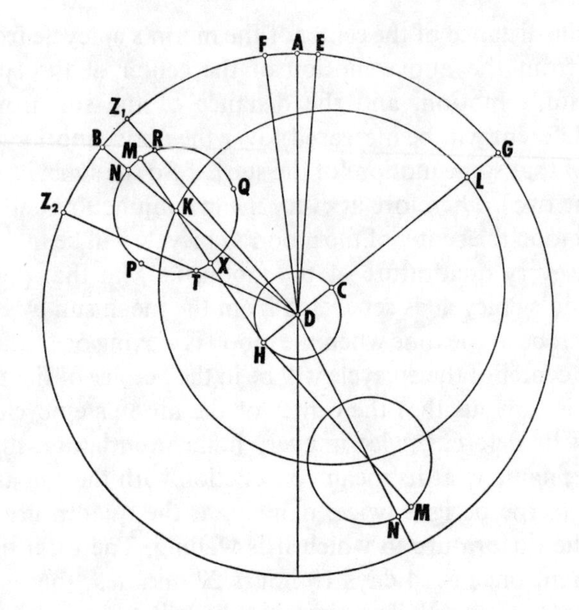

Fig. 11

occidentem a puncto *A*, centrum autem epicicli corpusque lune, si erat in opposito augis epicicli, et sol uersus orientem ab eodem puncto.

Esto itaque ut hodie in meridie NOVARIE peruenerit aux ad punctum *G* et nodus capitis ad punctum *E*, centrum autem epicicli ad punctum *B* corpusque lune ad punctum *Z* et sol ad punctum *F* et ad hec omnia puncta protrahantur linee recte a puncto *D* quod positum est esse centrum orbis signorum. Erit igitur arcus *AE* motus capitis unius diei et est 3 minuta et 11 secunda, et arcus *EG* erit proprius motus augis siue centri deferentis circa centrum orbis signorum et est 11 gradus 9 minuta et 7 secunda, totusque arcus *AG* erit motus augis diurnus compositus ex motu proprio et motu capitis eritque 11 gradus 12 minuta et 18 secunda. Erit quoque arcus *AB* motus centri epicicli diurnus siue motus lune medius et est 13 gradus 10 minuta et 34 secunda. Arcus autem *EB* erit motus latitudinis et est 13 gradus 13 minuta et 45 secunda, et arcus *AZ* erit motus corporis lune quem uerum motum uocamus; qui erit minor arcu *AB*, si corpus lune erat in auge epicicli, quantitate arcus orbis signorum qui subtenditur arcui epi-

177–78 autem...puncto: solis uersus orientem et centrum epycicli similiter β

181 puncta *om.* β

Legend for Fig. 11

Composite illustration (after Campanus) of the moon's motions in the Ptolemaic model (compare Figures 5 and 6, pp. 44 and 45). Circle *ABG* represents the ecliptic, with center *D*, the earth. Circle *LK* represents the moon's deferent, with center *C*. *C* moves on the "small circle" *CH* about *D*. The epicycle, with center *K*, is represented by circle *MN*. Initially the apogee of the deferent, the epicycle center, the moon's body (in either the apogee or the perigee of the epicycle), the ascending node, and the mean sun all have the same longitude, Aries 0°, which is marked by *A*; i.e., all those points lie along line *DA*. After 24 hours, all points have moved away from *A* (the amounts moved are exaggerated in the figure for clarity): the apogee and the node have moved westward, to *G* and *E*, respectively; the mean sun and epicycle center have moved eastward, to *F* and *K* (whose projection on the ecliptic is *B*), respectively. The moon has moved clockwise on the epicycle, to *R* if it was in the apogee initially, to *T* if it was in the perigee initially; and its projections on the ecliptic are Z_1 and Z_2, respectively. *N* represents the true apogee of the epicycle (being the point on the epicycle farthest from the earth *D* along line *DK*), while *M* represents the mean apogee of the epicycle (being the point on the epicycle farthest from *H*, the point opposite *C* on the small circle). The epicycle is drawn twice to illustrate how *M* and *N* change their relative positions on opposite halves of the deferent. *P* and *Q* are the easternmost and westernmost points of the epicycle, i.e., the points where tangents from *D* touch the epicycle (in which the moon's equation of anomaly is greatest).

ascending node toward the west of point *A*, but the center of the epicycle and the body of the moon, if it was in the perigee of the epicycle, and the sun toward the east of that point.

Let us suppose then that today at midday in NOVARA the apogee has reached point *G*; the ascending node, point *E*; the center of the epicycle, point *B*; the body of the moon, point *Z*; and the sun, point *F*: let straight lines be drawn to all these points from point *D*, which was defined as the center of the ecliptic. Therefore arc *AE* will be the motion of the node in one day, namely, 3 minutes and 11 seconds; arc *EG* will be the proper motion of the apogee[33]—in other words, the motion of the center of the deferent—about the center of the ecliptic, namely, 11 degrees 9 minutes 7 seconds; and the whole arc *AG* will be the daily motion of the apogee, the components of which are the proper motion and the motion of the node: it will be 11 degrees 12 minutes 18 seconds. Arc *AB* will be the daily motion of the center of the epicycle, in other words, the mean motion of the moon [in longitude], namely, 13 degrees 10 minutes 34 seconds. Arc *EB* will be the motion of [the argument of] latitude, namely, 13 degrees 13 minutes 45 seconds; arc *AZ* will be the motion of the body of the moon, which we call the true motion. If the body of the moon was in the apogee of the epicycle, [arc *AZ*] will be less than arc *AB* by the amount of that arc of the ecliptic which is subtended by the arc of the epicycle which the moon has

189 34 secunda: Et est compositus ex motu centri epicicli uersus orientem et motu capitis uersus occidentem *add. F mg.* Σ

cicli quem transiuit in isto tempore luna uersus occidentem et est 1 gradus 20 minuta et 45 secunda. Et hec est quantitas arcus *ZB* quem motus lune
195 uerus diminuit de motu suo medio, diciturque equatio argumenti lune diminuens motum. Si uero corpus lune erat in opposito augis epicicli, erit arcus *AZ* maior arcu *AB* quantitate arcus orbis signorum qui subtenditur arcui epicicli quem transiuit luna in isto tempore uersus orientem et est 1 gradus 35 minuta et 27 secunda et hec est quantitas arcus *BZ* quem motus
200 uerus lune addit super motum suum medium, diciturque equatio lune argumenti augens motum. Quia uero maxima motus diminutio fit centro epicicli existente in auge deferentis et luna existente in auge epicicli, nam cum centrum epicicli est in opposito augis deferentis et luna in auge epicicli maxima fit diminutio in die una de argumento lune ideoque multum
205 minuitur eius equatio que est diminuens motum, ideo nullatenus luna potest fieri retrograda nec etiam stationaria. Sed ista habet loco istarum proprietatum motum diminutum cum ipsa est in superiori parte epicicli, et motum augmentatum cum est in inferiori, et motum medium cum est super utraque latera ipsius.

210 Arcusque *AF* erit medius motus solis et est 59 minuta et 8 secunda, et *FB* arcus erit medius motus lune in distantia sui a sole, et arcus *FG* motus augis lune in distantia sui a sole, et unus eorum est equalis alteri et est uterque eorum 12 gradus 11 minuta et 26 secunda. Arcus autem *BG* est elongatio centri epicicli ab auge sui deferentis et hoc est duplum distantie
215 centri lune a sole quod in canonibus tabularum centrum lune siue duplex interstitium appellatur et est 24 gradus 22 minuta et 52 secunda.

Quoniam autem in linea *DG* est aux deferentis lune, oportet ut in ea quoque sit centrum ipsius deferentis. Sit itaque aux punctum *L* et centrum eius punctum *C*. Describam igitur super centrum *C* et secundum spatium
220 *CL* circulum *LK* qui necessario erit deferens lune, et ponam ut linea *DB*

196 erat: sit β
201 maxima: minima φ
202 in auge[1]: in opposito augis δ *om. B*
203 opposito...in *om. AJv*
204 argumento lune: uero *add.* αδφ *om. BP*
212 alteri: alii δ
217 in[1] *om.* δ

traversed toward the west in that time [i.e., one day], namely, 1 degree 20 minutes 45 seconds:[34] this is the size of the arc *ZB* by which the true motion of the moon falls short of its mean motion, and it is called the "subtractive equation of the moon's anomaly."[35] But if the body of the moon was in the perigee of the epicycle, arc *AZ* will be greater than arc *AB* by the amount of that arc of the ecliptic which is subtended by the arc of the epicycle which the moon has traversed toward the east in that time, namely, 1 degree 35 minutes 27 seconds: this is the size of the arc *BZ* by which the true motion of the moon exceeds its mean motion, and it is called the "additive equation of the moon's anomaly." Since the greatest subtractive effect on the mean motion occurs when the center of the epicycle is in the apogee of the deferent and the moon is in the apogee of the epicycle—for, when the center of the epicycle is in the perigee of the deferent and the moon in the apogee of the epicycle, the greatest subtraction from the moon's anomaly for one day occurs, and so its equation, which is [then] subtractive, is much diminished—therefore the moon can never become retrograde or even stationary.[36] But, in place of those properties [of station and retrogradation], it has a decreased [true] motion when it is in the upper part of the epicycle, and an increased motion when it is in the lower part, and a [true] motion [that is equal to its] mean when it is on either side of its epicycle.[37]

Arc *AF* will be the mean motion of the sun, namely, 59 minutes and 8 seconds, and arc *FB* will be the mean motion of the moon in elongation from the sun, and arc *FG* the motion of the apogee of the moon in elongation from the sun: each of the [latter] two is equal to the other, and both are 12 degrees 11 minutes 26 seconds. Arc *BG* is the elongation of the center of the epicycle from the apogee of its deferent; this is double the distance of the center of [the epicycle of] the moon from the sun, and, in the rules for the tables,[38] it is called the "centrum"[39] of the moon, or the "double interval"; it is 24 degrees 22 minutes 52 seconds.

Now, since the apogee of the moon's deferent is on the line *DG*, the center of the deferent too must be on that line. Let its apogee, then, be point *L* and its center point *C*. Therefore on center *C* with radius *CL* I draw a circle *LK*, which must be the deferent of the moon, and I suppose that line *DB* cuts circle

---

220 lune: Nota quod debes ymaginare quod epiciclus sit situs in una continuitate circulari et sperica que sit tante spissitudinis quante longitudinis est dyameter epicicly et ista continuitas debet ymaginari esse contenta infra concauitatem seu spissitudinem ecentrici seu defferentis et sic aux ecentrici poterit intelligi moueri ad partem unam et centrum epicicly ad aliam ecentricus enim lune mouetur ad occidentem et centrum epicicli ad orientem et nota quando respexeris hanc figuram ?dissertere faciem ad meridiem ita quod hec figura sit inter oculum aspicientem et meridiem *add.* Λ

secet circulum *LK* in puncto *K*. Erit ergo punctum *K* centrum epicicli lune. Describam igitur super ipsum circulum *MNPQ* qui sit epiciclus lune. Item circa centrum *D* et secundum spatium *DC* describam circulum *CH* qui est circulus paruus quem describit centrum deferentis proprio motu 225 uersus occidentem trahens secum longitudinem longiorem et totum circulum deferentem secundum situm suum. Producam autem lineam rectam *CD* usque ad *H* ut sit *CH* diameter illius circuli parui, et protraham lineam rectam ab *H* per *K* usque ad circumferentiam epicicli que sit *HXKM*, protraham etiam lineam a *D* per *K* usque ad circumferentiam eiusdem que 230 sit *DKN*. Eritque punctum *M* aux media epicicli et punctum *X* oppositio augis, punctum uero *N* aux uera; et a puncto *M* sumitur motus lune equalis in suo epiciclo qui dicitur argumentum medium, et est in superiori parte uersus occidentem, sicud dictum est supra, ut ab *M* uersus *Q*, esto enim quod *Q* sit punctus epicicli occidentalis, in inferiori autem uersus orientem 235 ut ab *X* ad *P*, esto enim quod *P* sit punctus orientalis in epiciclo; et est quantitas huius motus omni die naturali 13 gradus 3 minuta et 54 secunda. Tantus itaque erit uterque duorum arcuum qui sunt *MR* et *XT*. Arcus autem *MN* uocatur in canonibus tabularum equatio centri quoniam secundum distantiam centri ab auge uariatur: in auge enim et in eius oppo- 240 sito nullus est arcus quoniam *M* et *N* fiunt ibi punctus unus. In arcu autem qui est inter augem et eius oppositionem precedit uersus orientem aux uera augem mediam, in reliquo uero arcu econtrario, ut manifeste uidere poteris si in illo arcu descripseris epiciclum et protraxeris a punctis *H* et *D* lineas rectas per centrum epicicli usque ad eius circumferentiam. Sicud 245 autem puncta *M* et *N* fiunt punctus unus nichilque abinuicem distant centro epicicli existente in auge deferentis uel in eius opposito, ita centro epicicli perueniente ad utrasque longitudines medias quod est in distantia ab auge deferentis per 3 signa et 24 gradus est eorum distantia maxima. Recedente enim centro epicicli ab auge statim incipit descendere punctum 250 *M* a puncto *N*, et quanto amplius ipsum centrum elongatur ab auge tanto amplius elongatur *M* ab *N* et procedit huius additio in elongatione quousque distet centrum ab auge quantitate predicta. Tunc autem maxime distant abinuicem *M* et *N*. Deinceps uero sicud apropinquat centrum epicicli opposito augis, ita diminuitur distantia *M* ab *N*, ita quod cum centrum

229 lineam: aliam δ
231 augis: eius αφ*H*
232 parte: epycicli *add.* β
236 54: 58 *H*Λ
240 M et N: MN δ/fiunt: fuit δ sunt β fuerint *f*
243 punctis: puncto δ
245 M et N: MN δ
249 centro epicicli: epiciclo α
249–50 incipit...puncto N: incipit CL descendere et discedere punctum M a puncto preocupato δ

LK at point *K*. Then point *K* will be the center of the moon's epicycle. So I describe on that center circle *MNPQ*; let this be the moon's epicycle. Moreover, on center *D* with radius *DC* I draw circle *CH*, which is the small circle described by the center of the deferent [as it moves] with its proper motion toward the west, carrying with it the "greatest distance" and the whole deferent circle, with respect to its position. I produce the straight line *CD* to *H*, so that *CH* is the diameter of that small circle, and draw the straight line *HXKM* from *H* through *K* to the circumference of the epicycle. I also draw the line *DKN* from *D* through *K* to the circumference of the epicycle: point *M* will be the mean apogee of the epicycle and point *X* the [mean] perigee, while point *N* will be the true apogee. From point *M* the uniform motion of the moon on its epicycle, which is called the mean anomaly, is measured; and, as was stated above, in the upper part [of the epicycle] it has a westward direction, as from *M* toward *Q*, if we suppose *Q* to be the westernmost point of the epicycle, but in the lower part it has an eastward direction, as from *X* toward *P*, if we suppose *P* to be the easternmost point of the epicycle. The amount of this motion is 13 degrees 3 minutes 54 seconds every natural day. Thus, that [amount] will be the size of each of the two arcs *MR* and *XT*. In the rules for the tables arc *MN* is called the "equation of center," because it varies according to the distance of the center [of the epicycle] from the apogee [of the deferent];[40] for [when the epicycle is] in the apogee or the perigee the arc is zero, because there *M* and *N* are one point. But [when the epicycle is] on the arc which lies between the apogee and the perigee,[41] the true apogee [of the epicycle] precedes, [i.e., is] to the east of, the mean apogee, whereas on the other arc the opposite is true, as you can clearly see if you draw the epicycle on that arc and draw straight lines from points *H* and *D* through the center of the epicycle to its circumference.[42] And just as points *M* and *N* become one point and have no interval between them when the center of the epicycle is in the apogee or perigee of the deferent, so too when the center of the epicycle reaches either mean distance, which occurs 3 signs and 24 degrees[43] from the apogee of the deferent, the interval between them [*M* and *N*] is greatest. For as the center of the epicycle leaves the apogee, point *M* immediately begins to move down from point *N*, and the greater the elongation of the center [of the epicycle] from the apogee, the greater the elongation of *M* from *N*; this increase in elongation continues until the center is distant from the apogee by the aforesaid amount: then *M* and *N* are at their greatest distance from each other. From there on, as the center of the epicycle approaches the perigee, the interval between *M* and *N*

---

231 uera: et punctum X eius oppositio *add.* Π
244–45 Sicud... unus: a punctis autem M et N fiunt punctus unus alii prout centrum epycicli declinat a centro terre ad longitudinem longiorem ecentrici 360 partium erunt hee magnitudines. Preterea quidem sicud *f*

255 fuerit in oppositione augis nulla erit eorum distantia. Postea descendente centro ab oppositione augis incipit distare continuo *M* ab *N* et non cessat eorum distantia augeri quousque centrum epicicli distet ab oppositione augis per 2 signa et 6 gradus, ibique est eorum distantia maxima que rursus incipit minui et tanto amplius minuitur quanto magis centrum epicicli 260 apropinquat ad augem. Cumque illuc peruenerit nulla prorsus erit eorum distantia, idemque denuo et infinities fiet. Nolo autem ut mireris si maximam distantiam duarum augium predictarum posui in loco deferentis adeo distante ab auge sua et oppositioni augis apropinquante, quoniam ipsarum maximam distantiam oportet esse in loco deferentis ubi linea *HK* 265 ortogonaliter insistit linee *LH*. Ibi enim due linee *HK* et *DK* maximum angulum continebunt in puncto *K* ideoque maximus erit ibi angulus *MKN* qui sibi opponitur quare et arcus *MN* maximus.

Protraham rursus a centro *D* ad punctum *Z* qui antecedit punctum *B* itemque ad punctum *Z* qui sequitur ipsum lineas rectas. Sit ergo locus ubi 270 linea *DZ* prima secat epiciclum in parte superiori punctus *R* et locus ubi linea *DZ* secunda secat ipsum in parte inferiori punctus *T*. Erit igitur punctus *R* locus lune in epiciclo hodie ad meridiem predicti loci si heri fuit luna in auge epicicli aut punctus *T* si heri fuit in opposito augis ipsius.

Preter uero omnes prefatos motus habet luna adhuc motum alium 275 spere sue super polos orbis signorum secundum PTHOLOMEVM uel super capita arietis et libre secundum THEBITH qui est motus stellarum fixarum quemadmodum dictum est in sole.

Manifestum est igitur ex predictis quod luna 5 habet circulos. Habet enim deferentem et circulum paruum quem describit centrum deferentis 280 circa centrum orbis signorum et epiciclum et circulum descriptum super centrum orbis signorum et in superficie deferentis lune de cuius partibus mouetur centrum epicicli in deferente omni die naturali 24 gradus 22 minuta et 52 secunda. Nam centrum epicicli mouetur omni die de partibus ipsius uersus orientem 13 gradus 10 minuta et 34 secunda, aux uero defe-285 rentis mouetur de eisdem partibus uersus occidentem 11 gradus 12 minuta et 18 secunda. Habet quoque quintum circulum descriptum etiam super centrum orbis signorum et in eius superficie qui secat deferentem lune in

255 descendente: discedente *aBdgHqRwx*Π distendente *N* ascendente *f*
256 continue φ*BbCOW*/cessat: restat β
261 et *om.* αφ
263 oppositionem δ*A* oppositione β opposito *fWw*
269 Z *om.* β*A*ΠΨ
276 secundum: sicut ponit β/THEBITH: THEMO *p* TENEBIT *r*
284–86 aux…18 secunda *om.* δ

decreases, so that when the center is in the perigee the interval between them is zero. After that, as the center leaves the perigee, *M* immediately begins to move apart from *N*, and the interval between them continues to increase until the center of the epicycle is 2 signs and 6 degrees distant from the perigee: there the greatest interval between them is found; this [interval] immediately begins to decrease again, and decreases in proportion to the approach of the center of the epicycle to the apogee. When [the center of the epicycle] has reached that point [the apogee], there will be no interval at all between them [*M* and *N*], and the whole process will take place again and an infinite number of times. Now do not be surprised that I put the greatest interval between the two aforementioned apogees [of the epicycle] at a place on the deferent so far from its apogee and near to its perigee: the reason is that the greatest distance between them should be at that place on the deferent where line *HK* forms a right angle with line *LH*. For at that point the two lines *HK* and *DK* will contain the maximum angle at point *K*, and so *MKN*, which is the angle opposite [*HKD*], will be at a maximum then, and hence arc *MN* also [will be] at a maximum.[44]

Again, I draw straight lines from center *D* to the point *Z* [Z_1] which precedes point *B* and to the point *Z* [Z_2] which follows it. Then let the place where line DZ_1 cuts the epicycle in its upper part be point *R* and the place where line DZ_2 cuts it in its lower part be point *T*. Then point *R* will be the position of the moon on its epicycle today at midday in the place previously designated if the moon was in the apogee of the epicycle yesterday; or, if it was in its perigee yesterday, its position [today] will be point *T*.

Moreover, besides all the motions which we have mentioned, the moon has yet another motion, that of its sphere about the poles of the ecliptic according to PTOLEMY, or about the starting-points of Aries and Libra according to ṬĀBIT: this is the motion of the fixed stars, as was stated when we discussed the sun.[45]

Therefore it is clear from what we have said that the moon has five circles: it has a deferent, a small circle described by the center of the deferent about the center of the ecliptic, an epicycle, a circle described about the center of the ecliptic in the plane of the moon's deferent[46]—the center of the epicycle, [as it moves] on the deferent, traverses 24 degrees 22 minutes of this circle every natural day; for the center of the epicycle traverses 13 degrees 10 minutes 34 seconds of that [circle] toward the east, while the apogee of the deferent traverses 11 degrees 12 minutes 18 seconds of the same [circle] toward the west. It has also a fifth circle, likewise described about the center of the ecliptic but in its

---

280–81 circulum descriptum...lune: iste est equans *mg. o*
281 cuius: scilicet circuli descripti supra centrum orbis qui dicitur equans lune *add. fJΛ*

duobus locis qui dicuntur capud et cauda et mouetur et mouet secum istas duas sectiones uersus occidentem omni die 3 minutis et 11 secundis.

290 Manifestum quoque est quod luna habet 5 motus. Habet enim motum centri epicicli in deferente, motum quoque centri deferentis in circulo paruo et motum corporis sui in epiciclo et motum duarum sectionum qui dicitur motus capitis. Habet insuper motum sue spere continentis omnes hos motus.

295 Quid autem sit spera lune sic intelliges. Esto centrum epicicli in auge deferentis, luna uero in auge epicicli, et protrahatur linea recta a centro orbis signorum per centrum epicicli et centrum corporis lune quousque perueniat ad augem corporis lune. Posito ergo centro circuli centro ipsius orbis signorum, super ipsam lineam describatur circulus; cuius circuli 300 medietas manente fixa sua diametro ymaginetur moueri a quouis situ quousque redeat ad eundem, eritque corpus a uestigio motus huius circuli inclusum spera, et ipsa est que continet speram lune et speras omnium elementorum. Itemque esto centrum epicicli lune in opposito augis sui deferentis, luna quoque in opposito augis sui epicicli, ducaturque linea 305 recta a centro orbis signorum usque ad oppositionem augis corporis lune, et ipsa erit minor quam linea ducta ab eodem centro usque ad centrum epicicli in semidiametro epicicli et semidiametro corporis lune. Super ipsam igitur describatur circulus alius cuius centrum etiam sit centrum orbis signorum, istiusque circuli medietas sua diametro manente fixa ymagine-310 tur etiam circumduci a quouis situ usque dum redeat ad eundem. Eritque etiam ut prius corpus inclusum a uestigio huius motus spera et ipsa est que continet speras omnium elementorum diciturque spera elementaris. Corpus igitur quo spera descripta a primo circulo excedit speram descriptam a secundo dicitur spera lune et ipsum est corpus a uestigiis duorum pre-315 missorum motuum inclusum. Eodem autem modo intellige speras mercurii et ueneris et solis et trium superiorum qui sunt mars iupiter et saturnus. Est igitur concaua superficies spere lune locus spere elementaris; eius uero conuexa superficies peruenit usque ad superficiem concauam spere mercurii. Mercurii uero conuexa superficies peruenit usque ad concauam 320 ueneris; ueneris quoque conuexa ad concauam solis; solis conuexa ad concauam martis; martis quoque conuexa ad concauam iouis; et iouis

289 uersus occidentem *om.* β/omni die: naturali *add.* βφ
293 capitis: et caude *add.* φ/sue: octaue Λ
298 ipsius: ipso *AIl om.* β
310 usque dum: donec β
311 motus: circuli *BF*Σ *om. ACMpqRT*ΓΠΨ/spera: spere *ABqSTz*
313 primo: maiori β
314 secundo: minori β

plane, which cuts the deferent of the moon in two places, which are called ascending and descending nodes, and moves 3 minutes and 11 seconds a day toward the west,[47] carrying with it those two intersections.

It is also clear that the moon has five motions: it has the motion of the center of the epicycle on the deferent, the motion of the center of the deferent on the small circle, the motion of its body on the epicycle, the motion of the two intersections, which is called the motion of the node, and, in addition [to these], the motion of its sphere, which contains all the previous motions.

The meaning of the term "sphere of the moon"[48] can be understood as follows. Suppose the center of the epicycle to be in the apogee of the deferent, and the moon in the apogee of the epicycle. Let a straight line be drawn from the center of the ecliptic through the center of the epicycle and the center of the moon's body until it reaches the apogee [i.e., farthest point] of the moon's body. Then, with that line as radius and the center of the ecliptic as center, draw a circle. Imagine that half that circle, taken in any position, rotates [about] its diameter, which remains fixed in one place, until it returns to the same position. The body enclosed by the trace of the motion of this [semi]circle will be a sphere,[49] and this is the sphere that contains the sphere of the moon and the spheres of each of the elements.[50] Again, suppose that the center of the moon's epicycle is in the perigee of its deferent, and the moon also in the perigee of its epicycle, and let a straight line be drawn from the center of the ecliptic to the perigee [i.e., nearest point] of the moon's body: this line will be less than the line drawn from the same center to the center of the epicycle by [the amount of] the radius of the epicycle plus the radius of the moon's body. With this line as radius, then, and with the center of the ecliptic again as center, draw another circle, and imagine that half of that circle, too, taken in any position, is rotated [about] its diameter, which remains fixed in one place, until it returns to the same position. Then as before the body enclosed by the trace of this motion will be a sphere; this is the sphere that contains the spheres of all the elements and is called the "elemental sphere." Then the body by which the sphere described by the first circle exceeds the sphere described by the second is called the sphere of the moon: it is the body enclosed between the traces of the two motions described above. The spheres of Mercury, Venus, the sun, and the three outer planets, namely, Mars, Jupiter, and Saturn, must be understood in the same way. Therefore the concave surface of the sphere of the moon is the place of [i.e., contains][51] the elemental sphere, while its convex surface reaches the concave surface of Mercury's sphere; Mercury's convex surface reaches the concave surface of Venus; Venus's convex, the sun's concave; the sun's convex, Mars's concave; Mars's convex, Jupiter's concave; Jupiter's convex, Saturn's

---

305 lune: et hec uero indicabit (intrabis *Q* intrabit *T*) corpus lune *mg. A add. cQT*

conuexa ad concauam saturni; saturni uero conuexa superficies peruenit usque ad concauam superficiem stellarum fixarum; spere autem stellarum fixarum conuexa superficies peruenit usque ad concauam orbis noni.

325 Extra autem huius orbis conuexam superficiem utrum sit aliquid utpote alia spera necessitate rationis non cognoscimus. Fidei uero informatione sanctis ecclesie doctoribus assentientes reuerenter confitemur extra ipsam celum esse empireum in quo est bonorum spirituum mansio; ipsum quoque nonum uel esse celum cristallinum uel cristallinum esse extra ipsum et 330 sub empireo, hoc enim scriptura testatur diuina cui contradici fas non est. Si igitur cristallinum aliud sit a nono perueniet noni conuexa superficies usque ad concauam cristallini, cristallini autem conuexa perueniet usque ad concauam empirei. Empirei uero conuexa superficies nichil habet extra se; ipsa enim est supremum omnium corporalium rerum et maxime distans 335 a communi centro sperarum quod est centrum terre, unde ipsa est locus generalis et communis omnibus locatis, uelud omnia continens et a nullo alio contenta. Sic itaque patet celestes speras esse numero 11 si nonus orbis et celum cristallinum sunt diuersa uel tantum 10 si sunt idem. Corpusque ex omnibus illis constans dicitur quinta essentia preter quatuor elementa 340 supposita, uocatur etiam etherea regio sibique subiecta elementaris de cuius diuisione, situ et ordine aliquid est dicendum.

Spera igitur elementaris cuius supremum finis est rerum corruptibilium et principium incorruptibilium diuiditur in 4 speras 4 elementorum. Quarum prima est spera ignis quam orbiculariter circumdat spera lune et 345 ipsa orbiculariter circumdat tres inferiores. Secunda est spera aeris que etiam orbiculariter circumdat duas inferiores sicud ipsa orbiculariter circumdatur ab igne. Tertia est spera aque cuius orbicularis ambitus est diuino precepto decisus, terra in sue decisionis parte modicum consurgen-

325 aliquid utpote *om.* β
328 bonorum: beatorum α
329 extra: supra β
334 suprema δ
335 centro: omnium *add.* β
339 preter: propter α*H*
348 sue: sua β

concave; while Saturn's convex surface reaches the concave surface [of the sphere] of the fixed stars, and the convex surface of the sphere of the fixed stars reaches the concave of the ninth sphere.

Whether there is anything, such as another sphere, beyond the convex surface of this [ninth] sphere, we cannot know by the compulsion of rational argument [alone]. However, we are informed by faith, and in agreement with the holy teachers of the church we reverently confess that beyond it is the empyrean heaven in which is the dwelling-place of good spirits,[52] and furthermore that either the ninth sphere[53] itself is the crystal heaven,[54] or the crystal heaven is beyond it and below the empyrean. For this is attested by sacred scripture, which it is not lawful to contradict. If, then, the crystal heaven is a separate entity from the ninth sphere, the convex surface of the ninth sphere will reach the concave of the crystal heaven, and the crystal heaven's convex will reach the concave of the empyrean. The empyrean's convex surface has nothing beyond it. For it is the highest of all bodily things, and the farthest removed from the common center of the spheres, namely, the center of the earth; hence it is the common and most general "place" for all things which have position,[55] in that it contains everything and is itself contained by nothing. Thus, then, it is apparent that the heavenly spheres are 11 in number if the ninth sphere and the crystal heaven are different, or only 10 if they are identical. The body which they all go to make up is called the "fifth essence"[56]—["fifth" because it is] in addition to the four lower elements.[57] It is also called the "ethereal" region, while that below it is called the "elemental." We should say something about the division, position, and order of the latter.

The elemental sphere, then, whose outer limit is the end of corruptible things and the beginning of incorruptible ones, is divided into the four spheres of the four elements. The first of these is the sphere of fire, which is completely encircled by the sphere of the moon and itself completely encircles the three lower spheres. The second is the sphere of air, which in its turn completely encircles the two lower spheres, as it is itself completely encircled by the fire. The third[58] is the sphere of water, whose circular embrace is interrupted by divine decree, the earth rising up a little in the part where it is interrupted. The decree of

---

330 Nota auctor est catholicus et fidelis *mg. f*

340 etherea: Nam aether dicitur ab ἄει semper et θέω curro non autem ab αἴθω comburo ut uoluit ANAXAGORAS *mg. G* (*cf. Simplicium in* De caelo *270*ᵇ*16. 118. 22 et 119. 2*).

344 prima: Isti duo numeri qui sunt cubi illorum duorum qui secuntur et sunt illi scilicet qui secuntur in quibus reperitur 1635108092541000000 proportio diametri terre 1178100 et lune secundum fractiones positas et si diuiseris maiorem cubum per minorem exibunt 4or integra et 4or minuta fere qui sunt deno- 408024444725282821 63minatio proportionis corporis terre ad corpus lune *add.* Ψ

te. Fuit autem diuinum preceptum "congregentur aque que sub celo sunt
350 in locum unum et appareat arida" ut haberet homo qui quodammodo
finis est omnium locum sue habitationi congruentem. Ideoque rationabiliter credendum est solum locum illum terre detectum esse ab aquis qui humano usui fuit necessarius. Quia igitur sola quarta pars terre quam continent duo semicirculi quorum unus ab oriente in occidentem sub
355 equatore protenditur et alius ab oriente in occidentem per polum septentrionalem inhabitatur ut omnes aiunt, oportet alias tres quartas terre esse coopertas aquis. Quarta uero spera est spera terre cuius superficies apparens cum superficie aque est quasi superficies una. Et hec est naturaliter conuexa superficies spere aque, nam naturalis conuexa superficies spere
360 terre est naturalis concaua spere aque quod facile intelliges si ymagineris totam terre molem in formam uere spericam esse redactam eamque in medio aquarum esse omnium sepultam. Talis enim est eius naturalis situs, et centrum huius spere est centrum omnium sperarum predictarum. Ipse quidem sunt omnes concentrice, licet in speris celestibus circuli quibus
365 planetarum motus efficiuntur sint ecentrici.

Ex hiis autem 4 corporibus simplicibus et corruptibilibus omnium mixtorum corpora componuntur fiuntque ex eis diuerse mixtorum corporum species secundum sue commixtionis diuersitatem miscibiliumque proportionem. Sunt quoque omnia ex hiis composita corruptibilia et in elementa
370 prima resolubilia. Harum autem sperarum elementarium due propter sue soliditatem substantie prestant omnibus animantibus mansionem que sunt terra et aqua. In aliis uero duabus que sunt aer et ignis inpossibile est animal manere propter sue substantie raritatem quamuis ea que uolatilibus

349 preceptum: ut *add.* φ*Hz*
351–52 rationabiliter *om.* αφ
360 quod facile: quam faciliter δ quam facile *H*
362 omnino *D*/eius: forma et *add.* β
370 prima: priora αφ*H*
373 uolatilibus *acdHhStVWw* uolantibus *rz*

God was, "Let the waters which are beneath the heaven be assembled in one place, and let the dry land appear,"[59] so that man, who is in a certain sense the purpose of all things, might have a place suitable for him to live in. So on rational grounds we must believe that only that area of the earth which was necessary for human use was uncovered by the waters. Therefore, because only a fourth part of the earth, as is universally agreed, is inhabited, namely, that part contained between two semicircles, the one stretching from east to west along the equator and the other from east to west through the north pole, the other three quarters of the earth must be covered by the waters.[60] The fourth sphere is that of earth, whose visible surface is almost identical with the surface of the water. This [surface of the water] is, in its natural state, the convex surface of the sphere of water, for the natural convex surface of the sphere of the earth is the natural concave of the sphere of water. You will easily understand [the meaning of] this if you imagine the whole mass of earth transformed to a truly spherical shape and submerged in the midst of all the waters. For such is its natural position, and the center of that sphere [of the earth so situated] is the center of all the above-mentioned spheres. They are all concentric, although in the celestial spheres the circles by which the motions of the planets are effected are eccentric.

Now from these four simple and corruptible substances the bodies of all compound things are made up; from them come into being the different kinds of compound bodies, according to the differences in the mixture and the proportions of the elements which are mingled. All things which are compounded from these [elements] are also corruptible and can be resolved into the primary elements. Two of these spheres of the elements provide a habitation for all living things because of their solidity: these are earth and water. But in the other two, air and fire, it is impossible for an animal to live, because of the rarity of their substance, although those animals which are equipped with wings

---

349 preceptum: ut habetur Genesi 1 illud scilicet *add. dS*

353–57 Hic loquitur CAMPANVS ut uir catholicus non ut philosophus sicut superius demonstratione enim uel sensu apud nos nescitur ut ait PTHOLOMEVS et ALBATEGNI quod solum hec quarta habitetur uel sit habitabilis quare nichil concludit cum dicit alias 3 quartas esse sub aquis totas ymo est contra rationem secundum hoc enim spera terre esset necessario ecentrici ea uel spera aque uel ambe impossibile enim est quod circuli in eadem superficie facti qui sese intersecant sicut super idem centrum *mg. p*

373 raritatem: ubi non reuera ipsa sagax natura in nullo superflua nec in aliquo diminuta seu exorbata sed cuiuslibet perfectionis matrona trina de causa ipsa uolatilia tradidit quanta primo quidem ut eorum nociua fugerent secundo ut ipsorum necessaria utpote uictualia aquirerent tertio quod quatenus in eorum uolatu uel uolutatione gauderent, quare per aera frequenter uolatilia efferuntur et *add.* Λ

apta sunt frequenter per aerem efferantur uel ut utilia querant uel ut no-
375 ciua fugiant uel ut ipsa uolitatione letentur.

Iste quidem est numerus et ordo et situs sperarum omnium tam etheree quam elementaris regionis.

Postquam itaque sufficienter dictum est de orbibus et motibus lune et spera eius ceterisque speris omnibus, uolo nunc narrare magnitudines
380 orbium lune trium qui sunt deferens, epiciclus et circulus paruus quem describit centrum deferentis circa centrum orbis signorum et distantiam centri deferentis a centro orbis signorum longitudinemque lune longiorem et propiorem et magnitudinem corporis lune. Vt igitur que dicenda sunt plenius comprehendantur, describam orbem signorum qui sit circulus
385 *GH* super centrum *D* et circulum deferentem lune qui sit circulus *AB* circa centrum *C*; et supra centrum *D* quod est centrum orbis signorum describam secundum spatium *DC* circulum *CK* qui erit circulus quem describit centrum deferentis lune quod est punctum *C* circa centrum orbis signorum; et protraham per duo centra *C* et *D* diametrum communem
390 omnibus hiis circulis que sit linea *GDH*, sintque duo puncta in quibus hec diameter secat circulum deferentem lune *A* et *B*, *A* quidem punctum longitudinis longioris et *B* punctum longitudinis propioris. Describam quoque super duo puncta *A* et *B* que sunt longitudo deferentis lune longior et propior epiciclum lune. Sitque punctum *E* locus in quo linea *GH* secat epi-

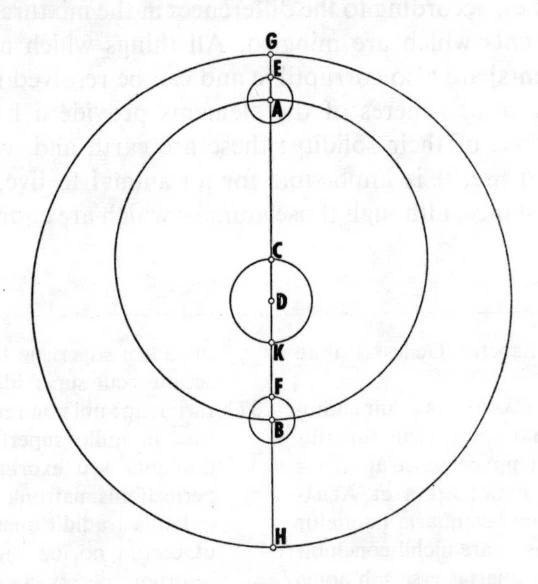

Fig. 12

380 deferens: equans et *add.* β
391 diameter: que est GDH *add.* δ/A quidem: ita quidem quod A sit δ

often travel through the air either to seek for necessaries [of life] or to escape harm or just to enjoy the act of flying itself.[61]

Such is the number, order, and position of all the spheres, both those in the ethereal region and those in the region of the elements.

Thus, now that sufficient has been said about the circles and motions of the moon and about its sphere and all the other spheres, I wish to state the sizes of three of the moon's circles, namely, the deferent, the epicycle, and the small circle which is described by the center of the deferent about the center of the ecliptic; and [I also wish to state] the distance of the center of the deferent from the center of the ecliptic, and the greatest and least distances of the moon,[62] and the size of the moon's body. Therefore, so that what I have to say may be more fully understood [see Fig. 12, p. 186], I describe the ecliptic —let this be circle *GH* on center *D*—and the moon's deferent circle—let this be circle *AB* on center *C*. On center *D*, which is the center of the ecliptic, with radius *DC* I describe circle *CK*, which will be the circle described by the center of the moon's deferent, namely, point *C*, about the center of the ecliptic. Through the two centers *C* and *D* I draw a diameter common to all these circles: let this be line *GDH*, and let the two points at which this diameter cuts the moon's deferent circle be *A* and *B*, *A* the point of greatest distance and *B* the point of least distance [on the deferent]. I also describe the moon's epicycle on the two points *A* and *B*, which are the greatest and least distances on the moon's deferent. Let point *E* be the place at which line *GH* cuts the moon's

Legend for Fig. 12

Schematic representation of the lunar model (after Campanus) to delineate the parts of its "sphere." *GH* represents the ecliptic about center *D*, the earth. *AB* represents the moon's deferent, about center *C*. *CK* is the "small circle." The epicycle is drawn twice, once about *A*, the farthest point from *D* on the moon's deferent, and once about *B*, the nearest point to *D* on the moon's deferent. *E* is the apogee of the epicycle at *A*, *F* the perigee of the epicycle at *B*. Then *DE* represents the "greatest possible distance of the moon" and *DF* its "least possible distance." These define the outer and inner boundaries, respectively, of the moon's "sphere."

392 propioris: secundum itaque spatium CA describam super centrum C circulum deferentem qui est circulus AB *add. F*Σ

395 ciclum lune in superiori parte ipsius cum ipse est in puncto *A*, et punctum *F* sit locus ubi eadem linea secat eundem in inferiori parte ipsius cum ipse est in puncto *B*. Erit ergo punctum *E* longior longitudo centri corporis lune que esse potest; punctum uero *F* erit propior que esse potest.

Hiis itaque sic descriptis narrabo magnitudines premissas, primo qui-
400 dem secundum partes illas de quibus linea *AC* que est semidiameter deferentis continet 60 partes; deinde secundum partes illas de quibus semidiameter terre est pars una; ultimo autem secundum partes uulgariter notas que sunt miliaria prout miliare unum constat ex 4000 cubitis. Dico ergo quod secundum partes de quibus linea *AC* que est semidiameter deferentis
405 est 60 partes erit linea *CD* que est distantia centri eius a centro orbis signorum 12 partes et 28 minuta. Linea uero *AE* que est semidiameter epicicli erit de eisdem partibus 6 partes et 20 minuta secundum propinquitatem. Linea uero *AD* que est longior longitudo ecentrici lune a centro terre erit 72 partes et 28 minuta; et linea *DB* que est longitudo propior ecentrici
410 lune a centro terre erit 47 partes et 32 minuta. Tota autem linea *ED* que est longior longitudo centri corporis lune a centro terre que unquam esse potest erit 78 partes et 48 minuta. Linea uero *DF* que est longitudo propior centri corporis lune a centro terre que esse potest erit 41 partes et 12 minuta.

415 At uero secundum partes illas de quibus semidiameter terre est pars una erunt predicte magnitudines ita: erit quidem linea *AC* 48 partes et 51 minuta; et linea *CD* 10 partes et 9 minuta; et linea *AE* 5 partes et 10 minuta uicinius; linea uero *AD* 59 partes sine aliqua fractione; et linea *DB* 38 partes et 43 minuta; linea quoque *ED* 64 partes et 10 minuta; et linea *DF*
420 33 partes et 33 minuta. Semidiameter autem corporis lune est secundum partes predictas uidelicet secundum quod semidiameter terre est pars una 17 minuta et 32 secunda. Semidiameter enim terre tripla est ad semidiametrum corporis lune et ulterius continens duas quintas eius fere.

De miliaribus autem prout unumquodque eorum constat ex 4000 cubi-
425 tis sic erunt predicte magnitudines. Linea quidem *AC* erit 158540 miliaria et 5 undecime unius miliaris; et linea *CD* 32941 miliaria et 4 undecime

399–400 quidem: enim narrabo premissas magnitudines δ *om. GPWw*
401 continet: habet αφ*H* / deinde: tertio quidem δ
413 a centro terre *om.* β
415 At uero: nunc autem restat easdem narrare δ
416 erunt...ita: secundum autem tales partes δ
417 linea CD: de eisdem partibus *add.* δ
423 ulterius *om.* αφ*H*

epicycle in its upper part when it [the center of the epicycle] is at point *A*, and let point *F* be the place where the same line cuts it [the epicycle] in its lower part when it is at point *B*. Then point *E* will be the greatest possible distance of the center of the moon's body, while point *F* will be [its] least possible distance.

Having thus set out this figure, then, I will state the magnitudes [of the] above [distances], first in those units of which line *AC*, which is the radius of the deferent, contains 60, then in those units of which the radius of the earth is one, and lastly in well-known units, namely, miles, according to the system in which one mile consists of 4,000 cubits.[63] I say then that, in the units of which line *AC*, which is the radius of the deferent, is 60, line *CD*, which is the distance of its [the deferent's] center from the center of the ecliptic, will be 12 units and 28 minutes; line *AE*, which is the radius of the epicycle, will be 6 units and 20 minutes approximately; line *AD*, which is the greatest distance of the moon's eccentric from the center of the earth, will be 72 units and 28 minutes; and line *DB*, which is the least distance of the moon's eccentric from the center of the earth, will be 47 units and 32 minutes. The whole line *ED*, which is the greatest distance that there can ever be between the center of the moon's body and the center of the earth, will be 78 units and 48 minutes, while *DF*, which is the least possible distance between the center of the moon's body and the center of the earth, will be 41 units and 12 minutes.[64]

But in those units of which the radius of the earth is one, the aforementioned magnitudes will be as follows: line *AC* will be 48 units and 51 minutes, line *CD* 10 units and 9 minutes, and line *AE* 5 units and 10 minutes approximately; line *AD* will be 59 units exactly, without any fraction, and line *DB* will be 38 units and 43 minutes, line *ED* 64 units and 10 minutes, and line *DF* 33 units and 33 minutes. Now the radius of the moon's body is, in the above units, namely, those in which the earth's radius is one unit, 17 minutes and 32 seconds.[65] For the earth's radius is in the ratio of 3 and about $\frac{2}{5}$ to the radius of the moon.

But in miles, according to the system in which a single mile consists of 4,000 cubits, the aforementioned magnitudes will be as follows: line *AC* will be 158,540$\frac{5}{11}$ miles, line *CD* 32,941$\frac{4}{11}$ miles, line *AE* 16,768$\frac{2}{11}$ miles, line *AD*

---

399 premissas: 4 modis diuersis et abinuicem distinctis *add.* δ

401 60 partes: secundo uero secundum partes de quibus linea AD que est a centro terre ad longitudinem longiorem ecentrici lune continet etiam 60 partes *add.* δ

403 Scito (Nota *Z*) quod CAMPANVS non posuit nisi tres modos sciendi magnitudines scilicet secundum primum modum tertium et quartum sed interponitur ille modus secundum lineam AD quia de illa posuit unam tabularum sequentium nec est ille secundus modus CAMPANVS (CAMPANI *Z*) sed additio scriptoris *mg. DZ*

414 12 minuta: *post hoc add.* δ *proportiones secundum quod AD continet 60 partes; uide appendicem A, 447*

420 33 minuta: et linea AB erit 97 partes et 42 minuta *add.* δ

unius miliaris; et linea *AE* 16768 miliaria et 2 undecime unius miliaris; et linea *AD* 191481 miliaria et 9 undecime unius miliaris; et linea *DB* 125653 miliaria et 2 undecime unius miliaris; et linea *ED* 208250 miliaria sine
430 aliqua fractione; et linea *DF* 108885 miliaria absolute. Semidiameter autem corporis lune erit 948 miliaria et 13 trigesime tertie unius miliaris. Et tota ipsius diameter erit duplum eius uidelicet 1896 miliaria et 26 trigesime tertie unius miliaris. Ambitus uero lunaris corporis in circuitu erit 5958 miliaria et 22 trigesime tertie unius miliaris. Et si super lineam *ED* addi-
435 derimus medietatem diametri corporis lune, erit quod agregabitur distantia superficiei conuexe spere lune a centro terre et ipsa est 209198 miliaria et 13 trigesime tertie unius miliaris. Rursus quoque si eandem semidiametrum corporis lune subtraxerimus de linea *DF*, remanebit distantia superficiei concaue spere lune a centro terre et ipsa est 107936 miliaria et
440 20 trigesime tertie unius miliaris. Est igitur spissitudo spere lune 101261 miliaria et 26 trigesime tertie unius miliaris. Ambitus autem conuexe superficiei spere sue est 1314961 miliaria et 11 trigesime tertie unius miliaris. Ambitus uero concaue est 678458 miliaria et 22 trigesime tertie unius miliaris. Ambitus quoque sui deferentis est 996540 miliaria. Et si ipsum
445 diuiseris per 27 dies et 8 horas qui sunt tempus unius reuolutionis centri epicicli in equante lune, exibit tibi motus eius in die una. Erit itaque dieta centri epicicli de partibus equantis 36458 miliaria et 47 minuta unius miliaris. Ambitus uero epicicli lune est 105400 miliaria; et si ipsum diuiseris per 27 dies et 13 horas qui sunt tempus unius reuolutionis eius in suo epi-
450 ciclo, exibit tibi dieta eius in epiciclo et est 3826 miliaria et 56 minuta unius miliaris.

Quantitas autem corporis lune comparata ad quantitatem corporis terre est quadragesima pars corporis eius fere. Corpus enim terre continet corpus lune 40 uicibus et insuper quintam decimam partem eius fere.
455 PTHOLOMEVS quidem quando posuit corpus terre trigintuplum nonuplum ad corpus lune et quartam eius partem fere processit secundum facilitatem que propinqua fuit ueritati. Ipse enim supposuit quod diameter terre contineret tribus uicibus diametrum lune et duas eius quintas quemadmodum ipse dicit et ita supposuit quod proportio diametri terre ad dia-
460 metrum lune esset quemadmodum proportio 17 ad 5. Cubi autem istorum

427–28 et linea AE...9 undecime unius miliaris *om. BCdgMm*
433 lunaris corporis: lune δ
442 sue: lune *BbCDfhMmPUXYy*ΓΛ
447 epicicli: lune *add.* βφ
453 quadragesima: quinquagesima *AdFiqUu vxy*ΠΣΨ alias 50ª *mg. O*
455 quando: ipse *add.* βδ
456 partem *om.* β/facilitatem: facultatem β*H*
458 continet αφ*H*
458–60 et duas...proportio: triplicatam (triplicata *o*Λ) et tunc proportio dyametri terre ad dyametrum lune esset quemadmodum proportio triplicata φ
459 dixit β
460 est β

191,481$\frac{9}{11}$ miles, line *DB* 125,653$\frac{2}{11}$ miles, line *ED* 208,250 miles without any fraction, and line *DF* 108,885 miles exactly. The radius of the moon's body will be 948$\frac{13}{33}$ miles.[66] The whole of its diameter will be double that, namely, 1,896$\frac{26}{33}$ miles. The circumference of the moon's body will be 5,958$\frac{22}{33}$ miles.[67] And if we add the radius of the moon's body to line *ED*, the sum will be the distance of the convex surface of the moon's sphere from the center of the earth: that is 209,198$\frac{13}{33}$ miles. Again, if we subtract that same radius of the moon's body from line *DF*, the remainder will be the distance of the concave surface of the moon's sphere from the center of the earth: that is 107,936$\frac{20}{33}$ miles. Therefore the thickness of the moon's sphere is 101,261$\frac{26}{33}$ miles;[68] the circumference of the convex surface of its sphere is 1,314,961$\frac{11}{33}$ miles; the circumference of its concave [surface] is 678,458$\frac{22}{33}$ miles. The circumference of its deferent is 996,540 miles; and if you divide the latter by 27 days and 8 hours,[69] which is the time of one revolution of the center of the epicycle on the moon's deferent,[70] your result will be its motion during one day. Thus the day's journey[71] of the center of the epicycle on the deferent will be 36,458 miles and 47 minutes of a mile.[72] The circumference of the moon's epicycle is 105,400 miles; and if you divide the latter by 27 days and 13 hours,[73] which is the time of one revolution of the moon on its epicycle, your result will be its daily journey on the epicycle: that is 3,826 miles and 56 minutes of a mile.[74]

The size of the moon's body is about a fortieth part of the size of the earth's body. For the earth's body contains the moon's body 40 times plus about $\frac{1}{15}$ part of it. Now PTOLEMY, when he stated that the body of the earth is approximately 39 and a quarter times the body of the moon, operated according to an easy procedure, which is [only] approximately correct.[75] For he supposed that the diameter of the earth is 3$\frac{2}{5}$ times the diameter of the moon, as he himself says, and thus he supposed that the ratio of the diameter of the earth to the diameter of the moon was 17 to 5. Now the cubes of the latter two numbers,

---

460 ad 5: hoc inuenies per multiplicationem 5 in se cubice et 17 in se cubice *add. BMΓ*

duorum numerorum qui sunt 4913 et 125 se habent in proportione predicta. Sed cum nos exquisita numeratione comparauerimus diametrum terre ad diametrum lune in numeris primis et minimis inueniemus esse proportionem ipsarum sicud est proportio duorum numerorum qui sunt 900 et 263. Isti enim duo numerant duos numeros in quibus proportio earum secundum fractiones premissas reperitur equaliter secundum numerum qui est 1309. Et numeri continentes proportionem predictarum diametrorum secundum fractiones premissas sunt isti duo numeri 1178100 et 344267. Si igitur duorum numerorum predictorum eorum uidelicet qui sunt primi et minimi in proportione duarum predictarum diametrorum inuenerimus cubos, inter eos erit proportio corporis terre ad corpus lune; proportio enim quarumlibet duarum sperarum adinuicem est sicud proportio duorum cuborum adinuicem quorum latera sunt diametri sperarum. Nam proportio cuborum adinuicem est sicud proportio suorum laterum triplicata; sperarum quoque sicud suarum diametrorum triplicata; quemadmodum proportio quadratorum est sicud proportio suorum laterum duplicata, et circulorum sicud suarum diametrorum similiter duplicata. Sunt autem illi duo cubi 729,000,000 et 18,191,447. Cum itaque diuiserimus maiorem eorum per minorem exibunt 40 integra et 4 minuta fere que sunt denominatio proportionis premissorum corporum terre scilicet et lune. Hee igitur predicte sunt uerissime numerationes magnitudinum premissarum extracte ex primis et ueris radicibus PTHOLOMEI demonstratoris nobilis.

Et ut magnitudines predicte promptius inueniri possint, supposui tres tabulas in quibus ipsas secundum suum ordinem descripsi; quibus etiam quartam tabulam preposui in qua easdem magnitudines annotaui secundum partes illas de quibus linea *AD* que est a centro terre ad longitudinem longiorem ecentrici lune habet 60 partes: nam ad hoc PTHOLOMEVS posuit principia et radices.

Restat autem ut ostendamus nunc qualiter ea que predicta sunt de luna ponenda sunt in instrumento nostro ut per ipsum possimus equare lunam

462 Sed: tamen *add.* α
463 minimis: numeris secundis δ
466 equaliter: quoniam *add.* αφ*H* quia *add.* *C*Γ quandoque *add.* *K*
467 continentes proportionem: conuenientes proportioni α
470 primi: scilicet de 900 et de 263 *add.* δθ *mg.* Σ
477 similiter duplicata: proportio equalis *Z*
482–83 demonstratoris nobilis *om.* β
484 supposui: suprascripsi δ presubposui *F*Σ supponam *rz*
486 quartam *om.* β/preposui: apposui δ
488 longiorem: a centro *add.* β

namely, 4,913 and 125, are in the proportion stated [i.e., about 39¼ to 1]. But when we use exact numbers and compare the diameter of the earth to the diameter of the moon in the smallest integers, we shall find that they are in the ratio of the two numbers 900 and 263. For those two divide the two numbers which give the ratio of the diameters, according to the above fractions, an equal number of times, the number of times being 1,309. The numbers giving the ratio of the aforementioned diameters, according to the above fractions, are the following two: 1,178,100 and 344,267. Therefore, if we find the cubes of the above two numbers [900 and 263], which are the smallest integers giving the ratio of the two diameters in question, they will give the ratio of the volume of the earth to the volume of the moon, because the ratio between [the volumes of] any two spheres is the same as the ratio between the two cubes whose sides are the diameters of the spheres. For the ratio between cubes is the same as the cube of the ratio between their sides, and similarly the ratio between spheres is the same as the cube of the ratio between their diameters, just as the ratio between squares is the same as the square of the ratio between their sides, and similarly [the ratio] between circles is the same as the square [of the ratio] between their diameters. Now those two cubes are 729,000,000 and 18,191,447. Thus, when we divide the greater by the less, the result will be 40 and about 4 minutes: this is the denominator of the ratio between the aforesaid volumes, namely, the earth's and the moon's. Therefore the aforementioned figures are the truest measurements of the magnitudes in question, derived from the basic and true parameters of PTOLEMY, that notable giver of proofs.

In order that the above magnitudes may be more easily found, I have set out below [Table 1, pp. 356–57] three tables in which I have written them down in their proper order. To these I have prefixed a fourth table,[76] in which I have noted the same magnitudes in the units of which *AD*, which is the distance from the center of the earth to the greatest distance on the moon's eccentric, contains 60: for that is the system of units in which PTOLEMY gave the basic distances and parameters.

It remains for us now to show how the foregoing description of the moon is to be represented on our instrument, so that by it we can equate the moon

---

465 et 263: scilicet quod una continet aliam 40 uicibus et quintam decimam partem eius (ipsius *y*) *add.* φ*HJ* Alibi habetur iste enim duo inuenerunt duos numeros in quibus proportio earum secundum fractiones premissas reperitur equaliter quoniam secundum numerum qui est 1309 et numerum continentes proportiones predictarum dyametrorum secundum fractiones sunt isti duo 1178100 et 344267 *add.* Λ

471 Hinc dicendum potius uidetur lunae proportionem ad corpus terrae esse ut unum ad 49 etenim cum luna est in perigio sui epicycli ?chordae min. 36 eius ergo semidiam. apparens est min. 18 terrae ?autem 66 lunae ergo ad terram est proportio qua 18 ad 66 seu 9 ad 33 siue 3 ad 11 omnino *mg.* Ψ_2

489 radices: Ideo hic demicant *add.* *J*

quandocumque opus fuerit. De premissis uero non studebimus aliquid ponere in hoc instrumento nostro nisi quod fuerit necessarium ponere in eo quantum attinet ad equationem cursus lune. Accipiam itaque unam ex 495 tabulis tribus rotundis quas dixi in principio que sit ualde plana et ualde pollita ex utraque parte, et in una parte eius scribam in medio ipsius punctum *D* supra quem describam circulum unum maiorem quem potero describere in ea, et sit iste circulus orbis signorum quem quadrabo duabus diametris, ipsumque diuidam in 12 signa et sub eo faciam alios circulos 500 tres concentricos sibi, ita quod secundus tantum distet a primo quod inter eos scribantur nomina signorum et tertius tantum a secundo quod inter eos ponatur diuisio graduum totius circuli et quartus tantum a tertio quod inter eos possint scribi numeri graduum distinctorum per 5 et 5, et fiant hec omnia prorsus eodem modo sicud in sole dictum est et sicud 505 uides in presenti figura. Postquam ista erunt ita disposita relinquam aliquod spatium modicum in una suarum diametrorum infra interiorem circulum tantum scilicet quantum uoluero distare circulationem maiorem epicicli lune ab ipso interiori circulo. Residuum uero spatium quod erit in eadem diametro usque ad punctum *D* diuidam in 96 partes quod leuiter 510 faciam per 12 et 8 cum suis partibus. De istis autem 96 partibus accipiam 12 partes et 28 minuta que ponam dimidium unius a parte centri *D*, et ubi terminabuntur ponam notam *C* eritque *C* centrum deferentis lune. Describam igitur circa centrum *D* et secundum spatium *CD* circulum *CK* qui erit circulus paruus quem describit centrum deferentis lune circa cen- 515 trum orbis signorum. Itemque a puncto *C* uersus circumferentiam orbis signorum accipiam de predictis 96 partibus 60 partes, et ubi terminabuntur ponam notam *A* eritque punctus *A* punctus augis deferentis et linea *AC* semidiameter deferentis. Lineabo igitur super centrum *C* et secundum spatium *AC* circulum *AB* qui erit ipse deferens lune. Rursus autem a 520 puncto *A* uersus circumferentiam orbis signorum de predictis 96 partibus accipiam 6 partes et tertiam partem unius que est 20 minuta, et ubi terminabuntur ponam notam *E* eritque punctus *E* punctus augis epicicli quando centrum epicicli est in auge deferentis. Et faciam super centrum *A* et secundum spatium *AE* circulum *EF* qui erit epiciclus lune quem opor- 525 tet diuidere in 12 signa et quodlibet signum in 30 gradus ut in eo possint prompte inueniri partes omnium argumentorum lune. Quia uero epiciclus

492 studebimus: studeas breuius *bDX* stude breuius *N* *om.* *AΛΠ*
493 ponere[1]: ponemus *AIoΛ*/ponere[2]: poni δ
494 unam *om.* β
495 in principio: nisi *add.* *B* fabricandas unam *add.* *MΓ*
499 circulos *om.* β
503 scribi: poni αφ
504–5 dictum...figura *om.* β
507 uolam α*H* uolo *AlT*Π uolueris Λ
509 leuiter: breuiter β linealiter *f*
518 deferentis: lune *add.* βφ
519 erit ipse: est circulus δ est *Z*
525 in eo *om.* β
526 omnes β

whenever it may be necessary. However, we will not trouble to put any of the above-mentioned features on this instrument of ours beyond what may be necessary with respect to equating the moon's motion. I take, then, one of the three circular boards which I mentioned at the beginning:[77] this should be very level and very smooth on both sides. On one side of it, in the middle, I mark point *D*, and with this as center describe a circle, the largest that I can draw on it [see Fig. 13, p. 196]: let that circle represent the ecliptic. I divide it into four quadrants by means of two diameters and [also] divide it into 12 signs. Inside it I draw three other circles concentric with it, such that the second is far enough away from the first for the names of the signs to be written between them, and the third far enough from the second for the division of the whole circle into degrees to be marked between them, and the fourth far enough from the third for the numbers of the degrees to be written between them, one number at each five-degree division. Let all this be done in exactly the same way as that described for the sun,[78] and as you see in the adjoining figure [Fig. 13]. After the above dispositions have been made, I mark off a short length along one of the diameters in from the inmost circle, namely, as much as I want the distance to be between the bigger circular representation of the moon's epicycle[79] and the inmost circle. The rest of the length of that diameter up to point *D* I divide into 96 parts:[80] I can easily do this by division into 12 parts and [further] division into 8 parts. Beginning from center *D* I take 12 parts and 28 minutes[81]—these [minutes] I take as half of one part—out of the 96 parts, and at their end I mark point *C*, which will be the center of the moon's deferent. So I describe on center *D*, with radius *CD*, circle *CK*, which will be the small circle which is described by the center of the moon's deferent about the center of the ecliptic. Again, from point *C* toward the circumference of the ecliptic I take 60 of the above 96 parts,[82] and at their end I mark point *A*; point *A* will be the apogee of the deferent, and line *AC* the radius of the deferent. Therefore I draw on center *C*, with radius *AC*, circle *AB*, which will be the moon's actual deferent. Again, from point *A* toward the circumference of the ecliptic I take 6 parts and $\frac{1}{3}$, i.e., 20 minutes,[83] of the above 96 parts, and at their end I mark point *E*: this will be the apogee of the epicycle when the epicycle is in the apogee of the deferent. I draw on center *A*, with radius *AE*, circle *EF*, which will be the moon's epicycle. This [circle] ought to be divided into 12 signs, and each sign into 30 degrees, so that every [possible] lunar anomaly might readily be found on it [expressed] in its [proper] units. But

---

524 epiciclus lune: et super idem centrum scilicet A describam circulum alium *add.* φ

-·-·-·- Tabula motus epicicli
·········Tabula motus centri ecentrici
- - - -Tabula motus centri epicicli

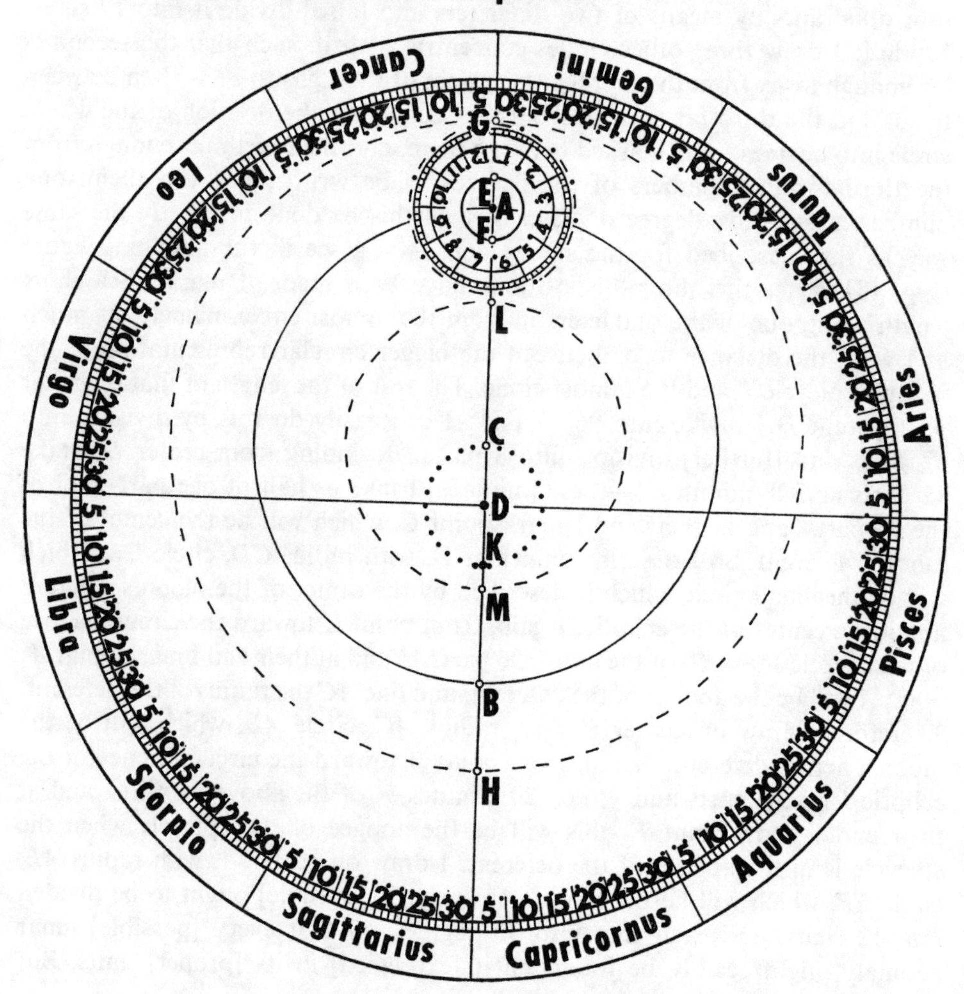

INSTRVMENTVM EQVATIONIS LVNE

Fig. 13

Legend for Fig. 13

Instrument for the moon (after Campanus). The outer graduated system of circles (with center *D*, the earth) represents the ecliptic. Circle *AB* represents the moon's deferent, *CK* the "small circle." *EF* represents the true epicycle, but because of its small size, the graduation is marked on the "epicycle in its convenient representation" drawn about the same center, *A*, but outside *EF*. The dotted and dashed lines represent the edges of the moving parts. The innermost of these (indicated by alternate dots and dashes) is the "disk of the motion of the epicycle" (merely the "epicycle in its convenient representation"); it revolves about center *A* within the "disk of the motion of the center of the epicycle" (indicated by the dashed lines). The latter revolves about center *C* within the "disk of the motion of the center of the eccentric" (indicated by the dotted lines), which revolves about center *D*. Threads are attached by one end to points *A*, *D*, and *K* (the thread at *K* is actually attached on the movable "disk of the motion of the center of the eccentric," not on the fixed circle *CK*). The moon's true position is found as follows. Stretch the thread at *D* through the point on the ecliptic whose distance from the mean sun is equal and opposite to that of the mean moon, and revolve the outermost movable disk (the "disk of the motion of the center of the eccentric") until point *A* coincides with the thread. (The moon's deferent is then in the correct position for the "crank mechanism" to operate.) Next stretch the thread at *D* through the point on the ecliptic marking the mean longitude of the moon, and, keeping the outermost disk fixed, move the "disk of the motion of the center of the epicycle" until point *A* again falls beneath the thread. (The epicycle is then in its correct position.) Stretch the thread at *K* through point *A*, and revolve the "disk of the motion of the epicycle" until the line between signs 1 and 12 lies along the thread. Stretch the thread at point *A* through the point on the graduated epicycle marking the moon's mean anomaly, and note where that thread cuts circle *EF*. Stretch the thread at *D* through that intersection of circle *EF*: where the thread at *D* now cuts the ecliptic is the moon's true longitude. Compare Figure 6 (p. 45).

EF est minoris quantitatis quam predicta diuisio requirat, nimis enim oporteret esse magnum instrumentum ad hoc quod epiciclus lune in eo posset diuidi secundum partes predictas, idcirco super centrum *A* descri-
530 bam circulum alium cuius circumferentia transeat per finem predictarum 96 partium eritque semidiameter istius circuli 23 partes et 32 minuta de predictis partibus. Dehinc uero infra istum circulum et super idem centrum scilicet *A* describam alium circulum tantum distantem a priori quod inter ipsos possint poni diuisiones graduum 360. Itemque infra istum circulum
535 et super idem centrum faciam circulum tertium tantum distantem a secundo quod inter ipsos possint poni numeri graduum distinctorum per 5 et 5. Rursus super idem centrum faciam circulum quartum tantum distantem a tertio quod inter eos possint scribi numeri signorum que omnia et etiam diuisiones ipsorum faciam penitus sicud feci superius in circulo ecentrico
540 solis. Sed in scribendo numerum graduum et signorum faciam contrario modo ei quod feci in sole. Incipiam enim ab auge epicicli et procedam scribendo numerum tam graduum quam signorum uersus dextram; nam luna in superiori parte sui epicicli mouetur uersus occidentem et in inferiori uersus orientem ut dictum est supra, omnes autem alii planete
545 econtrario.

Postquam autem hec ita perfecero faciam ita quod centrum deferentis quod est punctum *C* moueatur circa centrum orbis signorum quod est punctum *D* uersus occidentem et contra successionem signorum et centrum epicicli super circumferentiam deferentis uersus orientem et ad suc-
550 cessionem signorum. Sic enim oportet esse dispositionem huius instrumenti quemadmodum apparet ex predictis. Hoc autem faciam tali ingenio.

529 predictas: omnes *BMmOqy*Γ omnes *add. adhStU*
531 istius: parui *add.* αφ
533 tantum: solummodo *add.* δ
534 infra: intra δθφ
543 occidentem: dextram scilicet occidentem β
551 tali ingenio: hoc modo β

epicycle *EF* is too small for the requirements of the above division, for the instrument would have to be overly big to accommodate a lunar epicycle which could be divided in the above-mentioned way; therefore I draw on center *A* another circle,[84] whose circumference is to pass through the end of the above 96 parts: the radius of that circle will be 23 parts and 32 minutes of the above parts.[85] Next, inside the latter circle and on the same center, namely, *A*, I draw another circle far enough from the first for the divisions of the 360 degrees to be put between them. Again, inside that circle and on the same center, I draw a third circle far enough from the second for the numbers of the degrees to be written between them, one number every fifth degree. Again on the same center I draw a fourth circle, far enough from the third for the numbers of the signs to be written between them. All this, as well as the divisions, I carry out exactly as I did above for the sun's eccentric circle, except that, in writing the numbers of the degrees and signs, I do the opposite of what I did for the sun: I begin from the apogee of the epicycle and proceed toward the right [i.e., clockwise] in writing the numbers of degrees and signs alike. For the moon moves toward the west in the upper part of its epicycle and toward the east in the lower, as was stated above, while all the other planets do the opposite.

After I have finished doing this as described, I arrange that the center of the deferent, namely, point *C*, shall move about the center of the ecliptic, which is point *D*, toward the west, counter to the succession of the signs, and that the center of the epicycle shall move on the circumference of the deferent toward the east, following the succession of the signs; for such is the way in which this instrument should be arranged, as is clear from what has been said before. This I do by the following device: I hollow out the whole circular area between

---

545 Luna cadit supra; mercurius occidit infra *mg. o*

551 Hoc...ingenio: potest enim fieri in papiro aliquantule spisitudinis uel in pergameno uel in alia quacumque conuenienti materia *g*/Ingenium meliorem qua facere possit uel plus leuiorem hoc modo fit: Recipe puntum C quod est centrum deferentis et fac circulum breuem a punto A circa centrum D; post hoc pone circinum in punto C et fac circulum donec totum diametrum sit HCK; hoc facto acipe unam tabulam ualde politam in quantitate istius circuli. Postea fac unam aliam tabulam paulo maiorem prima deinde iunge rete due tabule simul bene certificati donec apareat in estremitate istius tabule in modum concauitatis; post hec faciamus unum circulum quod intret in illa concauitate donec ueniat ab estremitate circulationis secunde tabule et iste circulus circumuoluat in illa concauitate et isto circulo ponas epiciclum cum debito modo ut aparet in subiepta [*sic*] fiura [*sic*]. Et proprio modo facias in deferenti mercurio. Et nota quod est regula simmetrica quod in omni circulatione puntum circumu[o]luere super centrum ipsius; acipe extremi[ta]te illius circuli quam uis et acipe punto illo quod est per circulationem AHTK per centrum C tantum est dicere quantum accipere totidem quantitatis primi circuli ut aparet in parua fiura *mg. cum figura z*

Concauabo totum spatium in circuitu quod est intei circulum *CK* et interiorem illorum 4 circulorum qui descripti sunt propter diuisionem orbis signorum, et faciam concauitatem istam equalem ualde quantum possibile
555 fuerit uel ad tornum uel cum instrumento aliquo ad hoc apto uel superposito alicui tabule plane figure et magnitudinis oportune limbo aliquo in quo sit descriptio circuli signorum facta predicto modo, et in centro eius superposita tabula simili et equali circulo *CK*, ita uidelicet quod ista tabula et limbus sint equalis altitudinis. Et uolo quod concauitas ista sit
560 tante profunditatis quod corpus ab ea contentum cuius superficies sit ipsamet superficies tabule possit aliam concauitatem recipere in qua aliud corpus cuius superficies sit ipsamet superficies tabule possit collocari. Et fiat ista concauitas non perpendiculariter quidem descendens sed ex parte circuli *CK* declinet in profundo eius uersus centrum *D*, et ex parte circuli
565 signorum declinet etiam in eius profundo uersus circumferentiam exteriorem totius tabule ut corpus in ipsa concauitate locatum non possit inde exire. Postea faciam tabulam aliam isti concauitati similem quam uocabo tabulam motus centri ecentrici que cum posita fuerit in concauitate predicta moueatur in ea uniformiter et equaliter. In qua describam circulum
570 deferentem lune et epiciclum ipsius secundum utramque suam circulationem ueram uidelicet et oportunam quemadmodum uides in premissa figura. Et super centrum ipsius deferentis quod est punctum *C* describam in tabula ista duos circulos quorum unus supergrediatur aliquantulum epiciclum oportune circulationis qui sit *GH* et alius aliquantulum subsistat
575 eidem qui sit *LM*. Postquam autem istos circulos ita descripsero faciam in ista tabula que dicitur tabula motus centri ecentrici concauitatem aliam inter circulum *GH* et circulum *LM* et faciam eodem modo quo primam uel ad tornum uel cum instrumento ad hoc apto uel secundum uiam limbi secundum quam si processeris facies duos limbos lunaris forme quorum
580 unus erit sicud figura contenta inter circulum *GH* et interiorem circulum 4 extremorum et alius sicud figura contenta inter circulum *LM* et circulum *CK*. Sitque etiam ista concauitas adeo profunda quod corpus in ipsa locatum possit aliam concauitatem recipere in qua iaceat et moueatur epiciclus oportune circulationis, et non sit etiam ipsa perpendiculariter de-

554 equalem: planam δ
559 limbet δ*Af*
564–65 uersus...profundo *om.* *BdiMmqtY* ΓΨ
571–72 quemadmodum...figura *om.* β
577 quo: feci *add.* δ
579 facias β
582 etiam *om.* α*Z*
582–83 in ipsa locatum: ab ea contentum β

circle *CK* and the inmost of the four circles which were drawn for the purpose of representing the division of the ecliptic, and I make this cavity as level as possible, either on the lathe or with some instrument suitable for the purpose, or else by fastening, onto a flat board of suitable size and shape, a "limb"[86] on which the circle of the ecliptic has been drawn in the way described above and by attaching to its center a disk similar [in shape] and equal [in size] to circle *CK*, such that the disk and the limb are of equal thickness. And I require that that cavity be deep enough that a body contained in it whose surface is continuous with the surface of the [original] board can itself contain another cavity in which another body whose surface is continuous with the surface of the board can be placed.[87] The cavity should be made, not with vertical [sides], but sloping from top to bottom toward center *D* on the side of circle *CK*, and sloping from top to bottom toward the circumference of the whole board on the side of the circle of the ecliptic,[88] so that a body fitted into the cavity cannot slip out of it. Next I make another disk similar [in size and shape] to the cavity: I call this the "disk of the motion of the center of the eccentric"; when it has been fitted into the above-mentioned cavity, it should move uniformly and evenly inside it. On this disk I draw the moon's deferent circle and its epicycle in both forms of its circular representation, the true and the convenient, as you see in the foregoing figure. On the center of the deferent, which is point *C*, I draw on the disk two circles, of which one should pass just outside the epicycle in its convenient representation—let this be *GH*—and the other should pass a little inside it—let this be *LM*. After I have thus drawn those circles, I make, in that disk which is called the disk of the motion of the center of the eccentric, another cavity, between circle *GH* and circle *LM*, in the same way as I made the first, either on the lathe or with an instrument suited to this purpose or by means of a limb. If you use the latter method, you make two limbs in the form of "lunes,"[89] of which one will be of the dimensions of the figure contained between circle *GH* and the inmost of the four outer circles and the other [of the dimensions] of the figure contained between circle *LM* and circle *CK*. This cavity also should be deep enough so that a body placed in it can contain a further cavity in which the epicycle, in its convenient representation, may lie and move. It, too, should not have vertical [sides], but should

555–56 Nota quod mihi uidetur melius esse ut fiat concauitas ad tornum quia mouendo tabulam mouebuntur ambo deferentes simul et uniformiter uel si fiant duo limbi quod figantur ambo super aliqua tabula uel coniungatur cum cola ut ad motum tabule ambo moueantur *mg. S*

561 possit: in ea collocari et ipsamet superficies tabule possit *add. K* aliam concauitatem recipere in qua aliud corpus cuius superficies sit etiam ipsamet superficies tabule possit *mg. A*

571–72 figura: si libuerit uidere figuram ab autore primo et literaliter descriptam amoueas rotas et se ostendet *mg.* Γ

585 scendens sed declinans in interiori parte uersus centrum et in parte exteriori uersus circumferentiam ut corpus in ea locatum non possit inde exire. Postea faciam tabulam aliam isti concauitati similem quam uocabo tabulam motus centri epicicli in qua describam iterum circulum deferentem lune et epiciclum eius secundum utramque circulationem. Et faciam rur- 590 sus in ea concauitatem aliam secundum quantitatem que est inter extremum 4 circulorum epicicli oportune circulationis et intimum quod faciam uno predictorum modorum. Quam etiam faciam tali modo quod ipsa recuruetur ad interiora et exteriora ne corpus sibi impositum possit egredi ab ea. Deinde faciam aliam tabulam isti concauitati similem quam uocabo 595 tabulam motus epicicli in qua describam epiciclum oportune circulationis per 4 circulos et suas diuisiones subscribendo numeros graduum et signorum quemadmodum uides in figura. Quibus peractis componam tabulam motus epicicli in concauitate sua et tabulam motus centri epicicli in sua et tabulam motus centri ecentrici in sua. Et sint omnia ista ordinata ita quod 600 nulla istarum tabularum superemineat alteri sed omnium continentium et contentarum sit prorsus superficies una, et quod quelibet tabula moueatur in concauitate sua motu leui et equali ita quod tabula contenta contingat continentem ex omni parte et ex omni modo. Post ista uero faciam unum signum notabile in puncto augis deferentis et unum in puncto augis epicicli 605 oportune circulationis, et figam clauum unum perpendiculariter erectum cuius hasta sit tenuissima super centrum orbis signorum quod est punctum *D* et alium super centrum epicicli quod est punctum *A* et alium super punctum qui in circulo *CK* opponitur augi deferentis, hoc est in puncto *K*. Et ligabo ad pedem cuiuslibet istorum clauorum unum filum de seta 610 tenuissimum et equale.

Cum itaque uoluero equare lunam per illud instrumentum, intrabo cum tempore ad quod uolo ipsam equare quod uoco tempus datum in tabulam medii motus lune. Itemque cum eodem tempore intrabo in tabulam elongationis lune a sole. Rursus quoque cum eodem tempore intrabo in tabu- 615 lam medii argumenti lune et queram unumquodque eorum operando eodem modo, sicud docuimus operari superius in querendo argumentum solis. Postea accipiam medium motum lune et uidebo ubi terminetur in orbe signorum incipiendo ab ariete et faciam ibi notam uel materialem uel solum in intellectu. Considerabo quoque ubi terminetur numerus dis- 620 tantie lune a sole in orbe signorum incipiendo ab ariete ibique faciam

585 declinans: decliuis β
586 in ea: ab ea α*H*
593 recuruetur: retornetur *A*
598 et tabulam…epicicli in sua *om.* δ
602 leui: facili δ
604 augis[2] *om.* β
608 qui: quod δ
609 seta: serico $Aa_1hJryz\Gamma$ serica *cfj* seta serica Λ
612 tabula β tabula tempus *f*

slope [from top to bottom] toward the center at its inner edge and toward the circumference at its outer edge, so that a body fitted into it cannot slip out of it. Next I make another disk fitting that cavity and call it the "disk of the motion of the center of the epicycle." On it I draw again the moon's deferent circle and the epicycle in both its representations. I make in it, in turn, another cavity corresponding to the dimensions [of the area] between the outermost and inmost of the four circles of the epicycle in its convenient representation: I do so by one of the above-mentioned methods. This [cavity], too, I make in such a way that it is beveled toward the inside and outside, so that a body fitted into it cannot slip out of it. Then I make another disk fitting that cavity and call it the "disk of the motion of the epicycle." On it I draw the epicycle in its convenient representation, using the four circles and their divisions and writing in the numbers of the degrees and signs, as you see in the figure. Having completed these things, I fit the disk of the motion of the epicycle into its cavity, and the disk of the motion of the center of the epicycle into its, and the disk of the motion of the center of the eccentric into its. All these should be arranged in such a fashion that none of the disks projects above another, but the surface of all, whether containing or contained, is one and the same, and that every disk moves in its cavity with a smooth and even motion, so that the contained disk touches its container on all sides and in all ways. Next I make one clearly visible mark at the point of the apogee of the deferent, and one at the point of the apogee of the epicycle in its convenient representation, and affix a nail with a very thin spike upright on the center of the ecliptic, which is point *D*, and another on the center of the epicycle, which is point *A*, and another on the point on circle *CK* which is opposite the apogee of the deferent, that is, on point *K*.[90] I tie at the base of each of these nails a silk thread, very fine and of uniform thickness.

So when I want to equate the moon by this instrument, I enter with the time at which I want to equate it, which I call the datum time, into the table of the moon's mean motion. And I enter again with the same time into the table of the moon's elongation from the sun.[91] I enter once more with the same time into the table of the moon's mean anomaly. In searching out each of them, I operate in the same way as we instructed above[92] in seeking the sun's argument. Then I take the moon's mean motion and see where it brings me on the ecliptic, beginning [to count] from [the beginning of] Aries. There I make a mark, either on the actual instrument or just in my memory. I also observe where the number for the moon's elongation from the sun brings me on the ecliptic, beginning from Aries, and make a mark there: that is where the mean sun will

---

610 equale: undique. Sic ergo complete habetur compositio huius instrumenti quod uocatur instrumentum equationis lune *add. DFMNVZ*

notam eritque ibi sol secundum medium eius motum. Queram itaque locum qui tantum antecedat locum solis medium quantum locus solis medius antecedit locum lune ita quod sol sit in medio utriusque, ibique etiam faciam notam. Accipiam itaque filum ligatum ad punctum *D* et extendam
625 ipsum et faciam transire per hanc notam ultimam. Postea circumuoluam tabulam motus centri ecentrici quousque aux deferentis lune cadat sub isto filo. Deinde ponam illud filum extensum supra notam primam que terminat medium motum lune, et manente tabula motus centri ecentrici fixa secundum situm in quo posui augem ecentrici, circumducam tabulam
630 motus centri epicicli quousque centrum epicicli cadat sub filo isto, eruntque aux deferentis et centrum epicicli lune disposita in isto instrumento materiali quemadmodum sunt in celo. Est enim utrumque illorum sub illo puncto orbis signorum sub quo debet esse, sol quoque secundum medium motum eius cadit in medio eorum. Post hec autem sumam medium
635 argumentum lune et queram ipsum in epiciclo oportune circulationis, et ubi terminabitur numerus argumenti in epiciclo faciam notam. Tunc igitur accipiam filum ligatum ad punctum *K* et ipsum extensum faciam transire per centrum epicicli usque ad extremam circumferentiam ipsius, et circumducam tabulam motus epicicli quousque aux epicicli cadat directe
640 sub isto filo. Et tunc filum ligatum ad centrum epicicli quod est punctum *A* extendam ad notam prius signatam que terminat argumentum lune et considerabo ubi illud filum secat circumferentiam circuli *EF* qui est epiciclus lune secundum ueritatem, et per punctum illum extendam filum ligatum ad punctum *D*, et ubi filum illud secabit orbem signorum ibi erit
645 luna secundum uerum eius motum.

Si autem per hoc instrumentum uolueris scire utrum luna sit cursu tarda aut cursu uelox, facias quod filum quod ligatum est ad punctum *D* contingat epiciclum uere circulationis ex utraque parte. Et si fuerit luna in arcu superiori inter duos contactus, ipsa erit cursu tarda; si autem in in-
650 feriori, ipsa erit cursu uelox.

De motibus igitur et orbibus et spera lune ceterisque speris omnibus et de magnitudinibus orbium et totius corporis eius et de distantia quoque centri deferentis et longitudinis longioris et propioris et de spissitudine spere sue, de modo etiam compositionis instrumenti propositi et de arte
655 equandi lunam per ipsum, talia et tanta dixisse sufficiat.

Deinceps autem de mercurio sermo sequatur. Prius tamen patefaciemus causas operationum tabularum in equando lunam quemadmodum fecimus in sole. Dico ergo quod si luna semper esset in uera auge sui epicicli uel in

628 motus *om.* φ
633 esse *om.* β
641 extensum α / signatam: assignatam β / terminat: medium *add.* β
644 illud *om.* β
647 ligatum: alligatum αφ*H*
652 et totius corporis eius et: illorum trium corporum δ suorum trium et corporis *Hz*

be.[93] I look, therefore, for the point which precedes the mean sun by as much as the mean sun precedes the moon, so that the sun will be halfway between the two points; there, too, I make a mark. Then I take the thread attached to point *D* and, stretching it tight, make it pass through this last mark. Then I revolve the disk of the motion of the center of the eccentric until the apogee of the moon's deferent falls beneath the thread. Then I stretch the thread over the first mark, which indicates the position of the mean moon, and, leaving the disk of the motion of the center of the eccentric fixed in the position in which I put the apogee of the eccentric, I revolve the disk of the motion of the center of the epicycle until the center of the epicycle falls beneath the thread; then the apogee of the deferent and the center of the moon's epicycle will be arranged on the material instrument as they are in the heavens. For each of them is at that point of the ecliptic at which it ought to be, and the mean sun, too, falls halfway between them. Next I take the moon's mean anomaly and look for [that number] on the epicycle in its convenient representation, and make a mark where the number for the anomaly brings me on the epicycle. Then I take the thread attached to point *K*, and, stretching it tight, make it pass through the center of the epicycle to its far circumference, and I revolve the disk of the motion of the epicycle until the apogee of the epicycle falls directly beneath the thread. Then I stretch the thread attached to the center of the epicycle, which is point *A*, to the previously marked point indicating the position of the moon's anomaly and observe where the latter thread cuts the circumference of circle *EF*, which is the moon's epicycle in its true representation; through that point I stretch the thread attached to point *D*, and where the latter thread cuts the ecliptic will be the place of the moon in its true motion.

But if you want to know by this instrument whether the moon is on the slow or the fast part of its course,[94] make the thread attached to point *D* be tangent to the epicycle in its true representation on both sides: if the moon is on the upper of the arcs between the two points of tangency, it will be on the slow part of its course; but if on the lower, it will be on the fast part of its course.

Let so much suffice, then, [for us] to have said about the motions, the circles, and the sphere of the moon and all the other spheres, about the sizes of its circles and its whole bulk, about the distance of the center of its deferent and its greatest and least distances, about the thickness of its sphere, and also about the method of construction of the proposed instrument and the way to equate the moon by means of it.

Next should follow our account of Mercury. But first we will make clear the reasons for the operations with the tables for equating the moon, just as we did for the sun.[95] I say, then, that if the moon were always in the true apogee

---

636 argumenti in epiciclo: incipiendo a prin- cipio primi signi *add.* δ

eius oppositione, uerus motus lune semper esset idem quod medius. Nam
660 medius motus lune, ut dictum est prius, indicatur per lineam ductam a centro terre per centrum epicicli quam oportet transire per ueram augem epicicli et per eius oppositionem. Quare si ita esset, linea medii motus lune et linea ueri essent linea una, non esset igitur necessarium equare lunam sed solum querere medium eius motum. Nunc autem non est ita;
665 immo contingit lunam esse in qualibet parte epicicli quemadmodum manifestum est ex predictis. Necessarium igitur fuit scire in qua parte epicicli esset et quantum distaret ab auge uera epicicli; ubi autem esset in epiciclo, non potuit sciri nisi per motum eius uniformem in epiciclo qui sumitur ab auge media et dicitur argumentum medium, ideoque de isto
670 motu facte sunt tabule que ponunt distantiam eius ab auge media epicicli de quibus precessit sermo. Distantia autem eius ab auge uera non potest sciri nisi sciatur distantia centri epicicli ab auge sui deferentis, quoniam secundum hanc distantiam uariatur distantia augis medie epicicli ab auge uera. Idcirco sumunt tabularii medium motum lune et medium motum
675 solis et medium argumentum lune et subtrahunt medium motum solis de medio motu lune, et remanet eis distantia centri epicicli lune a sole. Quia uero sol semper distat equaliter a centro epicicli et ab auge deferentis lune, duplant residuum predictum et habent distantiam centri epicicli ab auge deferentis quam uocant centrum. In directo huius centri inuenitur in tabu-
680 lis distantia augis medie epicicli ab auge uera eiusdem quoniam hec distantia uariatur secundum uarietatem centri, sicud superius ostensum est, et intitulatur in tabulis hec distantia equatio centri. Hec equatio sumitur; et quia luna in epiciclo suo mouetur in superiori parte uersus occidentem, centro autem existente minus 6 signis, precedit aux uera augem mediam
685 uersus orientem; eo uero existente plus 6 signis sequitur, ut prius demonstrauimus, necesse est ut centro existente minus 6 signis, plus distet luna ab auge uera epicicli quam a media secundum quantitatem equationis centri; centro uero existente plus 6 signis, minus distet ab auge uera quam a media eadem quantitate. Idcirco quando centrum est minus 6 signis
690 addunt equationem centri argumento medio, et quando est maius minuunt, et habent argumentum uerum siue distantiam lune ab auge uera epicicli.

666 parte: sui *add.* α
677 lune: et hoc secundum medium motum solis ideo *add. DZ*
681 secundum: per β
682 in tabulis *om.* β
686 ut: in *add.* δ
688 distat αφ

or perigee of its epicycle, its true motion would always be the same as its mean motion. For the mean motion of the moon, as was stated before,[96] is defined by the line drawn from the center of the earth through the center of the epicycle: this line must pass through the true apogee and perigee of the epicycle. Therefore, if that were the case [i.e., if the moon were always in the apogee or perigee of the epicycle], the line defining the moon's mean motion and the line defining its true [motion] would be one and the same, and so it would not be necessary to equate the moon, but only to find its mean position. But in fact that is not the case: rather the moon may be in any part of the epicycle, as is clear from what was said above. Therefore it became[97] necessary to know in what part of the epicycle it was, and how far from the true apogee of the epicycle. However, it was not possible to know where it was on the epicycle except by its uniform motion on the epicycle, which is measured from the mean apogee and is called the mean anomaly. For that reason tables for that motion were constructed, giving its [the moon's] distance from the mean apogee of the epicycle; these are the tables we mentioned above.[98] Moreover, its distance from the true apogee cannot be known unless the distance of the center of the epicycle from the apogee of its deferent is known, since it is according to that distance that the distance between the mean and true apogees of the epicycle varies. For that reason the table-makers take the mean motion of the moon and the mean motion of the sun and the mean anomaly of the moon; they subtract the mean motion of the sun from the mean motion of the moon and get as remainder the distance between the center of the moon's epicycle and the sun. Now because the sun is always equidistant from the center of the moon's epicycle and the apogee of its deferent, they double the remainder we mentioned and get the distance between the center of the epicycle and the apogee of the deferent:[99] they call this distance the "centrum." Opposite this centrum in the tables is found the distance of the mean apogee of the epicycle from its true apogee, since this distance varies as the centrum varies, as was shown above:[100] this distance is entitled "equation of center" in the tables. This equation is taken. Now the moon moves on the upper part of its epicycle toward the west, and, when the centrum is less than six signs, the true apogee precedes, [i.e., is] to the east of, the mean apogee, but, when it [the centrum] is greater than six signs, [the true apogee] follows [the mean apogee], as we proved before;[101] therefore, when the centrum is less than six signs, the moon must necessarily be farther from the true apogee of the epicycle than from the mean by the amount of the equation of center, whereas when the centrum is greater than six signs, [the moon] must be nearer the true apogee than the mean by the same amount. For that reason, when the centrum is less than six signs, [the table-makers] add the equation of center to the mean anomaly, and, when it is greater, they subtract it, thus getting the true anomaly, or the distance of the moon from the true apogee of the epicycle.

Supposuerunt igitur tabularum compositores quod centrum epicicli esset in auge sui deferentis et inuestigauerunt quantus esset arcus orbis signorum interceptus inter lineam medii motus et lineam ueri ad omnem
695 distantiam lune ab auge uera epicicli; quoniam supposuerunt quod transiret aliqua linea a centro orbis signorum per centrum epicicli usque ad firmamentum, transirent quoque ab eodem centro ducte linee recte per principia quorumlibet graduum epicicli, et quesierunt quantum distarent termini istarum linearum omnium a termino linee prime. Et distantias
700 illas uocauerunt equationes argumentorum et scripserunt quamlibet in directo sui argumenti. Centro igitur epicicli existente in auge sui deferentis intrant cum argumento lune uero quod tunc idem est quod medium, et sumunt equationem argumenti sibi respondentem que est distantia ueri motus lune a medio eius. Considerant igitur si argumentum lune est maius
705 6 signis aut minus. Si enim fuerit maius 6 signis, tunc uerus motus addit supra medium eo quod luna mouetur in epiciclo in superiori parte eius uersus occidentem et in inferiori uersus orientem. Tunc igitur addunt equationem argumenti super motum medium et habent uerum. Si autem argumentum fuerit minus 6 signis, uerus motus erit minor medio quantita-
710 te equationis argumenti. Tunc ergo minuunt equationem argumenti de medio motu et relinquitur eis uerus motus. Ita quidem posuerunt artem equandi centro epicicli existente in auge deferentis quemadmodum est in omni coniunctione et oppositione.

Eo autem alibi existente maiores sunt omnes equationes quorumlibet
715 argumentorum et tanto maiores quanto plus centrum epicicli propinquum fuerit oppositioni augis, ibi enim sunt omnes maxime. Supposuerunt igitur iterum quod centrum epicicli esset in oppositione augis deferentis et inuestigauerunt quanta ibi esset equatio argumenti unius gradus et quanta argumenti duorum graduum, et ita de ceteris. Et subtraxerunt equationes
720 omnium argumentorum priores de suis comparibus modo repertis, et residua scripserunt in directo suorum argumentorum. In directo enim unius gradus scripserunt excessum equationis unius gradus centro epicicli existente in opposito augis deferentis, super equationem unius gradus eo existente in auge. Et huius residua uocauerunt equationem circuli breuis.

703 sumunt: assumunt α affirmant *f* 713 oppositione: media *add.* β*D*
711 eis: eius β

The table-makers supposed, then,[102] that the center of the epicycle was in the apogee of its deferent and calculated how great an arc of the ecliptic would be cut off between the line of the mean motion and the line of the true motion for every distance of the moon from the true apogee of the epicycle. For they supposed that a line passed from the center of the ecliptic through the center of the epicycle as far as the firmament and that other straight lines drawn from the same center passed through the first points of each degree [of anomaly] on the epicycle, and they calculated the distances of the end-points[103] of all of those lines from the end-point of the first line. They called those distances the equations of anomaly and wrote down each of them opposite its [corresponding] argument of anomaly. Therefore, when the center of the epicycle is in the apogee of its deferent, they enter with the moon's true anomaly, which, at that position, is the same as the mean, and take the equation of anomaly which corresponds to it: this [equation] is the difference between the moon's true motion and its mean. Then they observe whether the moon's anomaly is greater or less than six signs. For if it is greater than six signs, then the true motion is greater than the mean, because the moon moves on the upper part of its epicycle toward the west and on the lower toward the east. Therefore in that situation they add the equation of anomaly to the mean motion and get the true. But if the anomaly is less than six signs, the true motion will be less than the mean by the amount of the equation of anomaly. Therefore in that situation they subtract the equation of anomaly from the mean motion and get as remainder the true motion. Such is the method they established for equating [the moon] when the center of the epicycle is in the apogee of the deferent, as it is at every [mean] conjunction and opposition.

But when it is elsewhere, all the equations for each argument of anomaly are bigger, and the nearer the center of the epicycle is to the perigee, the bigger they are: for at the perigee they are all at their maximum. Therefore they supposed this time that the center of the epicycle was in the perigee of the deferent and calculated for that situation the size of the equation of 1 degree of anomaly, the size for 2 degrees of anomaly, and so forth. Then they subtracted each of the previous equations of anomaly from the corresponding one of those which they had just found, and they wrote down the differences, each opposite its corresponding argument of anomaly: [e.g.,] opposite 1 degree they wrote down the excess of the equation of 1 degree when the center of the epicycle was in the perigee over the equation of 1 degree when it was in the apogee. They called these differences the "equation of the small circle."[104] Again, they considered

---

697 firmamentum: et hic est medius motus et quod *add.* φ et hic est medius motus *add. J*

713 oppositione: nam in omni coniunctione et oppositione mediis solis et lune centrum epycicli est in auge ut expositum est *vel sim. add.* φ*J*

725 Rursus quemlibet istorum excessuum posuerunt 60 partes quas uocauerunt minuta proportionalia et inuestigauerunt in qualibet distantia centri epicicli ab auge deferentis quantum ex illis excessibus addiderant equationes argumentorum centro epicicli existente in quouis loco super suas compares prius positas que sunt centro epicicli existente in auge deferentis. Hanc ergo 730 partem scripserunt in directo cuiuslibet centri et uocauerunt eam minuta proportionalia eo quod demonstrent quantum de excessibus equationum argumentorum centro epicicli existente in oppositione augis super equationes eorumdem centro existente in auge debeat sumi centro epicicli existente in quouis loco. Intrant itaque cum centro et sumunt minuta proportionalia 735 et intrant cum argumento uero et sumunt equationem argumenti et equationem circuli breuis. Per equationem argumenti sciunt quantum distaret uerus motus a medio si centrum epicicli esset in auge. Per equationem circuli breuis sciunt quod tantum excederet distantia ueri motus a medio si centrum esset in opposito augis distantiam eorum si idem centrum esset 740 in auge. Per minuta proportionalia sciunt quod centro epicicli existente ubi est tantum distantia ueri motus a medio excedit distantiam eorundem si centrum esset in auge quantum illa minuta sunt de 60. Sumunt igitur totam partem de equatione circuli breuis quota pars sunt minuta proportionalia de 60 et partem illam addunt super equationem argumenti prius 745 sumptam, et habent equationem argumenti equatam que est distantia ueri motus a medio, centro epicicli ubi est existente. Hanc igitur equationem equatam addunt super medium motum si argumentum est maius 6 signis uel minuunt si est minus, et habent uerum locum lune. Hec est causa totius operationis tabularum in luna. De mercurio uero amodo dicam.

727 addiderunt *Aadgkoux*$_1$Π adderent *cefG HLlPrsvWwz*
731 demonstrant *D*Ψ demonstret *cGHIlNP svWwx* demonstrarent *adFgLMOprStU Yz*Σ
734 sumunt: assumunt α
738 quod tantum: quantum *fgLM*Γ
740–42 Per...auge *om. BMy*Γ
741 tantum: debet excedere *add.* α*H*Λ / excedit *om.* α*H*Λ
743 totam: tantam α
744–45 prius...argumenti: equationem dico β appellantes causa (eam *y*) equationem φ et habebunt equationem Π
748 minus: 6 signis *add.* α
749 operationis: operis θ*H* uel operis *add.* φ / De...dicam *om. BbcgJMP*Γ

each of those differences to consist of 60 units, which they called "minutes of proportion," and they calculated, for every distance of the center of the epicycle from the apogee of the deferent, by what fraction of those differences the equations of anomaly [for the situation] when the center of the epicycle was in the given place exceeded the corresponding equations which had previously been calculated for the center of the epicycle in the apogee of the deferent.[105] This fraction they wrote down opposite each [degree of] centrum and called it the "minutes of proportion" because the minutes show what proportion one ought to take, for a given position of the center of the epicycle, of the excess of the equation of anomaly for the center of the epicycle in the perigee over the equation for the center of the epicycle in the apogee. Thus they enter with the centrum and take the minutes of proportion, and then enter with the true anomaly and take the equation of anomaly and the equation of the small circle: the equation of anomaly tells them how much the true motion would differ from the mean if the center of the epicycle were in the apogee; the equation of the small circle tells them that the difference of the true motion from the mean, if the center [of the epicycle] were in the perigee, would be that much greater than their difference if it were in the apogee. The minutes of proportion tell them that, the center of the epicycle being in its actual position, the difference between mean and true motions exceeds what their difference would be, if the center [of the epicycle] were in the apogee, by an amount [of the equation of the small circle] proportional to the ratio between those minutes and 60. Therefore they take the fraction of the equation of the small circle corresponding to the ratio between the minutes of proportion and 60, and they add that fraction to the equation of anomaly which they had previously found, and get the equated equation of anomaly, which is the difference of the true motion from the mean for the actual position of the center of the epicycle. Then they add this equated equation to the mean motion if the anomaly is greater than six signs, or subtract it if it is less than six signs, and get the true place of the moon. That is the explanation for the whole process of operation with the tables for the moon.

V

[De mercurio]

Mercurius uero 4 habet circulos quibus suorum motuum in longitudine complet uarietates. Habet enim epiciclum in cuius circumferentia mouetur corpus eius, in superiori quidem parte uersus orientem, in inferiori autem uersus occidentem; in quo quidem contrarius est lune. Habet etiam ecentricum deferentem in cuius circumferentia mouetur centrum epicicli uersus orientem. Habet quoque circulum paruum in cuius circumferentia mouetur centrum deferentis uersus occidentem equaliter, super centrum ipsius circuli parui describendo in temporibus equalibus equales angulos et de ipsius circumferentia equales arcus. Nam deferens mercurii non est fixus sicud nec deferens lune. Habet rursus circulum equantem super cuius centrum describit centrum epicicli in temporibus equalibus equales angulos et de circumferentia ipsius equales arcus, unde quia ad centrum ipsius refertur equalitas motus centri epicicli ideo dicitur equans. Sunt autem deferens et equans et circulus paruus predictus in superficie una. Superficies uero epicicli declinat ab ea modis diuersis quemadmodum loco suo dicetur. Equans uero circulus equalis ponitur deferenti licet eo inequali posito nullum sequatur inconueniens. Qui etiam est ecentricus habens habitudinem fixam quoniam aux sua inseparabilis est a loco spere stellarum fixarum; cui adheret etiam circulus paruus predictus fixe quoque habitudinis.

Sunt autem centrum orbis signorum et centrum equantis et centrum predicti circuli parui in una linea recta quam terminant ex parte augis

2 longitudine: plene *add.* β et latitudine plene *add.* *M*Γ

9 describendo: super ipsum centrum *add. codd. plerique om. MP*

15 una: sed *add.* β

16–17 quemadmodum...dicetur *om. BMm OUY*Γ

19 habitudinem: etiam longitudinem β latitudinem *FfL*ΠΣΨ al. latitudinem *mg. G*

V

[Theory of Mercury]

Next I will speak of Mercury. Mercury has four circles[1] by which it brings about the variations of its motions in longitude:[2] it has an epicycle on the circumference of which its body moves, toward the east on the upper part and toward the west on the lower; in this respect it goes contrary to the moon. It also has an eccentric deferent on the circumference of which the center of the epicycle moves toward the east. It has, too, a small circle on the circumference of which the center of the deferent moves toward the west uniformly, describing in equal times equal angles about the center of the small circle and equal arcs on its circumference. For Mercury's deferent, like the deferent of the moon, is not fixed. Furthermore, it has an equant circle about the center of which the center of the epicycle describes in equal times equal angles, and equal arcs on its circumference; hence it is called the "equant,"[3] because equality [i.e., uniformity] of motion of the center of the epicycle takes place with respect to its center. Now the above-mentioned deferent and equant and small circle are in one plane,[4] but the plane of the epicycle is inclined to [the latter plane] in various ways, as will be explained in the proper place.[5] The equant circle is assumed to be equal to the deferent, even though no awkward consequences would ensue if it were assumed unequal to it.[6] Moreover, it [the equant] is eccentric and has a fixed position, since its apogee[7] never moves from a [given] point on the sphere of the fixed stars. Attached to it [the equant] is the above-mentioned small circle, which is also fixed in position.

The center of the ecliptic, the center of the equant, and the center of the above small circle lie on one straight line, the end-points of which are 17 degrees

---

1 De circulis mercurii *mg. A*; Sequitur de mercurio *mg. a tit. lRs*; *tit.* Incipit theorica de mercurio *b*; *tit.* Incipit tractatus de mercurio *dt*; Mercurius *mg. G*; *tit.* Incipit tractatus de mercurio et primo de quatuor circulis quibus motuum suorum complet uarietates *g*; *tit.* De equatorio et theorica mercurii *j*; *tit.* Tractatus tertius de mercurio. capitulum primum *k*; *tit.* sequitur capitulum de mercurio Γ *mg. M*; *tit.* Sequitur theorica mercurii *mY*; *tit.* De mercurio *rz*; *tit.* Incipit tractatus de mercurio feliciter *S*; *tit.* De mercurio, capitulum tertium de motu mercurii *T*; *tit.* De motibus mercurii secundum diuersos circulos *Z*

2 Notantur dicit in longitudine quia habet etiam motum in latitudine quo declinat ab ecliptica qui non docetur in hoc instrumento cum instrumentum habet superficiem planam et non spericam *mg. o*

17 gradus et 30 minuta libre, et ex parte opposita 17 gradus et 30 minuta
25 arietis. Centrum uero equantis est in medio aliorum duorum distans equaliter ab utroque eorum. Linea uero que est inter centrum circuli parui predicti et centrum equantis est semidiameter ipsius circuli parui unde manifestum est quod circumferentia istius parui circuli diuidit per equalia lineam que est inter centrum eius et centrum orbis signorum punctusque
30 diuisionis est ipsum centrum equantis. Cumque centrum deferentis erit in auge istius parui circuli, erunt tunc 4 centra in linea una recta: centrum scilicet deferentis, centrum parui circuli predicti, centrum equantis et centrum orbis signorum. Eruntque distantie eorum ab inuicem equales. Sunt enim tres linee interiacentes equales. Cum uero centrum deferentis
35 fuerit in opposito augis predicti circuli parui, erunt tunc deferens et equans idem circulus; habebunt enim idem centrum. Nam oppositio augis circuli parui est centrum equantis. Dictum est autem quod ipsi sunt in superficie una quare cum centrum deferentis mouetur in circumferentia predicti parui circuli oportet etiam ipsius augem moueri. Cum enim aux omnis
40 circuli sit in linea recta que egreditur a centro orbis signorum et transit per centrum eius, centro autem illius circuli moto inpossibile est hanc lineam manere fixam, oportet ut cuiuscumque circuli ecentrici centrum mouetur eius aux necessario moueatur et cuiuscumque circuli ecentrici centrum est fixum eius aux necessario sit fixa. Mouebitur itaque aux
45 ecentrici deferentis mercurii, non tamen circulariter sicud lune. In luna quidem aux deferentis sui mouetur circulariter quoniam centrum deferentis eius mouetur in circumferentia circuli cuius centrum est centrum orbis signorum. Tota nempe distantia augis deferentis lune ubicumque fuerit componitur ex semidiametro deferentis et semidiametro sui circuli parui
50 in cuius circumferentia centrum deferentis mouetur. In mercurio autem non est ita; nam centrum sui circuli parui longe distat a centro orbis signorum.

Quis igitur erit motus augis deferentis mercurii? Motus eius erit mirabilis et pulcre considerationis habens certas metas quas non egreditur, ad

25 est *om.* α*H*
29 eius: equantis φ
38 quare: quia δφ/cum: ergo θφ*H* *om.* *Z*
40 et transit *om.* φ
44 aux: huius *add.* β
49 deferentis et semidiametro *om.* *hiMV*Γ
53 mercurii: est dicendum *add.* δ dico reuera *add.* φ ueneri elicio reuera *add.* Λ/Motus: enim *add.* δ

and 30 minutes of Libra[8] on the side of the apogee and 17 degrees and 30 minutes of Aries on the opposite side. The center of the equant is in between the other two, equidistant from both. The line between the center of the above small circle and the center of the equant is the radius of the small circle; hence it is clear that the circumference of that small circle divides the line between its center and the center of the ecliptic into two equal parts, and the point of division is the center of the equant. And when the center of the deferent is in the apogee of that small circle, then four centers will lie on one straight line, namely, the center of the deferent, the center of the above small circle, the center of the equant, and the center of the ecliptic. And the distances of each from its neighbors will all be equal; for the three intervening lines are equal. But when the center of the deferent is in the perigee of the above small circle, then the deferent and the equant will be one and the same circle, since they will have the same center: for the perigee of the small circle is the center of the equant. It was stated that they lie in one plane; therefore, when the center of the deferent moves on the circumference of the above small circle, its [the deferent's] apogee must also move. For, since the apogee of every [eccentric] circle lies on a straight line proceeding from the center of the ecliptic and passing through its own center, when that circle's center moves, it is impossible for that line to remain motionless. So the apogee of any eccentric circle whose center moves must necessarily move, and the apogee of any eccentric circle whose center is fixed must necessarily remain fixed. Therefore the apogee of the eccentric deferent of Mercury will move, but will not move through a whole circle as the moon's does: for the moon, the apogee of its deferent moves through a whole circle, since the center of its deferent moves around the circumference of a circle whose center is the center of the ecliptic. For the total distance [from the center of the ecliptic] of the apogee of the moon's deferent, whatever its position, is the sum of the radius of the deferent and the radius of its small circle, on the circumference of which the center of the deferent moves. However, that is not so for Mercury, since the center of its small circle is far removed from the center of thc ecliptic.

What, then, will be the motion of Mercury's apogee? Its motion will be wonderful and beautiful to contemplate: it has fixed bounds which it does not

---

37 equantis: Nota quod conueniens uidetur quod equans et deferens sunt inequales quam equales sed difficile est ymaginari quod duo circuli equaliter cum unus alium non possit in se comprehendere sint in una superficie nisi foret dicamus quod deferens sit circulus realis equans uero ymaginatus *add. J*

45 scilicet quia non circa centrum mundi *mg. K*

51 nam: Quia aliquando distantia augis est maior aliquando minor unde quando maxima est componitur ex semidiametro deferentis et tota diametro circuli parui et distantia centri equantis a centro orbis (circuli *Q*) signorum unde *FΣ mg. Q*

55 quarum unam cum uenerit illico conuertetur ad alteram nullamque possibile est in loco conuersionis esse quietem; si qua enim foret necessarium quoque esset motum centri deferentis solui quiete: hoc autem est inpossibile. Erit itaque motus iste cuiusdam accessionis et recessionis eiusque mete erunt due linee recte egresse a centro orbis signorum contingentes 60 circulum paruum predictum. Nam inter istas duas lineas semper uagabitur aux deferentis et cum ad unam earum uenerit continuo conuertetur ad alteram. Cumque peruenerit ad punctum quod inter eas diuidit equaliter, coniungetur augi equantis; nec erit finis motui eius de una ad alteram. Inpossibile autem est ut eas egrediatur, et necesse est ut in omni plena cir- 65 culatione semel ad utramque earum perueniat et bis augi equantis coniungatur. Cum enim centrum deferentis fuerit in alterutro duorum contactuum predictorum, erit aux necessario in linea contingente quoniam aux et centrum circuli cuius est aux semper sunt in una linea recta cum centro orbis signorum. Foras autem eam egredi est inpossibile quoniam si cen- 70 trum deferentis non fuerit in alterutro duorum punctorum contactus oportebit ipsum esse uel in arcu superiori qui est inter predictos contactus uel in inferiori. Est enim semper in aliquo puncto circumferentie ipsius parui circuli. Vtrolibet autem horum dato linea recta que a centro orbis signorum egredietur et per ipsum centrum transibit intercipietur inter predictas duas 75 lineas contingentes. Quia igitur in hac linea est aux deferentis, constat eam non posse egredi duas predictas lineas contingentes.

Augi uero equantis coniungetur hec aux in omni circulatione bis; transiens enim ab una contingente ad alteram cum ad locum medium peruenerit augi equantis coniungetur. Nec erit unus idemque punctus deferentis 80 aux eius in omni instanti; nam quilibet mouetur circulariter secundum motum centri sui deferentis et contingit quemlibet ipsorum esse in qualibet parte equantis quod inpossibile est de auge deferentis sicud apparet ex predictis. Non enim egreditur duas predictas lineas contingentes. Sicud autem aux deferentis predictas contingentes non egreditur, ita oppositio 85 augis easdem contingentes in continuum directumque protractas in partem oppositam quousque circumferentie circuli equantis obuient non egreditur

56–57 enim...quiete: esset (est *N*) foret necessarium quod esset centrum deferentis similiter in quiete δ
57 solui quiete: solum quiescere Λ
65 semel: solis *JP*
77 coniungetur: iungetur αφ
84 predictas contingentes: has duas lineas β / ita: sic nec β

pass; when it reaches one of them, it will immediately turn about [and move] toward the other, and there can be no period of rest in the place where it turns about; for if there were any, it would be necessary for the motion of the center of the deferent, too, to be replaced by[9] rest; but that is impossible.[10] Thus that motion will consist of a kind of progression and regression,[11] and its bounds will be formed by the two straight lines proceeding from the center of the ecliptic tangent to the above small circle. For the apogee of the deferent will always wander between those two lines, and, when it reaches one of them, it will immediately turn about toward the other. When it reaches the point which divides [the arc] between them into two equal parts, it will coincide with the apogee of the equant. And there will be no end of its motion from one to the other [of the bounding lines]. It is impossible for it to pass outside them, and in every complete revolution [of the center of the deferent on the small circle] it must necessarily reach each of them once and coincide with the apogee of the equant twice. For when the center of the deferent is in either of the two above-mentioned points of tangency, the apogee will necessarily be on the tangent line, since the apogee and the center of the circle whose apogee it is are always on the same straight line as the center of the ecliptic. It is impossible for it to go outside [those bounds], for, if the center of the deferent is not in either of the two points of tangency, it will have to be either on the upper arc between the two above-mentioned points of tangency or on the lower one. For it is always on some point of the circumference of the small circle. But whichever of these two [arcs] you suppose [it to be on], the straight line proceeding from the center of the ecliptic and passing through the center [of the deferent] will be in the area cut off between the two aforementioned tangents. Therefore, since the apogee of the deferent lies on the above line, it is certain that it cannot go outside the two above-mentioned tangents.

This apogee will coincide with the apogee of the equant twice in every complete revolution. For in its passage from one tangent to the other, it will coincide with the apogee of the equant when it reaches the middle point. No single point of the deferent can be its apogee at all times, for each point moves in a circle following the motion of the center of the deferent, and so it comes about that each point is in every part of the equant [in turn], and that [situation] is impossible for the apogee of the deferent, as is clear from what was said above; for it does not pass outside the above-mentioned tangents. And just as the apogee of the deferent does not pass outside the above-mentioned tangents, similarly the perigee does not pass outside the [lines resulting when the] same tangents [are] produced in a straight line on the opposite side until they meet the circumference of the equant circle, but, like the apogee, [the perigee] moves

---

79 coniungetur: id est erit (in *add. y*) eadem linea *add.* φJ

sed inter eas de una ad alteram ad similitudinem augis mouetur et utrique earum semel et oppositioni augis equantis bis in omni integra circulatione coniungitur. Erit igitur motus augis maior quam motus sue oppositionis 90 quoniam predicte contingentes maiorem arcum equantis includunt ex parte augis quam ex parte sue oppositionis. Quantum uero arcus excedit arcum tantum motus augis motum sue oppositionis excedet.

Iste est dictorum circulorum motus et situs in genere. Quantitates autem ipsorum motuum sunt ita. Mouetur quidem corpus mercurii in circum- 95 ferentia sui epicicli ab eo puncto qui maxime elongatur a centro equantis equaliter quidem omni die naturali 3 gradus 6 minuta et 24 secunda de circumferentia epicicli. Motus enim ipsius equalis in epiciclo sumitur ab illa diametro epicicli que secundum rectitudinem opponitur centro equantis. Punctus autem predictus dicitur aux media epicicli et habetur per li- 100 neam rectam ductam a centro equantis per centrum epicicli usque ad eius circumferentiam. Punctus enim in circumferentia epicicli qui terminat hanc lineam est punctus ille a quo mouetur corpus mercurii equaliter et qui dicitur aux media; nam aux uera epicicli est punctus ille in ipsius circumferentia qui maxime elongatur a centro orbis signorum et est ille qui 105 terminat lineam ductam a centro orbis signorum per centrum epicicli usque ad eius circumferentiam. Arcum uero qui separat inter augem ueram et mediam uocant tabularii equationem centri.

Centrum autem epicicli mouetur per circumferentiam deferentis super centrum equantis uersus orientem equaliter omni quidem die pertransiens 110 de partibus equantis 59 minuta et 8 secunda secundum equalitatem motus solis medii. Nam iste motus quem medium mercurii dicunt et motus solis medius equales sunt et simul. Semper enim idem circulus per polos orbis

92 excedet: excedit δθφ
95 qui: quo *AaBgMOpY*ΓΠ
102 corpus mercurii *Toomer* centrum epicicli *codd. plerique* centrum corporis Π corporis planete *mg. o*/a quo mouetur corpus mercurii: in circumferentia ipsius qui maxime elongatur a centro equantis super circumferentiam cuius mouetur centrum eius β
112 simul: similes α

between them from one to the other and coincides with each of them once and with the perigee of the equant twice in every complete revolution. Therefore the motion of the apogee will be greater than the motion of the perigee, since the above tangents cut off a greater arc of the equant on the side of the apogee than on the side of the perigee.[12] The excess of the motion of the apogee over that of the perigee will be proportional to the excess of the arc [cut off on the side of the apogee] over the arc [cut off on the side of the perigee].

That is a general [qualitative] description of the motion and position of the circles we enumerated. The particular quantities involved are as follows: the body of Mercury moves with a uniform motion on the circumference of its epicycle 3 degrees 6 minutes 24 seconds[13] of the epicycle's circumference every natural day, [measured] from that point which is farthest removed from the center of the equant. For its uniform motion on the epicycle is measured from that diameter of the epicycle which lies along a straight line opposite [i.e., drawn to its center from] the center of the equant. The above-mentioned point is called the "mean apogee of the epicycle" and is obtained by a straight line drawn from the center of the equant through the center of the epicycle to its circumference. The point on the circumference of the epicycle which is the end-point of that line is the point from which the uniform motion of the body of Mercury is [measured][14] and which is called the "mean apogee." For the true apogee of the epicycle is that point on its circumference which is farthest removed from the center of the ecliptic, and that is [the point] which is the end of the line drawn from the center of the ecliptic through the center of the epicycle as far as its circumference. The arc between the true and mean apogees is called the "equation of center" by the table-makers.

The center of the epicycle moves on the circumference of the deferent toward the east with a motion, uniform with respect to the center of the equant, of 59 minutes and 8 seconds of the equant every day, in accordance with its equality to the mean motion of the sun.[15] For that motion, which they call the mean motion of Mercury, and the mean motion of the sun are equal and [take place] together. For the same circle passing through the poles of the

---

92 excedet: Quoniam predicte contingentes linee maiorem arcum equantis includunt ex parte augis (minorem uero *add.* Γ) quam ex parte sue oppositionis *add.* *Rz*Γ

101 circumferentiam: Punctus in epiciclo qui terminat augem mediam semper est idem punctus et certus secundum mentem actoris. Ipse enim intelligit quod epiciclus in loco suo non mouetur sed planeta in epiciclo sicut inferius patet *add.* *J*

102 ille: qui maxime elongatur a circumferentia eius (eius *om.* *F*) equantis uel centro (circumferentia *F*) eiusdem *add.* *F*Σ

signorum transiens utrosque ex una parte terminat. Quapropter semper in eodem puncto longitudinis orbis signorum esse dinoscuntur.

115 Centrum autem deferentis mouetur per circumferentiam parui circuli predicti super eius centrum uersus occidentem equaliter omni quidem die de partibus circumferentie ipsius parui circuli quantitate predicta uidelicet 59 minutis et 8 secundis secundum equalitatem motus solis medii. Transibit igitur omni die centrum epicicli ex circumferentia sui deferentis de 120 partibus equantis quasi duplum motus solis medii scilicet 1 gradum 58 minuta et 16 secunda. Circuibit itaque centrum epicicli suum deferentem bis in eo tempore in quo semel circuibit equantem quod quidem oportet esse in anno solari cum motus iste et motus solis medius equales sint et simul. Fuit autem ex creatoris beneplacito ut centro deferentis existente 125 in auge parui circuli esset centrum epicicli in auge deferentis. Fuerunt itaque simul centrum epicicli aux deferentis et aux equantis. Fuit itaque centrum epicicli in loco sui deferentis qui maxime elongatur uel elongari potest a centro orbis signorum. Necesse est ergo propter equalitatem motuum ut centrum epicicli sit in auge equantis quandocumque centrum 130 deferentis fuerit in auge parui circuli. Esto enim ita; recedente ergo centro epicicli ab auge equantis uersus orientem, recedit centrum deferentis ab auge parui circuli uersus occidentem. Et quia isti motus sunt equales, cum peruenerit centrum epicicli ad oppositionem augis equantis, perueniet centrum deferentis ad oppositionem augis parui circuli quam dictum est 135 esse centrum equantis. Eruntque tunc idem deferens et equans eritque tunc etiam centrum epicicli in oppositione augis deferentis et in loco sui deferentis qui tunc maxime apropinquat centro orbis signorum. Ad eum autem qui maxime apropinquare potest numquam perueniet, quoniam oporteret ut centro deferentis existente in auge parui circuli esset centrum 140 epicicli in opposito augis deferentis quod esse non potest. Possibile uero

122 eo: eodem *AoP* omni δ
123 solis: solaris α
124 simul: similes α
130 Est *ABfgkxy*ΛΠΨ / Esto...ita *om. Fhij*Σ
130–32 Esto...circuli *om. M*Γ
134 quam: quod δ
135 Eruntque: Et quia *BMmqY*Γ Eritque *fz*
135–36 eritque tunc etiam: erit *BMmUY* erit etiam Γ eritque etiam Π eritque Ψ
138 quoniam: si sic *add.* δ tunc *add. P*

ecliptic always passes through the end-points of [the lines defining] both [mean motions].[16] For that reason they are known to be always in the same point of the ecliptic in longitude.

The center of the deferent moves on the circumference of the above small circle toward the west with a uniform motion, with respect to its center, of the same amount, namely, 59 minutes and 8 seconds, of the circumference of the small circle every day, in accordance with its equality to the mean motion of the sun. Therefore the center of the epicycle will traverse about double[17] the mean motion of the sun, namely, 1 degree 58 minutes 16 seconds every day, measured with respect to the center of the equant, on the circumference of its deferent. Thus the center of the epicycle will go twice around its deferent in the time it takes it to go once around the equant; this time must be one solar year, since that motion and the mean motion of the sun are equal and [take place] together. Now it was the will of the Creator[18] that, when the center of the deferent was in the apogee of the small circle, the center of the epicycle should be in the apogee of the deferent. Therefore the apogee of the deferent and the apogee of the equant were [coincident with] the center of the epicycle at the same time. Thus the center of the epicycle was in that place on its deferent which is the farthest removed that it becomes or ever can become from the center of the ecliptic. It necessarily follows, then, from the equality of the motions[19] that the center of the epicycle is in the apogee of the equant whenever the center of the deferent is in the apogee of the small circle. For suppose that to be the situation; then, as the center of the epicycle moves away to the east from the apogee of the equant, the center of the deferent moves away to the west from the apogee of the small circle. And because those motions are equal, when the center of the epicycle reaches the perigee of the equant, the center of the deferent will reach the perigee of the small circle, which was stated to be the center of the equant. At that point deferent and equant will coincide, and then, too, the center of the epicycle will be in the perigee of the deferent and at that point on its deferent which is at that moment nearest the center of the ecliptic. But it will never reach that point which is the nearest that [any point on the deferent] can come [to the center of the ecliptic], since [for that to happen] the center of the epicycle would have to be in the perigee of the deferent when the center of the deferent was in the apogee of the small circle, and that is not possible. However, it is possible for it to approach closer to it in a dif-

---

113 terminat: ex parte longitudinis licet non semper ex parte latitudinis propter mercurium qui habet aliquando latitudinem ab orbe signorum tam ipse quam suus epiciclus et suus deferens qui intersecat orbem signorum in duobus punctis qui dicuntur caput et cauda draconis *add. J*

123 solari: una uice motu proprio alia uice motu augis *add. T mg. A*

129 equantis: directe super augem equantis in eadem linea *add. A*

est ut magis alibi quam in hoc loco eidem apropinquet. Centro enim deferentis existente in contactu occidentali circumferentie parui circuli, fuit centrum epicicli uicinius centro orbis signorum quam sit in hoc loco, immo etiam uicinius quam unquam esse potest. Non tamen tunc erat in opposito
145 augis deferentis. Fuit autem centrum deferentis in contactu occidentali cum ipsum distabat ab auge parui circuli quantitate 4 signorum. In tali enim puncto contingunt due linee ipsum paruum circulum quia ab alterutro contactuum usque ad augem eius sunt 4 signa, et usque ad oppositionem augis duo. Cum ergo centrum epicicli distabat ab auge equantis
150 per 4 signa de partibus equantis, tunc fuit centrum epicicli in maxima uicinitate sui ad centrum orbis signorum.

Recedet ergo rursus centrum epicicli ab hac oppositione augis et mouebitur uersus augem equantis, et recedet etiam centrum deferentis ab oppositione augis parui circuli et mouebitur uersus augem suam. Cumque
155 peruenerit centrum deferentis ad contactum orientalem, erit iterum centrum epicicli maxime uicinum centro orbis signorum eritque hec uicinitas equalis priori. Centrum quoque deferentis distabit ab auge parui circuli per 4 signa et centrum epicicli ab auge equantis per tantumdem, mouebiturque iterum centrum deferentis uersus augem parui circuli et centrum
160 epicicli uersus augem equantis, ita ut cum centrum deferentis peruenerit ad augem parui circuli perueniet quoque centrum epicicli ad augem equantis. Sic itaque manifestum est quod centrum epicicli est in auge equantis quandocumque centrum deferentis est in auge parui circuli, et quod centrum epicicli describit bis in anno suum deferentem. Non tamen est in
165 auge sua nisi semel quoniam dum centrum epicicli est in medietate orientali sui deferentis est aux eius inter augem equantis et lineam contactus occidentalis quod est in medietate opposita, et dum centrum epicicli est in medietate occidentali sui deferentis est aux eius inter augem equantis et lineam contactus orientalis quod est in medietate opposita. Numquam
170 igitur erit in auge deferentis nisi cum fuerit etiam in auge equantis; hoc autem solum semel in una reuolutione contingit. Cumque hoc fuerit, erit in puncto qui maxime potest elongari a centro orbis signorum; in opposito

144 erat: erit βδ

146 cum ipsum: tamen ipsum tunc δ/distabit *AadeFgikPSt*ΠΣΨ distat *M*Γ

147–48 alterutro: duorum *add.* β

149 distabit β distat *A*

158 tantumdem: totidem δ eundem *fp*

ferent place from this, for, when the center of the deferent was in the western point of tangency on the circumference of the small circle, the center of the epicycle was nearer to the center of the ecliptic than it is in the present position [i.e., in the perigee of the small circle], and furthermore nearer than it can be in any [other] position.[20] However, at that point it was not in the perigee of the deferent: the center of the deferent was in the western point of tangency when it was four signs away from the apogee of the small circle; such is [the location of] the point[s] at which the two lines are tangent to the small circle, because [the distance] from each of the two points of contact to its apogee is four signs and to its perigee two signs.[21] Therefore, when the center of the epicycle was distant from the apogee of the equant by four signs, measured along the equant, that was the moment when the center of the epicycle was at its closest to the center of the ecliptic.

The center of the epicycle will, then, move away again from this position in the perigee and travel toward the apogee of the equant, and the center of the deferent too will move away from the perigee of the small circle and travel toward its apogee. And when the center of the deferent reaches the eastern point of tangency, the center of the epicycle will again be at its closest to the center of the ecliptic, and this nearest distance will be equal to the previous one. The center of the deferent will, [here] too, be four signs distant from the apogee of the small circle, and the center of the epicycle the same distance from the apogee of the equant, and the center of the deferent will again travel toward the apogee of the small circle, and the center of the epicycle toward the apogee of the equant, so that, when the center of the deferent reaches the apogee of the small circle, the center of the epicycle also will reach the apogee of the equant. And thus it is clear that the center of the epicycle is in the apogee of the equant whenever the center of the deferent is in the apogee of the small circle, and that the center of the epicycle traverses its deferent twice a year. However, it is in its apogee only once [a year], because, while the center of the epicycle is on the eastern half of its deferent, the apogee [of the deferent] is between the apogee of the equant and the line of the western tangent, which is on the opposite half; and, while the center of the epicycle is on the western half of its deferent, the apogee [of the deferent] is between the apogee of the equant and the line of the eastern tangent, which is on the opposite half. Therefore it will never be in the apogee of the deferent except when it is also in the apogee of the equant; but this happens only once in one revolution. And when this happens, it will be at that point which is the farthest possible removed

---

151 signorum: Illud refertur non ad immediate dicta sed ad aliud supra ubi dicitur eritque tunc centrum epicicli in oppositione augis equantis et defferentis qui tunc sunt idem *add. J et sim. mg. o*

quoque augis deferentis erit solum semel, sed non erit tunc in puncto qui maxime potest apropinquare centro orbis signorum. In illo enim numquam
175 erit. Erunt quoque in reuolutione una duo loca in quibus maxime eidem apropinquabit quando uidelicet ab auge equantis distiterit hinc inde per 4 signa. Aux etiam deferentis bis coniungetur augi equantis in reuolutione una.

Quoniam autem mercurius quando est in superiori parte sui epicicli
180 mouetur uersus orientem et cum centro epicicli; in inferiori uero uersus occidentem et contra centrum epicicli; arcus uero sui epicicli quem pertransit in una die dum est in inferiori parte ipsius subtenditur arcui orbis signorum maiori arcu quem pertransit in eodem tempore centrum epicicli, necesse est ut mercurio accidat processus statio et retrogradatio. Intelli-
185 gantur enim due recte linee egredientes a centro orbis signorum contingere epiciclum. Cum igitur fuerit in arcu eius superiori qui est a contactu occidentali usque ad orientalem, uidebitur semper moueri uersus orientem motu duplici uidelicet motu eius in epiciclo et motu centri epicicli in deferente eritque semper processiuus. Cum autem fuerit in arcu inferiori
190 qui est a contactu orientali usque ad occidentalem, uidebitur moueri motu centri epicicli uersus orientem, motu uero suo in epiciclo uersus occidentem; in quo quamdiu arcus orbis signorum quem pertransit centrum epicicli erit maior arcu eiusdem orbis cui subtenditur arcus quem pertransit mercurius de epiciclo, semper erit processiuus. Cum uero equabuntur, erit
195 stationarius diceturque locus epicicli in quo huius motuum equalitas accidet ex parte contactus orientalis statio prima. Post hoc incipiet arcus orbis signorum cui subtenditur arcus epicicli quem pertransit mercurius una

190 a: in δ
192 signorum: cui subtenditur arcus deferentis *add.* δ
194 semper: etiam *add.* θ*HZ*
195–96 accidit φ fiet β

from the center of the ecliptic. It will also be only once in the perigee of the deferent, but then it will not be at the point which is the nearest possible approach to the center of the ecliptic. For it will never be at that point. In one revolution there will be two places where it will approach nearest, namely, when it is four signs distant from the apogee of the equant on either side.[22] Furthermore, the apogee of the deferent will coincide with the apogee of the equant twice in one revolution.

Now Mercury, when on the upper part of its epicycle, moves toward the east and in the same direction as the center of the epicycle, but on the lower part toward the west and in the opposite direction to the center of the epicycle; moreover, the arc of its epicycle which it traverses in one day while it is on the lower part subtends a greater arc of the ecliptic than the center of the epicycle traverses in the same time. Therefore Mercury must exhibit direct [i.e., forward] motion, station, and retrogradation. For let us suppose two straight lines proceeding from the center of the ecliptic tangent to the epicycle: when [Mercury] is on the upper arc, which extends from the western point of tangency to the eastern, it will always appear to move toward the east with a double motion, namely, its motion on the epicycle and the motion of the center of the epicycle on the deferent, and it will always be moving direct. But when it is on the lower arc, which extends from the eastern point of tangency to the western, it will appear to move toward the east in terms of the motion of the center of the epicycle, but in its proper motion on the epicycle [it will appear to move] toward the west. [When it is] on this [arc], as long as the arc of the ecliptic traversed by the center of the epicycle is greater than the arc of the ecliptic subtended by the arc of the epicycle traversed by Mercury [in the same time], its motion will continue to be direct. But when they [i.e., these two ecliptic arcs] become equal, it will be stationary, and the place on the epicycle in which its equality of motions occurs near the eastern point of tangency will be called the "first station." After that the arc of the ecliptic subtended by the arc of the epicycle traversed by Mercury in one day will begin to be greater than the arc

---

173 Apparet per ALFRAGANVM differentia quarta decima (*cf. al-Farġānī* 14. 9. [*Alfraganus, Carmody*, pp. 27–28]) quod centrum deferentis mercurii mouetur uniformiter in circumferentia circuli parui quod conuenit cum CAMPANO et cum ROGERO HERFORDENSI. Communis tamen theorica ponit quod mouetur equaliter ratione centri equantis (*cf. Theorica planetarum Gerardi, sec. 52*: "*Mouetur autem deferens ita ut in temporibus equalibus equales angulos describat super centrum equantis*") *mg. D*

177–78 Aux...una: Augi etiam deferentis bis coniunguntur ille due auges in una reuolutione epicicli quia semel coniunguntur ille igitur sunt in eodem puncto alia uice quia sunt in eadem linea recta protensa a centro mundi per ambas auges uel melius dic quod bis coniungitur et sunt in eodem puncto in firmamento ut patet in figura *J*

183 centrum epicicli: quam dum est in superiori parte ipsius epicicli *add. V mg. K*

die esse maior arcu eiusdem orbis quem pertransit una die centrum epicicli, eritque tunc mercurius retrogradus quousque perueniat ad locum epi-
200 cicli in quo isti duo arcus iterum fiunt equales. Cum ergo equabuntur erit
iterum stationarius, diceturque locus epicicli in quo huius motuum equalitas accidet ex parte contactus occidentalis statio secunda. Post istum uero
locum incipiet iterum arcus orbis signorum quem pertransit centrum
epicicli excedere arcum eiusdem orbis cui subtenditur arcus epicicli quem
205 pertransit mercurius, eritque rursus processiuus usque ad locum stationis
prime; inde uero retrogradus usque ad locum stationis secunde.

Hec autem duo loca stationis prime et secunde semper distant equaliter
ab auge uera epicicli centro quidem epicicli existente in auge deferentis 4
signa 27 gradus et 14 minuta et hec est maxima distantia utriusque statio-
210 nis ab auge epicicli. Centro uero epicicli equato distante ab auge equantis
per 4 signa, in quo quidem loco maxime apropinquat centro orbis signorum sicud dictum est supra, distant duo loca duarum stationum ab auge
epicicli 4 signis 24 gradibus et 29 minutis, et hec est minima distantia
utriusque stationis ab auge epicicli. Centro autem epicicli existente in op-
215 posito augis equantis distant ambo loca stationum ab auge epicicli 4 signis
24 gradibus et 42 minutis. Centro quoque epicicli equato distante ab auge
equantis 2 signis et 4 gradibus quod dicitur in mercurio longitudo media,
distant duo loca duarum stationum ab auge epicicli 4 signis 25 gradibus
et 9 minutis. Erit autem tempus totius retrogradationis eius centro quidem
220 epicicli existente in loco primo 21 dies 2 hore et 16 minuta; eo uero existente in loco secundo 22 dies 20 hore et 45 minuta. In loco autem tertio
erit 22 dies 17 hore et 24 minuta. In loco uero quarto erit 22 dies 5 hore et
18 minuta. Quia uero mercurius percurrit totum epiciclum in 115 diebus 21
horis et 5 minutis si quodlibet predictorum temporum ex eo subtraxeris
225 remanebit tempus totius directionis eius centro epicicli existente in eisdem
locis. Eritque tempus illud in loco primo 94 dies 18 hore et 49 minuta; in
loco autem secundo 93 dies nulla hora et 20 minuta; in loco uero tertio 93
dies 3 hore et 41 minuta; in loco autem quarto 93 dies 15 hore et 47 minuta.
Hec sunt tempora retrogradationum et directionum mercurii centro epicicli
230 existente in predictis locis. Arcum autem directionis aut retrogradationis
si libuerit sic inuenies: distantiam stationis prime ab auge dupla, eritque
arcus directionis; quem si subtraxeris de 12 signis quod remanserit erit
arcus retrogradationis.

Vt autem que dicta sunt figurali exemplo comprehendantur describam

198 eiusdem orbis: cui subtenditur arcus *add.* δ
198–99 centrum epicicli: de deferente *add.* δ
203 signorum: cui subtenditur arcus deferentis *add.* δ
209 et hec est: sumpta ab auge epicicli δ
227 nulla hora: absque horis β*DZ*
229 directionum: durationum *f*

of the ecliptic traversed by the center of the epicycle in one day, and Mercury will then be retrograde until it reaches the place on the epicycle at which those two arcs become equal again. Then, when they are equal, [Mercury] will again be stationary, and the point on the epicycle at which the equality of its motions occurs near the western point of tangency will be called the "second station." After that point the arc of the ecliptic which the center of the epicycle traverses [in one day] will again begin to exceed the arc of the ecliptic subtended by the arc of the epicycle traversed by Mercury [in the same time], and its motion will again be direct until [it reaches] the place of the first station, but retrograde from there as far as the place of the second station.[23]

Now these two places of the first and second stations are always equidistant from the true apogee of the epicycle:[24] when the center of the epicycle is in the apogee of the deferent, they are 4 signs 27 degrees 14 minutes distant from it, and this is the greatest distance of both stations from the apogee of the epicycle. When the equated center of the epicycle[25] is 4 signs away from the apogee of the equant, in which place it approaches nearest to the center of the ecliptic, as was stated above,[26] the places of the two stations are both 4 signs 24 degrees 29 minutes distant from the apogee of the epicycle, and this is the least distance of both stations from the apogee of the epicycle. When the center of the epicycle is in the perigee of the equant, the places of both stations are 4 signs 24 degrees 42 minutes distant from the apogee of the epicycle. When the equated center of the epicycle is 2 signs and 4 degrees distant from the apogee of the equant, [a situation] which is given the name of "mean distance" for Mercury,[27] the places of the two stations are both 4 signs 25 degrees 9 minutes distant from the apogee of the epicycle. The period of its total retrogradation, when the center of the epicycle is in the first of the [above-mentioned] places, will be 21 days 2 hours 16 minutes; when it is in the second, 22 days 20 hours 45 minutes; in the third, it will be 22 days 17 hours 24 minutes; and in the fourth, 22 days 5 hours 18 minutes.[28] Since Mercury traverses the whole of its epicycle in 115 days 21 hours 5 minutes,[29] if you subtract each of the above periods from the latter, the remainders will be the periods of its total direct motion when the center of the epicycle is in the respective places. That period will be for the first place 94 days 18 hours 49 minutes, for the second place 93 days 0 hours 20 minutes, for the third place 93 days 3 hours 41 minutes, and for the fourth place 93 days 15 hours 47 minutes.[30] Those are the periods of retrogradation and direct motion of Mercury when the center of the epicycle is in the above positions. However you may find the arc of retrogradation or of direct motion, if you wish, as follows: double the distance of the first station from the apogee, and [the result] will be the arc of direct motion. If you subtract this from 12 signs, the remainder will be the arc of retrogradation.

But in order that what has been stated may be understood by an illustrating

235 orbem signorum qui sit circulus *AB* circa centrum *D* in quo protraham diametrum *ADB*, sitque punctus *A* in 17 gradibus et 30 minutis libre. Erit igitur punctus *A* aux equantis; itaque centrum equantis erit in linea *AD*. Ponam ergo ut sit punctum *E* circa quem lineabo circulum equantem qui sit *GHZ*, sintque duo puncta *G* et *Z* in linea *ADB*, *G* quidem ex parte *A*,
240 *Z* uero ex parte *B*. Eritque *G* aux equantis; *Z* uero oppositio augis. Sumam autem ex linea *DA* lineam *EF* equalem linee *ED* eritque punctus *F* centrum circuli parui circa cuius circumferentiam mouetur centrum deferentis eiusque semidiameter linea *FE*. Lineabo itaque circa punctum *F* et secundum longitudinem *FE* circulum *CE* qui erit circulus paruus predic-
245 tus. Ponam quoque ut *C* sit punctus in quo circumferentia huius circuli secat diametrum *ADB*. Erit ergo *C* centrum deferentis in illa hora in qua centrum epicicli est in auge deferentis idemque punctus *C* erit aux istius parui circuli. Sumam quoque ex linea *AD* lineam *CK* equalem linee *EG* que erit semidiameter deferentis eiusque aux punctum *K*. Describam igitur
250 circa centrum *C* circulum *HK* qui erit deferens mercurii. Quia igitur quando centrum deferentis est in auge parui circuli, est centrum epicicli in auge deferentis, erit punctum *K* centrum epicicli. Circa ipsum itaque describam circulum *ST* qui sit epiciclus mercurii sintque duo puncta *S*, *T* loca in

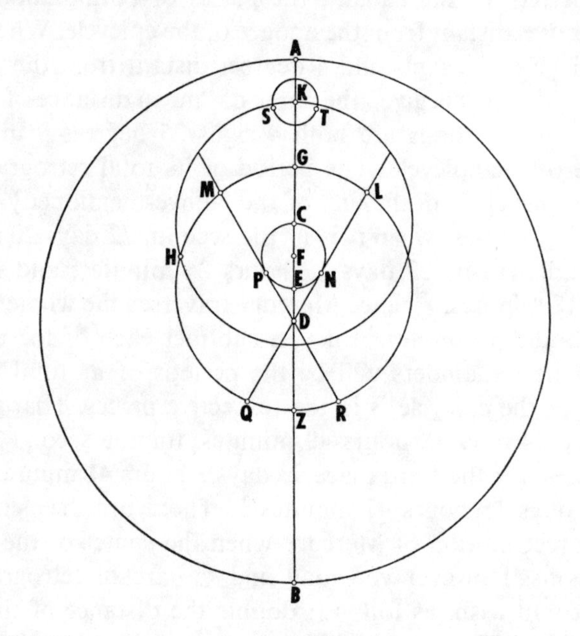

Fig. 14

239 ADB: AB αφ*H* 247 punctus: scilicet α*H* *om.* φ
243 semidiameter: sit *add.* δ erit *add.* Π

diagram [see Fig. 14, p. 228], I describe on center *D* the ecliptic—let this be circle *AB*—and draw in it the diameter *ADB*. Let point *A* be in 17 degrees and 30 minutes of Libra. Thus point *A* will be the apogee of the equant, and therefore the center of the equant will be on line *AD*. I suppose [that center], then, to be point *E*, about which I draw the equant circle—let this be *GHZ*, and let the two points *G* and *Z* lie on line *ADB*, *G* on the side of *A*, and *Z* on the side of *B*. *G* will be the apogee of the equant, and *Z* the perigee. I cut off from line *DA* line *EF* equal to line *ED*, and point *F* will be the center of the small circle on the circumference of which moves the center of the deferent, while line *FE* will be its radius. So I draw about point *F* with radius *FE* a circle *CE*, which will be the above small circle. I also put *C* as the point at which the circumference of this small circle cuts the diameter *ADB*. Then *C* will be the center of the deferent at that moment when the center of the epicycle is in the apogee of the deferent, and the same point *C* will be the apogee of that small circle. I also cut off from line *AD* line *CK* equal to line *EG*. *CK* will be the radius of the deferent, and its apogee will be point *K*. So I describe about center *C* circle *HK*, which will be the deferent of Mercury. Then, since the center of the epicycle is in the apogee of the deferent when the center of the deferent is in the apogee of the small circle, point *K* will be the center of the epicycle. So I describe about it circle *ST* to be the epicycle of Mercury, and let the two points

Legend for Fig. 14

Diagram of the Ptolemaic model for Mercury (after Campanus; compare Figure 8, p. 50). *AB* represents the ecliptic, with center *D*, the earth. *GZ* represents the equant circle, with center *E*, the equant point. Circle *KH* represents the deferent, with center *C*. The epicycle, *ST*, is here drawn at *K*, the apogee of the deferent. *NEP* is the small circle, with center *F*, about which *C* revolves clockwise. The points *M* and *L* on the tangents *MPDR* and *LNDQ* from the earth *D* to the small circle are the intersections of the tangents with the equant circle and mark the limits of the to-and-fro motion of the apogee of the deferent. Similarly points *Q* and *R* mark the limits of the motion of the perigee of the deferent. *S* and *T* are the points where tangents from *D* meet the epicycle: on the upper arc between *T* and *S* Mercury's motion on the epicycle is from west to east, while on the lower arc between *S* and *T* its motion on the epicycle is from east to west.

quibus due linee protracte a centro orbis signorum contingent epiciclum:
255 punctus quidem *S* in contactu orientali, *T* uero in contactu occidentali. Protraham autem a puncto *D* usque ad circumferentiam equantis duas lineas contingentes circulum paruum in duobus punctis *P* et *N* que sint *DPM* et *DNL*, sitque punctus *P* in orientali contactu et punctus *N* in occidentali. Producam quoque eas in partem oppositam usque ad circum-
260 ferentiam equantis fiantque *MDR* et *LDQ*.

Hiis igitur sic dispositis dico quod mercurius mouetur in circumferentia epicicli a puncto *T* per augem epicicli ad punctum *S* totusque iste motus est uersus orientem. Ab *S* uero mouetur per oppositum augis epicicli ad *T* totusque iste motus est uersus occidentem, et est ille motus in die naturali
265 3 gradus 6 minuta et 24 secunda. Centrum quoque epicicli mouetur uersus *H* describens super punctum *E* in temporibus equalibus equales angulos et de circumferentia circuli *GH* equales arcus, et est iste motus in die naturali 59 minuta et 8 secunda de circumferentia circuli secundum equalitatem motus solis medii. Quia uero dum mercurius in suo epiciclo mouetur
270 uersus occidentem ab *S* ad *T* contingit alicubi post punctum *S* quod centrum epicicli non plus mouetur de partibus orbis signorum uersus orientem quam mercurius uersus occidentem, necesse est ut mercurius sit ibi stationarius eritque ibi statio prima. Quia quoque mercurio magis accedente ad oppositum augis epicicli incipit plus moueri de partibus orbis signorum
275 uersus occidentem quam centrum epicicli uersus orientem, necesse est ut mercurius tunc fiat retrogradus et augebitur hec retrogradatio quousque mercurius sit in opposito augis epicicli. Exinde uero incipiet minui minueturque usque dum rursum equetur motus mercurii uersus occidentem motui centri epicicli uersus orientem quod erit ante punctum *T* eritque ibi
280 statio secunda. Post quam rursus incipiet dirigi cum motus centri uersus orientem superet motum mercurii uersus occidentem.

Centrum insuper deferentis quod est punctum *C* mouetur uersus *N* describens super punctum *F* quod est centrum parui circuli in temporibus equalibus equales angulos et de circumferentia circuli *CN* equales arcus et
285 est iste motus equalis motui centri epicicli prius dicto, scilicet 59 minuta et 8 secunda de circumferentia circuli *CN*. Transibit igitur omni die centrum epicicli in circumferentia deferentis de partibus equantis 1 gradum 58 minuta et 16 secunda uicinius. Cum autem centrum deferentis erit in puncto *N* qui distat a puncto *C* per 4 signa parui circuli, erit centrum

254 contingerent δ
258 orientali: occidentali *AFfMsY*
259 occidentali: orientali *AFfM*
259–60 circumferentiam: etiam *add.* δ
261 dispositis: descriptis β
266 punctum: centrum βδ/E: quod scilicet est centrum equantis *add. AQT*/angulos: Ideo enim dicitur circulus ille cuius (C *add. T*) est centrum equantis *add. AQT*
267 arcus: equaliter describit *add. A*

S and *T* be the places in which the two lines drawn from the center of the ecliptic are tangent to the epicycle, point *S* at the eastern point of tangency and point *T* at the western point of tangency. I draw from point *D* to the circumference of the equant two lines tangent to the small circle at the two points *P* and *N*—let the lines be *DPM* and *DNL*, and let point *P* be at the eastern point of tangency and point *N* at the western. I also produce them on the opposite side [of *D*] as far as the circumference of the equant—let [these extended tangents] be *MDR* and *LDQ*.

Having made these dispositions, I say that Mercury moves on the circumference of the epicycle from point *T* through the apogee of the epicycle to point *S*, and the whole of that motion is toward the east. But from *S* it moves through the perigee of the epicycle to *T*, and the whole of that motion is toward the west. [The amount of] that motion is 3 degrees 6 minutes 24 seconds in a natural day. The center of the epicycle moves [from *K*] toward *H*, forming in equal times equal angles at point *E* and equal arcs of the circumference *GH*. That motion is 59 minutes and 8 seconds of the circumference of circle *GH* in a natural day, in accordance with its equality to the mean motion of the sun. But, while Mercury is moving westward on its epicycle from *S* toward *T*, at some point beyond point *S* it comes about that the center of the epicycle is traversing, toward the east, no more of the ecliptic than is Mercury toward the west; hence at that point Mercury must be stationary, and the first station must be there. Then, as Mercury approaches the perigee of the epicycle, it begins to traverse more of the ecliptic toward the west than does the center of the epicycle toward the east; so Mercury must then become retrograde, and this retrogradation will increase [in speed] until Mercury is in the perigee of the epicycle.[31] From there on, however, it will begin to decrease, and will decrease until the motion of Mercury toward the west is once again equal to the motion of the center of the epicycle toward the east, which will be before point *T*; there the second station will be. After that it will again begin to have a direct motion, since the motion of the center toward the east will be greater than the motion of Mercury toward the west.

Furthermore, the center of the deferent, which is point *C*, moves toward point *N*, describing in equal times equal angles at point *F*, which is the center of the small circle, and equal arcs on the circumference of circle *CN*; that motion is equal to the above-mentioned motion of the center of the epicycle, namely, 59 minutes and 8 seconds of the circumference of circle *CN* [per day]. Therefore the center of the epicycle will traverse on the circumference of the deferent about[32] 1 degree 58 minutes 16 seconds of the equant every day. When the center of the deferent is at point *N*, which is four signs of the small circle

290 epicicli maxime propinquum puncto *D*. Cum uero centrum deferentis erit in puncto *E*, fiet idem circulus deferens cum equante centrumque epicicli cadet in equante in puncto *Z* qui est oppositio augis. Cum iterum centrum deferentis ascendet ab *E* per *P* centrum epicicli mouebitur per medietatem oppositam uersus *K* et cum centrum deferentis peruenerit ad *P* qui distat 295 a puncto *C* per 4 signa parui circuli, erit rursus centrum epicicli maxime propinquum puncto *D*. Cumque centrum deferentis peruenerit a puncto *P* ad *C* centrum epicicli fiet in puncto *K*, et sic omnia procedent denuo sicud prius. Ex quo patet quod centrum epicicli in quo tempore semel circuit equantem bis circuit deferentem et tempus illud est annus solaris ut prius 300 dictum est.

At uero aux deferentis non egreditur arcum *MGL* nec eius oppositio arcum *QZR*; oportet enim ut aux eius sit in linea que egreditur a puncto *D* et transit per centrum eius. Quia ergo centrum eius semper est in circumferentia circuli *CN*, nulla autem linea protracta a puncto *D* per aliquod 305 punctum circumferentie circuli *CN* cadit extra arcum *MGL*, inpossibile est augem predictam cadere extra arcum *MGL*; neque igitur eius oppositio cadet extra arcum *QZR*. Quamdiu uero erit centrum deferentis in semicirculo *CNE* erit aux predicta in arcu *GL*; et quamdiu erit centrum deferentis in semicirculo *EPC*, erit aux ipsa in arcu *GM*. Erunt igitur aux et 310 centrum deferentis semper ex eadem parte diametri *AB*. Quia igitur dum centrum deferentis est in alterutro duorum semicirculorum *CNE,EPC*, est centrum epicicli ex parte opposita diametri *AB*, inpossibile est centrum epicicli esse in auge deferentis alibi quam in puncto *K*.

Si uero posuerimus centrum epicicli in circumferentia deferentis alibi 315 quam in punctis *K*, *Z*; est enim in puncto *Z* cum centrum deferentis est in puncto *E*; posuerimus quoque centrum deferentis in circumferentia circuli *CN* in puncto ubi conuenit, fient diuerse linee *EK* et *DK*. Quas si protraxerimus usque ad circumferentiam epicicli erit punctus terminans *EK* aux epicicli media a qua mouetur mercurius in circumferentia epicicli 320 equaliter, diciturque a tabulariis distantia eius ab ea argumentum medium. Punctus uero qui terminat *DK* dicitur aux epicicli uera; distantia quoque mercurii ab ea dicitur argumentum equatum. Arcus autem epicicli qui est

291 cum equante: et equans δ
293 ab E: in puncto Z *add. AQ* / per P: ad C et *add. codd. plerique sed om. M* et *add. aBCY*
296 Cumque centrum: Cumque punctum uel centrum β
301 egreditur: egredietur *AfHhLlrtxΛ*
315–16 K, Z...posuerimus: K, Z, erit centrum deferentis alibi quam in punctis C, E. Est enim in puncto K cum centrum deferentis est in puncto C et est (est *om. V*) etiam in puncto Z cum centrum deferentis est in puncto E. Cum posuerimus δ
317 in puncto ubi conuenit: alibi quam in punctis C uel E δ / DK: si K sit centrum epicicli ubicumque fuerit *add.* δ

distant from point *C*, the center of the epicycle will be at its nearest to point *D*. But when the center of the deferent is at point *E*, the deferent will become the same circle as the equant, and the center of the epicycle will coincide with point *Z* of the equant, which is the perigee.[33] When the center of the deferent moves up again from *E* through *P*, the center of the epicycle will move through the opposite half [of the deferent] toward *K*, and, when the center of the deferent reaches *P*, which is four signs of the small circle distant from point *C*, the center of the epicycle will again be at its nearest to point *D*. And when the center of the deferent reaches *C* from point *P*, the center of the epicycle will reach point *K*, and thus the whole process will be repeated anew just as before. From this it is plain that the center of the epicycle goes around the deferent twice in the time in which it goes around the equant once, and that time is a solar year, as was stated before.

But the apogee of the deferent will not pass outside arc *MGL*, nor its perigee outside arc *QZR*. For its apogee must lie on the line proceeding from point *D* and passing through its [the deferent's] center. Therefore, since its center is always on the circumference of circle *CN*, and [since] no line drawn from point *D* through any point on the circumference of circle *CN* falls outside arc *MGL*, it is impossible for the above apogee to fall outside arc *MGL*. Neither will its perigee, then, fall outside arc *QZR*. But as long as the center of the deferent is on semicircle *CNE*, the above apogee will be on arc *GL*, and as long as the center of the deferent is on semicircle *EPC*, the apogee will be on arc *GM*. Therefore the apogee and the center of the deferent will always be on the same side of the diameter *AB*. Since, therefore, as long as the center of the deferent is on either of the two semicircles *CNE* and *EPC* the center of the epicycle is on the opposite side of diameter *AB*,[34] it is impossible for the center of the epicycle to be in the apogee of the deferent anywhere except at point *K*.

But if we put the center of the epicycle [at a point] on the circumference of the deferent other than points *K* or *Z*—it is at point *Z* when the center of the deferent is at point *E*—and [if] we also put the center of the deferent at a suitable point on the circumference of circle *CN*, *EK* and *DK*[35] will become different lines. If we produce these lines as far as the circumference of the epicycle, the end-point of *EK* will be the mean apogee of the epicycle, from which the uniform motion of Mercury on the circumference of the epicycle is measured; the distance of Mercury from this [mean apogee] is called the "mean anomaly" by the table-makers. The end-point of *DK* is called the "true apogee" of the epicycle, and Mercury's distance from it the "equated anomaly." The

---

306 caue ne huius alicubi ponatur S pro D quia in exemplari erant similis figure *mg. V*

inter eas dicitur equatio centri quoniam ipse uariatur secundum distantiam centri epicicli ab auge equantis. Hec enim distantia centrum appellatur.
325 Precedit autem aux media ueram in medietate que descendit a puncto *K* ad *Z*. In medietate uero altera que ascendit a puncto *Z* ad *K* aux uera precedit mediam. Ideoque in una medietate addunt tabularii equationem centri argumento medio; in alia uero medietate subtrahunt, fitque eis argumentum uerum.

330 Spera etiam mercurii continens predictos motus habet motum proprium equalem et similem motui sperarum solis et lune de quo motu quidem et diffinitione et situ et ordine sperarum omnium sufficit quod supra dictum est.

Habet igitur mercurius 4 circulos: epiciclum, equantem, deferentem et
335 circulum paruum. Quatuor quoque motus, scilicet motum eius in epiciclo, centri epicicli in deferente, centri etiam deferentis in circulo paruo et motum spere sue. Omnesque hii motus sunt in longitudine. Vocamus autem longitudinem distantiam que est ab occidente uersus orientem. Sed sunt ei alii motus in latitudine qui sunt diuersitatis plurime quibus mouetur
340 uersus meridiem aut septentrionem; nam latitudinem uocamus distantiam que est a septentrione uersus meridiem; de quibus nichil ad propositum cum ad instrumentum illud talium motuum doctrina non pertineat.

Consequens autem est hiis que dicta sunt ut narremus magnitudines orbium mercurii qui sunt deferens, equans, epiciclus et circulus paruus
345 quem describit centrum deferentis suo motu et distantiam centrorum equantis et circuli parui necnon etiam deferentis a se inuicem et a centro orbis signorum, longitudinem quoque mercurii a centro terre longiorem et propiorem et magnitudinem sue spere et sui corporis quemadmodum fecimus in luna.

350 Ad hoc autem ut plane comprehendantur ea que dicenda sunt describam predictos 4 circulos mercurii in superficie una. Protraham etiam per augem equantis et centrum orbis signorum diametrum *ADM*. Sitque *A* locus augis, *D* centrum orbis signorum. Erit ergo centrum equantis in linea *AD*. Sit ergo punctum *K* et describatur circa ipsum circulus *LM* qui
355 sit equans. Sumaturque ex linea *AK* linea *HK* equalis linee *KD* eritque *H*

331 equalem et: scilicet in 100 annis uno gradu equaliter (equalem et *y*) *fJy*

340 aut septentrionem: a septentrione et a meridie uersus septentrionem δ a septentrione *BMO*Γ

341 meridiem: aut a meridie uersus (ad *A*) septentrionem *add.* φ*AJ*

342 pertineat: perueniat *CDMNRTVZ*

347 a centro terre *om.* δ

350 dicenda: dicta *AW*

arc of the epicycle which lies between them [the two apogees] is called the "equation of center," since it varies according to the distance of the center of the epicycle from the apogee of the equant. For the latter distance is called the "centrum." The mean apogee precedes [i.e., is to the east of] the true [when the epicycle is] on the half [of the deferent] which goes from point *K* down to point *Z*. But on the other half, which goes from point *Z* up to *K*, the true apogee precedes the mean. Therefore on the one half[36] the table-makers add the equation of center to the mean anomaly, whereas on the other half they subtract it, to get the true anomaly.

The sphere of Mercury which encloses the motions described above has a motion of its own as well, which is equal to and like the motion of the spheres of the sun and moon.[37] What was said above[38] about this motion and about the definition, position, and order of all the spheres is sufficient.

Mercury has, then, four circles: an epicycle, an equant, a deferent, and a small circle, and also four motions, namely, its own motion on the epicycle, the center of the epicycle's on the deferent, the center of the deferent's on the small circle, and the motion of its sphere. All of these motions are in longitude. By "longitude" we mean the direction from west to east. But it has other motions in latitude,[39] of great diversity, by which it moves to the south or north. For by "latitude" we mean the direction from north to south. However, [to talk] about these is pointless, since the theory of such motions has no relevance to the present instrument.

The appropriate sequel to what we have described is for us to relate the sizes of the circles of Mercury, which are the deferent, the equant, the epicycle, and the small circle which the center of the deferent describes in its proper motion, and [to relate] the distances of the centers of the equant, the small circle, and the deferent from each other and from the center of the ecliptic, and also the greatest and least distances of Mercury from the center of the earth, and the size of its sphere and its body, [all] as we did for the moon.[40]

But in order that what we have to say may be fully understood, I draw the above-mentioned four circles of Mercury in one plane [see Fig. 15, p. 236]. I also draw the diameter *ADM* through the apogee of the equant and the center of the ecliptic: let *A* be the place of the apogee[41] and *D* the center of the ecliptic. Then the center of the equant will be on line *AD*. Let it be point *K*, and let there be described about it circle *LM* to be the equant. Let line *HK* be cut off from line *AK* equal to line *KD*: then *H* will be the center of the small

338 orientem: hoc in figuram mercurii circa quam ymaginor tam frontem debere uertere uersus meridiem intuentem ita quod figura sit inter oculum et meridiem eam intuentem *add.* Λ

342 pertineat: sequitur figura prima mercurii *add. f*/CAMPANVS ut hic apparet non demonstrat de motibus planetarum in latitudine *mg. D*

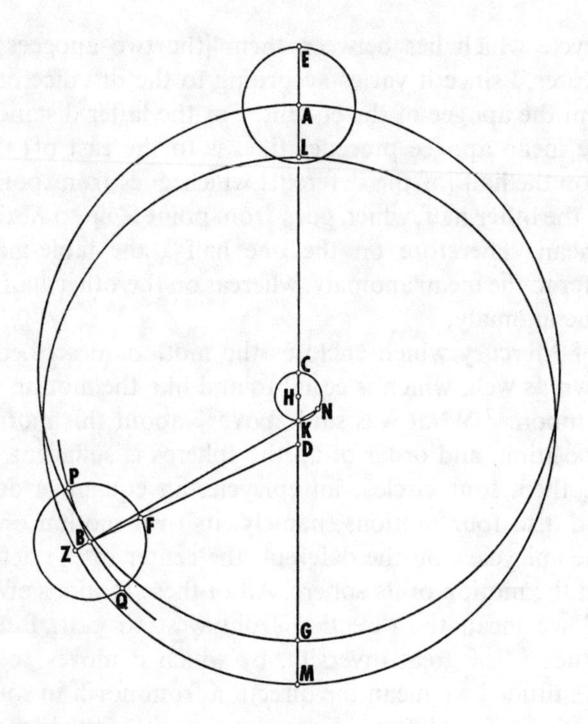

Fig. 15

centrum parui circuli eiusque semidiameter linea *HK*. Describatur igitur super punctum *H* et secundum spatium *HK* circulus *CK* qui erit ipse paruus circulus. Sitque *C* locus in quo circumferentia huius circuli secat diametrum *ADM*. Erit ergo *C* centrum deferentis in illa hora in qua
360 centrum epicicli est in auge deferentis. Describam itaque circa punctum *C* circulum *AG* equalem circulo *LM*. Eritque circulus iste deferens eiusque aux punctum *A* quod etiam erit centrum epicicli. Describatur igitur circa ipsum *A* epiciclus mercurii cuius aux sit punctum *E* eritque *E* locus supremus ad quem possit peruenire mercurius. Sumam itaque ex circulo paruo
365 arcum *CN* quem ponam 4 signa, ex opposita uero parte sumam de equante arcum *LZ* quem ponam etiam 4 signa, et ponam ut centrum deferentis sit in puncto *N*. Circa ipsum etiam describam sub puncto *Z* unum arcum deferentis qui sit arcus *PBQ* et protraham a puncto *K* lineam *KZ* que secet arcum predictum in puncto *B*. Oportet itaque propter ea que dicta sunt
370 de equalitate motuum ut centrum epicicli sit in puncto *B*. Eruntque etiam tria puncta *N*, *K*, *B* in linea una recta et linea *NB* semidiameter deferentis.

357 erit: est α
362 Describam φ
363 epyciclum φ
368 secat δ

Legend for Fig. 15

Schematic representation of Mercury's model (after Campanus) to delineate the parts of its "sphere." *D* represents the earth, and *LM* represents the equant circle, with center *K*, the equant point. *AG* represents the deferent when its center is at *C* on the small circle *CNK* (when the epicycle center is in the apogee *A* of the deferent). The epicyclic apogee is marked at *E*. The epicycle is drawn again at the point of its closest approach to the earth *D*, namely, when the center of the deferent has moved 120° from *C* on the small circle, to *N*. An arc of the deferent, *PBQ*, is drawn for this situation ($\angle LKZ = \angle CHN = 120°$, $NB = CA$), and the epicycle drawn about *B* on the line *NKZ*. Point *F*, where line *DB* cuts the epicycle in this situation, is thus the perigee of the epicycle at *B*. *DE*, the "greatest possible distance" of Mercury, and *DF*, the "least possible distance," define the outer and inner boundaries, respectively, of Mercury's "sphere."

circle, and its radius will be line *HK*. Therefore let there be described about point *H*, with radius *HK*, a circle *CK*, which will be the small circle itself. Let *C* be the place in which the circumference of this circle cuts the diameter *ADM*. Then *C* will be the center of the deferent at the moment at which the center of the epicycle is in the apogee of the deferent. So I draw about point *C* circle *AG* equal to circle *LM*. That circle will be the deferent, and its apogee will be point *A*, which will also be the center of the epicycle. Therefore let there be drawn about *A* the epicycle of Mercury; let its apogee be point *E*. Then *E* will be the uppermost point which Mercury can reach. Now I cut off from the small circle arc *CN*, which I make four signs, and on the opposite side [of *ADM*] I cut off from the equant arc *LZ*, which I also make four signs, and I suppose the center of the deferent to be at point *N*. With that [point *N*] as center I draw under point *Z* an arc of the deferent—let this be arc *PBQ*—and draw from point *K* line *KZ* to cut the above arc at point *B*. From what was said about the equality of motions[42] it follows that the center of the epicycle must be at point *B*. Also the three points *N*, *K*, and *B* will lie on a straight line,[43] and

Protraham autem lineam *DB* que secet epiciclum in inferiori parte in puncto *F*. Erit igitur *F* locus infimus in quo possit esse mercurius.

Hiis itaque sic dispositis narrabo magnitudines premissas; primo qui-
375 dem secundum partes illas de quibus semidiameter deferentis que est linea *AC* habet 60 partes; deinde secundum partes illas de quibus semidiameter terre est pars una; ultimo uero secundum partes uulgariter notas que sunt miliaria, prout miliare unum constat ex 4000 cubitis, ut sit processus noster hic similis ei quem habuimus in luna: sic enim intendimus in omnibus.
380 Dico ergo quod secundum partes de quibus linea *AC* que est semidiameter deferentis est 60 partes est unaqueque linearum trium que sunt *CH*, *HK*, *KD* tres partes. Et linea *AE* que est semidiameter epicicli 22 partes et 30 minuta. Et linea *AD* que est longitudo longior centri epicicli 69 partes. Et linea *DB* que est longitudo propior centri epicicli 55 partes et
385 34 minuta. Et linea *ED* que est maior elongatio centri corporis mercurii que possit esse a centro orbis signorum est 91 partes et 30 minuta. Et linea *DF* que est maior propinquitas centri corporis mercurii que possit esse ad centrum orbis signorum est 33 partes 4 minuta. Et linea *DG* que est maior apropinquatio deferentis centro orbis signorum que esse possit est 51 par-
390 tes. Et linea *DM* que est longitudo propior equantis est 57 partes.

At uero secundum partes illas de quibus semidiameter terre est pars una possibile est nobis easdem magnitudines inuenire si supponamus supremum eius ad quod peruenit luna esse infimum eius ad quod peruenit mercurius sicud dictum est in precedentibus. Nisi enim ita fuerit necesse esset
395 planetas proprias speras exire aut inter eas locum esse uacuum aut eas esse maiores quam ipsorum planetarum motus requirat; quorum duo prima inpossibilia, tertium uero superfluum esse uidetur. Quod quidem si quis posuerit erunt tamen hee magnitudines quibus minores inpossibile est esse. Modus autem inueniendi eas erit secundum quod narrabo. Quo-
400 niam supponam supremum lune esse infimum mercurii, hoc enim conueniens et sufficiens esse uidetur, supremum autem ad quod peruenit centrum corporis lune ut dictum est in precedentibus distat a centro orbis signorum per 64 partes et 10 minuta de partibus illis de quibus semidiameter terre est pars una; semidiameter autem corporis lune est 17 minuta et 32 secunda
405 de partibus eisdem; semidiameter quoque corporis mercurii est quasi 2

372 secat δ
373 mercurius: centrum mercurii δ
374 dispositis: ut in presenti figura patet *add.* δ
388–90 Et linea DG...51 partes *om. B*
389–90 51 partes: sine minutis *add. F*
394–97 Nisi...superfluum: Quia si non uel erunt planetarum spere ex omni parte minores quam planetarum motus requirat uel alicubi maiores et alicubi minores uel ex omni parte maiores quarum positionum due prime sunt inpossibiles tertia uero superflua δ
397 Quod: Quam δ
398 minores: maiores δ

line *NB* will be the radius of the deferent. I draw line *DB* to cut the lower part of the epicycle at point *F*. Then *F* will be the lowest point in which Mercury can be.[44]

Having made these dispositions, I will relate the above magnitudes, first in the units of which the radius of the deferent, which is line *AC*, contains 60, then in the units of which the radius of the earth is one, and finally in the well-known units, namely, miles, according to the system in which one mile consists of 4,000 cubits, so that our procedure here may be like that which we followed for the moon. For it is our intention so to proceed for all [the bodies].

I say, then, that in the units of which line *AC*, which is the radius of the deferent, contains 60, each of the three lines *CH*, *HK*, and *KD* is 3 units; line *AE*, which is the radius of the epicycle, is 22 units and 30 minutes; line *AD*, which is the greatest distance of the center of the epicycle, is 69 units; line *DB*, which is the least distance of the center of the epicycle, is 55 units and 34 minutes; line *ED*, which is the greatest possible distance of the center of Mercury's body from the center of the ecliptic, is 91 units and 30 minutes; line *DF*, which is the closest possible approach of the center of Mercury's body to the center of the ecliptic, is 33 units and 4 minutes. Line *DG*, which is the closest possible approach of the deferent to the center of the ecliptic, is 51 units; and line *DM*, which is the least distance of the equant, is 57 units.[45]

It is possible for us to determine the same distances in the units of which the radius of the earth is one, if we suppose that the uppermost point which the moon reaches is the lowest point which Mercury reaches, as was stated previously.[46] For if that were not so, it would necessarily follow either that the planets go outside their own spheres, or that there is empty space between them [the spheres], or that they [the spheres] are larger than the motion of the planets requires; the first two of these alternatives are impossible, while the third appears to be superfluous.[47] But even if someone does suppose [the third] to be so, these magnitudes will still be the smallest possible.[48] The way of computing them is as I shall explain. Since I suppose the moon's uppermost point to be Mercury's lowest—for this is a convenient and sufficient supposition—and the highest point which the center of the moon reaches, as was stated previously, is 64 units and 10 minutes[49] distant from the center of the ecliptic, if we use the units of which the radius of the earth is one; and [since] the radius of the moon's body is 17 minutes and 32 seconds[50] in the same units, while the radius of Mercury's body is about 2 minutes and 8 seconds in the same

---

373 in...mercurius: id est propinquissimus terre siue centro orbis signorum *o*Λ id est propinquissimus centro terre *y* quia D est centrum orbis signorum *add. QT* quia D esset signorum. Hic ponitur figura circularis *mg. A*

minuta et 8 secunda de eisdem partibus; est enim quasi octaua uicesima pars semidiametri terre; quia itaque si predictis 64 partibus et 10 minutis addiderimus semidiametrum lune et semidiametrum mercurii, proueniet nobis infimum centrum corporis mercurii quod est linea *DF*, oportet ut
410 linea *DF* sit 64 partes 29 minuta et 40 secunda de partibus illis de quibus semidiameter terre est pars una. Ponam eam igitur 64 partes et 30 minuta. Nam de 40 secundis faciam unum minutum. Eritque hec linea medium ad inueniendum omnes alias. Inueniuntur itaque secundum hunc modum quoniam multiplicabo cuiuscumque linee numerationem sumptam secun-
415 dum partes illas de quibus semidiameter deferentis habet 60 partes per *DF* linee numerationem sumptam secundum partes illas de quibus semidiameter terre est pars una, et productum diuidam per numerationem linee *DF* sumptam secundum partes illas de quibus semidiameter deferentis habet 60 partes et exibit numeratio earum que assumpte sunt ad multi-
420 plicationem sumpta secundum partes illas de quibus semidiameter terre est pars una. Erunt itaque linea quidem *AC* 117 partes et 2 minuta, et unaqueque trium linearum *CH*, *HK*, *KD* 5 partes et 51 minuta et linea *AE* 43 partes et 53 minuta et linea *AD* 134 partes et 35 minuta et linea *DB* 108 partes et 23 minuta et linea *ED* que est supremum mercurii 178 partes et
425 28 minuta et linea *DG* 99 partes et 29 minuta et linea *DM* 111 partes et 11 minuta.

De miliaribus autem erunt predicte magnitudines ita. Linea quidem *AC* 379826 miliaria et 4 undecime unius miliaris. Quelibet linearum trium *CH*, *HK*, *KD* 18985 miliaria et 10 undecime unius miliaris. Linea *AE*
430 142421 miliaria et 4 undecime unius miliaris. Linea *AD* 436784 miliaria et 1 undecima unius miliaris. Linea *DB* 351753 miliaria et 2 undecime unius miliaris. Linea *ED* 579205 miliaria et 5 undecime unius miliaris. Linea *DF* 209331 miliaria et 9 undecime unius miliaris. Linea *DG* 322868 miliaria et 7 undecime unius miliaris. Linea *DM* 360840 miliaria et 5 undecime unius
435 miliaris.

Volo autem nunc inuenire distantiam utriusque superficiei spere mercurii, scilicet concaue et conuexe, a centro terre. Per hoc enim manifesta erit spissitudo spere ipsius. Quia uero concaua superficies huius spere est eadem cum conuexa superficie spere lune, ut concordet distantia concaue mercurii
440 a centro terre distantie quam superius inuenimus de conuexa lune, ab eadem uerificabimus quantitatem linee *DF* quam posuimus maiorem sua uera quantitate in 20 secundis prout semidiameter terre est pars una. Tali enim

409 oportet: igitur *add.* δ
414 linee: aliarum *AB* aliam *M* alius *f*
415 per: in *ABfM*
417 producta β*H* productam *Ak*
420 sumpta *Hz*Π sumptam *codd. plerique*
428 379826: 379862 *adeGgHhIkLlmNorsUu vWwXxYyZz*ΛΠΣΨ 179862 *M*Γ 379863 *b*
439 concaue: superficiei spere *add.* φ
440 eodem δθφ

units—for it is about a twenty-eighth part of the radius of the earth[51]—then, if we add the radius of the moon and the radius of Mercury to the above 64 units and 10 minutes, the result will be the lowest position reached by the center of Mercury's body, namely, line *DF*; line *DF* must then be 64 units 29 minutes 40 seconds, expressed in those units of which the radius of the earth is one. Therefore I take it as 64 units and 30 minutes, making the 40 seconds 1 minute. This line will serve as the intermediary for finding all the others. They are found in the following way: I multiply each line's measurement, expressed in the units of which the radius of the deferent contains 60, by the measurement of line *DF*, expressed in the units of which the radius of the earth is one, and divide the product by the measurement of line *DF* expressed in the units of which the radius of the deferent contains 60:[52] the result will be the measurement of the lines to which multiplication was applied, expressed in the units of which the radius of the earth is one. Then line *AC* will be 117 units and 2 minutes, each of the three lines *CH*, *HK*, and *KD* will be 5 units and 51 minutes, line *AE* will be 43 units and 53 minutes, line *AD* will be 134 units and 35 minutes, line *DB* will be 108 units and 23 minutes, line *ED*, which is the uppermost point reached by Mercury, will be 178 units and 28 minutes, line *DG* will be 99 units and 29 minutes, and line *DM* 111 units and 11 minutes.

In miles the above distances will be as follows:[53] line *AC* will be 379,826$\frac{4}{11}$ miles, each of the three lines *CH*, *HK*, and *KD* 18,985$\frac{10}{11}$ miles, line *AE* 142,421$\frac{4}{11}$ miles, line *AD* 436,784$\frac{1}{11}$ miles, line *DB* 351,753$\frac{2}{11}$ miles, line *ED* 579,205$\frac{5}{11}$ miles, line *DF* 209,331$\frac{9}{11}$ miles, line *DG* 322,868$\frac{7}{11}$ miles, and line *DM* 360,840$\frac{5}{11}$ miles.

Now I want to find the distance of each surface of the sphere of Mercury, namely, the concave and the convex, from the center of the earth. For from these the thickness of the sphere itself will be apparent. Since the concave surface of this sphere is the same as the convex surface of the moon's sphere, so that the distance of Mercury's concave surface from the center of the earth agrees with the distance which we found above for the moon's convex [surface], we will use the latter to verify the size of line *DF*, which we made greater than its true amount by 20 seconds, in the units of which the radius of the earth is

---

425 28 minuta: et linea DF 64 partes et 30 minuta *add. c mg. Aa*

falso posito non accidit in magnitudinibus predictis error sensibilis propter eius paruitatem relatam ad eas. Quia tamen uolumus ut infimum mercurii 445 precise conueniat supremo lune, idcirco ponemus lineam predictam in uera eius quantitate. Est autem quantitas uera eius 64 partes 29 minuta et 40 secunda de partibus illis de quibus semidiameter terre est pars una. Si igitur istas partes et minuta et secunda reduxerimus ad miliaria, inueniemus 209313 miliaria et 26 tertias trigesimas unius miliaris que est uera 450 quantitas linee predicte. Semidiameter uero corporis mercurii est 115 miliaria et 13 tertie trigesime unius miliaris. Dictum est enim quod ipsa est 2 minuta et 8 secunda prout semidiameter terre est pars una. Subtraham itaque semidiametrum corporis mercurii de linea *DF* et remanebit distantia superficiei concaue spere mercurii a centro terre que est 209198 miliaria 455 et 13 tertie trigesime unius miliaris, et hanc eandem distantiam inuenimus in superficie conuexa spere lune a centro terre. Eandem quoque semidiametrum corporis mercurii addemus super lineam *ED* eritque distantia superficiei conuexe spere mercurii a centro terre et est 579320 miliaria et 5 undecime et 13 tertie trigesime unius miliaris. Erit igitur spissitudo spere 460 mercurii 370122 miliaria et 5 undecime unius miliaris totusque ambitus superficiei conuexe spere mercurii erit 3641446 miliaria et 2 septime unius miliaris. In hoc enim posui 5 undecimas et 13 tertias trigesimas predictas unum miliare. Ambitus autem concaue ipsius dictus est in luna. Ipse enim est ambitus conuexe lune ipsumque inuenimus 1314961 miliaria et 11 465 tertias trigesimas unius miliaris. Ambitus uero corporis mercurii est 725 miliaria et 11 tertie trigesime unius miliaris. Ambitus autem deferentis aut equantis ipsius cum sint equales erit 2387480 miliaria. Et si ipsum diuiseris per 365 dies et quartam qui sunt tempus unius reuolutionis centri epicicli in ecentrico equante, exibit tibi motus eius in die una. Erit itaque dieta 470 ipsius 6536 miliaria et 34 minuta unius miliaris. Ambitus autem epicicli mercurii erit 895220 miliaria. Et si ipsum diuiseris per 116 dies qui sunt tempus reuolutionis eius in suo epiciclo, exibit tibi dieta ipsius in epiciclo et est 7717 miliaria et 24 minuta unius miliaris. Corpus autem terre continet corpus mercurii 22247 uicibus et insuper 161 duodecimas quingentesi- 475 mas corporis eius. Est enim proportio diametrorum terre et corporis mercurii sicud proportio 225 ad 8. Si itaque cubos istorum duorum numerorum inueneris, inter eos erit proportio corporis terre et corporis mercurii. Sunt autem illi cubi 11390625 et 512. Sic igitur sunt quantitates predicte quas posui per ordinem in subscripta tabula.

480 Modum autem faciendi instrumentum nostrum per quod equemus mer-

446 64: 46 φ
455 inueniemus β
461 3641446: 2641446 a_1*BdFfgikMmopUuv*
ΥΓΠΛΣΨ
469 equante: circa equantem δ equantem *Z*
480 nostrum: ipsum δ *om. Aefky*

one. This false assumption does not produce any sensible error in the above magnitudes, because of its smallness in relation to them. But because we want the lowest point of Mercury to coincide exactly with the uppermost point of the moon, we will give the above line [*DF*] its true dimensions. Its true dimensions are 64 units 29 minutes 40 seconds in those units of which the radius of the earth is one. If, then, we convert those units, minutes, and seconds to miles, we get 209,313$\frac{26}{33}$ miles; that is the true size of the above line. But the radius of Mercury's body is 115$\frac{13}{33}$ miles, for we said that it is 2 minutes and 8 seconds when the radius of the earth is one unit. Thus I subtract the radius of Mercury's body from line *DF*, and the remainder will be the distance of the surface of the concave sphere of Mercury from the center of the earth: this [distance] is 209,198$\frac{13}{33}$ miles, which is the same distance as that we found for the convex surface of the moon's sphere from the center of the earth.[54] We also add the same radius of Mercury's body to line *ED*, and the result will be the distance of the convex sphere of Mercury from the center of the earth: it is 579,320$\frac{5}{11}$ plus $\frac{13}{33}$ miles. Therefore the thickness of Mercury's sphere will be 370,122$\frac{5}{11}$ miles,[55] and the circumference of the convex surface of Mercury's sphere will be 3,641,446$\frac{2}{7}$ miles—in this calculation I took the above $\frac{5}{11}$ plus $\frac{13}{33}$ as one mile.[56] The circumference of its concave [surface] was stated when we were dealing with the moon, for it is the circumference of the moon's convex, and we found that to be 1,314,961$\frac{11}{33}$ miles.[57] The circumference of Mercury's body is 725$\frac{11}{33}$ miles.[58] The circumference of its deferent—or of its equant, since they are equal—will be 2,387,480 miles.[59] If you divide this by 365 days and a quarter, which is the period of one revolution of the center of the epicycle on the equant, the result will be its motion in one day. Thus its day's journey will be 6,536 miles and 34 minutes of a mile.[60] The circumference of Mercury's epicycle will be 895,220 miles;[61] if you divide this by 116 days,[62] which is the period of one revolution of Mercury on its epicycle, the result will be its day's journey on the epicycle, and that is 7,717 miles and 24 minutes of a mile.[63] The body of the earth contains the body of Mercury 22,247 times plus $\frac{161}{512}$ of its body.[64] For the ratio between the diameters of the earth and of the body of Mercury is 225 to 8.[65] Thus, if you find the cubes of the latter two numbers, the ratio between them will be that of the earth's body to Mercury's body. Those cubes are 11,390,625 and 512. Such are the above quantities, which I have arranged in order in the table below [Table 2, p. 358].

Next I shall describe the way to make the instrument with which we may

curium ammodo dicam. Nec laborabo ponere in eo aliquid eorum que dicta sunt nisi fuerit necessarium ad inueniendum uerum locum ipsius in longitudine. Sumam igitur tabulam illam in cuius una parte ordinaui instrumentum lune et partem alteram eius faciam ualde planam et ualde
485 pollitam quanto plus possibile fuerit. Et scribam in medio eius punctum *D* super quem describam circulum unum maiorem quem potero describere in ea, ut sit iste circulus orbis signorum; sub quo faciam alios tres super idem centrum. Distet autem secundus tantum a primo quod inter eos possint conuenienter scribi nomina signorum; et tertius tantum a secundo
490 quod inter eos possint poni singule diuisiones graduum; et quartus tantum a tertio quod inter eos possint scribi numeri graduum distinctorum per 5 et 5. Et quadrabo istos circulos duabus diametris ortogonaliter se secantibus super centrum eorum. Quarum una transeat per 17 gradus et 30 minuta libre et per gradum oppositum eritque in ea sub libra aux mercurii.
495 Et diuidam eos in 12 signa. Et omnia circa eorum diuisionem et intitulationem expediam sicud in sole et luna dictum est.

Quibus expeditis relinquam infra interiorem istorum 4 circulorum in una dictarum diametrorum, in illa scilicet que transit per 17 gradus et 30 minuta libre tantum spatium quantum uolo distare circulum unum cum suis
500 diuisionibus descriptum supra centrum equantis ab eo. Faciam enim in instrumento isto equantem maiorem deferente ut in eius circumferentia possit numerari medium centrum mercurii ita quod sit totus equans cum suis diuisionibus extra deferentem. Posset etiam fieri minor deferente essetque idem. Ego uero faciam ipsum maiorem quoniam oportebit nos facere
505 unum alium circulum cum diuisione signorum et graduum super centrum *H* quod est centrum parui circuli ut in eius circumferentia possit numerari motus centri deferentis uersus occidentem. Et istum cum omnibus circulis ad eius diuisionem necessariis faciam minorem deferente et totum intra ipsum. Residuum uero spatium quod erit in predicta diametro usque ad
510 *D* diuidam in 32 partes. Et quamlibet earum intelligam esse diuisam in alias tres ut sint in uniuerso 96, sed subdiuisionem istam in tres non oportet actu efficere quoniam omnia expediuntur per diuisionem primam. Sit igitur linea *DK* una predictarum 32 partium. Erit ergo *K* centrum equantis. Itemque sumam a puncto *K* uersus circumferentiam de eadem diametro
515 lineam *HK* equalem linee *KD*, eritque *H* centrum circuli parui quem describit centrum deferentis motu suo uersus occidentem. Super ipsum ergo describam circulum *CK* qui erit circulus paruus predictus sitque punctus

496 et luna *om. bDjX* et luna dictum est *om. BM*Γ

500 ab eo: scilicet interiori 4 circulorum *add.* β

503–4 essetque: accideretque δ accidetque *Z* accideret quod *N* adque *f*

504 maiorem: deferente *add.* β

equate Mercury. But I shall not trouble to put in it any of the features we have mentioned which are not necessary for finding its true position in longitude. I take, then, that board on one side of which I set out the instrument of the moon, and make the other side very flat and smooth, as much as is possible. In the middle of it I mark point *D* and, on this point [as center], describe a circle, the largest that I can describe on that [board]: let this circle be the ecliptic [see Fig. 16, p. 246]. Inside it, and on the same center, I draw three more [circles]: the second should be far enough from the first for the names of the signs to be conveniently written between them; the third far enough from the second for the single divisions of the degrees to be marked between them; and the fourth far enough from the third for the numbers of the degrees to be written between them, one number being marked every five degrees. I divide these circles into four by means of two diameters cutting one another at right angles at their center. Let one of these diameters pass through 17 degrees and 30 minutes of Libra[66] and the degree opposite it: on that [diameter] on [the side of] Libra will be the apogee of Mercury. I divide the circles into 12 signs, and perform everything concerned with the division and labeling of them in the way explained for the sun[67] and moon.

Having done all this, I cut off below the innermost of the four circles along one of the above-mentioned diameters, namely, the one which passes through 17 degrees and 30 minutes of Libra, the amount that I want to establish as the distance between it [the innermost of the four circles] and a circle described about the center of the equant and divided in the appropriate way.[68] For in this instrument I shall make the equant greater than the deferent, so that the mean centrum[69] of Mercury can be counted off on its circumference, in such a way that the whole of the equant, with all its divisions, lies outside the deferent. [The equant] could also be made smaller than the deferent, and [the result] would be the same.[70] But I shall make it greater, since we shall have to draw another circle, divided into signs and degrees, on center *H*, which is the center of the small circle, so that the motion of the center of the deferent toward the west can be counted on its circumference. This [small circle], together with all the circles necessary for its division, I shall draw smaller than the deferent and completely inside it.[71] I divide the interval remaining on the above diameter up to *D* into 32 parts,[72] and I imagine each of them to be divided into three smaller parts, so that there are 96 altogether; however, one should not actually carry out the subdivision into three, since everything can be accomplished by means of the primary division.[73] Let line *DK*, then, be one of the above 32 parts. Then *K* will be the center of the equant. Again, I take on the same diameter, from point *K* toward the circumference, line *HK* equal to line *KD*; then *H* will be the center of the small circle which is described by the center of the deferent in its motion toward the west. Upon [*H*], then, I describe circle *CK*, which will be the above small circle; let point *C* be on the above-

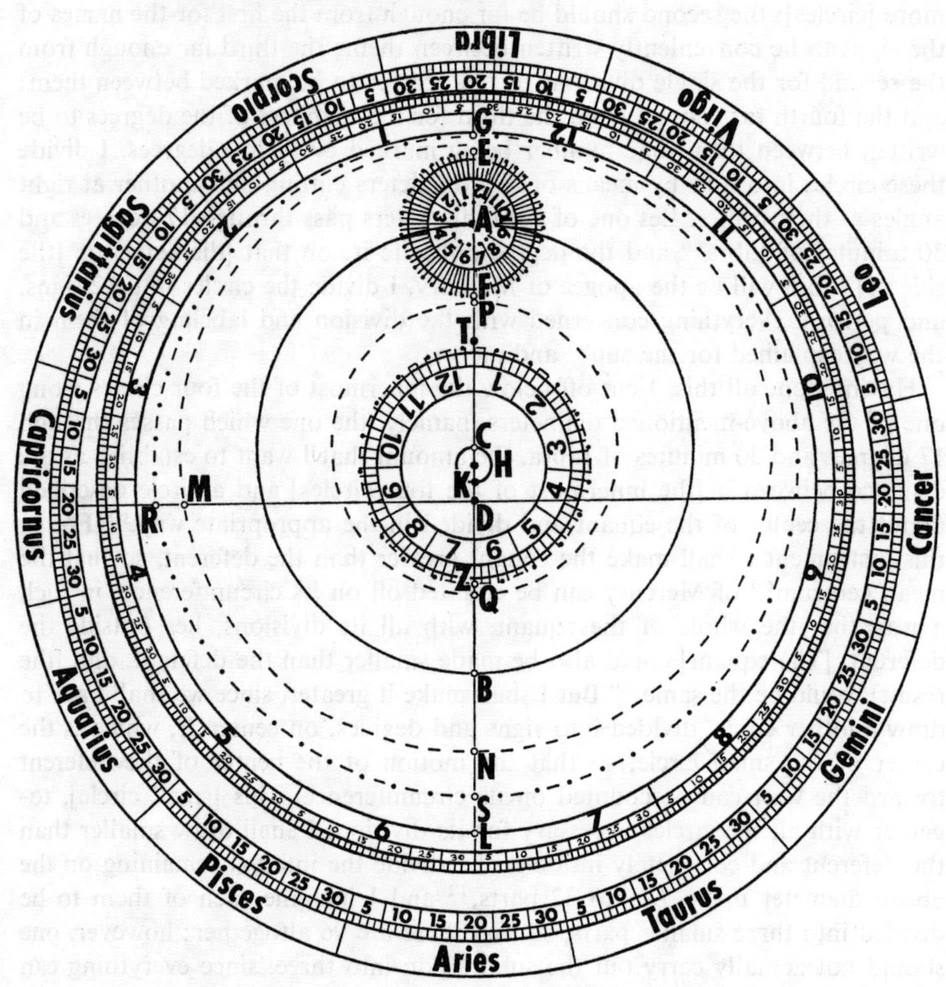

INSTRVMENTVM EQVATIONIS MERCVRII

Fig. 16

Legend for Fig. 16

Instrument for Mercury (after Campanus). The outer graduated system of circles (with center *D*, the earth) represents the ecliptic. The graduated system of circles immediately within that (with center *K*) represents the equant. The zero point of the equant is aligned with Mercury's apogee, in Libra 17;30°. *CK* represents the small circle, with center *H*; the graduation of the small circle is marked outside the small circle itself (because of its size), just inside circle *TZ*, which has the same center. *AB* represents the deferent, and *EF* represents the epicycle, on center *A*. The dotted and dashed lines represent the edges of the moving parts. The smallest of these (indicated by the dotted lines) is the "disk of the motion of the epicycle"; it revolves about center *A* within the "disk of the motion of the center of the epicycle" (indicated by the dashed lines). The latter revolves about center *C* and is in turn contained in the "disk of the motion of the center of the eccentric" (indicated by alternate dots and dashes), which revolves about center *H*. Threads are attached by one end to points *H*, *K*, and *D*, and marks are made at points *E* (the apogee of the epicycle) and *T* (on the edge of the "disk of the motion of the center of the eccentric"). Mercury's true position is found as follows. Stretch the thread at *H* through the point on the inmost graduated system of circles marking Mercury's mean centrum (its mean longitude from apogee), and revolve the outermost movable disk (the "disk of the motion of the center of the eccentric") until the mark at *T* coincides with the thread. (Mercury's deferent is then in the correct position.) Stretch the thread at *K* through the point on the equant circle marking Mercury's mean centrum and, keeping the outermost disk fixed, move the "disk of the motion of the center of the epicycle" until the epicycle center *A* coincides with the thread; then revolve the "disk of the motion of the epicycle" until the epicycle apogee too (point *E*) coincides with the thread. (The epicycle is then in its correct position.) Mark Mercury's mean anomaly on the edge of the epicycle and stretch the thread at *D* through this mark. Where the thread cuts the ecliptic is Mercury's true longitude. Compare Figure 8 (p. 50) and Plate 2.

C in diametro predicta. Erit ergo *C* centrum deferentis in illa hora in qua centrum epicicli est in auge deferentis. Rursus a puncto *C* uersus circum-
520 ferentiam sumam de predicta diametro lineam *CA* que sit 20 partes de predictis 32 partibus, eritque *A* punctus centrum epicicli et aux deferentis et linea *CA* semidiameter deferentis. Describam igitur super centrum *C* et secundum longitudinem linee *CA* circulum *AB* qui erit deferens mercurii. Itemque a puncto *A* uersus circumferentiam sumam lineam *AE* in predicta
525 diametro que sit 7 partes et medietas partis de predictis 32 partibus. Eritque punctus *E* aux epicicli, eiusque semidiameter linea *AE*. Describam itaque super centrum *A* et secundum longitudinem *AE* circulum *EF* qui erit epiciclus. Infra quem describam tres alios circulos sibi concentricos et diuidam ipsum in 12 signa et quodlibet signum in 30 gradus et scribam
530 numerum graduum et signorum. Que omnia expediam sicud feci in epiciclo lune, excepto quod numerum signorum et graduum scribam contrario modo. Incipiam enim ab auge epicicli et procedam scribendo uersus sinistram sicud feci in circulo ecentrico solis.

Remanent autem de predictis 32 partibus pars una et medietas partis.
535 Est enim linea *ED* 30 partes et medietas partis. Super centrum igitur equantis quod est punctum *K* describam circulum *GL* qui transeat per ultimam illarum 32 partium. Et extra ipsum super idem centrum describam circulum alium qui tantum distet a priori quod inter illos possit scribi numerus signorum. Item extra ipsum secundum et super idem centrum faciam circu-
540 lum tertium tantum distantem a secundo quod inter eos possit scribi numerus graduum distinctorum per 5 et 5. Rursus extra ipsum tertium et super idem centrum faciam circulum quartum tantum solummodo distantem a tertio quod inter eos possint fieri singule diuisiones graduum. In diuisione autem istius circuli protraham omnes lineas rectas quibus fiet
545 diuisio signorum et graduum uersus centrum ipsorum quod est punctum *K*, non autem uersus centrum orbis signorum. Hoc enim erit nobis generale in omni diuisione quorumlibet circulorum quod ad centrum circuli illius qui diuiditur respiciant omnes recte linee quibus ipse diuiditur. Hac igitur diuisione facta scribam numerum graduum et signorum incipiendo ab
550 auge et procedendo uersus sinistram sicud feci in ecentrico solis.

Postmodum oportet ita facere quod centrum deferentis moueatur in circumferentia circuli *CK* super centrum *H* uersus occidentem et centrum epicicli in circumferentia deferentis uersus orientem. Hoc autem faciam eo ingenio quo processi in luna. Describam enim super centrum *C* duos cir-
555 culos concentricos deferenti quorum unus excedat epiciclum medietate et

518 C[1] *om.* δ
538 alium: secundum *add.* α*H*
548 qui: quod δ

mentioned diameter. Then *C* will be the center of the deferent at the time at which the center of the epicycle is in the apogee of the deferent. Again I take on the above diameter, from point *C* toward the circumference, line *CA*, which is to contain 20 of the above 32 parts: point *A* will be the center of the epicycle and the apogee of the deferent, and line *CA* will be the radius of the deferent. So I describe on center *C*, with radius of the length of line *CA*, circle *AB*, which will be Mercury's deferent. Again I take on the same diameter, from point *A* toward the circumference, line *AE*, which is to contain $7\frac{1}{2}$ of the above 32 parts. *E* will be the apogee of the epicycle, and line *AE* its radius. So I describe on center *A*, with radius *AE*, circle *EF*, which will be the epicycle. Inside it and concentric with it I describe three more circles, and I divide it [the epicycle] into 12 signs and every sign into 30 degrees, and I write in the numbers of the degrees and signs. I do all this as I did for the moon's epicycle,[74] except that I write the numbers of the signs and degrees in the opposite direction: beginning from the apogee of the epicycle I proceed toward the left [i.e., counterclockwise] as I write, just as I did for the sun's eccentric circle.

There remain $1\frac{1}{2}$ of the above 32 parts. For line *ED* is $30\frac{1}{2}$ parts. Therefore I describe on the center of the equant, which is point *K*, circle *GL* to pass through the end of those 32 parts. Outside it, on the same center, I describe a second circle far enough from the first for the numbers of the signs to be written between them. Again outside the second and on the same center, I draw a third circle far enough from the second for the numbers of the degrees, marked off in groups of five, to be written between them. Again, outside the third and on the same center, I draw a fourth circle just far enough from the third for the single divisions of the degrees to be marked between them. In dividing that circle, I draw all the straight lines with which the division into signs and degrees is carried out toward its center, namely, point *K*, and not toward the center of the ecliptic. For it will be a general principle of ours that in the division of any circle all straight lines by which the circle is divided shall be directed toward the center of the circle which is being divided. Having, then, carried out this division, I write in the numbers of the degrees and signs, beginning from the apogee and proceeding toward the left [i.e., counterclockwise], as I did for the sun's eccentric.

Next it is necessary to arrange that the center of the deferent shall move on the circumference of circle *CK* about center *H* toward the west, and that the center of the epicycle shall move on the circumference of the deferent toward the east. This I do by means of the device which I used for the moon:[75] I draw on center *C* two circles concentric with the deferent, of which one should pass

---

548 diuiditur[2]: *tit.* Figura instrumenti mercurii habet hic locum sed propter magnitudinem eius ponitur in franceno *add. j*

quarta de predictis 32 partibus et alius subsistat ei tantumdem. Sitque maior istorum duorum circulorum *MN*; minor uero circulus *PQ*. Super centrum quoque parui circuli quod est punctum *H* describam duos alios circulos quorum unus excedat circulum *MN* alia medietate et quarta qui sit circulus
560 *RS*, alius uero subsistat circulo *PQ* tantumdem qui sit circulus *TZ*. Eritque circulus *RS* contingens interiorem quatuor circulorum qui facti sunt propter diuisionem equantis. Diuidam quoque circulum *TZ* in signa et gradus propter motum centri deferentis. Nam super idem centrum describam alios circulos tres infra ipsum ut sint quatuor qui diuidantur in 12 signa et
565 quodlibet signum in suos gradus sicud modo fecimus de equante. Et scribatur numerus signorum et graduum, sed in scribendo incipiemus a puncto qui supponitur augi et procedemus uersus dextram contrario modo ei quod fecimus in equante. In isto enim circulo mensurabimus motum centri deferentis et in equante motum centri epicicli. Vocabo quoque istum
570 circulum circulum motus centri deferentis.

Concauabo igitur totum spatium quod est inter duos circulos *RS* et *TZ* ita quod in concauitatem illam ingrediatur tabula una plana ita spissa quod ipsa possit etiam concauari. Fiat quoque sic ista concauitas quod tabula sibi inposita non possit inde egredi. Post hec faciam tabulam isti
575 concauitati similem quam uocabo tabulam motus centri ecentrici, et lineabo in ea deferentem qui est *AB* cum epiciclo qui est *EF* et duos equedistantes deferenti qui sunt *MN* et *PQ* secundum magnitudines suas. Postea concauabo in ista tabula totum spatium quod est inter duos circulos *MN* et *PQ* ita quod in concauitatem illam ingrediatur tabula alia plana similis illi
580 concauitati, adeo etiam spissa quod ipsa possit adhuc concauari. Deinde faciam aliam tabulam isti secunde concauitati similem quam uocabo tabulam motus centri epicicli, et lineabo in ea deferentem et epiciclum cum diuisionibus et inscriptionibus epicicli. Postea iterum in ista tabula concauabo totum illud spatium quod continet diuisiones et inscriptiones epi-
585 cicli. Et faciam tabulam tertiam isti tertie concauitati similem quam uocabo tabulam motus epicicli. Et diuidam eam et inscribam sicud diuiditur et inscribitur epiciclus. Postea ordinabo unamquamque earum in concauitate sua ita quod quelibet interior contingat exteriorem ex omni parte et omni

560 subsistet *AadfIKnpstu*
565 signum: signorum δθ
588–89 omni modo: ex omni modo δ

outside the epicycle by a distance of $\frac{3}{4}$ of one of the above 32 parts, and the other pass beneath it by the same amount. Let the greater of these two circles be *MN* and the lesser *PQ*. I also describe about the center of the small circle, namely, point *H*, two other circles, of which one should exceed circle *MN* by another $\frac{3}{4}$ [of a part]—let this be circle *RS*—and the other should be less than circle *PQ* by the same amount—let this be circle *TZ*.[76] Circle *RS* will touch the inmost of the four circles which were drawn for the division of the equant. I also divide circle *TZ* into signs and degrees, to represent [on it] the motion of the center of the deferent:[77] I draw inside it on the same center three more circles, so that there are four [in all] to be divided into 12 signs, and every sign into its constituent degrees, as we did just before for the equant. Let the numbers of the signs and degrees be written in, but in doing so we begin from the point beneath the apogee and proceed toward the right [clockwise], contrary to what we did for the equant. For on this circle we shall measure the motion of the center of the deferent, whereas on the equant [we measure] the motion of the center of the epicycle. I call this circle the "circle of the motion of the center of the deferent."

I hollow out the whole area between the two circles *RS* and *TZ* in such a way that the cavity will receive a flat disk thick enough to be hollowed out itself.[78] This cavity should also be formed in such a way that a disk placed in it cannot slip out.[79] Next I make a disk fitting that cavity and call it the "disk of the motion of the center of the eccentric," and I draw on it the deferent *AB*, the epicycle *EF*, and the two [circles] equidistant from the deferent, *MN* and *PQ*, [all] with the proper dimensions. Then I hollow out in that disk the whole area between the two circles *MN* and *PQ* in such a way that the cavity will receive another flat disk of the same size as the cavity, [this disk], too, being thick enough to be hollowed out. Then I make another disk fitting this second cavity and call it the "disk of the motion of the center of the epicycle," and I draw on it the deferent and the epicycle, with the divisions and inscriptions of the epicycle. Then again in this disk I hollow out the whole area containing the divisions and inscriptions of the epicycle. I make a third disk fitting this third cavity and call it the "disk of the motion of the epicycle." I divide it and inscribe it as the epicycle is divided and inscribed. Then I arrange each of them in its cavity, so that each inside [disk] touches each outer [container] on

---

562 equantis: et TZ (TZ *om.* *N*) etiam erit contingens PQ circulum (circuli *N*) *add.* δ

575 Nota ?quod est centrum ecentrici in mercurio scilicet uerum et illud in instrumento non mouetur sed centrum ecentrici quod ymaginatur in circulo oportune circulationis mouetur cum aux eius moueatur *mg.* *o*
Si uidere uolueris huius instrumenti figuram literaliter descriptam remoueas tunc rotas et ostendet se quod petis etc. *mg.* *M*Γ

modo et moueatur in ea motu leui et equali et non possit egredi ab ea.
590 Sitque omnium istarum tabularum continentium et contentarum superficies una quemadmodum dictum est in luna. Postea uero faciam unum signum notabile in tabula motus centri ecentrici sub auge deferentis, cuius extremitas contingat circulum *TZ*, et aliud in auge epicicli. Figam quoque unum clauum in puncto *H* et alium in puncto *K*, itemque alium in puncto
595 *D*. Et ligabo ad pedem cuiuslibet istorum clauorum unum filum de seta tenuissimum et equale sicud feci in luna.

Cum igitur uoluero equare mercurium per istud instrumentum, intrabo cum tempore ad quod uoluero ipsum equare quod uoco tempus datum in tabulam medii argumenti solis. Nam motus equalis centri epicicli mer-
600 curii et motus equalis solis idem motus sunt sicud diximus in hiis que premissa sunt. Sumam igitur ipsum et demam ex eo tria signa 29 gradus et 40 minuta quoniam aux ecentrici mercurii precedit augem solis ista quantitate. Si autem argumentum solis fuerit minus quantitate predicta, addam super ipsum 12 signa, ex agregato uero subtraham ipsam quantitatem;
605 quod autem relinquitur erit distantia centri epicicli ab auge equantis de partibus equantis et hoc uocatur medium centrum mercurii. Intrabo quoque cum eodem tempore in tabulam medii argumenti mercurii et sumam ipsum. Est autem medium argumentum mercurii distantia corporis eius ab auge media epicicli ut prius dictum est. Inuentis igitur medio centro et
610 medio argumento queram numerum medii centri in circulo equante et in circulo motus centri deferentis. Et faciam utrobique notam unam uel materialem uel solum in intellectu. Numerum quoque medii argumenti queram in diuisione epicicli et faciam etiam ibi notam. Postea accipiam filum ligatum ad punctum *H* et extendam ipsum super notam factam in circulo
615 motus centri deferentis. Et circumuoluam uersus occidentem tabulam motus centri deferentis quousque signum factum in ea cadat sub isto filo et perueniat ad notam que terminat numerum medii centri. Deinde accipiam filum ligatum ad punctum *K* et extendam ipsum super notam factam in equante que terminat numerum centri medii. Et manente hac tabula fixa
620 circumuoluam tabulam motus centri epicicli ab ea contentam quousque centrum epicicli cadat directe sub isto filo. Tabulam quoque epicicli ab

589 mouetur δ
594 alium²: tertium δ
598 uoluero: uolam β uolo *Af* uelim Λ
602 ecentrici: equantis δ
603 minus: in *add.* β
605 relinquetur δ
615–16 Et...deferentis *om. Morz*Γ
621–23 Tabulam...filo *om. BbiKW*

all sides and in all ways, and moves in it with a smooth and even motion, and cannot slip out of it. The surface of all those containing and contained disks should be one and the same, as we stated for the moon. Next I make one clearly visible mark on the disk of the motion of the center of the eccentric below the apogee of the deferent: the edge of this mark should touch circle *TZ*. [I make] another [mark] at the apogee of the epicycle.[80] I also affix a nail on point *H*, another on point *K*, and yet another on point *D*. I fasten a very fine thread of silk of uniform thickness to the base of each of these nails, as I did for the moon.

Therefore, when I want to equate Mercury by this instrument, I enter with the time at which I want to equate it, which I call the "datum time," into the table of the sun's mean anomaly.[81] For the uniform motion of the center of Mercury's epicycle and the uniform [i.e., mean] motion of the sun are the same motion, as we stated in what went before.[82] Therefore I take it and subtract from it 3 signs 29 degrees 40 minutes, since the apogee of Mercury's eccentric precedes [i.e., is to the east of] the apogee of the sun by that amount.[83] But if the anomaly of the sun is less than the above quantity, I add 12 signs to it [the anomaly] and subtract that amount [i.e., 3^s 29;40°] from the result; the remainder will be the distance of the center of [Mercury's] epicycle from the apogee of the equant, measured along the equant; this is called the "mean centrum" of Mercury. I also enter with the same time into the table of Mercury's mean anomaly and take that. The mean anomaly of Mercury is the distance of its body from the mean apogee of the epicycle, as was stated before.[84] Therefore, when the mean centrum and the mean anomaly have been found, I look on the equant circle and on the circle of the motion of the center of the deferent[85] for the number corresponding to the mean centrum. In each place I make a mark either on the actual instrument or just in my memory. I also look on the divided part of the epicycle for the number corresponding to the mean anomaly, and I make a mark there, too. Next I take the thread attached to point *H* and stretch it over the mark I made on the circle of the motion of the center of the deferent. I revolve the disk of the motion of the center of the deferent[86] toward the west [i.e., clockwise] until the mark made on it [at point *T*] falls beneath that thread and meets the mark indicating the amount of the mean centrum. Then I take the thread attached to point *K* and stretch it over the mark I made on the equant indicating the amount of the mean centrum. Letting this disk [the disk of the motion of the center of the deferent] remain fixed, I revolve the disk of the motion of the center of the epicycle, which it contains, until the center of the epicycle falls directly beneath that [second] thread. I also revolve

---

616 Alii habent quousque aux circuli defferentis maioris distantie cadat sub isto filo *mg*. Λ

ista contentam circumuoluam quousque aux epicicli cadat directe sub isto eodem filo. Eritque huius dispositio circuli deferentis et centri epicicli et augis ipsius medie ad tempus datum. Corpus quoque mercurii erit in loco 625 illo epicicli ubi facta est nota que terminat medium argumentum. Per hanc ergo notam faciam transire filum ligatum ad punctum *D* extendamque ipsum quousque secet orbem signorum. Et ubi secauerit ibi erit uerus locus mercurii ad tempus datum.

Si autem per illud instrumentum scire uolueris utrum mercurius sit 630 stationarius, directus aut retrogradus, fac filum ligatum ad punctum *D* contingere epiciclum ex utraque parte. Considera itaque utrum nota facta in epiciclo cadat in arcu superiori inter duos contactus aut in inferiori. Si in superiori uel etiam in puncto contactus, directus est. Si autem in inferiori, fac transire filum ligatum ad punctum *D* per notam factam in epiciclo 635 que designat locum mercurii et nota locum in orbe signorum ubi filum illud secat ipsum. Deinde permuta filum in epiciclo uersus dextram per 3 gradus epicicli et uide iterum ubi secat orbem signorum. Si igitur inter duas sectiones intercipitur de orbe signorum quasi gradus unus, stationarius est. Et erit in statione prima, si est prope contactum sinistrum; in 640 secunda uero, si est prope contactum dextrum. Si autem quod intercipitur inter duas sectiones de orbe signorum minus est uno gradu, directus est; si autem plus uno gradu, retrogradus est. Quod si subtiliter perscrutatus fueris poteris scire quando incipiet dirigi si est retrogradus, et quando erit in statione secunda, aut quando incipiet retrogradari si est directus, et 645 quando erit in statione prima. Similiter quoque utrum sit ascendens uel descendens tam in ecentrico quam in epiciclo, adeo facile scire poteris quod non est necesse ponere doctrinam ad hoc. Laus igitur sit omnipotenti Deo cuius misericordia donauit nobis intellectum et uiam ad comprehendendum per materiale instrumentum tantam uarietatem et tam admirabi-650 lem diuersitatem motuum mercurii.

Nunc autem consequens est ad premissa ut patefaciamus causas operis tabularum in mercurio quemadmodum fecimus in sole et luna. Hoc autem manifestum erit per illud instrumentum nobis memoriter tenentibus quecumque superius de orbibus et motibus mercurii dicta sunt. Quia uero 655 operatio tabularum in eo difficilis admodum esse uidetur, precipue autem propter minuta proportionalia que diuersificantur in mercurio et uenere et tribus superioribus, idcirco studium nostrum erit ut sermone claro et facili

623 huius: hec δ
633 erit δ
637 3 gradus: in arcu inferiori *add.* φ*J*/secet *BMY*Γ
639 est[1]: erit β
643 incipiet: incipit β incipiat *M*
644 incipiet: incipit β incipiat *M*
649 uarietatem: ueritatem δ
656 et[1]: a θ

the disk of the motion of the epicycle, which is contained by that [disk of the motion of the center of the epicycle], until the apogee of the epicycle falls directly beneath that same thread. This disposition will be [the situation] of the deferent circle and the center of the epicycle and its mean apogee at the datum time. Furthermore, the body of Mercury will be in that place on the epicycle where the mark indicating the mean anomaly was made. Thus I make the thread attached to point *D* pass through this mark and stretch it out until it cuts the ecliptic. And the place where it cuts [the ecliptic] will be the true position of Mercury at the datum time.

But if you want to know by that instrument whether Mercury is stationary, in direct motion, or retrograde, make the thread attached to point *D* touch the epicycle on either side. Then observe whether the mark you made on the epicycle falls on the upper of the arcs between the points of tangency or the lower. If it is on the upper, or actually at a point of tangency, it is in direct motion. But if it is on the lower, make the thread attached to point *D* pass through the mark you made on the epicycle which defines the position of Mercury, and note the place on the ecliptic where that thread intersects it. Then change the position of the thread on the epicycle [by moving it] three degrees of the epicycle toward the right [i.e., adding three degrees of anomaly], and, once again, observe where it intersects the ecliptic. If, then, the amount of the ecliptic cut off between the two points of intersection is about one degree, [Mercury] is stationary; and it will be in the first station if it is near the left point of tangency, but in the second [station] if it is near the right point of tangency. But if the amount of the ecliptic cut off between the two points of intersection is less than one degree, [Mercury] is in direct motion, whereas if [that amount is] more than one degree, [Mercury] is retrograde.[87] And if you make a careful examination, you will be able to tell when it will begin to be in direct motion and when it will be in the second station, if it is retrograde, or when it will begin to be retrograde and when it will be in the first station, if it is in direct motion. Similarly you will be able to tell whether it is ascending or descending,[88] both on the eccentric and on the epicycle, so easily that it is unnecessary to give instructions for that. Praise, then, to God omnipotent, whose mercy has granted us understanding and a means of comprehending through a material instrument so great a variety and so wonderful a diversity of motions [as we find] in Mercury.

The appropriate sequel to the foregoing is for us now to explain the reasons for the method of operation with the tables for Mercury as we did for the sun and moon.[89] These will become clear through that instrument, if we bear in mind everything that was said before about the circles and motions of Mercury. But since the method of operation with the tables for it seems rather difficult, especially on account of the minutes of proportion, which are different for Mercury from [those for] Venus and the three superior planets, our object will

omnes difficultates eius faciliter explanemus et, ne breuitas sermonis obscuritatem pariat, non uitabimus prolongare sermonem nostrum secun-
660 dum quod opus erit explanationi difficultatis proposite. Si enim plene comprehenderimus causas operis in mercurio, facilimum erit intelligere eas in uenere et tribus superioribus.

Dico ergo quod centrum epicicli mercurii mouetur uersus orientem in circumferentia sui deferentis super centrum sui equantis motu equali, ita
665 quod super ipsius centrum describit in temporibus equalibus equales angulos et in eius circumferentia equales arcus. Vnde linea recta protracta a centro equantis ad centrum epicicli mouet ipsum epiciclum uersus orientem equaliter, ideoque hec linea dicitur linea motus medii in equante. Quia uero ad orbem signorum omnem motum referimus, oportunum est querere que
670 sit linea medii motus mercurii in orbe signorum. Dico autem quod linea medii motus mercurii in orbe signorum est linea recta que egreditur a centro orbis signorum protensa equedistanter linee medii motus in equante modo dicte quousque ad orbem signorum perueniat. Necesse est enim quod hec linea moueatur in orbe signorum equaliter sicud prima mouetur equaliter
675 in equante. Nam cum ipse sint equedistantes sequitur ut anguli duo con-

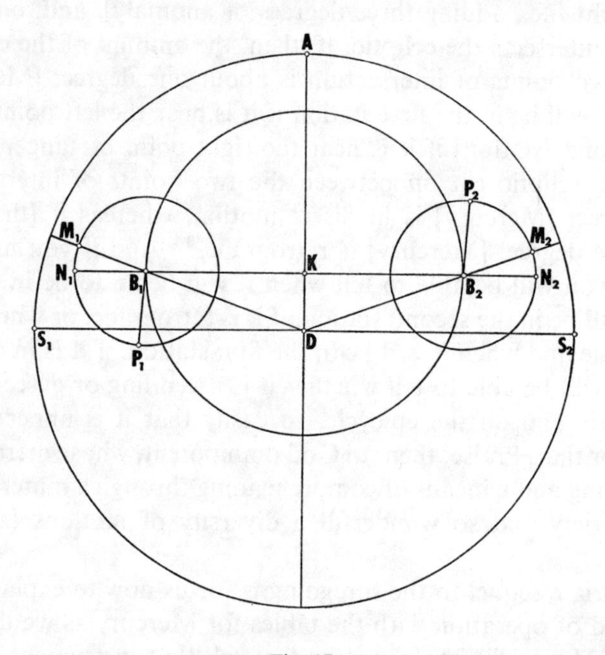

Fig. 17

661 causas operis: causas operationis *D* eas operationes *M*Γ
668 ideoque: idcirco β
672–73 modo dicte: modo dicto φ*JWw* modo ducte *rz* directa δ

be to explain all the difficulties simply in clear and easy language; and, lest brevity of explanation should give rise to obscurity, we will not hesitate to make our account as long as may be necessary for the elucidation of the difficulty confronting us. For, once we have fully understood the reasons for the operations [of the tables] for Mercury, it will be very easy to understand them for Venus and the three superior planets.

I say, then, that the center of Mercury's epicycle moves toward the east on the circumference of its deferent with a motion uniform with respect to the center of its equant, in such a way that it describes in equal times equal angles about its [the equant's] center and equal arcs on its circumference. Hence the straight line drawn from the center of the equant to the center of the epicycle carries the epicycle toward the east with uniform motion [see Fig. 17, p. 256], and for that reason this line is called the "line of mean motion in the equant." But since we refer all motions to the ecliptic, it is pertinent to ask what line in the ecliptic [system] is the line of Mercury's mean motion. I say that the line of Mercury's mean motion in the ecliptic is the straight line proceeding from the center of the ecliptic and produced parallel to the line of mean motion in the equant, which we just mentioned, until it reaches the ecliptic.[90] For this line must move uniformly in the ecliptic in the same way as the first-mentioned [line] moves uniformly in the equant. For, since they are parallel, it follows that

Legend for Fig. 17

Illustration of the relationship between mean and true centrum and mean and true anomaly for Mercury (based on a figure found in some manuscripts, but not mentioned in the text). AS_1S_2 represents the ecliptic, about center D, the earth ("centrum orbis signorum": MSS). Circle B_1B_2 represents the equant circle about center K, the equant ("centrum equantis": MSS). The epicycle is schematically represented at B_1 and B_2 on opposite sides of the equant (although in fact the epicycle center moves on the deferent, here we are concerned only with the relationship between angles and with the speed of the epicycle, not its distance, so the schematic placement on the equant is legitimate). The epicycle center moves uniformly about K. Thus Mercury's "mean position" with respect to D, the earth, is given by a line moving about D parallel to the radius vector KB. This line (DS in the figure) Campanus calls the "line of mean motion [in the ecliptic]" and also the "line of the mean centrum." The line DB joining the earth and the center of the epicycle he calls the "line of the true centrum." The difference between the two is the equation of center, angle KBD. M and N are the true and mean apogees of the epicycle, respectively ("aux uera" and "aux media": MSS). The mean anomaly is measured from the mean apogee N, but the equation of anomaly is a function of the true anomaly ($\angle MBP$). Clearly the difference between the two, arc MN ("equatio centri": MSS), is equal to the equation of center ($\angle KBD$). Two positions of the epicycle are shown to illustrate a rule of signs. On the first half of the equant (beginning from apogee) the equation of center has to be subtracted from the mean centrum to get the true centrum ($\angle AKB_1 - \angle KB_1D = \angle ADB_1$) but added to the mean anomaly to get the true anomaly ($\angle N_1B_1P_1 + \angle M_1B_1N_1 = \angle M_1B_1P_1$, and $\angle M_1B_1N_1 = \angle KB_1D$); on the second half of the equant the converse is true ($\angle AKB_2 + \angle KB_2D = \angle ADB_2$, where both $\angle AKB_2$ and $\angle ADB_2$ are greater than 180°; and $\angle N_2B_2P_2 - \angle M_2B_2N_2 = \angle M_2B_2P_2$).

tenti ab eis et a diametro transeunte per ambo centra predicta que sunt centrum equantis et centrum orbis signorum sint equales. Quoniam si duabus lineis equedistantibus tertia superuenerit facit angulum extrinsecum equalem intrinseco quemadmodum geometre probant. Due autem
680 linee equedistantes sunt que predicte sunt, tertia uero eis superueniens est diameter transiens per ambo centra predicta, anguli autem extrinsecus et intrinsecus sunt duo anguli quos continent predicte due linee equedistantes cum predicta diametro super duo centra predicta. Sunt autem hec in mercurio sicud in sole dictum est de medio motu eius. Quia igitur hec
685 linea equaliter mouetur in orbe signorum, idcirco dicitur linea medii motus mercurii et indicat eius medium motum et medium centrum. Sed medium motum indicat relata ad initium arietis; medium uero centrum, relata ad augem equantis. Est igitur medius motus mercurii arcus interceptus inter hanc lineam et principium arietis. Medium autem centrum eius est arcus
690 interceptus inter ipsam et augem equantis. Sed uerum centrum ipsius ostendit linea recta egressa a centro orbis signorum transiens per centrum epicicli. Arcus quidem interceptus inter ipsam et augem equantis dicitur centrum uerum. Arcus igitur interceptus inter lineam centri medii et lineam centri ueri est differentia eorumdem, quoniam in illo arcu excedit quando-
695 que centrum medium centrum uerum et quandoque in ipso exceditur ab eodem; medium quidem centrum excedit uerum in predicto arcu dum centrum epicicli est in medietate equantis que est ab auge eius usque ad oppositionem augis quod est dum centrum medium est minus 6 signis; uerum etiam centrum excedit medium in predicto arcu dum centrum epici-
700 cli est in alia medietate equantis que est ab oppositione augis sue usque ad eius augem quod est dum centrum medium est plus 6 signis. Vt igitur habito medio centro habeamus uerum necesse est eorum differentiam cognoscere, et si centrum medium est minus 6 signis eandem differentiam a centro medio remouere, si uero est maius 6 signis eandem differentiam
705 centro medio addere. Et quod erit post diminutionem uel augmentum istius differentie erit centrum uerum.

Hanc autem differentiam ueri centri et medii uocauerunt equationem centri, quoniam ipsa crescit et minuitur secundum distantiam centri epicicli ab auge equantis. Si enim fuerit centrum epicicli in auge equantis,

685 equaliter: essentialiter *V*
690–91 ostendit: erit φ*Bb*
694 ueri: non *add.* *BMmOqy*Γ
707 uocauerunt: tabularii *add.* *MWw*Γ

the two angles formed by each line and the diameter passing through the two above centers, namely, the center of the equant and the center of the ecliptic,[91] are equal. For, if two parallel lines are intersected by a third [line], [that line] forms an exterior angle equal to the interior [opposite one], as is proved by the geometricians.[92] Now the two parallel lines are the ones mentioned above, while the third [line] intersecting them is the diameter which passes through the two above centers, and the exterior and interior angles are the two angles formed by the above two parallel lines and the above diameter through the two centers. These [details] are the same for Mercury as [those] we described for the sun[93] [when we discussed] its mean motion. Since, then, this line moves uniformly in the ecliptic, it is called the "line of Mercury's mean motion," and it marks its [Mercury's] mean motion and mean centrum. But it marks its mean motion with respect to the beginning of Aries and its mean centrum with respect to the apogee of the equant. Therefore the mean motion of Mercury is the arc [of the ecliptic] cut off between this line and the beginning of Aries, while its mean centrum is the arc cut off between it [this line] and the apogee of the equant. However, its true centrum is marked by the straight line proceeding from the center of the ecliptic and passing through the center of the epicycle:[94] the arc cut off between this [second line] and the apogee of the equant is called the "true centrum." Therefore the arc cut off between the line of the mean centrum and the line of the true centrum is the difference between them [the mean centrum and the true centrum], because that is the arc by which the mean centrum is at times greater, and at other times less, than the true centrum: the mean centrum is greater than the true by the amount of the above arc as long as the center of the epicycle is on the half of the equant from its apogee to its perigee, i.e., as long as the mean centrum is less than six signs; the true centrum is greater than the mean by the above arc as long as the center of the epicycle is on the other half of the equant, namely, [that] from its perigee to its apogee, i.e., as long as the mean centrum is more than six signs. Therefore, to get the true centrum from the mean, we must discover the difference between them and, if the mean centrum is less than six signs, subtract that difference from the mean centrum, but, if it is more than six signs, add that difference to the mean centrum. The result of the subtraction or addition of that difference will be the true centrum.

They called this difference between the mean and true centra the "equation of center" because it increases and decreases according to the distance of the center of the epicycle from the apogee of the equant. For, if the center of the epicycle is in the apogee of the equant, the line of the true centrum and the

---

678 superuenerit: per 29 primi *add.* Λ *Joy mg. f*
679 probant: et per 29 principium Evclidis

710 linea ueri centri et linea medii erunt linea una. Nulla igitur erit inter eas distantia, ideoque etiam equatio centri nulla. Cum autem centrum epicicli recesserit ab auge equantis, diuerse fient linee predicte. Eritque alia que indicat centrum medium et alia que indicat centrum uerum. Fiet igitur inter eas aliqua distantia que quidem distantia tanto amplius crescet quan-
715 to plus centrum epicicli elongabitur ab auge equantis usque dum distet ab ea secundum medium eius motum 3 signis et 13 aut 14 gradibus. Ibi enim erit maior distantia earum que esse possit. Abinde autem quanto plus centrum epicicli mouebitur uersus oppositionem augis equantis tanto plus minuetur illa distantia usque dum centrum epicicli fiat in oppositione augis
720 equantis. Ibique nulla erit earum distantia quoniam ibi quoque linea medii centri et linea ueri fiunt linea una. Centro uero epicicli inde recedente et ad augem equantis accedente incipiunt fieri diuerse linee medii centri et ueri, et quanto amplius centrum epicicli mouetur uersus augem equantis tanto amplius predicte linee distant abinuicem usque dum centrum epicicli distet
725 iterum ab auge equantis 3 signis et 13 aut 14 gradibus secundum medium eius motum. Ibique erit iterum earum distantia maxima et equalis maxime priori. Centro autem epicicli plus apropinquante augi equantis minuetur distantia earum quousque centrum epicicli ueniat ad augem equantis ubi nulla erit earum distantia que omnia plane uidere potes in instrumento
730 premisso.

Quia igitur uiderunt tabularum compositores distantiam que est inter uerum centrum et medium uariari secundum diuersitatem distantie centri epicicli ab auge equantis, idcirco conati sunt inuenire in omni distantia centri epicicli ab auge equantis quanta sit distantia linee medii centri a
735 linea ueri et uariauerunt distantiam centri epicicli ab auge equantis per augmentum unius gradus. Nec oportuit eos laborare circa inuentionem huius distantie que est inter uerum centrum et medium nisi penes distantias quas habet centrum epicicli ab auge equantis in una medietate equantis in ea scilicet que est ab auge usque ad eius oppositionem; quoniam distantie
740 ueri centri et medii quas facit centrum epicicli in medietate altera equantis que est ab oppositione augis eius usque ad augem suam equales sunt suis comparibus quas facit centrum epicicli in medietate prima. Nam centrum epicicli cum distat ab auge equantis in duabus medietatibus predictis equaliter facit distantias ueri centri et medii equales. Nec est aliqua diuer-
745 sitas inter ista nisi quod in medietate prima centrum medium est maius uero quantitate illius distantie; in secunda autem minus quemadmodum

710 erit: ibi *add.* δθ
734 medii centri: medii motus *BMmOU*v_2*Y*Γ
735 uariauerunt: narrauerunt δ
736 oportet βφ
742 comparibus: distantiis *add.* θφ
744 aliqua: alia *AaBMVYZ*Γ

line of the mean will be one and the same. Therefore the interval between them will be nil, and so the equation of center will be nil also. But when the center of the epicycle moves away from the apogee of the equant, the above lines will become separate, and the one marking the mean centrum will be different from the one marking the true centrum. Therefore there will be an interval between them, and this interval will increase the farther the center of the epicycle moves away from the apogee of the equant, until it is 3 signs and 13 or 14 degrees[95] away from it in mean motion. For at that point the greatest possible interval between them will be found. From there on, however, the nearer the center of the epicycle moves toward the perigee of the equant, the smaller that interval will become, until the center of the epicycle reaches the perigee of the equant. There the interval between them will be nil, for there, too, the line of the true centrum and the line of the mean become one and the same. But, as the center of the epicycle moves away from that point and approaches the apogee of the equant, the lines of the mean and true centra begin to become separate, and the nearer the center of the epicycle moves to the apogee of the equant, the greater the interval between the two above lines grows, until the center of the epicycle is again 3 signs and 13 or 14 degrees distant from the apogee of the equant in mean motion. There the interval between them will again be a maximum, and equal to the previous maximum. Then as the center of the epicycle approaches the apogee of the equant, the interval between them will decrease until the center of the epicycle reaches the apogee of the equant, and there the interval between them will be nil. You can see this whole process clearly on the instrument described above.

Therefore, because the table-makers saw that the interval between true and mean centra varies according to the distance of the center of the epicycle from the apogee of the equant, they tried to find, for every distance of the center of the epicycle from the apogee of the equant, the size of the interval between the line of the mean centrum and the line of the true [centrum]; and they made the distance of the center of the epicycle from the apogee of the equant vary by steps of one degree. They did not have to bother to find this interval, namely, that between the true and mean centra, for the distances of the center of the epicycle from the apogee of the equant on more than one half of the equant, namely, the one from the apogee to the perigee; for the intervals between the true and mean centra produced by the center of the epicycle on the other half of the equant, from the perigee to the apogee, are equal to the corresponding ones produced by the center of the epicycle on the first half. For, when the center of the epicycle is at equal distances from the apogee of the equant on either of the two above halves, it produces equal intervals between the true and mean centra. The only difference between those [situations] is that on the first half the mean centrum is greater than the true by the amount of that interval, whereas on the second it is less [than the true by the same

prius dictum est. Vnde necesse est quod eadem sit distantia inter uerum centrum et medium quando centrum epicicli distat ab auge per unum gradum et quando distat per 359 gradus. Item necesse est quod sit eadem 750 distantia quando distat per duos gradus et quando distat per 358 gradus, et ita deinceps in medietate prima augmentando et in medietate secunda equaliter diminuendo. Processerunt igitur tabularii hoc modo. Scripserunt enim in linea una descendente gradus medietatis prime equantis incipientes ab auge ipsius. Et processerunt per augmentum gradus et gradus usque 755 ad complementum totius medietatis quam terminauerunt in 6 signis. In linea uero altera descendente etiam scripserunt gradus medietatis secunde equantis incipientes ab auge ipsius, et processerunt per diminutionem gradus et gradus usque ad complementum totius medietatis quam terminauerunt necessitate exigente in sex signis. Super has duas lineas scrip- 760 serunt titulum, linee numeri. Quesiuerunt igitur per uias geometricas associatas uiis arismetricis quanta sit distantia ueri centri a medio in qualibet positarum distantiarum centri epicicli ab auge equantis. Et huius distantias repertas scripserunt in linea tertia descendente in directo eius centri a quo quelibet causatur, scripseruntque similiter super eas titulum, equatio 765 centri. Ista sufficiant de centro medio et centro uero et eorum distantia que equatio centri dicitur.

Nunc autem de argumento mercurii et operationum que sunt circa ipsum causas scrutemur. Dictum est supra quod mercurius in suo epiciclo mouetur dum est in superiori parte uersus orientem et iste motus, prout 770 incipit ab auge media epicicli, est equalis et uniformis et dicitur argumentum medium; prout autem incipit ab auge uera ipsius, inequalis est et difformis et dicitur argumentum uerum. Quid autem sit aux media epicicli et quid uera dictum est prius. Est autem aux media eius punctus in circumferentia ipsius qui terminat lineam que egreditur a centro equantis et tran- 775 sit per centrum epicicli. Et hec linea est illa cui prius diximus equedistare lineam medii motus siue medii centri. Aux autem uera est punctus in circumferentia epicicli qui cadit in linea ueri centri. Et est hec linea illa que egreditur a centro orbis signorum et transit per centrum epicicli. Excessus igitur inter uerum argumentum et medium est distantia inter augem ueram 780 et mediam epicicli. Dico igitur hunc excessum esse similem equationi centri, similem autem sic intelligo quoniam tantus est iste excessus de tota circumferentia epicicli quanta est equatio centri de tota circumferentia

764 super eas: eis δ

767 operationibus α*H*/sunt: fiunt β*H* fuerunt *A*

777 qui: que δ

amount],[96] as was stated before. Hence, the interval between the true and mean centra must be the same when the center of the epicycle is 1 degree distant from the apogee as when it is 359 degrees distant. Again, the interval must be the same when [the epicycle center] is 2 degrees distant as when it is 358 degrees distant, and so forth, with parallel increase on the first half and decrease on the second half. Therefore the table-makers proceeded as follows:[97] they wrote down the degrees of the first half of the equant, beginning from its apogee, in a single column; and they increased the numbers one degree at a time, until they had completed the whole half, ending at six signs. They also wrote down the degrees of the second half of the equant, beginning from its apogee, in a second column; and they decreased the numbers one degree at a time until they had completed the whole half, ending, necessarily, at six signs. Above these two columns they wrote the heading "Lines of Numbers."[98] They then investigated by geometrical methods, combined with arithmetical ones,[99] the size of the interval between the mean centrum and the true at each of the tabulated distances of the center of the epicycle from the apogee of the equant. When they had found these intervals, they wrote them in a third column, each opposite the centrum [i.e., distance of epicycle center from apogee] which produces it, and they gave these, too, a heading, "Equation of Center." Let that suffice for the mean and the true centrum, and for the interval between them, which is called the equation of center.

Now, however, let us examine the anomaly of Mercury, and the reasons for the operations concerning it. We said above[100] that Mercury moves on its epicycle toward the east when it is on the upper part, and that that motion, when measured from the mean apogee of the epicycle, is uniform and is called the mean anomaly, but, when measured from its true apogee, is not uniform[101] and is called the true anomaly. The meanings of "mean apogee" and "true apogee" of the epicycle were explained above: its mean apogee is the point on its circumference which is the end of the line proceeding from the center of the equant and passing through the epicycle's center. This line is the one to which we said above[102] the line of mean motion or the mean centrum was parallel. Its true apogee is the point on the circumference of the epicycle which falls on the line of the true centrum.[103] This [line of the true centrum] is the line proceeding from the center of the ecliptic and passing through the center of the epicycle. Therefore the difference between the true and mean anomalies is the distance between the true and mean apogees of the epicycle. I say, then, that this difference is similar[104] to the equation of center; by "similar" I understand the following: that the proportion of that difference to the whole circumference of the epicycle is the same as the proportion of the equation of center to the

---

753 enim: sicut patet in tabula equationis mercurii *add. JoΛ*

orbis signorum. Ita enim dicuntur a geometris arcus similes quorum uidelicet ad suas circumferentias est una proportio. Constat quidem quod linea
785 ueri centri in qua est aux uera epicicli secat duas equedistantes predictas: illam quidem que egreditur a centro equantis super centrum epicicli et illam que egreditur a centro orbis signorum super centrum eius. Faciet igitur hec linea cum illis duabus equedistantibus angulum extrinsecum, et est ille qui est in centro epicicli, equalem intrinseco, et est ille qui est in
790 centro orbis signorum. Hoc enim probant geometre quod linea recta cadens super duas equedistantes facit angulum extrinsecum equalem intrinseco quemadmodum prius dictum est. Angulus igitur in centro orbis signorum qui subtenditur equationi centri equalis est angulo in centro epicicli qui subtenditur distantie que est inter augem ueram et augem mediam
795 epicicli. Erit igitur equatio centri similis isti distantie, sic enim dicunt geometre quod arcus circulorum oportet esse similes quibus in suis centris subtenduntur anguli equales. Tanta igitur est diuersitas inter centrum uerum et centrum medium de partibus orbis signorum quanta est inter argumentum medium et argumentum uerum de partibus epicicli. Quantum
800 igitur necesse est subtrahere uel addere medio centro ad habendum uerum tantum quoque necesse est addere uel subtrahere medio argumento ad habendum uerum, sed contrarie. Nam in prima medietate equantis que est ab auge ad eius oppositionem, sicud centrum medium est maius uero, ita argumentum uerum est maius medio. Itemque in medietate equantis se-
805 cunda que est ab oppositione usque ad augem sicud centrum uerum est maius medio, ita argumentum medium est maius uero. Ideoque necesse est semper ut quidquid subtrahitur a centro medio ad habendum uerum, idem addatur argumento medio ad habendum uerum, et quidquid additur centro medio ad habendum uerum idem subtrahatur ab argumento medio ad
810 habendum uerum.

Reuertamur nunc ad opera tabularum et explanemus ea usque ad uerificationem centri et argumenti. Precipiunt igitur autores tabularum ut uolentes equare mercurium ad aliquod tempus intremus cum tempore dato et sumamus medium motum solis et medium argumentum mercurii. Me-
815 dium autem motum solis sumimus quoniam ipse est medius mercurii. Dictum est enim supra quod lineam indicantem medium motum solis et lineam indicantem medium motum mercurii terminat idem circulus tran-

784 quidem: ex premissa figura *add.* δ
787 super centrum eius: equedistanter priori δ
789–90 epicicli...centro *om. BmStY*
790 geometri δ
790–92 quod...intrinseco *om.* δ
805 oppositione: augis *add.* β
808–10 et quidquid...uerum *om.* δ
811 Reuertamur: igitur *add.* δθ ergo *add.* Σ
811–12 uerificationem: operationem *BgMOQqY*Γ
815 sumimus: sumamus δ sumemus *f*

whole circumference of the ecliptic. For those arcs that have the same proportion to their respective circumferences are called similar[105] by the geometricians. It has been established that the line of the true centrum, on which lies the true apogee of the epicycle, cuts the two above-mentioned parallel lines, [intersecting] the one proceeding from the center of the equant at the center of the epicycle and the one drawn from the center of the ecliptic at the center of the ecliptic. Therefore this line, as it meets those two parallel lines, will form an exterior angle, the one at the center of the epicycle, equal to the interior [angle], the one at the center of the ecliptic.[106] For the geometricians prove that a straight line intersecting two parallel lines forms an exterior angle equal to the interior [opposite] one, as was stated before.[107] Therefore the angle which is subtended at the center of the ecliptic by the equation of center[108] is equal to the angle subtended at the center of the epicycle by the interval between the true and mean apogees of the epicycle. So the equation of center will be similar to that interval, for the geometricians say that those arcs of circles must be similar which subtend equal angles at their centers.[109] Therefore the difference between mean and true centra contains as many parts of the ecliptic as the difference between mean and true anomalies contains parts of the epicycle. Therefore the same amount must be added to or subtracted from the mean anomaly to obtain the true as has to be subtracted from or added to the mean centrum to obtain the true, but the operations are opposite: for on the first half of the equant, from the apogee to the perigee, the mean centrum is greater than the true, but, on the other hand, the true anomaly is greater than the mean. And again, on the second half of the equant, from the perigee to the apogee, the true centrum is greater than the mean, but, on the other hand, the mean anomaly is greater than the true. And thus, whatever amount is subtracted from the mean centrum to obtain the true, the same must be added to the mean anomaly to obtain the true, and, whatever is added to the mean centrum to obtain the true, the same must be subtracted from the mean anomaly to obtain the true.

Let us now return to the operations with the tables and explain them as far as the equating of the centrum and the anomaly.[110] The table-makers, then, instruct[111] [us] that, if we wish to equate Mercury for a given time, we should enter with the datum time and take the mean motion of the sun and the mean anomaly of Mercury. We take the mean motion of the sun because it is the mean motion of Mercury [also]. For we stated above[112] that the line marking the mean motion of the sun and the line marking the mean motion of Mercury are bounded by one and the same circle passing through the poles of the

---

783–84 in principio 3 EVCLIDIS *mg.* *F*Σ
790 geometre: per 29 principium *add.* *Jo* per 29 primi *add.* Λ
796 geometre: in 12ª diffinitione terty EVCLIDIS *add.* Λ

siens per polos orbis signorum et ex eadem parte. Itaque per medium motum solis habemus distantiam linee medii motus mercurii ab ariete in
820 orbe signorum. Per medium uero argumentum mercurii habemus distantiam corporis mercurii in suo epiciclo ab auge media epicicli uersus orientem. Precipiunt postmodum ut a medio motu solis quem prediximus esse medium mercurii subtrahamus augem mercurii que est 6 signa 17 gradus et 30 minuta: hoc ideo quoniam aux equantis mercurii cadit sub 17 gradibus
825 et 30 minutis libre a quo puncto usque ad arietem sunt 6 signa 17 gradus et 30 minuta. Cum igitur quantitatem istam subtraxerimus de medio motu solis qui est distantia linee medii motus mercurii ab ariete, remanebit nobis distantia linee medii motus mercurii ab auge equantis quam prius diximus esse medium centrum mercurii. Et hoc uolebant habere tabularii
830 per hanc subtractionem unde etiam residuum uocant centrum medium. Precipiunt postmodum ut cum hoc centro intremus in lineas numeri et sumamus equationem centri: hoc ideo quoniam uolunt ex centro medio extrahere centrum uerum et ex argumento medio argumentum uerum. Precipiunt postmodum ut si centrum est minus 6 signis eandem equatio-
835 nem minuamus a centro medio et relinquitur uerum; si uero centrum est maius 6 signis, eandem equationem addamus centro medio et habebitur uerum: hoc ideo quoniam cum centrum epicicli distat ab auge equantis minus 6 signis centrum medium maius est uero in quantitate equationis centri, quando autem distat ab ea plus 6 signis centrum medium est minus
840 uero in quantitate eadem. Precipiunt postmodum ut addamus equationem centri argumento medio si eam subtraxerimus a centro, et quod subtrahamus eandem ab argumento medio si eam addiderimus centro: hoc ideo quoniam quantum minuit aut addit centrum uerum a centro medio uel super centrum medium tantum addit aut minuit argumentum uerum super
845 argumentum medium uel ab argumento medio contrarie. Per hanc ergo additionem uel subtractionem contrarie factam habemus argumentum uerum. Precipiunt postmodum ut si hanc equationem addiderimus centro, superscribamus ei, addatur; si uero ipsam subtraxerimus ab eo, superscribamus ei, minuatur: hoc ideo quoniam in fine operis erit hec eadem
850 equatio addenda medio motui si fuerit addita centro, aut minuenda ab eo si fuit diminuta a centro. Et ipsi timuerunt ne obliuisceremur utrum esset addita centro uel subtracta ab eodem propter multa opera que oportebat nos interponere antequam ipsam adderemus medio motui uel subtraheremus ab eo. Per ista igitur que dicta sunt habemus uerum centrum et uerum

833 et...uerum *om. BMmYΓ*
835 relinquetur *eVxz* relinquatur δ
842 addiderimus: addidimus δ
845–47 Per...uerum *om. BMmYΓ*
848–49 superscribamus: superscribantur δ superscribatur *HΣ*
853 adderemus: deberemus addere δ
853–54 subtraheremus: subtrahere δ detrahere *H*

ecliptic, and [that this circle intersects them] on the same side [of the ecliptic]. Thus in the mean motion of the sun we have the distance of the line of Mercury's mean motion from [the beginning of] Aries, [measured] along the ecliptic; while in the mean anomaly of Mercury we have the distance of Mercury's body on the epicycle from the mean apogee of the epicycle, [measured] toward the east. Next they instruct us to subtract the apogee of Mercury, which is 6 signs 17 degrees 30 minutes, from the mean motion of the sun, which we have already stated to be the mean motion of Mercury. The reason for this is that the apogee of Mercury's equant lies at 17 degrees and 30 minutes of Libra, and from this point to Aries there are 6 signs 17 degrees 30 minutes. Therefore, when we have subtracted the latter amount from the mean motion of the sun, which is the distance of the line of Mercury's mean motion from Aries, the remainder will be the distance of the line of Mercury's mean motion from the apogee of the equant, which distance we previously said was the mean centrum of Mercury. This was what the table-makers wished to obtain by the above subtraction; hence they call the remainder the mean centrum. Next they instruct us to enter the columns of argument[113] with this centrum and to take the equation of center. The reason for this is that they want to derive the true centrum from the mean centrum and the true anomaly from the mean anomaly. Next they instruct us to subtract that equation from the mean centrum if the centrum is less than six signs: then the remainder is the true [centrum]; but if the centrum is greater than six signs, [they instruct us] to add that equation to the mean centrum, and we will get the true. The reason for this is that, when the center of the epicycle is less than six signs distant from the apogee of the equant, the mean centrum exceeds the true by the amount of the equation of center, but, when [the epicycle center] is more than six signs distant from it [the apogee], the true centrum exceeds the mean by that amount. Next they instruct us to add the equation of center to the mean anomaly if we subtracted it from the centrum, but to subtract it from the mean anomaly if we added it to the centrum. The reason for this is that the true anomaly exceeds or is exceeded by the mean anomaly by the same amount as the true centrum is exceeded by or exceeds the mean centrum, the difference being of opposite sign [for the anomaly to what it is for the centrum]. Thus, by performing this opposite operation of addition or subtraction, we have the true anomaly.[114] Next they instruct us to write "add" above the equation if we added it to the centrum, but, if we subtracted it from it [the centrum], to write "subtract" above it. The reason for this is that, at the end of the operations,[115] this same equation will have to be added to the mean motion if it was added to the centrum or subtracted from it if it was subtracted from the centrum. And they were afraid that we might forget whether it had been added to the centrum or subtracted from it, because of the many operations which we have to go through before adding it to or subtracting it from the mean motion. By the process thus described, then, we

855 argumentum, distantiam quoque ueri centri a medio, hoc est, equationem centri.

Ad causas itaque consequentis operis accedamus. Si mercurius esset semper in auge uera sui epicicli uel in eius opposito, non oporteret nos amplius laborare in eius uero loco inueniendo. Nam linea ueri centri esset 860 linea ueri loci. Quia uero potest esse in qualibet parte sui epicicli, ideo alia potest esse linea ueri loci quam linea ueri centri. Harum igitur duarum linearum distantiam oportet nos inuenire. Si enim cognouerimus distantiam ueri loci a uero centro et cognouerimus utrum illa distantia sit addens super uerum centrum quod erit cum uerum argumentum erit minus 6 sig-865 nis aut sit minuens a uero centro quod erit cum argumentum uerum erit plus 6 signis, et nos iam prius cognouimus per equationem centri distantiam ueri centri a medio motu, et sciuimus etiam utrum ista distantia sit minuens a medio motu quod est cum centrum est minus 6 signis aut addens super ipsum quod est cum centrum est plus 6 signis; nos per ista sciemus distan-870 tiam ueri loci a medio motu. Quoniam si ambe distantie predicte erunt addentes, eas coniunctas addemus super medium motum et habebimus uerum locum. Si autem ambe erunt minuentes, eas coniunctas minuemus de medio motu et relinquitur uerus locus. Si uero una earum sit addens et alia minuens, detrahemus minorem de maiori, de residuo autem facie-875 mus quod faciendum erat de maiori et sic habebimus etiam uerum locum.

Laborandum est igitur circa inuentionem distantie que est inter lineam ueri loci et lineam ueri centri. Hec autem distantia dicitur equatio argumenti quoniam secundum quod uariatur argumentum uerum uariatur etiam hec distantia ut continuo demonstrabimus. Et quamuis etiam centri 880 uarietas hanc eandem distantiam uariet ut patebit ex sequentibus, non tamen propter hoc debuit dici equatio centri duplici ratione. Prima quidem quoniam centrum iam habuit suam equationem propriam de qua paulo ante tractauimus; secunda autem quoniam licet argumentum et centrum hanc distantiam uarient, argumentum tamen maxime, sicud indicabunt 885 sequentia, ideoque equatio argumenti dicitur et non equatio centri. Dico igitur quod distantia duarum linearum, uidelicet ueri loci et ueri centri, que dicitur equatio argumenti, diuersificatur et propter distantiam mercurii ab auge uera epicicli et propter distantiam centri epicicli ab auge equantis. Primum quidem horum manifestum est: pone enim centrum epi-890 cicli in quouis loco et esto gratia exempli quod sit in auge equantis. Intel-

857 accedamus: attendamus δ
865 aut sit: at si *B* aut si *AFiY om. f*
867 sciuimus etiam *om.* φ*A*Π / ista distantia *om.* φ*A*
881 Primo δ
882 propriam: primam δ
883 secundo δ*A*Ψ secundum *N*
885 ideoque: etiam *add.* β

have the true centrum and the true anomaly and also the distance between true and mean centra, that is, the equation of center.

Let us then turn to the reasons for the succeeding operations. If Mercury were always in the true apogee or perigee of its epicycle, we should not have to expend any further effort in finding its true position. For the line of its true centrum would be the line of its true position. But because it can be at any point on its epicycle, the line of the true position can be different from the line of the true centrum. Therefore we must find the distance between these two lines. For, if we discover the distance between [the line of] the true position and [the line of] the true centrum and also whether that distance is additive with respect to the true centrum, which it will be when the true anomaly is less than six signs, or subtractive with respect to the true centrum, which it will be when the true anomaly is more than six signs, and since we have already found the distance of [the line of] the true centrum from [the line of] the mean motion by the equation of center and also whether that distance is subtractive with respect to the mean motion, which it is when the centrum is less than six signs, or additive, which it is when the centrum is more than six signs, [then] by the above we will know the distance of [the line of] the true position from [the line of] the mean motion. For, if both the above distances[116] are additive, we will add their sum to the mean motion and get the true position; if both are subtractive, we will subtract their sum from the mean motion, and the remainder will be the true position; whereas, if one of them is additive and one subtractive, we will subtract the lesser from the greater and apply the remainder [to the mean motion] with the same sign as we should have applied the greater, and thus, too, we shall get the true position.

Therefore we have to work to find the distance between the line of the true position and the line of the true centrum. This distance is called the "equation of anomaly," because it varies as the true anomaly varies, as we shall presently prove. And although the variation of the centrum also affects the variation of this distance, as will become clear from what follows, nevertheless it should not, on account of this, be called the "equation of center," for two reasons. The first is that the centrum has already been assigned its own equation, which we discussed just above; the second is that, although both anomaly and centrum affect the variation of this distance, it is the anomaly which has the greatest effect, as will be shown in what follows.[117] For those reasons it is called the "equation of anomaly" and not the "equation of center." I say, then, that the distance between the two lines, namely, [the line] of the true position and [the line] of the true centrum, which is called the equation of anomaly, varies both because of the distance of Mercury from the true apogee of the epicycle and because of the distance of the center of the epicycle from the apogee of the equant. The first of these [factors] is obvious: for suppose the center of the epicycle to be in any place you please, for example, in the apogee of the equant.

ligas ergo rectas lineas duci a centro orbis signorum usque ad eius circumferentiam per singulos gradus epicicli. Manifestum est igitur quod linea transiens per primum gradum minus distat a linea ueri centri quam transiens per secundum; et transiens per secundum minus quam transiens per
895 tertium; et sic de ceteris usque ad lineam contingentem epiciclum que inter omnes maxime distat ab ea. Cetere uero transeuntes per gradus qui sunt inter contactum et oppositionem augis tanto minus distant ab ea quanto gradus per quos transeunt plus distant a contactu. Idemque erit de lineis que transeunt per gradus qui sunt in alia medietate epicicli. Nam
900 tantum distabunt singule in hac medietate a linea ueri centri quantum distant in medietate prima sue compares ab eadem. Compares autem uoco lineas transeuntes per gradus equaliter hinc et inde distantes ab auge uera epicicli. Nec est aliqua differentia inter distantias duarum comparium a linea ueri centri nisi quod in medietate prima epicicli est distantia addens
905 et in medietate secunda est distantia minuens. Patet igitur ex predictis quod equatio argumenti que est distantia linee ueri loci a linea ueri centri uariatur propter diuersitatem ueri argumenti. Patet insuper quod duo argumenta uera quorum unum terminatur in medietate prima et aliud in secunda distantque ambo termini equaliter ab auge uera epicicli habent
910 equales equationes. Faciunt enim equales distantias linee ueri loci a linea ueri centri. Sed primum habet equationem addentem et secundum habet equationem minuentem, unde eandem equationem habet uerum argumentum unius gradus et uerum argumentum 359 graduum. Item eandem equationem habet argumentum uerum 2 graduum et argumentum uerum 358
915 graduum et sic de ceteris quousque ueniamus ad finem semicirculi, ex parte ista addendo gradum et gradum, ex parte uero altera minuendo gradum et gradum. Quod autem diuersitas distantie centri epicicli ab auge equantis uariet predictas equationes non est difficile uidere. Centro enim epicicli ubicumque alibi existente quam in auge equantis, manifestum est quod
920 premisse distantie maiores fient quam sint sue relatiue eo existente in auge. Eruntque iste distantie maxime que esse possunt quando centrum epicicli propinquissimum erit centro orbis signorum, quod erit quando centrum medium est 4 signa ut dictum est superius. Ideoque ibi erunt equationes omnium argumentorum maxime. Due quoque distantie centri epicicli ab
925 auge equantis compares facient equationes relatiuorum argumentorum equales. Voco autem distantias centri epicicli compares que in duabus

901 eodem δ
903 inter: istas *add.* β
909 distantque *hoXZ*Γ distentque *codd. plerique* distatque *DN* distantesque *V*
919 ubique α*H*

Then imagine straight lines drawn from the center of the ecliptic to the circumference of the epicycle, one to each degree of [anomaly on] the epicycle: it is obvious that the line passing through the first degree is nearer the line of the true centrum than the one passing through the second, and the one passing through the second nearer than the one passing through the third, and so on up to the line tangent to the epicycle, which is at the greatest distance of all from it [the line of the true centrum]. The other lines, which pass through the degrees between the tangent and the perigee, are nearer to it [the line of the true centrum] according as the degrees through which they pass are farther from the tangent. The same will be true of lines passing through the degrees on the other half of the epicycle. For each line on this half will be the same distance from the line of the true centrum as the corresponding one on the first half is from the same line. By "corresponding" lines I mean those lines which pass through degrees equally removed from the true apogee of the epicycle on opposite sides of it. The only difference between the distances of two corresponding lines from the line of the true centrum is that on the first half of the epicycle [i.e., for an anomaly of 0°–180°] the distance is additive, while on the second half it is subtractive. Therefore it is clear from what we have said that the equation of anomaly, which is the distance of the line of the true position from the line of the true centrum, varies because of the variation of the true anomaly. It is clear, furthermore, that two true anomalies, one of which has the point determining it lying on the first half [i.e., between 0° and 180°] and the other on the second [between 180° and 360°], both points being the same distance from the apogee of the epicycle, have equal equations. For the lines of the true position are at equal distances from the line of the true centrum. But the first has an additive equation and the second a subtractive equation. Hence the same equation belongs to a true anomaly of 1 degree and a true anomaly of 359 degrees; again the same equation belongs to a true anomaly of 2 degrees and a true anomaly of 358 degrees, and so forth, until we come to the end of the semicircle, increasing by one degree at a time on one side and decreasing by one degree at a time on the other side. Furthermore, it is not difficult to see that the variation in the distance of the center of the epicycle from the apogee of the equant produces a variation in the above equations. For when the center of the epicycle is anywhere outside the apogee of the equant, it is clear that the above distances[118] will be greater than those at the same anomalies[119] when it [the center of the epicycle] is in the apogee. And those distances will be the greatest possible when the center of the epicycle is closest to the center of the ecliptic; that will be when the mean centrum is four signs, as was stated above.[120] And so at that position the equations of all the anomalies will be at a maximum. Furthermore, two corresponding distances of the center of the epicycle from the apogee of the equant will produce equal equations for the same anomalies. I call those distances of the center of the epicycle "corresponding" which are

medietatibus sumuntur equales ab auge equantis. Centrum igitur uerum unius gradus et 359 graduum faciunt equationes relatiuorum argumentorum equales. Ita quoque faciunt centrum uerum duorum graduum et 358
930 et sic de ceteris, ex hac parte augmentando et ex illa diminuendo gradum et gradum usque ad complementum semicirculi. Quia igitur 180 gradus epicicli qui sunt ab auge sua usque ad eius oppositionem uariant istam distantiam que dicitur equatio argumenti, itemque 180 gradus equantis qui sunt ab auge sua usque ad oppositionem eius uariant eandem distantiam,
935 oportebat ad habendam plenam et promptam cognitionem de equatione argumenti inuenire 32580 equationes, ut uidelicet centro epicicli existente in auge equantis quereremus equationes 180 argumentorum; itemque centro epicicli distante ab auge per unum gradum quereremus alias 180 equationes eorumdem argumentorum. Idem quoque faceremus centro epicicli
940 distante ab auge equantis per duos gradus et per tres gradus et sic de ceteris usque ad 180 gradus equantis, essentque facte 180 equationes 180 uicibus exceptis illis que facte fuerunt centro epicicli existente in auge. Quare essent equationes argumenti in uniuerso 32580.

Hoc autem erat ualde longum et habens laborem multum ideoque ad
945 istarum equationum facilem comprehensionem usi sunt quodam subtili ingenio. Supposuerunt enim quod centrum epicicli secundum uerum eius motum distaret ab auge equantis per 2 signa et 4 aut 5 gradus. Et supposuerunt quod gradus positi in duabus primis lineis descendentibus in tabula mercurii super quas scribitur titulus, linee numeri, quos diximus supe-
950 rius in inuentione equationis centri numerare equantem numerauerunt etiam epiciclum in inuentione equationis argumenti. Et quesiuerunt per uias geometricas associatas uiis arismetricis quanta esset equatio argumenti unius gradus et quanta argumenti duorum graduum et quanta argumenti trium et ita de ceteris usque ad 180. Et quamlibet equationem inuentam
955 scripserunt in quadam linea descendente, ponentes quamlibet equationem in directo sui argumenti. Et huic linee superposuerunt titulum, equatio argumenti. Rursus supposuerunt quod centrum epicicli esset in auge equantis et quesiuerunt equationes eorumdem 180 argumentorum que fuerunt omnes minores primis 180 prius inuentis, quelibet uidelicet minor
960 sua relatiua. Fueruntque minime que esse possent propter hoc quod centrum epicicli maxime distabat a centro orbis signorum. Subtraxerunt igi-

929 358: graduum *add.* α
930 augmentando: augendo δ/illa: alia δ hac *B*
933–34 que dicitur...distantiam *om.* *BhKMY*Γ
935 habendum δθ
938 distante: existente β
940 distante: existente β
941–42 180 uicibus: uidelicet *bDX*
949 quos: quas δ*ABM*Γ
950 numerare equantem *om.* δ/numerauerunt: numerarent δ*cek* numerabunt *A*
955 quadam: alia *add.* δ
960 possint θ*H* possunt *AdekrSt*Λ

measured as equal [arcs] from the apogee of the equant on the two [opposite] halves. Therefore a true centrum of 1 degree and a true centrum of 359 degrees produce equal equations for the same anomalies. So do the true centra of 2 degrees and 358 degrees, and so forth, with an increase of one degree at a time on the one side and a decrease of one degree at a time on the other, until the semicircle is completed. Since, therefore, the 180 degrees of the epicycle from its apogee to its perigee produce a variation in that distance which is called the equation of anomaly, and the 180 degrees of the equant from its apogee to its perigee also produce a variation in the same distance, one ought to find 32,580 equations[121] to have complete and immediate information about the equation of anomaly. That is, we should calculate the equations of the 180 anomalies for the center of the epicycle in the apogee of the equant, and again, for the center of the epicycle at 1 degree of distance from the apogee, we should calculate another 180 equations for the same anomalies, and do the same for the center of the epicycle at distances from the apogee of the equant of 2 degrees, of 3 degrees, and so on up to 180 degrees of the equant. And thus 180 equations would have been calculated 180 times, not counting that set which was calculated for the center of the epicycle in the apogee. Therefore the total number of equations of anomaly would be 32,580.

However, this [process] would have been very long and involved much labor, and so [the table-makers] used an ingenious device for the easy determination of these equations: they supposed the center of the epicycle to be at a distance of 2 signs and 4 or 5 degrees from the apogee of the equant in true motion,[122] and they posited that the degrees found in the first two columns of the table for Mercury, above which is written the heading "Lines of Numbers," which we said above represent the [degrees of the] equant when one is finding the equation of center, also represent [the degrees of] the epicycle when one is finding the equation of anomaly. They then calculated by geometrical methods, combined with arithmetical ones, the size of the equation for an anomaly of 1 degree, the size for an anomaly of 2 degrees, the size for an anomaly of 3, and so forth up to 180, and wrote down each equation that they found in a special column, putting each equation opposite the proper anomaly. Above this column they wrote the heading "Equation of Anomaly."[123] Again, they supposed the center of the epicycle to be in the apogee of the equant, and calculated the equations of the same 180 anomalies; these equations were all less than the first 180 which had already been found—each, that is, less than the one for the same anomaly—and were furthermore the least possible, since the center of the epicycle was at a maximum distance from the center of the ecliptic. Therefore they subtracted

tur minores secundo inuentas de maioribus primo inuentis, quamlibet uidelicet de sua relatiua. Singula uero residua scripserunt in quadam alia linea descendente, quodlibet quidem residuum in directo sui argumenti, et
965 superposuerunt huic linee titulum, longitudo longior. Amplius autem supposuerunt quod centrum epicicli secundum eius medium motum distaret ab auge equantis per 4 signa. Et quesiuerunt equationes eorumdem 180 argumentorum que fuerunt omnes maiores primis, quelibet uidelicet maior sua relatiua. Fueruntque maxime que esse possunt propter hoc quod cen-
970 trum epicicli uicinissimum erat ibi centro orbis signorum ut dictum est supra. Subtraxerunt igitur minores primo repertas de hiis maioribus tertio repertis, quamlibet uidelicet de sua relatiua. Singulos uero excessus harum maiorum super illas minores scripserunt in quadam alia linea descendente, quemlibet quidem excessum in directo sui argumenti. Et superposuerunt
975 huic linee titulum, longitudo propior.

Habuerunt igitur tabularii per predicta equationes omnium argumentorum in tribus sitibus centri epicicli: primo quidem eo existente in longitudine media a centro orbis signorum, secundo eo existente in longitudine maxima ab eodem, tertio eo existente in longitudine minima ab eodem.
980 Habuerunt quoque quantum equationes longitudinis maxime diminuunt a suis relatiuis longitudinis medie; hec enim diminutio intitulatur, longitudo longior. Habuerunt etiam quantum equationes longitudinis minime addunt super suas relatiuas longitudinis medie; hec enim additio intitulatur, longitudo propior. Sciuerunt itaque per hec omnia quod si centrum
985 epicicli est in longitudine media quod est cum centrum uerum est duo signa et 4 gradus debent intrare cum argumento uero in lineas numeri et sumere equationem argumenti in directo eius in linea illa cui superscribitur, equatio argumenti; et hec ipsa sine aliquo augmento uel diminutione est distantia linee ueri loci a linea ueri centri, addens quidem super uerum centrum si
990 argumentum uerum est minus 6 signis, minuens uero ab eo si argumentum uerum est maius 6 signis. Cum autem centrum epicicli est in sua longitudine maxima quod est cum cadit in auge equantis, sciuerunt per predicta quod debent intrare cum uero argumento in lineas numeri et sumere equationem argumenti in directo eius in linea equationis argumenti, et
995 debent etiam cum eodem argumento sumere longitudinem longiorem in linea cui superscribitur, longitudo longior, et demere longitudinem longiorem inuentam de equatione argumenti inuenta. Et residuum erit distan-

965 superposuerunt: supposuerunt θ
969 possint δ possent *V*Π
974 superposuerunt *ABSUuvWy*ΠΣ supposuerunt *codd. plerique*
979 tertio: uero *add.* β
983 addunt: similiter *add.* δ/relatiuas: equationes *MmQqY*Γ
994 in linea equationis argumenti *om.* δ
997 erit *B* est *codd. plerique*

the lesser equations, the ones which had been calculated second, from the greater, which had been calculated first, that is, subtracted each from the one for the same anomaly. Then they wrote down the remainders in another special column, each remainder opposite the proper anomaly, and put above this column the heading "Greatest Distance."[124] Next they supposed the center of the epicycle to be four signs distant from the apogee of the equant in mean motion, and they calculated the equations of the same 180 anomalies: all these were greater than the first—that is, each was greater than the one for the same anomaly—and they were the greatest possible, since the center of the epicycle was at its closest to the center of the ecliptic at that point, as was stated above. Therefore they subtracted the lesser [equations], the ones that were found first, from these greater [equations], which were found third—each one, that is, from the one for the same anomaly. They wrote each of the differences between these greater and those lesser [equations] in another special column, each opposite the proper anomaly. And above this column they wrote the heading "Least Distance."[125]

By the foregoing procedure, then, the table-makers had established the equations of all the anomalies for three positions of the center of the epicycle, first for its mean distance from the center of the ecliptic, second for its greatest distance from it, and third for its least distance from it. They had also established the differences by which the equations at greatest distance fall short of the equations for the same anomalies at mean distance; for this is the difference [column] with the heading "Greatest Distance." They had furthermore established the differences by which the equations at least distance exceed the equations for the same anomaly at mean distance. For this is the difference [column] entitled "Least Distance." Thus, from all this they knew that, if the center of the epicycle is at mean distance, which it is when the true centrum is 2 signs and 4 degrees,[126] they ought to enter the columns of argument with the true anomaly and take the equation of anomaly opposite it in the column with the heading "Equation of Anomaly": this by itself, without any increase or decrease,[127] is the distance between the line of the true position and the line of the true centrum. It is additive with respect to the true centrum if the true anomaly is less than six signs, but subtractive with respect to it [the true centrum] if the true anomaly is greater than six signs. But when the center of the epicycle is at maximum distance, which it is when it falls in the apogee of the equant, they knew by the above [process] that they should enter the columns of argument with the true anomaly and take the equation of anomaly opposite it in the column of equation of anomaly, and that they should also take the greatest distance corresponding to the same anomaly in the column headed "Greatest Distance" and subtract the value they found for this greatest distance from the value they found for the equation of anomaly: the remainder will be

tia linee ueri loci a linea ueri centri, addens aut minuens secundum regulam modo dictam. Cum autem centrum epicicli est in sua longitudine minima 1000 quod est cum centrum medium est 4 signa, sciuerunt per predicta quod debent intrare sicud prius cum argumento uero in lineas numeri et sumere equationem argumenti in linea sua, et debent etiam cum eodem argumento sumere longitudinem propiorem in linea cui superscribitur, longitudo propior, et addere longitudinem propiorem inuentam equationi argumenti 1005 inuente. Agregatumque erit distantia linee ueri loci a linea ueri centri, addens aut minuens secundum regulam predictam.

Si autem centrum epicicli fuerit alibi quam in tribus predictis locis, si quidem fuerit inter longitudinem mediam et longitudinem maximam, sciuerunt quod cum argumento debeat sumi equatio argumenti ut prius et 1010 etiam longitudo longior, et quod aliquid de longitudine longiori debeat diminui de equatione argumenti ad habendam distantiam linee ueri loci a linea ueri centri, eo quidem plus quo centrum epicicli uicinius erat longitudini maxime et eo minus quo centrum epicicli uicinius erat longitudini medie. Certam autem partem longitudinis longioris quam deberent dimi- 1015 nuere de equatione argumenti non potuerunt habere per predicta. Si uero centrum epicicli fuerit inter longitudinem mediam et longitudinem minimam aut etiam inter longitudinem minimam et oppositionem augis equantis, sciuerunt quod cum argumento sicud prius debeat sumi equatio argumenti et etiam longitudo propior, et quod aliquid de longitudine propiori 1020 addendum sit equationi argumenti ad habendum distantiam linee ueri loci a linea ueri centri, eo quidem plus quo centrum epicicli uicinius erit longitudini minime, eo autem minus quo propinquius fuerit longitudini medie aut etiam oppositioni augis equantis. Crescit enim pars longitudinis propioris addenda equationi argumenti a longitudine media usque ad 1025 minimam, ita quod in longitudine minima tota longitudo propior additur ut dictum est prius. A longitudine autem minima usque ad oppositionem augis equantis decrescit pars addenda. Sciuerunt itaque quod centro epicicli existente in omnibus hiis locis que sunt inter longitudinem mediam et minimam aut minimam et oppositionem augis equantis aliquid de longi- 1030 tudine propiori deberent addere super equationem argumenti. Certam autem partem longitudinis propioris quam deberent addere non potuerunt scire per predicta.

Sollicitati igitur sunt studio magno qualiter in omni loco ubi est aliquid

1003–4 in linea...propiorem *om. BKMUY*Γ
1006 regulam: uel rationem *add.* β
1009 debeat: debebat φ*ADHVz*
1010–11 debeat diminui de: diminuendum erat ab δθφ
1011 habendum δ
1017 aut etiam...minimam *om. deFghlMmt WwY*Γ
1018 debebat θ
1020 habendam φ*M*Γ
1021 erit: est *BFgiMSVY*ΓΣΨ erat *Acz*Π
1030 Certam: aut totam β

the distance between the line of the true position and the line of the true centrum, additive or subtractive according to the rule just enunciated. But when the center of the epicycle is at minimum distance, which it is when the mean centrum is four signs, they knew by the above [process] that they should enter the columns of argument with the true anomaly, as before, and take the equation of anomaly in its column, and that they should also take the least distance corresponding to the same anomaly in the column headed "Least Distance" and add the value they found for the least distance to the value they found for the equation of anomaly: the sum will be the distance between the true position and the true centrum, additive or subtractive according to the above rule.

But in the case that the epicycle was in a position other than the three abovementioned: if it was between the mean distance and the greatest distance, they knew that with the anomaly [as argument] they should take the equation of anomaly, as before, and also the greatest distance, and that some fraction of [the value for] the greatest distance should be subtracted from the equation of anomaly to get the distance between the line of the true position and the line of the true centrum; moreover, the nearer the center of the epicycle was to the maximum distance, the bigger [that fraction should be], and the nearer the center of the epicycle was to the mean distance, the smaller [it should be]. However, they could not tell from the foregoing [process] the exact fraction of the greatest distance which they should subtract from the equation of anomaly. On the other hand, if the center of the epicycle was between the mean distance and the minimum distance and also [if it was] between the minimum distance and the perigee of the equant, they knew that with the anomaly [as argument] they ought to take the equation of anomaly, as before, and also the least distance, and that some fraction of [the value for] the least distance should be added to the equation of anomaly to get the distance between the line of the true position and the line of the true centrum; moreover, the nearer the center of the epicycle is to the minimum distance, the greater [that fraction should be], but the nearer it is to the mean distance or to the perigee of the equant, the smaller [it should be]. For the fraction of the nearest distance which has to be added to the equation of anomaly increases [as the center of the epicycle moves] from the mean distance to the minimum, so that at the minimum distance the whole nearest distance is added, as was stated previously. But from the minimum distance to the perigee of the equant the fraction to be added decreases. They knew, then, that, when the center of the epicycle is in any of these positions between the mean and minimum distances or between the minimum and the perigee of the equant, they should add some fraction of the least distance to the equation of anomaly. However, they could not tell from the [process] described above the exact fraction of the least distance which they should add.

They were therefore extremely exercised [by the problem of] how they could discover, for every position where some fraction of the greatest distance has to

minuendum de longitudine longiori aut addendum de longitudine propiori
1035 possent cognoscere certam partem huius minuende et huius addende. Inuenerunt autem hoc arte delectabili et pulcra secundum quod dicam. Manifestum enim est ex predictis quod tota longitudo longior quecumque ipsa fuerit est minuenda de sua relatiua equatione argumenti centro epicicli existente in sua longitudine maxima quod est in auge equantis. Centro
1040 autem epicicli existente in sua longitudine media nichil de ipsa longitudine longiori minuendum est a sua relatiua equatione argumenti. At uero inter hec duo loca semper aliquid eius minuendum est ab ea: plurimum quidem si plurimum apropinquat longitudini maxime, minimum autem si plurimum ab ea remouetur. Quamlibet igitur longitudinem longiorem posue-
1045 runt 60 minuta et diuiserunt hec 60 minuta per illum arcum qui est inter longitudinem maximam et mediam proportionaliter, dantes uidelicet unicuique gradui illius arcus portionem que sibi secundum uiam proportionis inuente debetur, tanto maiorem quidem quanto gradus quilibet uicinior fuerit longitudini maxime, tanto minorem autem quanto fuerit ab ea
1050 remotior. Supposuerunt igitur rursus quod linee numeri predicte numerarent distantias centri ueri a longitudine maxima siue ab auge equantis. Et quesiuerunt per uias geometricas associatas uiis arismetricis quantum de predictis 60 minutis debetur centro uero in qualibet sua distantia ab auge in toto arcu qui est a longitudine maxima usque ad mediam. Et posuerunt
1055 unamquamque istarum partium de predictis 60 minutis inuentam in quadam alia linea descendente in directo uidelicet sui centri. Et superposuerunt huic linee titulum, minuta proportionalia: item etiam alium titulum, minuatur. Minuta quidem proportionalia intitulauerunt ea quia ipsa indicant nobis partem longitudinis longioris minuendam ex equatione argu-
1060 menti. Minuatur etiam superscripserunt quoniam pars quam indicant semper minuenda est et numquam addenda cum ipsa pars sit ex longitudine longiori que semper minuenda est. In hac autem linea minutorum proportionalium descendente posuerunt in principio in directo trium graduum qui immediate sequuntur longitudinem maximam 60, quoniam cen-
1065 tro epicicli ibi existente tota quelibet longitudo longior minuenda est ex sua relatiua equatione argumenti. In directo autem aliorum trium graduum consequentium posuerunt in dicta linea minutorum proportionalium 59, quoniam centro epicicli ibi existente non debet tota longior longitudo minui ex sua relatiua equatione argumenti, sed solum 59 minuta, id est, 59
1070 sexagesime illius longitudinis debent minui ex sua relatiua equatione argumenti. Ita quoque in directo aliorum trium sequentium graduum posue-

1040 ipsa: sua βφ
1047 illius: unius δ
1053 debeatur δ deberetur *MSt*Γ
1061 pars *om.* φ*BFgIMY*ΓΠΣΨ

be subtracted or some fraction of the least distance to be added, the exact size of this fraction to be subtracted or added.[128] They found it by an entertaining and beautiful method, as I shall describe. It is obvious from what was said above that the whole of the greatest distance, whatever its value, has to be subtracted from the equation of anomaly for the same anomaly when the center of the epicycle is at maximum distance, which it is [when it is] in the apogee of the equant. But when the center of the epicycle is at mean distance, no part of the greatest distance has to be subtracted from the equation of anomaly corresponding to it. However, between these two positions some fraction of that [greatest distance] always has to be subtracted from that [equation]: a very large fraction if it is very near the maximum distance, but a very small one if it is very far from it. Therefore they made each greatest distance equal to 60 minutes and divided these 60 minutes in due proportion along the arc which lies between the maximum distance and the mean; that is, they assigned to each degree of that arc the fraction [of 60] which is due to it according to the proportion they had discovered: the nearer that degree was to the maximum distance, the greater the fraction, but the farther it was from it, the lesser.[129] [I.e.,] they posited that the above-mentioned columns of argument [at this juncture] represent the distances between the [line of the] true centrum and the maximum distance, namely, the apogee of the equant, and computed by geometrical methods, combined with arithmetical ones, what fraction of the above 60 minutes is to be assigned to the true centrum for every [degree of] distance from the apogee along the whole arc from the greatest distance to the mean distance. They wrote these fractions of 60 minutes which they found in another special column, each opposite its proper centrum, and wrote above this column the heading "Minutes of Proportion," as well as the further heading "Subtract."[130] They called them "minutes of proportion" because they tell us what fraction of the greatest distance has to be subtracted from the equation of anomaly. They wrote "Subtract" also above them because the fraction they indicate always has to be subtracted, and never added, since it is a fraction of the greatest distance, which always has to be subtracted. In this column of minutes of proportion they put "60" at the beginning opposite [each of] the three degrees immediately following the maximum distance, for, when the center of the epicycle is there [i.e., in one of those degrees], the whole of [the value of] any greatest distance has to be subtracted from the equation of anomaly corresponding to it. Opposite the next three degrees they put "59" in the aforementioned column of minutes of proportion, because, when the center of the epicycle is there, it is not the whole of the greatest distance which should be subtracted from the equation of anomaly for the same anomaly, but only 59 minutes [of it], that is, $\frac{59}{60}$ of that distance should be subtracted from the equation of anomaly for the same anomaly. Similarly they put "58" opposite the next three

runt 58, eo quod ibi existente centro epicicli 58 sexagesime longitudinis longioris debeant ex sua relatiua equatione diminui. In directo autem aliorum trium sequentium posuerunt 57 et in directo duorum sequentium
1075 56 propter similem rationem. Et ita processerunt semper diminuendo hec minuta proportionalia secundum quod magis elongabantur a longitudine maxima donec peruenerunt ad duo signa et 4 gradus ubi posuerunt unum minutum quoniam centro epicicli ibi existente sola una sexagesima longitudinis longioris est diminuenda ex sua relatiua equatione argumenti,
1080 quod quasi idem est ac si nichil diminueretur. Ita igitur negotiati sunt de longitudine longiori, diuidendo eam per 60 minuta et in directo cuiuslibet centri ab auge usque ad longitudinem mediam ponendo minuta sibi respondentia, faciendo ea decrescere gradatim secundum quod oportet a 60 positis in directo unius gradus usque ad unum positum in directo duorum
1085 signorum et 4 graduum.

Sequitur de longitudine propiori. De hac quoque manifestum est ex predictis quod tota ipsa quecumque fuerit addenda est super suam relatiuam equationem argumenti centro epicicli existente in sua longitudine minima, quod est quando centrum uerum est 3 signa et 28 gradus. Tunc
1090 enim est centrum medium 4 signa. Centro autem epicicli existente in sua longitudine media nichil de ipsa longitudine propiori addendum est super suam relatiuam equationem argumenti. At uero eo existente inter hec duo loca aut etiam inter longitudinem minimam et oppositionem augis equantis semper aliquid ex ea addendum est super ipsam: plurimum quidem si
1095 plurimum apropinquat longitudini minime, minimum autem si plurimum ab ea remouetur. Quamlibet igitur longitudinem propiorem similiter diuiserunt in 60 minuta. Et hec 60 minuta diuiserunt proportionaliter per illum arcum qui est inter longitudinem mediam et minimam et per illum qui est inter longitudinem minimam et oppositionem augis equantis, dan-
1100 tes uidelicet unicuique gradui duorum arcuum predictorum portionem que sibi debetur secundum uiam proportionis inuente: tanto minorem quidem quanto quisquis dictorum graduum remotior fuerit a longitudine minima, tanto maiorem autem quanto fuerit ei propinquior. Supposuerunt igitur sicud prius quod linee numeri numerarent distantias centri ueri ab auge
1105 equantis. Et quesiuerunt similiter per uias geometricas associatas uiis arismetricis quantum de predictis 60 minutis debetur centro uero in qualibet sua distantia ab auge in toto illo arcu qui est a longitudine media ad mini-

1072 58[1]: minuta *add.* αφ
1077 peruenirent β
1081 minuta *om.* δ
1082–83 respondentia: correspondentia φ
1093 etiam: cum est θ est *BmuY*Π
1095 minime: medie *ABFMpQYz*ΓΣΨ
1098–99 mediam...longitudinem *om.* *StV*/ et per illum...minimam *om.* *rz*
1100 portionem: proportionem α
1105 uiis *om.* βδ
1106 debeatur θ*H* debebatur δ

degrees, because, when the center of the epicycle is there, $\frac{58}{60}$ of the greatest distance should be subtracted from the equation of anomaly for the same anomaly. They put "57" opposite the three following [degrees] and "56" opposite the two following [those], for the same reason, and proceeded in this fashion, continuously decreasing the minutes of proportion as they moved farther from the maximum distance, until they reached 2 signs and 4 degrees, where they put "1 minute,"[131] because, when the center of the epicycle is there, only $\frac{1}{60}$ of the greatest distance has to be subtracted from the equation of anomaly for the same anomaly, which is almost the same as if nothing were being subtracted.[132] This, then, was the procedure they followed for the greatest distance, dividing it into 60 minutes and putting opposite each [value of the] centrum from apogee to mean distance the minutes corresponding to it, making them decrease gradually in the proper way from the "60" opposite 1 degree to the "1" opposite 2 signs and 4 degrees.

We next deal with the least distance. It is clear from the foregoing that this, too, whatever its value, has to be added in its entirety to the equation of anomaly for the same anomaly when the center of the epicycle is at minimum distance, which is when the true centrum is 3 signs and 28 degrees;[133] for then the mean centrum is four signs. But, when the center of the epicycle is at mean distance, no part of the least distance should be added to the equation of anomaly for the same anomaly. However, when it is between these two positions and also when it is between the minimum distance and the perigee of the equant, some fraction of it [the least distance] always has to be added to it [the equation of anomaly]: a very large one if it is very near the minimum distance and a very small one if it is very far from it. Therefore they divided each least distance in the same way [as they had the greatest distances] into 60 minutes, and they divided those 60 minutes in due proportion along the arc which lies between the mean distance and the minimum, and [also] along the arc which lies between the minimum distance and the perigee of the equant; that is, they assigned to each degree of the two above-mentioned arcs the fraction [of 60] due to it according to the proportion they had found: the farther that degree was from the minimum distance, the smaller the fraction, and the nearer [that degree] was to it [the minimum distance], the greater the fraction.[134] [I.e.,] they posited, as before, that the columns of argument [at this juncture] represent the distances between the [line of the] true centrum and the apogee of the equant, and they computed in the same way, by geometrical methods, combined with arithmetical ones, what fraction of the above 60 minutes was to be assigned to the true centrum for every [degree of] distance from the apogee along the whole of the arc from the mean distance to the minimum,

mam et in toto illo qui est a minima usque ad oppositionem augis equantis. Et posuerunt unamquamque istarum partium de predictis 60 minutis
1110 inuentam in predicta linea minutorum proportionalium descendente in directo sui centri. Et superscripserunt istis minutis titulum, addatur, quoniam ipsa indicant nobis partem longitudinis propioris que semper est addenda et numquam minuenda. Inceperunt igitur ponere ista minuta longitudinis propioris ab illo gradu ubi defecerunt minuta longitudinis
1115 longioris: hoc est a longitudine media. Posuerunt itaque in directo duorum signorum et 5 graduum unum minutum, quoniam centro epicicli ibi existente unum minutum longitudinis propioris, id est, sexagesima pars eius, addenda est super suam relatiuam equationem argumenti. Titulum autem predictum, scilicet, addatur, scripserunt inter illud minutum quod est
1120 primum longitudinis propioris et minutum precedens quod est ultimum longitudinis longioris. Et uoluerunt quod iste titulus responderet omnibus minutis longitudinis propioris sequentibus. In directo quoque duorum signorum et 6 graduum posuerunt duo minuta que sunt tricesima pars de 60, eo quod centro epicicli ibi existente tricesima pars longitudinis pro-
1125 pioris addenda est supra suam relatiuam equationem argumenti. Item in directo duorum signorum et 7 graduum posuerunt 4 minuta et in directo duorum signorum et 8 graduum posuerunt 6 propter rationem consimilem. Et ita processerunt semper augmentando hec minuta proportionalia secundum quod magis apropinquabant longitudini minime donec peruene-
1130 runt ad ipsam. Ibique posuerunt 60 minuta quoniam centro epicicli ibi existente tota quelibet longitudo propior debet addi supra suam relatiuam equationem argumenti. Non solum autem in directo ipsius gradus longitudinis minime posuerunt 60 minuta, immo etiam in directo 5 graduum antecedentium et in directo 5 sequentium, quoniam centro epicicli in quo-
1135 libet illorum 11 graduum existente tota longitudo propior est addenda super suam relatiuam equationem argumenti. Abinde autem usque ad oppositionem augis equantis inceperunt hec minuta diminuere secundum quod centrum epicicli elongabatur a longitudine minima, unde in directo 3 graduum qui consequuntur 5 predictos posuerunt 59, eo quod centro
1140 epicicli ibi existente debent 59 sexagesime longitudinis propioris addi super equationem argumenti. In directo quoque trium sequentium posuerunt 58 propter rationem consimilem. Et ita processerunt semper diminuendo hec minuta proportionalia secundum quod magis elongabantur a longitudine minima quousque peruenerunt ad oppositionem augis equantis. Ibique
1145 posuerunt 40 minuta que sunt due tertie de 60, eo quod centro epicicli ibi

1115 a: in α/itaque: inquam *BMmqY*Γ
1119 scripserunt: superscripserunt *AbFI*ΣΨ supra *add.* δ
1140 super: suam relatiuam *add.* φ

and along the whole of the arc from the minimum distance to the perigee of the equant. They put these fractions of the above 60 minutes which they found in the above-mentioned column of minutes of proportion, each opposite the proper centrum, and wrote above those minutes the heading "Add," because they [the minutes of proportion] tell us the fraction of the least distance, and this [fraction] always has to be added, never subtracted. Therefore they began to write these minutes of least distance at the degree [of centrum] where the minutes of greatest distance left off, namely, at the mean distance. Thus they put 1 minute opposite 2 signs and 5 degrees, because, when the center of the epicycle is there, one minute of the least distance, that is, $\frac{1}{60}$ of it, must be added to the equation of anomaly for the same anomaly. They wrote the above-mentioned heading, namely, "Add," between the first minute of least distance and the preceding minute, which is the last minute of greatest distance, intending that heading to apply to all the following minutes of least distance. Opposite 2 signs and 6 degrees they put 2 minutes, which is $\frac{1}{30}$ of 60, because, when the center of the epicycle is there, $\frac{1}{30}$ of the nearest distance must be added to the equation of anomaly for the same anomaly. Again, opposite 2 signs and 7 degrees they put 4 minutes, and opposite 2 signs and 8 degrees they put 6 minutes, for the same reason;[135] and they proceeded in this fashion, continuously increasing these minutes of proportion as they moved nearer the minimum distance, until they reached it, and there they put 60 minutes, for, when the center of the epicycle is there, the whole of any value of the least distance must be added to the equation of anomaly for the same anomaly. However, they put 60 minutes opposite not only the actual degree of minimum distance, but also opposite the five preceding degrees and the five following ones, because, when the center of the epicycle is in any of those eleven degrees, the whole of the least distance must be added to the equation of anomaly for the same anomaly. From there on, however, until the perigee of the equant, these minutes began to decrease, [decreasing the more] as the center of the epicycle moved away from the minimum distance. Hence, opposite the three degrees following the above five they put "59," because, when the center of the epicycle is there, $\frac{59}{60}$ of the least distance must be added to the equation of anomaly, and opposite the three following they put "58," for the same reason. And they proceeded in this fashion, continuously decreasing these minutes of proportion as they moved farther from the minimum distance, until they reached the perigee of the equant, and there they put "40 minutes," which is $\frac{2}{3}$ of 60, because, when the center

existente solum due tertie longitudinis propioris debent addi super suam
relatiuam equationem argumenti.

Ita igitur negotiati sunt de longitudine propiori, diuidendo eam per 60
minuta sicud fecerant longitudinem longiorem et ponendo in directo cuius-
1150 libet centri a longitudine media usque ad oppositionem augis equantis
minuta sibi respondentia, faciendo quidem ea crescere gradatim secundum
quod oportet usque ad longitudinem minimam, abinde uero usque ad
oppositionem augis equantis faciendo ea decrescere; in quo quidem diuersificantur minuta longitudinis longioris in mercurio a minutis longitudinis
1155 propioris, quod minuta longitudinis longioris semper decrescunt, minuta
autem longitudinis propioris non semper crescunt, immo a longitudine
minima usque ad oppositionem augis semper decrescunt. Cuius causa est
quoniam locus longitudinis maxime est in auge equantis, locus autem
longitudinis minime non est in opposito augis. Vnde a longitudine minima
1160 usque ad oppositionem augis equantis ita oportet minuta proportionalia
gradatim decrescere sicud a loco ante longitudinem minimam in quo distantia centri epicicli a centro orbis signorum equalis erat distantie eiusdem
ab eodem quam habet in opposito augis equantis gradatim creuerunt usque ad longitudinem minimam. Manifestum est enim quod quelibet duo
1165 loca in quibus centrum epicicli distat equaliter a centro orbis signorum
debent habere equalia minuta proportionalia. Quia igitur necesse est ut
cuilibet distantie centri epicicli a centro orbis signorum que est eo existente
inter longitudinem minimam et oppositionem augis equantis sit alia equalis eo existente inter longitudinem mediam et minimam, nam in oppositio-
1170 ne augis minus distat a centro orbis signorum quam in longitudine media,
oportet ut hinc inde in locis in quibus hoc contingit ponantur minuta proportionalia equalia. Vnde si in duobus locis quorum unus sit ante longitudinem minimam et citra mediam et alius post minimam usque ad oppositionem augis inueneris minuta proportionalia equalia, scias centrum epi-
1175 cicli in illis duobus locis equaliter distare a centro orbis signorum, ut quando uerum centrum est 3 signa et 5 gradus et quando est 4 signa et 25
gradus. Vtrobique enim reperies in directo 50 minuta proportionalia. Sic
igitur intellige de ceteris. In mercurio itaque non sequitur, minuta proportionalia decrescunt procedendo inferius, ergo debet sumi longitudo longior
1180 et aliquid ex ea diminui ab equatione argumenti; quoniam a longitudine
minima usque ad oppositionem augis equantis decrescunt, et tamen debemus sumere longitudinem propiorem et aliquid ex ea addere super equa-

1149 fecerunt δ
1151 respondentia: correspondentia β
1156 immo: incipiendo *add.* β
1161 in quo: esset *add.* *ABdiKkMmu*ΓΨ
1162 equalis erat distantie: qualis erat distantia β
1167 a centro orbis: ab orbe θ*H* ab orbe a centro orbis δ ab orbe longitudinem a centro orbis signorum *Z*

of the epicycle is there, only $\frac{2}{3}$ of the least distance should be added to the equation of anomaly for the same anomaly.[136]

That is the way they dealt with the least distance, dividing it into 60 minutes, as they had done for the greatest distance and putting opposite each [degree of] centrum from the mean distance to the perigee of the equant the corresponding minutes, increasing them by steps in the proper way as far as the minimum distance, but from there decreasing them up to the perigee of the equant; in this last respect the minutes of greatest distance differ from the minutes of least distance for Mercury, because the minutes of greatest distance decrease continuously, but the minutes of least distance do not increase continuously: on the contrary, they decrease continuously from the minimum distance to the perigee of the equant. The reason for this is that the position of maximum distance is in the apogee of the equant, but the position of minimum distance is not in the perigee. Hence, from the minimum distance to the perigee of the equant the minutes of proportion must decrease by steps in the same way as they increased by steps up to the minimum distance from that place before the minimum distance in which the distance of the center of the epicycle from the center of the ecliptic was equal to its distance from it [the center of the ecliptic when the center of the epicycle is] in the perigee.[137] For it is obvious that any two positions in which the center of the epicycle is equidistant from the center of the ecliptic must have equal minutes of proportion. Since, then, to every distance of the center of the epicycle from the center of the ecliptic when the epicycle is between the minimum distance and the perigee of the equant there necessarily corresponds an equal distance [of the center of the epicycle from the center of the ecliptic] when it is between the mean and the minimum distances—for in the perigee it is less far from the center of the ecliptic than [it is] in the mean distance—the minutes of proportion must be assumed to be equal at the positions on either side [of the minimum distance] in which this [equality of distances] is found. Therefore, if you find equal minutes of proportion in two places, of which one lies before the minimum distance but beyond the mean [distance] and the other [lies] between the minimum and the perigee, you may be sure that the center of the epicycle is equidistant from the center of the ecliptic in those two places, as, for instance, when the true centrum is 3 signs and 5 degrees and when it is 4 signs and 25 degrees:[138] opposite each position you will find 50 minutes of proportion. The same conclusion should be drawn in other [similar situations]. For Mercury, then, it is incorrect to infer as follows: "The minutes of proportion decrease as one goes down; therefore we should take the greatest distance and subtract some fraction of it from the equation of anomaly."[139] For [the minutes of proportion] decrease from the minimum distance to the perigee of the equant, and yet [there] we should take the least distance and add some fraction of it to the equation of anomaly,

tionem argumenti, tantum scilicet quantum minuta illa sunt de 60. Sed
bene sequitur, minuta proportionalia crescunt procedendo inferius, igitur
1185 debemus sumere longitudem propiorem et aliquid ex ea addere super
equationem argumenti. Propter istam itaque diuersitatem uoluerunt ut
super minuta proportionalia in mercurio scriberetur titulus, super minuta
quidem longitudinis longioris que semper decrescunt, minuatur; super minuta autem longitudinis propioris que quandoque crescunt quandoque
1190 decrescunt, addatur, sicud superius dictum est. Per hos autem titulos sciemus quando intrabimus cum centro uero ad accipiendum minuta proportionalia, utrum postmodum cum argumento uero debeamus accipere longitudinem longiorem aut propiorem et utrum debeat aliquid addi supra
equationem argumenti aut minui ab ea. Si enim super minuta proportio-
1195 nalia scribatur, minuatur, scimus quoniam equatio argumenti debet minui.
Oportebit ergo ut cum argumento sumamus longitudinem longiorem que
est diminuens. Si autem super minuta proportionalia scribatur, addatur,
scimus quoniam equatio argumenti debet recipere additionem. Oportebit
igitur ut cum argumento sumamus longitudinem propiorem que est addens.
1200 In uenere autem et tribus superioribus non fuit ista diuersitas quoniam
longitudo minima fuit in eis centro epicicli existente in opposito augis deferentis; media autem centro uero existente 2 signa et 28 gradus uicinius;
maxima uero centro epicicli existente in auge sicud in mercurio. In quolibet igitur illorum 4 planetarum quesiuerunt equationes 180 argumento-
1205 rum in hiis tribus locis, et inuentas in longitudine media posuerunt in
directo suorum argumentorum superscribentes eis titulum, equatio argumenti. Excessus autem istarum 180 equationum super illas 180 que fuerunt
centro epicicli existente in auge scripserunt etiam in directo suorum argumentorum intitulantes eas, longitudo longior. Diminutiones quoque
1210 earumdem 180 equationum ab illis 180 que sunt centro epicicli existente in
opposito augis scripserunt similiter in directo suorum argumentorum intitulantes ipsas, longitudo propior. Et hec omnia expediuerunt sicud de
mercurio dictum est. Postea supposuerunt quod quelibet longitudo longior
et quelibet longitudo propior diuideretur in 60 minuta. Et diuiserunt pro-
1215 portionaliter 60 minuta longitudinis longioris per illum arcum qui est ab

1187 scribatur β scribantur *fh*
1195 scribitur θφ*H*
1197 scribitur θφ*H*
1200 fuit: est δ
1201 fuit in eis: est *DX*
1202 uero: epicicli *ABFMmUY*Γ uero epicicli Σ
1203 uero: fuit *add.* θ*H* uero *add.* *B* est *add.* δ
1206–7 argumenti: centri uel argumenti *BmOpQqUY*ΓΣ_1 centri Π
1208–9 argumentorum: et *add.* δ
1214 diuideretur: diuidatur β diuidentur *D*

namely, the fraction that those minutes [the minutes of proportion at the centrum in question] are of 60. But it is correct to infer as follows: "The minutes of proportion increase as one goes down; therefore we must take the least distance and add some fraction of it to the equation of anomaly." Thus, because of that difference, they decided that a heading should be written above the minutes of proportion for Mercury, namely, as was stated above, "Subtract" above the minutes of greatest distance, which decrease continuously, and "Add" above the minutes of least distance, which sometimes increase and sometimes decrease. By these headings we shall know, when we enter with the true centrum to take the minutes of proportion, whether subsequently, when we enter with the true anomaly, we should take the greatest distance or the least distance, and whether we should add something to the equation of anomaly or subtract something from it. For if "Subtract" is written above the minutes of proportion, we know that the equation of anomaly should be decreased. Then we will have to take the greatest distance corresponding to the equation, for that is subtractive. But if "Add" is written above the minutes of proportion, we know that the equation of anomaly must be added to. Therefore we will have to take the least distance corresponding to the anomaly, for that is additive.

For Venus and the three superior [planets], however, that difference[140] did not occur, because, for them, the minimum distance was found to occur when the center of the epicycle is in the perigee of the deferent, the mean [distance] when the true centrum is approximately 2 signs and 28 degrees,[141] and the maximum when the center of the epicycle is in the apogee, [the latter] as for Mercury. Therefore, for each of these four planets, they computed the equations of the 180 anomalies in these three places, and put [the 180] which they found for the mean distance opposite the corresponding anomalies, writing above them the heading "Equation of Anomaly." They also wrote down the amounts by which [each of] those 180 equations exceeded [the corresponding one of] the 180 [equations] for the center of the epicycle in the apogee, [putting each] opposite the corresponding anomaly, and gave them the heading "Greatest Distance." In the same way they also wrote down the amounts by which [each of] the same 180 equations fell short of the 180 [equations] for the center of the epicycle in the perigee, [putting each] opposite the corresponding anomaly, and gave them the heading "Least Distance." They carried out all these operations in the way we described for Mercury. Next they supposed each [value of the] greatest distance and each [of the] least distance to be divided into 60 minutes, and they divided the 60 minutes of greatest distance in due proportion

---

1195 non intellige quod equatio argumenti ab alio subtrahatur sed quid ab ea erit aliquid minuendum scilicet pars longitudinis longioris secundum proportionem minutorum proportionalium *mg. o*

auge usque ad longitudinem mediam. Et processerunt ista minuta semper decrescendo a 60 usque ad unum. Diuiserunt etiam proportionaliter 60 minuta longitudinis propioris per arcum illum qui est a longitudine media usque ad oppositionem augis. Et processerunt ista minuta semper cres-
1220 cendo ab uno usque ad 60. In istis igitur 4 planetis bene sequitur, minuta proportionalia decrescunt, ergo debet sumi longitudo longior et aliquid ex ea diminui ab equatione argumenti; aut minuta proportionalia crescunt, ergo debet sumi longitudo propior et aliquid ex ea addi equationi argumenti. In uenere igitur et tribus superioribus non fuit necessarium intitulare
1225 minuta proportionalia superscribendo, minuatur aut addatur; quoniam si decrescunt inferius ubi intramus cum centro scimus quod ipsa sunt de longitudine longiori et ideo longitudinem longiorem debemus sumere cum argumento et in proportione minutorum inuentorum ad 60 diminuere de ea ex equatione argumenti. Si uero crescunt inferius, scimus quod ipsa sunt
1230 de longitudine propiori et ideo longitudinem propiorem debemus sumere cum argumento et in proportione minutorum inuentorum ad 60 addere de ea super equationem argumenti. In hoc itaque differt mercurius a uenere et tribus superioribus quod magis patebit postquam modum orbium et motuum ueneris et trium superiorum determinauerimus. Sed idcirco hec
1235 predicta tetigimus quoniam equatio mercurii et predictorum 4 secundum tabulas habet eundem modum in omnibus hoc excepto.

In luna uero ut dictum est supra minuta proportionalia diuisa sunt per totum semicirculum qui est ab auge deferentis usque ad eius oppositionem. Quesiuerunt enim equationes 180 argumentorum centro epicicli existente
1240 in auge deferentis. Rursus quesiuerunt equationes eorumdem 180 argumentorum centro epicicli existente in opposito augis. Et subtraxerunt 180 equationes primo repertas que minores erant de 180 equationibus secundo repertis que fuerunt maiores, quamlibet uidelicet de sua relatiua. Et residua posuerunt in linea una descendente, quodlibet quidem residuum in
1245 directo sui argumenti. Et superposuerunt isti linee titulum, circulus breuis: unde quod in aliis planetis appellatur longitudo propior, in luna appellatur circulus breuis. Quemlibet quoque circulum breuem diuiserunt in 60 minuta proportionalia. Et distribuerunt ea proportionaliter per totum semicirculum, ponentes in directo cuiuslibet centri minuta proportionalia eidem
1250 debita. Et inceperunt a cifris ponentes cifram sub minutis proportionalibus in directo primorum 11 graduum; quoniam centro epicicli existente in auge

1220 igitur *om.* β
1222 ab: ex α
1235 equatio...predictorum 4: equatio argumenti omnium predictorum δ
1247 breuem *om.* φ
1249 directo: opposito β
1249–50 eidem debita: eiusdem β
1250 cifris: tytulis Γ / ponendo β

along the arc which lies between the apogee and the mean distance:[142] those minutes decreased in a continuous progression from 60 to 1. They also divided the 60 minutes of least distance along the arc which lies between the mean distance and the perigee: those minutes increased in a continuous progression from 1 to 60. Thus, for those four planets it is correct to make the following inference: "The minutes of proportion are decreasing; therefore we must take the greatest distance and subtract some part of it from the equation of anomaly"; or "The minutes of proportion are increasing; therefore we should take the least distance and add some part of it to the equation of anomaly." For Venus and the three superior [planets], then, it was not necessary to write above the minutes of proportion the headings "Subtract" or "Add." For, if they decrease going down [the column] when we enter with the centrum [as argument], we know that they are minutes of the greatest distance, and so we must take the greatest distance corresponding to that anomaly and subtract from the equation of anomaly that fraction of it [the greatest distance] which is the same as the ratio between the minutes we found and 60. But if [the minutes of proportion] increase going down [the column], we know that they are minutes of the least distance, and so we must take the least distance corresponding to the anomaly, and add to the equation of anomaly that fraction of it [the least distance] which is the same as the ratio between the minutes we found and 60. In this respect, then, Mercury differs from Venus and the three superior [planets]; this [difference] will become clearer when we have explained the particulars of the circles and motions of Venus and the three superior [planets]. But we have touched on the above matters [here] for the reason that the equation of Mercury in the tables behaves in the same way as that of the four above-mentioned planets in every respect except this one.

For the moon, however, as was stated above,[143] the minutes of proportion are divided along the whole of the semicircle between the apogee and the perigee of the deferent.[144] For [the table-makers] calculated the equations of the 180 anomalies for the center of the epicycle in the apogee of the deferent. They calculated the equations of the 180 anomalies again for the center of the epicycle in the perigee, and they subtracted the 180 equations which they found first, which were smaller, from the 180 which they found second, which were larger; that is, they subtracted each one from the one for the same anomaly. They put the differences in a column, each opposite the corresponding anomaly, and they wrote above that column the heading "Small Circle":[145] hence [the function] which is called "least distance" for the other planets is called "small circle" for the moon. They also divided each [value of the] small circle into 60 minutes of proportion. They distributed these in due proportion along the whole semicircle, putting opposite each [degree of] centrum the minutes of proportion due to it. They began with the zeros,[146] putting "0" in the column of minutes of proportion opposite the first eleven degrees, because, when the

deferentis uel in 11 gradibus sequentibus nichil de circulo breui debet addi super relatiuam equationem argumenti positam. In directo autem 8 graduum sequentium posuerunt unum minutum; quoniam centro epicicli
1255 existente in quolibet illorum 8 graduum debet una sexagesima de circulo breui addi super suam relatiuam equationem positam. In directo autem 7 sequentium posuerunt duo minuta propter causam consimilem. Et ita processerunt augmentando minuta proportionalia usque ad oppositionem augis deferentis ubi posuerunt in directo 9 graduum ultimorum totius
1260 semicirculi 60 minuta proportionalia; quoniam centro epicicli in quolibet illorum existente totus circulus breuis debet addi super suam relatiuam equationem argumenti positam.

Patet itaque causa minutorum proportionalium in omnibus planetis que in luna se habent simpliciter; in uenere et tribus superioribus dupliciter; in
1265 mercurio autem tripliciter. Et patet quid sit longitudo longior et quid etiam propior in mercurio et uenere et tribus superioribus, et quod circulus breuis in luna est sicud longitudo propior in aliis. Item patet quod minuta proportionalia in luna et in aliis 5 sunt sumenda cum centro uero. Equatio autem argumenti in luna et aliis quinque circulusque breuis in luna et
1270 longitudo longior et propior in aliis quinque sunt sumenda cum argumento uero. Item patet quod in luna semper tantum est addendum de circulo breui super equationem argumenti quantum sunt minuta proportionalia de 60. Idemque in aliis quinque de longitudine propiori, sed de longitudine longiori semper tantum est subtrahendum ab equatione argumenti quan-
1275 tum sunt minuta proportionalia de 60.

Hiis igitur ita declaratis reuertamur ad opera tabularum et reddamus causam de omnibus. Precipiunt rursus autores tabularum ut habitis centro uero et argumento uero intremus cum centro uero in lineas numeri et sumamus in directo eius minuta proportionalia: hoc ideo ut secundum
1280 proportionem eorum ad 60 augmentemus aut diminuamus equationem argumenti per partem conuenientis longitudinis. Precipiunt iterum ut in mercurio uideamus titulum minutorum proportionalium; in aliis uero consideremus utrum crescant aut decrescant: hoc ideo quia per ista scimus quam longitudinem debeamus accipere cum argumento. Precipiunt post-
1285 modum ut cum argumento uero intremus rursus in lineas numeri et sumamus equationem argumenti in directo eius et longitudinem longiorem aut propiorem secundum quod in mercurio docet titulus, in aliis uero cremen-

1266 quod: quid δ quid sit *AdfgtΣ* quidem *BY*

1267 in luna: et quod circulus breuis in luna *add.* δ

1268–71 Equatio...uero *om. ABdFghikMS tuYΓΠΣΨ*

1276 declaratis: determinatis αφ

center of the epicycle is in the apogee of the deferent or in the eleven degrees following it, no part of the small circle should be added to the corresponding equation of anomaly taken. Opposite the next eight degrees they put "1 minute," because, when the center of the epicycle is in any of those eight degrees, $\frac{1}{60}$ of the small circle should be added to the corresponding equation of anomaly taken. Opposite the next seven they put "2 minutes" for a similar reason. And they proceeded in this way, increasing the minutes of proportion as far as the perigee, where they put "60 minutes of proportion" opposite the last nine degrees of the whole semicircle,[147] because, when the center of the epicycle is in any of those [nine degrees], the whole of the small circle should be added to the corresponding equation of anomaly taken.

Thus the reason for the minutes of proportion is clear for all the planets:[148] for the moon they have a single trend,[149] for Venus and the three superior [planets] a double one, for Mercury a triple one. It is also clear what is the meaning of "greatest distance" and "least distance" for Mercury, Venus, and the three superior [planets], and that the "small circle" for the moon is the same as the "least distance" for the others. Furthermore, it is clear that the argument of the minutes of proportion for the moon and the other five planets is the true centrum, while the true anomaly is the argument of the equation of anomaly for the moon and the other five [planets], and also of the small circle for the moon and of the greatest and least distances for the other five [planets]. Again, it is clear that for the moon there must be added to the equation of anomaly that fraction of the small circle which is the same as the ratio between the minutes of proportion and 60. For the other five [planets] the same [holds true] for the least distance, but the fraction of the greatest distance equal to the ratio between the minutes of proportion and 60 must always be subtracted from the equation of anomaly.

Having thus made these matters clear, let us return to the operations with the tables and give the reasons for all of them. The table-makers next instruct us, when we have obtained the true centrum and the true anomaly, to enter the columns of argument with the true centrum and take the minutes of proportion opposite it; this is in order that we may increase or decrease the equation of anomaly by a fraction of the appropriate distance[150] according to the ratio between those minutes and 60. They then tell us to observe the heading of the minutes of proportion[151] for Mercury, but for the other [planets] to observe whether [the minutes] are increasing or decreasing [at this centrum];[152] this [we do] because by that [heading or trend] we know which distance we ought to take with the anomaly [as argument]. Next they tell us to enter the columns of argument again with the true anomaly and to take the equation of anomaly opposite it, along with either the greatest or the least distance, depending on what the heading says for Mercury, or depending on whether [the minutes of proportion] are increasing or decreasing for the other [planets]; this [they do]

tum uel decrementum: hoc ideo quoniam uolunt scire distantiam ueri loci a linea ueri centri. Et hec distantia aut est equatio argumenti sola ut si
1290 centrum epicicli est in longitudine media, aut est ipsa equatio augmentata uel diminuta per longitudinem propiorem uel longiorem secundum proportionem minutorum proportionalium ad 60. Precipiunt postmodum ut multiplicemus minuta proportionalia in longitudinem sumptam, siue longior fuerit siue propior, et productum diuidatur per 60: hoc ideo quoniam uolunt
1295 habere partem minuendam uel addendam super equationem argumenti. Precipiunt postmodum ut si longitudo est longior, subtrahatur pars inuenta ab equatione argumenti; si propior, addatur: hoc ideo quoniam longitudo longior minuit equationes argumenti positas, propior uero auget. Precipiunt postmodum ut huic equationi si argumentum est minus 6 signis superscriba-
1300 tur, addatur; si uero est plus 6 signis superscribatur ei, minuatur: hoc ideo quia si argumentum uerum est minus 6 signis, equatio ipsius est addens super uerum centrum; si autem est plus 6 signis, equatio eius est minuens a uero centro. Precipiunt postmodum ut si super utramque equationem uidelicet centri et argumenti superscribitur, addatur, coniungantur ambe et agre-
1305 gatum addatur medio motui et habebitur uerus; aut si super utramque scribatur, minuatur, coniungantur ambe et dematur agregatum de medio motu et remanebit uerus. Quod si super unam scribitur, addatur, et super aliam, minuatur, subtrahatur minor de maiori; de residuo uero fiat secundum titulum maioris, id est, si maior erat addenda, addatur illud residuum
1310 medio motui, si minuenda, minuatur ab eo, et remanebit uerus locus: hoc ideo quoniam per istas duas equationes uolunt habere uerum locum. Certum est autem quod linea ueri loci distat a linea medii motus aut per ambas equationes predictas quod est cum ambe sunt addentes aut ambe minuentes, aut per excessum unius earum super alteram quod est cum una earum
1315 est addens et alia minuens.

Completus est igitur sermo noster de causis eorum que fiunt in equatione mercurii, ueneris, trium superiorum et lune. Indulgeat autem lector si sermo noster uideatur concepta mensuratione prolixior. Malui enim cum quodam ipsius honere pro dicendorum facilitate prolongare sermonem
1320 quam cum sue mentis angustia sub uerborum breuitate ueritatem in occulto relinquere.

Non est autem pretereundum silentio quod omnia que prius dicta sunt de istorum planetarum equationibus possunt per instrumentum premis-

1294 et productum...60 *om.* α/uoluerunt δ uolumus *V*
1304 superscribatur *AltV* scribitur *BdFfgIi KNUu*ΛΣΨ scribatur *MoYy*Γ/coniungantur: iungantur α
1309 erat: est β
1318 concepta: contenta *AFgV* esse contenta *M*Γ ei contempta δ ei concepta *Hz*
1319 honere: honore *CQRTu* genere δ/dicendorum: discendi Σ

because they want to know the distance between [the line of] the true position and the line of the true centrum: this distance either is the equation of anomaly unmodified, as when the center of the epicycle is at mean distance, or is that equation increased by [a part of] the least distance or decreased by [a part of] the greatest distance according to the ratio between the minutes of proportion and 60. Next they tell us to multiply the minutes of proportion by the distance we have taken, whether greatest or least, and to divide the product by 60; this [they do] because they want to obtain the fraction to be subtracted from or added to the equation of anomaly. Next they tell us to subtract the fraction we have found from the equation of anomaly if the distance [taken] is the greatest, but to add it if it is the least; this [they do] because the greatest distance diminishes the equations of anomaly taken, while the least [distance] increases them. Next they tell us to write "add" above this equation if the anomaly is less than six signs, but to write "subtract" over it if [the anomaly] is more than six signs; this [they do] because the equation of anomaly is additive with respect to the true centrum if the true anomaly is less than six signs, but, if [the anomaly] is more than six signs, the equation is subtractive with respect to the true centrum. Next they tell us that, if "add" is written above both equations, namely, the equation of center and the equation of anomaly,[153] they should be added together and the sum added to the mean motion: the result will be the true motion; or if "subtract" is written above both, they should be added together and the sum subtracted from the mean motion: the remainder will be the true motion. But if "add" is written above one and "subtract" above the other, the lesser should be subtracted from the greater, and the difference should be applied according to the heading of the greater; that is, if the greater was to be added, that difference should be added to the mean motion, but if it was to be subtracted, [the difference] should be subtracted: the result will be the true position. This [they do] because they want to obtain the true position by those two equations, and it is a fact that the line of the true position differs from the line of the mean motion either by the sum of the two above-mentioned equations, as is the case when both are additive or both subtractive, or by the excess of one over the other, as is the case when one of them is additive and the other subtractive.

Thus our discussion of the reasons for the operations concerning the equation of Mercury, Venus, the three superior [planets], and the moon is completed. Let the reader be indulgent if our discussion seems more diffuse than the concise length [he requires].[154] For I have preferred to make the discussion longer for the easier comprehension of what I have to say, though this may be something of a burden to him, than to leave the truth hidden by employing brevity of expression, should his intelligence be limited.

However, we must not omit to mention that all the elements of the planetary equations which we list above can be found by means of the instrument pre-

sum inueniri. Etenim equationes centrorum argumentorum, longitudines
1325 longiores, propiores et circulum breuem in luna, minuta quoque proportionalia per hoc instrumentum reperire promptum est. Posito namque centro epicicli mercurii in quouis loco deferentis, centro quoque deferentis sic ordinato ut illi loco epicicli conueniat, si filum ligatum ad centrum equantis extenderis super centrum epicicli, filum autem ligatum ad cen-
1330 trum orbis signorum semel extenderis usque ad orbem signorum equedistanter priori, itemque semel per centrum epicicli, notauerisque duo loca in orbe signorum in quibus filum illud secundum secat ipsum: habebis equationem centri ad situm propositum. Arcus enim inter duo loca predicta est equatio centri. Itemque in uenere et tribus superioribus, excepto
1335 quod in hiis 4 centrum deferentis immobile est ut infra dicetur. In luna autem posito similiter centro sui epicicli in quouis loco, extende filum ligatum ad punctum quod in circumferentia parui circuli opponitur diametraliter centro deferentis et fac ipsum transire per centrum epicicli usque ad ipsius circumferentiam et nota locum in circumferentia epicicli ubi secat
1340 ipsam. Filum quoque ligatum ad centrum orbis signorum extende per centrum epicicli usque ad eius circumferentiam et uide ubi secat ipsam. Arcus enim epicicli inter istas duas sectiones interceptus est equatio centri lune ad situm propositum.

Equationes autem argumentorum quorumlibet in planetis omnibus sic
1345 inuenies. In sole quidem transeat a centro sui deferentis filum usque ad solem; a centro autem orbis signorum transeat primo equedistanter isti, deinde per solem: eritque arcus orbis signorum interceptus inter duo loca istius fili ultimi equatio argumenti solis ad situm propositum. In luna autem et ceteris sic: centrum quidem epicicli lune sit in auge sui deferentis, alio-
1350 rumque quinque sit in sua longitudine media. Transeatque filum a centro orbis signorum semel per centrum epicicli, deinde per quoduis uerum argumentum. Eritque arcus orbis signorum inter duo loca fili huius equatio argumenti sumpti.

Circulus autem breuis sic inuenitur: posito centro epicicli lune in oppo-
1355 sito augis sui deferentis sume equationem cuiusuis argumenti modo predicto et uide quantum hec equatio excedit equationem huius eiusdem ar-

1324 argumentorum: et argumentorum *j* argumentorumque *BmNY* argumentorum quia Γ
1325 circulus breuis θφ*H* circulos breues A_1
1340 ipsam: ipsum β*HZ*
1346 transeat: filum *add.* φ*JV*Σ
1348 fili: circuli *ABFgikOpSuY*ΓΣ *om. MQ*
1352 inter: hec *add.* β

viously described:[155] for the equations of center and anomaly, the greatest and least distances, and the small circle for the moon, and also the minutes of proportion, can readily be found by that instrument [see Fig. 16, p. 246]. For suppose the center of Mercury's epicycle to be located at any point on the deferent and the center of the deferent, too, so positioned as to correspond to that position of the epicycle;[156] then, if you stretch the thread attached to the center of the equant over the center of the epicycle and stretch the thread attached to the center of the ecliptic to meet the ecliptic, first parallel to the first thread and a second time through the center of the epicycle, and note the two points on the ecliptic in which the second thread cuts it, you will have the equation of center for the given situation:[157] for the arc between the two above points is the equation of center. The same [procedure is valid] for Venus and the three superior [planets], except that for these four the center of the deferent does not move, as will be stated below.[158] Similarly for the moon [see Fig. 13, p. 196], if the center of its epicycle is located in any position, stretch the thread attached to the point on the circumference of the small circle diametrically opposite the center of the deferent, and make it pass through the center of the epicycle to its circumference, and note the place which it cuts on the circumference of the epicycle. Also stretch the thread attached to the center of the ecliptic through the center of the epicycle to its circumference, and note where it cuts it [the circumference]: the arc of the epicycle cut off between those two points of intersection is the moon's equation of center for the given situation.[159]

You can find the equation of anomaly of any degree of anomaly for all the planets as follows: for the sun [see Fig. 10, p. 152], let the thread at the center of the eccentric[160] pass through the sun, and let the thread at the center of the ecliptic first be stretched parallel to the latter, then through the sun: the arc of the ecliptic cut off between the two positions of the second thread will be the sun's equation of anomaly for the given situation.[161] For the moon and the other [planets the procedure is] as follows: let the center of the moon's epicycle be in the apogee of its deferent, and let [the epicycles] of the other five [planets] be at their [respective] mean distances;[162] let the thread at the center of the ecliptic pass once through the center of the epicycle and then through any true anomaly you care to take: the arc of the ecliptic between the two positions of this thread will be the equation of the anomaly you took.

[The values of] the small circle are found as follows: put the center of the moon's epicycle in the perigee of its deferent, and take the equation of any anomaly by the above method, and note by how much this equation exceeds

1326 per...reperire: possunt inueniri. Inuentio equationis centri per dictum instrumentum in mercurio uenere et tribus superioribus inuenire $F\Sigma$

gumenti centro epicicli existente in auge. Eritque hic excessus circulus breuis argumenti sumpti. Longitudinem autem longiorem et propiorem in aliis 5 sic inuenies. Sume equationem cuiusuis argumenti modo predicto
1360 centro epicicli cuiusuis eorum existente in sua longitudine media, itemque eiusdem argumenti centro epicicli existente in sua longitudine maxima, rursus eiusdem argumenti centro epicicli existente in sua longitudine minima; eritque excessus equationis prime super secundam longitudo longior illius argumenti; excessus autem tertie super primam erit eius longitudo
1365 propior.

Minuta autem proportionalia sic inuenies: posito centro epicicli cuiusuis planete in quouis loco extra loca predicta sume equationem cuiusuis argumenti modo predicto et accipe differentiam huius equationis ad equationem eiusdem argumenti centro epicicli existente in auge si est in luna, uel
1370 eo existente in longitudine media si est in aliis quinque. Et uide quota pars circuli breuis illius argumenti sit hec differentia si est in luna, aut quota pars sit longitudinis longioris aut propioris eiusdem argumenti si est in aliis quinque: longioris quidem si locus in quo posuisti esse centrum epicicli sit inter longitudinem mediam et augem equantis; propioris autem si sit
1375 inter longitudinem mediam et oppositionem augis. Quota enim pars hec differentia fuerit alicuius predictorum, tota minuta ex 60 sunt que respondent centro posito, ut si fuerit medietas, minuta proportionalia centri positi erunt 30; si autem tertia, erunt 20; idemque in ceteris parte aut partibus. Liquet igitur quod per hoc instrumentum nobile cum fuerit exquisite
1380 factum et magnitudinis oportune, non solum poterimus equare planetas uerum etiam tabulas equationum suarum componere, compositasque probare et corruptas uerificare. De predictis itaque tanta dicta sufficiant.

1359 Sume: etiam *add.* δθ
1376 ex 60: est 60 δ
1377 posito: proposito δ
1378 positi *FSY* propositi *codd. plerique*
1379 nobile: mobile δ notabile *fi*

the equation of that same anomaly when the center of the epicycle is in the apogee: this excess will be [the value of] the small circle for the anomaly you took. [The values of] the greatest distance and the least distance for the other five planets can be found as follows: take the equation of any anomaly by the above method with the center of the epicycle at mean distance, whatever the planet; then take it again for the same anomaly, with the center of the epicycle at maximum distance, and once again for the same anomaly, with the center of the epicycle at minimum distance: the amount by which the first equation exceeds the second will be the greatest distance for that anomaly, and the amount by which the third exceeds the first will be the least distance for it.

The minutes of proportion can be found as follows: put the center of the epicycle of any planet in any position outside the above-mentioned points,[163] take the equation of any anomaly[164] by the above method, and find the difference between this equation and the equation of the same anomaly when the center of the epicycle is in the apogee for the moon, or at mean distance for the other five [planets]. Note what fraction of the small circle for that anomaly this difference constitutes for the moon, or what fraction it constitutes of the greatest or least distance for the same anomaly for the other five [planets]; [you should compare it to] the greatest distance if the point at which you put the center of the epicycle lies between the mean distance and the apogee of the equant, but to the least distance if it lies between the mean distance and the perigee. For, whatever fraction of one of the above [i.e., small circle, greatest distance, or least distance] that difference constitutes, the same expressed as a number of minutes out of a total of 60 corresponds to [i.e., is the value for] the centrum taken. For instance, if it is half, the [number of] minutes of proportion for the centrum taken will be 30; if it is a third, it will be 20, and similarly for other fractions, both unit and compound ones.[165] Thus it is clear that by this remarkable instrument, if it has been constructed with great accuracy and of suitable dimensions, we will be able not only to equate the planets but also to construct tables of the equations, and to test tables that have been constructed, and to correct corrupt ones.[166] Let so much suffice, then, for what we have to say on the above matters.

1382 corruptas: correctas seu probatas *f*

VI

[De uenere et tribus superioribus planetis]

De uenere uero et tribus superioribus qui sunt mars, iupiter et saturnus deinceps dicatur. Volo autem istorum quatuor planetarum tractatum simul coniungere, quoniam licet in multis proprietatibus differant, in pluri-
5 bus tamen conueniunt et precipue in hiis que ad faciendum instrumentum propositum sunt necessaria. Etenim idem modus eademque doctrina erit faciendi instrumentum nostrum in quolibet eorum. Nec erit prorsus aliqua diuersitas nisi in distantia centrorum et magnitudine aut paruitate epiciclorum et situ augium. Quapropter si sufficientem doctrinam habuerimus
10 faciendi instrumentum in uno eorum sciuerimusque distantias centrorum et magnitudines epiciclorum et loca augium in reliquis tribus, habebimus sufficienter propositum.

Dico igitur quod unusquisque 4 planetarum qui sunt uenus, mars, iupiter et saturnus tres habet circulos quibus suorum motuum in longitudine
15 complet uarietates. Habet enim epiciclum in cuius circumferentia mouetur corpus eius in superiori quidem parte uersus orientem, in inferiori uero uersus occidentem in quo quidem conueniunt cum mercurio et differunt a luna. Habet etiam circulum deferentem in cuius circumferentia mouetur centrum epicicli uersus orientem. Habet insuper equantem respectu centri
20 cuius mouetur equaliter centrum epicicli per circumferentiam deferentis. Nam super centrum equantis describit centrum epicicli uniuscuiusque dictorum planetarum in temporibus equalibus equales angulos et de ipsius circumferentia equales arcus quemadmodum dictum est de mercurio. Vnde quia ad centrum ipsius refertur equalitas motus centri epicicli idcirco dici-

2 qui...saturnus *om.* β
4 multis: multiplicatis *bDX*
6 modus: motus *AeM*Γ
8 aut: et β
9 situ: in situ δ secundum situm *AgiS*ΠΨ
19 insuper equantem: in superficie equantis *DX*

VI

[Theory of Venus and the Three Superior Planets]

Let us now speak of Venus and the three superior [planets], namely, Mars, Jupiter, and Saturn. I wish to treat these four planets together in my discussion, because, although they differ [from each other] in many particulars, nevertheless they are the same in most respects, especially in those [details] which are of importance in the construction of the proposed instrument.[1] For the method of construction of our instrument, and the instructions for it, will be the same for each of them. The only differences will be in the distances between the centers[2] and the greater or lesser size of the epicycle and the positions of the apogee. Hence, if we have adequate instruction in how to construct the instrument for one of these [planets], and we know the distances between the centers, the size of the epicycle, and the position of the apogee for [each of] the other three, we will have sufficient for our purpose.

I say, then, that each of the four planets, Venus, Mars, Jupiter, and Saturn, has three circles by which it brings about the variations of its motions in longitude: it has an epicycle on the circumference of which its body moves, toward the east on its upper part and toward the west on the lower. In this respect they resemble Mercury and differ from the moon.[3] It also has a deferent circle, on the circumference of which the center of the epicycle moves toward the east. It has, furthermore, an equant with respect to whose center the center of the epicycle, [as it travels] on the circumference of the deferent, moves uniformly. For the center of the epicycle of each of the above planets describes in equal times equal angles with respect to the center of the equant and equal arcs on its circumference, as we stated for Mercury. Hence it is called the "equant," because the equality of the motion of the center of the epicycle is

---

1 *tit.* de coniunctione 4 planetarum *A*; Capitulum quartum de motu et theorica trium superiorum planetarum et de uenere Rubrica *mg. A*; *tit.* De uenere et tribus superioribus *dGglRStv mg. bx*; sequitur theorica ueneris et trium superiorum *C et tit. Y*; Sequitur de uenere *f*; *tit.* De equatorio ueneris et trium superiorum planetarum et de aliquibus communioribus extractis ex theorica eorumdem *j*; Tractatus quartus de uenere et tribus superioribus Capitulum primum *k*; *tit.* Capitulum de saturno ioue marte et uenere coniunctim pertractans *MΓ*; *tit.* Theorica ueneris et trium superiorum *m*; *tit.* De uenere et tribus superioribus planetis *rz*; Capitulum quartum de motu trium planetarum superiorum et de uenere *T*; De uenere marte ioue saturno *Z*

25 tur equans. Sunt autem deferens et equans in predictis 4 planetis in superficie una. Superficies uero epicicli eorum declinat ab ea modis pluribus sicud postea dicetur. Sunt quoque ambo circuli, deferens et equans, cuiusque eorum ecentrici et centra amborum sunt in linea una recta cum centro orbis signorum: centrum quidem equantis remotius ab eo, centrum uero 30 deferentis diuidens per equalia lineam que continuat inter centrum equantis et centrum orbis signorum. Vnde centrum deferentis distat equaliter a centro equantis et a centro orbis signorum. Hec autem linea si protrahatur hinc inde quousque ad orbem signorum ex utraque parte perueniat erit diameter mundi; transit enim per centrum orbis signorum; eruntque in ea 35 aux deferentis et aux equantis ex parte centrorum suorum; ex parte autem opposita erit in ipsa oppositio augis. Necesse est itaque in hiis 4 planetis ut locus augis deferentis et equantis cuiusuis eorum in orbe signorum sit locus unus. Et est ille locus in uenere quidem 17 gradus et 50 minuta geminorum in quo quidem loco est aux solis. In marte autem unus gradus et 50 minuta 40 leonis. In ioue uero 14 gradus et 30 minuta uirginis. In saturno etiam 5 minuta sagittarii. Oppositiones uero augium predictorum circulorum sunt in locis condiametralibus locis predictis.

Mouetur autem corpus planete in circumferentia sui epicicli ab eo puncto qui maxime elongatur a centro equantis equaliter. Nam motus equalis 45 planete in suo epiciclo sumitur ab illa diametro epicicli que secundum rectitudinem opponitur centro equantis. Punctus autem predictus, et est ille qui terminat istam diametrum, dicitur aux media epicicli. Nam aux uera ipsius est punctus ille in eius circumferentia qui maxime distat a centro orbis signorum, et est ille qui terminat diametrum epicicli que secundum 50 rectitudinem opponitur centro orbis signorum. Et secundum distantiam planete ab istis duabus augibus epicicli sumitur argumentum medium et

29–31 centrum quidem...signorum *om. V* 47 diametrum: et *add.* θ
46 et *om.* θ / est *om. BMmPUY*Γ

taken with reference to its center. The deferent and the equant of the above four planets lie in one plane, but the plane of the epicycle is inclined to it [that plane] in several [different] ways, as will be explained later.[4] Both circles, deferent and equant, of each [planet] are eccentric, and their centers lie on a straight line with the center of the ecliptic;[5] the center of the equant is the farther of the two from it [the center of the ecliptic], and the center of the deferent bisects the line joining the center of the equant and the center of the ecliptic. Hence the center of the deferent is equidistant from the center of the equant and the center of the ecliptic. Now, if that line is produced in both directions until it meets the ecliptic on both sides, it will be a diameter of the world, for it passes through the center of the ecliptic; and on it will lie the apogee of the deferent and the apogee of the equant, on the same side [of the center of the ecliptic] as their centers lie, while the perigee will lie on that [line] on the opposite side. Thus for each of these four planets the position in the ecliptic of the apogee of the deferent must be one and the same as that of the apogee of the equant.[6] That position is, for Venus, Gemini 17 degrees 50 minutes, which is the same position as the sun's apogee; for Mars, Leo 1 degree 50 minutes; for Jupiter, Virgo 14 degrees 30 minutes; for Saturn, Sagittarius [0 degrees] 5 minutes.[7] The perigees of the above circles are diametrically opposite the above positions.

The uniform motion of the body of the planet on the circumference of its epicycle begins from the point which is farthest removed from the center of the equant. For the uniform motion of the planet on its epicycle is measured from that diameter of the epicycle which lies along a straight line opposite [i.e., drawn to its center from] the center of the equant.[8] The above-mentioned point, namely, the point at the end of that diameter, is called the mean apogee of the epicycle. For its true apogee is the point on its circumference which is farthest removed from the center of the ecliptic, namely, the point at the [far] end of that diameter of the epicycle which lies along a straight line opposite the center of the ecliptic. The mean anomaly and true anomaly are taken according to the distance of the planet from these two apogees of the epicycle, the mean anomaly

---

38–39 Aux ueneris et solis nunc uero id est 1473 est in primo gradu cancri; Aux martis nunc est in 14 gradus minuta 36 secunda 10 leonis. Aux iouis est ?15 gradus 0 minuta 57 secunda uirginis. Aux saturni est 12 gradus 45 minuta 25 secunda sagittarii *mg. G* in uenere 17 gradus et 50 minuta geminorum sed 27 gradus et 50 minuta est locus augis solis secundum autorem (*cf. lineeas III. 124–25*) *mg. V*/ solis: Nunc uero scilicet anno 1473 est aux ueneris et solis similiter in primo gradu cancri *add. Ww*

40 leonis: Nunc uero scilicet anno 1473 est in 14 gradus 36 minuta et 10 secunda leonis *add. Ww*/ uirginis: Nunc autem scilicet anno 1473 est 23 gradus minuta 0 57 secunda uirginis *add. Ww*

41 sagittarii: sed aux saturni nunc scilicet anno 1473 est 12 gradus 45 minuta 25 secunda sagittarii *add. Ww*

argumentum uerum: argumentum quidem medium per distantiam eius ab auge media, uerum autem per distantiam eius ab auge uera. Differentia uero inter argumentum medium et argumentum uerum, et ipsa est distan- 55 tia duarum augium predictarum, dicitur a tabulariis equatio centri. Demonstrauimus enim supra quod iste arcus epicicli est similis illi arcui orbis signorum quo differt centrum uerum a centro medio. Ideoque uterque eorum, cum similes sint, cum etiam a centro uarientur et cum a centro sumantur, dicitur equatio centri: unus quidem effectiue et finaliter, alius 60 autem effectiue solum. Finaliter uero dixi, quoniam qui in orbe signorum est accipitur ut per ipsum uerificetur centrum; qui autem in epiciclo ut per ipsum uerificetur argumentum. Ideoque equatio centri in luna dicitur centri effectiue solum. Per ipsam enim argumentum uerificatur; centrum autem non. Omnia uero hec sunt in istis 4 quemadmodum dictum est de 65 mercurio. Quantitates autem motuum equalium in suis epiciclis in istis quatuor planetis sunt ita. Venus quidem in omni die naturali mouetur 36 minuta et 59 secunda; mars uero 27 minuta et 42 secunda; iupiter autem 54 minuta et 9 secunda; saturnus etiam 57 minuta et 8 secunda. Centrum autem epicicli mouetur in istis planetis quemadmodum in mercurio per 70 circumferentiam deferentis uersus orientem super centrum equantis equaliter. In uenere quidem omni die naturali 59 minuta et 8 secunda de partibus equantis secundum equalitatem motus solis medii. Nam motus iste quem medium ueneris dicunt et motus solis medius equales sunt et simul quemadmodum de mercurio et sole dictum est. Sunt itaque tres planete 75 qui sunt sol, uenus et mercurius semper coniuncti quantum ad medium motum. Vnus enim et idem circulus transiens per polos orbis signorum semper terminat lineas mediorum motuum istorum trium planetarum et ex eadem parte. Dicuntur itaque semper esse in eodem puncto orbis signorum. In marte autem 31 minuta et 26 secunda; in ioue uero 4 minuta et 80 59 secunda; in saturno autem 2 minuta.

Quod si in planetis tribus qui sunt mars iupiter et saturnus coniunxeris motum equalem planete in epiciclo cum motu equali centri epicicli in deferente, erit agregatum equale motui solis medio. Quilibet enim eorum tantum mouetur isto duplici motu quantum sol suo simplici. Ideoque quanto 85 minus mouetur quis eorum in ecentrico eo plus mouetur in epiciclo et econ-

53 ab auge uera: a uera θ*H*
58 cum a centro: cum centro δ a centro *BMmPUY*Γ
62 dicitur: equatio *add.* δ
64 sunt: fiunt *BFgiMPY*ΓΣΨ / quemadmodum: superius *add.* δθφ
66 mouetur *om.* βδφ
69 in mercurio: in luna uel mercurio β
73 simul: similes *bdgPZ* similis *Fh*
84 quanto: quo δ quod *AIT*Ψ
85 quis: unus *BMmUVY*Γ quilibet *FiWw*Σ

from its distance from the mean apogee and the true [anomaly] from its distance from the true apogee. The difference between the mean anomaly and the true anomaly, which is the distance between the two above apogees, is called the "equation of center" by the table-makers. For we have previously shown[9] that that arc of the epicycle is similar to the arc of the ecliptic by which the true centrum differs from the mean. Therefore, since these two arcs are similar, and since, moreover, each of them varies as the centrum varies and has the centrum as its argument, each is called the equation of center: one of them as efficient and final cause, the other as efficient cause only. I said "as final cause" because the [arc] of the ecliptic [which is called equation of center] is taken in order that it may be used to correct the [actual] centrum; whereas the [arc] of the epicycle [is taken] in order that it may be used to correct the anomaly. Therefore for the moon the equation of center is so called only as efficient cause; for it is used to correct the anomaly, but not the centrum.[10] All the above particulars are the same for these four [planets] as we stated them for Mercury. The amounts of the mean motions of these four planets on their epicycles are as follows:[11] Venus moves 36 minutes and 59 seconds every natural day; Mars, 27 minutes and 42 seconds; Jupiter, 54 minutes and 9 seconds; and Saturn, 57 minutes and 8 seconds. The center of the epicycle moves for these planets, as for Mercury, on the circumference of the deferent toward the east, and uniformly with respect to the center of the equant. For Venus [the epicycle moves] 59 minutes and 8 seconds of the equant every natural day, in accordance with its equality to the mean motion of the sun. For that motion, which is called the mean motion of Venus, and the mean motion of the sun are equal and [take place] together, as we stated for Mercury and the sun. Therefore the three planets, namely, the sun, Venus, and Mercury, are always together, as far as mean motion is concerned. For the same circle passing through the poles of the ecliptic always passes through the end-points of the lines of mean motion of these three planets, and on the same side [of the center of the ecliptic].[12] Thus they are said to be always in the same point of the ecliptic. For Mars, however, [the mean daily motion of the epicycle is] 31 minutes and 26 seconds; for Jupiter, 4 minutes and 59 seconds; and for Saturn, 2 minutes.

But for the three planets, Mars, Jupiter, and Saturn, if you add the mean motion of the planet on the epicycle to the mean motion of the center of the epicycle on the deferent, the sum will be equal to the mean motion of the sun.[13] For each of them moves the same amount by that compound motion as the sun does by its simple [motion]. And therefore the slower any of them moves on the eccentric, the faster it moves on the epicycle, and vice versa. Thus, at

---

62 argumentum: Et illud patet in tabulis respicienti canonem qui incipit si certum locum et canonem qui incipit cum quemlibet *add. J*

trario quoque. Necesse est ergo ut in omni coniunctione eorum cum sole secundum motus medios sint semper in eadem habitudine epicicli quam inpossibile est mutari in coniunctione aliqua. Idem quoque in oppositione cuius hec est demonstratio. Sit modo sol coniunctus cuiuis eorum secun-
90 dum motum medium. Quia ergo sol est uelocior, coniungetur ei denuo antequam ille perficiat reuolutionem unam in ecentrico; non quidem in loco coniunctionis prime, sic enim fuisset quies in motu alterius aut retrogradatio quam non habet motus in ecentrico. In tempore itaque quod est inter duas coniunctiones perficit sol reuolutionem unam, et preter hoc arcum
95 qui est inter locum coniunctionis prime et secunde. Planeta autem non describit de ecentrico nisi arcum qui est inter locum coniunctionis prime et secunde. Tantum uero mouetur planeta in epiciclo quantum motus solis addit super motum eius in ecentrico. Relinquitur itaque quod in epiciclo fecerit reuolutionem unam. Est itaque in eodem puncto epicicli in hora
100 coniunctionis secunde in quo fuit in hora coniunctionis prime quod est propositum. Eodem quoque modo conuinces ipsum esse in eodem puncto epicicli in omni sui et solis oppositione. Amplius autem oportet punctum epicicli in quo est planeta in coniunctione eius cum sole secundum motum medium diametraliter opponi puncto ipsius in quo est in oppositione.
105 Constat enim quod a coniunctione solis et planete usque ad oppositionem superaddit sol semicirculum super motum planete in ecentrico. Tantum autem mouetur planeta in epiciclo, itaque a coniunctione ad oppositionem describit planeta in suo epiciclo semicirculum. Punctus igitur ubi est planeta in epiciclo in oppositione diametraliter opponitur puncto ubi erat in
110 coniunctione. Eodem quoque modo probabis ipsum distare per quartam epicicli a puncto in quo fuerat in coniunctione quando sol secundum medium motum distabit ab eo per quartam, quoniam constat quando sol distat ab eo secundum medium sui motum per quartam quod ipse super medium planete motum ab hora coniunctionis addit quartam. Igitur pla-
115 neta in epiciclo suo motus est per quartam. Arcus igitur orbis signorum interceptus inter solem et quemlibet trium superiorum secundum motum medium ipsorum similis est arcui epicicli intercepto inter punctum ad quem peruenit et punctum ubi erat in coniunctione eorum; quoniam tantum oportet moueri quemlibet trium superiorum in suo epiciclo quantum su-
120 peraddit medius motus solis super medium motum cuiusque ipsorum.

87 secundum motus medios: motus medii *DXZ*
90 est *om.* α*H*
93 motus: epycicli *add.* θ
99 fecit δ
101 conuinces: communices *AMNOp* conueniens est *rz* communicies uel inuenies *F* conicies uel inuenies Σ oportet Λ
102 epicicli *om.* β
111 quando: quoniam *ABhMmPrsVYz*ΓΨ
111–12 quando...constat *om.* δ
113 sui: utriusque θφ*H* uniuscuiusque *rz*/motum: similiter *add.* βδ/quod ipse: ipse enim tunc δ

every mean conjunction of [one of] these [planets] with the sun, [the planet] must be in the same position on the epicycle,[14] and it is impossible that this be changed at any conjunction. The same is true for opposition, and the proof of it is as follows. Suppose the sun to be in mean conjunction with any one of them.[15] Then, since the sun is swifter, it will come into conjunction with the planet again before that [planet] has completed one revolution on the eccentric; not indeed at the position where the first conjunction took place, for, if it did, there would have been rest or retrogradation in the motion of the planet; whereas the motion on the eccentric does not exhibit [either of] these. Thus, in the interval between two conjunctions, the sun completes one revolution and [traverses] in addition the arc between the positions of first and second conjunction;[16] the planet, however, in its motion on the eccentric, traverses only the arc between the positions of first and second conjunction; but on the epicycle the planet moves as much as the motion of the sun [in that interval] exceeds its [the planet's] motion on the eccentric. It follows, then, that [the planet] has made one revolution on the epicycle. Thus it is in the same point on the epicycle at the moment of the second conjunction as it was at the moment of the first conjunction, and this is what we stated. In the same way you can prove that it is in the same point of the epicycle at every opposition to the sun. Furthermore, the point of the epicycle at which the planet lies at [every] mean conjunction with the sun must be diametrically opposite to the point at which it is at opposition. For it is established that, from conjunction of sun and planet to opposition, the sun gains on the planet in its motion on the eccentric by a semicircle. But the planet moves the same amount on the epicycle; so the planet describes a semicircle on its epicycle from conjunction to opposition. Therefore the point of the epicycle at which the planet lies at opposition is diametrically opposite to the point at which it lay at conjunction. In the same way you can prove that it will be one-fourth of the epicycle distant from the point in which it was at conjunction when the sun is a quarter-circle removed from it in mean motion, because it is established that, when the sun is a quarter-circle removed from it in mean motion, it [the sun] has gained a quarter-circle on the planet with respect to its uniform motion [on the eccentric] from the time of conjunction. Therefore the planet has moved a quarter of the way around its epicycle. Therefore the arc of the ecliptic contained between the sun and the mean position of any of the three superior [planets] is similar to the arc of the epicycle contained between the point which the planet has [now] reached and the point at which it was at their [last] conjunction; for each of the three superior [planets] must move on its epicycle by an amount equal to the excess of the mean motion

Superaddit autem ab hora coniunctionis tantum quantum distat. Fuit autem ex creatoris beneplacito ut in prima coniunctione solis cum quolibet trium superiorum esset planeta in auge sui epicicli; oportuit itaque ut in oppositione sequente esset in opposito augis epicicli. Quapropter sem-
125 per fuit semperque erit motu durante quilibet eorum in coniunctione sui et solis in auge sui epicicli et in oppositione etiam in opposito augis. Eritque in quolibet tempore distantia planete ab auge media epicicli similis distantie medii motus solis a medio motu planete.

Manifestum est igitur ex hiis que dicta sunt de singulis planetis quod
130 omnes in aliquo communicant cum sole. Videturque sol esse quasi quoddam commune speculum ipsorum in quod omnes intuentes mutuantur inde quedam exemplaria sui motus. Luna enim sic centrum epicicli mouet uersus orientem et sic augem deferentis uersus occidentem ut semper utrumque distet hinc inde equaliter a sole. Venus autem et mercurius cen-
135 tra suorum epiciclorum semper mouent equaliter cum sole, sic eundem assidue lateraliter commitantes exceptis duobus instantibus in quibus eorum ecentrici secant ecentricum solis, ideoque in istis duobus sitibus eidem diametraliter subsunt quod ipsum numquam quantolibet spatio preueniunt nec sequuntur. Saturnus uero, iupiter et mars, quod in suis ecentricis
140 diminuunt a sole, in suis epiciclis punctualiter restaurant, eidem per agregationem istorum duorum motuum equati.

Vt autem que dicta sunt de istis quatuor planetis figurali exemplo comprehendantur, describam orbem signorum qui sit circulus *MN* supra centrum *D* et protraham diametrum *MDN*. Et ponam ut punctus *M* sit locus
145 augis cuiusque eorum. Erit itaque punctus *M* in uenere quidem 17 gradus et 50 minuta geminorum; in marte uero unus gradus et 50 minuta leonis; in ioue autem 14 gradus et 30 minuta uirginis; in saturno uero 5 minuta sagittarii. Hec enim sunt loca augium in planetis istis ut dictum est prius. Eruntque centra deferentium et equantium in linea *MD*. Ponam itaque ut
150 punctus *C* sit centrum deferentis et lineabo super ipsum circulum *AGB* qui sit ipse deferens. Et sumam ex linea *CM* lineam *CK* equalem linee *DC*. Eritque punctus *K* centrum equantis. Super ipsum ergo lineabo circulum *LGH* equalem circulo *AGB*. Eritque circulus iste equans. Ponam quoque

123 epicicli: media *add. DXZ*
124 augis: medie *add. DXZ* sui *add. BfgkSy*Λ
126 etiam *om.* α*H*
131 spectaculum β*V*
131–32 mutuantur inde: ipsum imitantur mutuantque ab eo β*V*
132 sui motus *om.* β*V*/mouet: tendit βφ*V*
136 lateraliter: linealiter *FMP*ΓΣ specialiter *V om. B*
137 ecentrici: deferentes *add.* δθφ
138 subsunt: subsistunt *BJMOPqV*Γ
141 equati: coequati β
142 autem: ea *add.* φ
143 MN: MH αφ*H*
144 MDN: MD θ*H* MDH *BfMVYz*
152 K: super *add.* β
153 LGH: LGM *F*Σ

of the sun over the mean motion of the planet in question. This excess, [measured] from the moment of conjunction, is the distance [of the sun from the mean planet]. Now it was the will of the Creator[17] that, at the first conjunction of the sun with each of the three superior [planets], the planet should be in the apogee of its epicycle. It necessarily followed, then, that at the following opposition it was in the perigee of its epicycle. And thus each of them always was, and always will be as long as their motion endures, in the apogee of its epicycle at conjunction with the sun and also in the perigee of its epicycle at opposition. And at any moment the distance of the planet from the mean apogee of its epicycle will be equal to the distance of the mean sun from the mean planet.

It is clear, then, from what has been said about the individual planets, that all of them are connected with the sun in some way; and the sun seems to be, as it were, a common mirror in which all look and from which all borrow some patterns for their own motion:[18] the moon moves the center of its epicycle toward the east and the apogee of its deferent toward the west in such a way that each on its own side is always equidistant from the sun; Venus and Mercury always move the centers of their epicycles to keep pace with the sun and thus accompany it continuously on one side or the other, except for two instants at which their eccentrics cut the sun's eccentric,[19] and thus in those two positions they are on a straight line below it, because [then] they are no distance either in front of or behind it; as for Saturn, Jupiter, and Mars, they make up exactly in their motion on the epicycle the amount by which they fall short of the sun in their motion on the eccentric and are [thus] made equal [in motion] to it [the sun] by the addition of those two motions.

Now in order that what has been said about these four planets may be understood by an illustrative diagram [see Fig. 18, p. 308], I describe the ecliptic—let this be circle *MN*—about center *D*, and draw the diameter *MDN*. I put point *M* as the position of the apogee for each of them [the planets]. Then point *M* will be Gemini 17 degrees and 50 minutes for Venus, Leo 1 degree and 50 minutes for Mars, Virgo 14 degrees and 30 minutes for Jupiter, and Sagittarius [0 degrees and] 5 minutes for Saturn. For those are the positions of the apogee for these planets, as was stated above.[20] The centers of the deferents and of the equants will lie on line *MD*. Thus I make point *C* the center of the deferent and on it [as center] draw circle *AGB* to be the actual deferent. From line *CM* I cut off *CK* equal to line *DC*: then point *K* will be the center of the equant. Therefore on it [as center] I draw circle *LGH* equal to circle *AGB*; this circle will be the

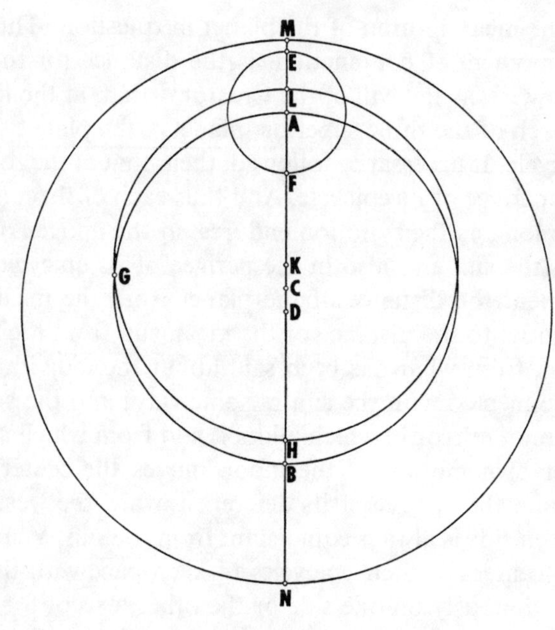

Fig. 18

punctum *A* centrum epicicli. Et super ipsum lineabo circulum *EF* qui sit
155 epiciclus eiusque diameter *EAF*. Sintque puncta *E*, *F*, *K* in linea una recta. Eritque *E* aux media, *F* uero oppositio eius in epiciclo. Nam aux eius uera erit in linea *DA*. Dico itaque quod planeta in suo epiciclo mouetur semper equaliter a puncto *E* uersus orientem. Centrum autem epicicli quod est punctus *A* mouetur etiam uersus orientem equaliter super punctum *K*
160 describens super ipsum in temporibus equalibus equales angulos et de circumferentia circuli *LGH* equales arcus. Quippe linea *KFE* mota semper equaliter super suam extremitatem que est punctus *K* uersus orientem defert secum centrum epicicli. Suntque puncta *E* et *F* signata et certa et semper eadem in epiciclo inseparabiliter adherentia huic linee. Linee uero
165 *DA* adherent quandoque secundum accidens ut centro epicicli existente in auge deferentis aut in eius opposito solum; alibi autem diuersificantur ab ea. Ideoque fuit necessarium ut equalitas motus planete in epiciclo sumeretur a puncto *E* qui terminat lineam *AK*. A puncto autem qui in circumferentia epicicli terminat lineam *DA* inpossibile fuit sumi equalitatem ip-
170 sius. Cum enim punctus ille diuersificetur a puncto *E* qui est signatus et

155 Sintque: Suntque *AI* Fiantque *DXZ* eruntque *f*
156 media: et etiam aux uera et *add.* δ / uero *om.* δθ / aux eius: media erit in linea CE et aux eius *add.* δ
166 alibi: aliquando φ
169 sumi: sumere β

Legend for Fig. 18

Illustration of the Ptolemaic model for Venus and the three outer planets (after Campanus). *MN* represents the ecliptic, with center *D*, the earth ("centrum mundi": MSS). The deferent is represented by circle *AGB*, with center *C* ("centrum deferentis": MSS). The epicycle *EF* is here represented in the apogee of the deferent, point *A*. The equant point, *K* ("centrum equantis": MSS), is the center of an equant circle *LGH*; by mere convention equant *LGH* is equal to deferent *AGB*. *D*, *C*, and *K* lie on one straight line, and $DC = CK$. Thus *M* represents the longitude of the apogee and *N* the longitude of the perigee. In the situation represented, *E* is the apogee of the epicycle and *F* the perigee of the epicycle (since the epicycle is in the apogee of the deferent, the mean and true apogees of the epicycle coincide). Compare Figure 7 (p. 48), Plate 4, and Figure 26 (p. 428).

equant. Furthermore, I make point *A* the center of the epicycle, and draw about it circle *EF* as epicycle, with diameter *EAF*. Let points *E*, *F*, and *K* lie on one straight line. Then *E* will be the mean apogee in the epicycle and *F* its perigee; for the true apogee will lie on line *DA*.[21] I say, then, that the planet always moves on the epicycle with a uniform motion from point *E* toward the east. The center of the epicycle, which is point *A*, also moves toward the east uniformly with respect to point *K*, describing in equal times equal angles about it [*K*] and equal arcs on the circumference of circle *LGH*. For line *KFE*, always moving uniformly toward the east about its extremity, point *K*, carries the center of the epicycle with it; and points *E* and *F* are definite, fixed, unchanging points on the epicycle, which are inseparably attached to this line [*KFE*]. Sometimes they lie on line *DA*, but contingently, namely, only when the center of the epicycle is in the apogee or perigee of the deferent; in other positions [of the epicycle] they do not lie on it. For that reason it was essential for the mean motion of the planet on the epicycle to be measured from point *E*, which lies at the end of line *AK*;[22] and it was impossible for its mean [motion] to be measured from the point on the circumference of the epicycle at the end of line *DA*. For that point is different from the definite, fixed point *E* and lags behind

certus sequens ipsum quantitate equationis centri in toto semicirculo *AGB*, in alio autem semicirculo antecedens ipsum quantitate eadem, necessarium fuit ut in temporibus equalibus elongaretur planeta inequaliter a predicto puncto, si in circumferentia epicicli equaliter mouetur. Suppo-
175 namus igitur planetam esse in puncto *E* epicicli et centrum epicicli esse in auge. Erit ergo punctus *E* secundum accidens terminans lineam *DA*. Mouebitur igitur planeta a puncto *E* uersus orientem centrumque epicicli in eandem partem. Punctusque terminans *DA* continuo diuersificabitur a puncto *E* mouebiturque ab eo uersus occidentem usque dum *DA* stet orto-
180 gonaliter super *MD*. Ibique erit equatio centri maxima et terminabitur motus iste uersus occidentem. Mouebiturque amodo ad punctum *E* uersus orientem usque dum *DA* continuetur secundum rectitudinem cum *DM*. Ibique erit punctus terminans *DA* ipse punctus *E* secundum accidens quia centrum epicicli erit in opposito augis. Igitur ab auge deferentis usque
185 dum *DA* stet ortogonaliter super *MD* elongabitur planeta a puncto terminante *DA* motu proprio uersus orientem et motu huius termini uersus occidentem. Abinde autem usque dum *DA* continuetur secundum rectitudinem cum *DM* elongabitur planeta a puncto predicto minus quantitate sui motus in tantum quantum punctus predictus mouetur uersus orientem
190 ad punctum *E*. Idemque erit in alia medietate deferentis modo contrario. Constat itaque planetam in suo epiciclo elongari a predicto puncto qui est aux uera epicicli in temporibus equalibus inequaliter. Quantitates uero istorum motuum equalium, planete uidelicet in suo epiciclo, et centri epicicli in deferente, dicte sunt prius.

195 Quoniam autem 4 planete predicti de quibus nunc specialiter est sermo qui sunt uenus mars iupiter et saturnus mouentur ut dictum est in superiori parte suorum epiciclorum uersus orientem cum centro epicicli, in inferiori autem uersus occidentem contra centrum epicicli, estque arcus orbis signorum cui subtenditur arcus inferioris partis epicicli quem pertranseunt
200 una die uersus occidentem in aliquo situ equalis ei quem pertransit centrum epicicli uersus orientem et in aliquo situ maior eo; necesse est ut unicuique etiam istorum 4 accidant processio, statio et retrogradatio. Processio quidem in toto superiori arcu epicicli intercepto inter duas rectas lineas egressas a centro terre contingentes epiciclum utrimque, et extende-
205 tur ista processio licet cum diminutione motus usque ad locum inferioris arcus epicicli post contingentem orientalem ubi primo duo arcus orbis

174 si: sed β/moueatur *DXZ* moueretur *F*Σ
174–75 supponemus α
180 Ibique: quia *add.* δ*aeIlw*/et *om.* δ*CfIW wx*Ψ
188 elongatur βδ
190–210 ad punctum E...orientem *om. B*
190 Idemque: Ideoque βφ*H* idem *FV*ΣΨ eritque *M*Γ
195 nunc *om.* β
202 et *om. AfimPy*
206 ubi: usque δ

[i.e., is to the west of] the latter by the amount of the equation of center [while the epicycle is] on the whole of the semicircle *AGB*, while on the other semicircle [that point] precedes it [*E*] by that amount. Therefore the planet must necessarily travel unequal distances in equal times from the above point, if it moves on the circumference of the epicycle with uniform speed. Let us suppose, then, that the planet is at point *E* of the epicycle, while the center of the epicycle is in the apogee. Then point *E* will be the end of line *DA*, [but only] contingently. Then the planet will move from point *E* toward the east, and the center of the epicycle will move in the same direction. And the point at the end of line *DA* will immediately become different from point *E* and will move away from it toward the west until *DA* is at right angles to *MD*. There the equation of center will be a maximum,[23] and there will be the end of the motion of that point toward the west. From that time on it will move eastward toward point *E*, until *DA* coincides with *DM* along a straight line. There the point at the end of *DA* will be point *E* itself, contingently, because the center of the epicycle will be in the perigee. Therefore from the apogee of the deferent until *DA* is at right angles to *MD* the planet will move away from the point at the end of *DA* [by the combination of] its own motion toward the east and the motion of this end-point toward the west. But from there until *DA* coincides with *DM* along a straight line, the amount the planet has moved away from the aforementioned point will be less than the amount of its proper motion by the amount which the aforementioned point moves eastward toward point *E*. The same will be true on the other half of the deferent, but with the opposite sign.[24] Thus it is established that the planet moves unequal distances on the epicycle in equal times with respect to the above point, which is the true apogee of the epicycle. The amounts of these mean motions, namely, of the planet on the epicycle and of the center of the epicycle on the deferent, were stated above.[25]

Now since, as was stated, the four above-mentioned planets with which we are now dealing in particular, namely, Venus, Mars, Jupiter, and Saturn, move toward the east and in the same direction as the center of the epicycle in the upper part of their epicycles, but toward the west and in the opposite direction to the center of the epicycle in the lower [part], and since the arc of the ecliptic subtended by the arc of the lower part of the epicycle which the planet traverses toward the west in one day is in some situations equal to the arc [of the ecliptic] traversed by the center of the epicycle toward the east [in one day] and in some situations greater than it,[26] each of those four planets must exhibit forward motion, station, and retrogradation. The forward motion takes place along the whole upper arc of the epicycle contained between the two straight lines proceeding from the center of the earth tangent to the epicycle on either side; and this forward motion will continue, though decreased in speed, up to that point on the lower arc of the epicycle past the eastern tangent where the two above-

---

184 Igitur: cor *mg. f*

signorum predicti erunt equales, ibique erit planeta stationarius existens in sua statione prima. Post uero augebitur arcus orbis signorum quem transit planeta in suo epiciclo uersus occidentem super arcum eius quem tran-
210 sit centrum epicicli uersus orientem, eritque propter hoc retrogradus. Durabitque eius retrogradatio usque ad locum istius inferioris arcus epicicli ante contingentem occidentalem ubi iterum duo predicti arcus orbis signorum erunt equales et ibi erit iterum planeta stationarius existens in sua statione secunda. Postea incipiet dirigi et fiet rursus processiuus sicud
215 prius dictum est.

Hec autem omnia sunt etiam in istis 4 sicud dictum est supra de mercurio. Vnde etiam in hiis 4 oportet ut duo loca stationis prime et secunde distent semper equaliter ab auge uera epicicli. Hec autem distantia in hiis planetis centro epicicli cuiuslibet istorum existente in auge deferentis aut
220 in sua longitudine media aut in opposito augis est de partibus epicicli cuiuslibet eorum secundum quod dicam. In uenere quidem centro epicicli existente in auge 5 signa 15 gradus et 51 minuta; eo uero existente in longitudine media 5 signa 17 gradus et 9 minuta; in opposito autem augis 5 signa 18 gradus et 21 minuta. In marte autem centro quidem epicicli exis-
225 tente in auge 5 signa 7 gradus et 28 minuta; eo uero existente in longitudine media 5 signa 13 gradus et 9 minuta; in opposito autem augis 5 signa 19 gradus et 15 minuta. In ioue autem centro quidem epicicli existente in auge 4 signa 4 gradus et 5 minuta; eo uero existente in longitudine media 4 signa 5 gradus et 40 minuta; in opposito autem augis 4 signa 7 gradus et 11
230 minuta. In saturno autem centro quidem epicicli existente in auge 3 signa 22 gradus et 44 minuta; eo uero existente in longitudine media 3 signa 24 gradus et 8 minuta; in opposito autem augis 3 signa 25 gradus et 30 minuta.

Tempora autem medietatis retrogradationis cuiusque ipsorum centro
235 epicicli existente in tribus locis predictis erunt secundum hunc modum. In uenere quidem centro epicicli existente in auge 21 dies 12 hore et 58 minuta; eo autem existente in longitudine media 20 dies 20 hore et 6 minuta; eo uero existente in opposito augis 20 dies 4 hore. In marte autem centro quidem epicicli existente in auge 40 dies una hora 24 minuta; eo autem
240 existente in longitudine media 36 dies 11 hore 57 minuta; eo uero existente in opposito augis 32 dies 6 hore et 35 minuta. In ioue autem centro quidem epicicli existente in auge 61 dies 10 hore et 38 minuta; eo autem existente in longitudine media 60 dies 4 hore et 53 minuta; eo uero existente in opposito augis 59 dies una hora et 35 minuta. In saturno autem centro

212 iterum: secundo *add.* αδ
218 semper *om.* β*V*
228 4²: *lacuna BPV*ΠΨ
235–47 In uenere...una hora *om. BCMOP V*Γ (*habent separatim in tabula BCMO P*Γ)

mentioned arcs of the ecliptic first become equal: there the planet will be stationary, being in the first station. After [that position] the arc of the ecliptic which the planet [by its motion] on the epicycle traverses toward the west [in one day] will become greater than the arc which the center of the epicycle traverses toward the east [in one day], and hence it will be retrograde. Its retrogradation will continue up to the place on that lower arc of the epicycle before the western tangent where the two above-mentioned arcs of the ecliptic once more become equal, and there the planet will once more become stationary, being in its second station. After that it will begin to move forward, and its motion will again be in a forward direction, in the way which we described above.

All these features are the same in these four planets as we described them above for Mercury.[27] Hence for these four, too, the positions of first and second station must always be equidistant from the true apogee of the epicycle. This distance is as I shall now state for each of these planets, when the center of its epicycle is in the apogee of the deferent, at mean distance, and in the perigee [of the deferent, respectively], measured as an arc of the epicycle of the planet in question: Venus, center of the epicycle at apogee, 5 signs 15 degrees 51 minutes; at mean distance, 5 signs 17 degrees 9 minutes; at perigee, 5 signs 18 degrees 21 minutes. Mars, center of the epicycle at apogee, 5 signs 7 degrees 28 minutes; at mean distance, 5 signs 13 degrees 9 minutes; at perigee, 5 signs 19 degrees 15 minutes. Jupiter, center of the epicycle at apogee, 4 signs 4 degrees 5 minutes; at mean distance, 4 signs 5 degrees 40 minutes; at perigee, 4 signs 7 degrees 11 minutes. Saturn, center of the epicycle at apogee, 3 signs 22 degrees 44 minutes; at mean distance, 3 signs 24 degrees 8 minutes; at perigee, 3 signs 25 degrees 30 minutes.[28]

The periods of half the retrogradation of each of them when the center of the epicycle is in the three above positions will be as follows: Venus, center of the epicycle at apogee, 21 days 12 hours 58 minutes; at mean distance, 20 days 20 hours 6 minutes; at perigee, 20 days 4 hours. Mars, center of the epicycle at apogee, 40 days 1 hour 24 minutes; at mean distance, 36 days 11 hours 57 minutes; at perigee, 32 days 6 hours 35 minutes. Jupiter, center of the epicycle at apogee, 61 days 10 hours 38 minutes; at mean distance, 60 days 4 hours 53 minutes; at perigee, 59 days 1 hour 35 minutes. Saturn, center of the epicycle

245 quidem epicicli existente in auge 70 dies 8 hore et 30 minuta; eo uero existente in longitudine media 69 dies 4 hore et 7 minuta; eo autem existente in opposito augis 68 dies et una hora.

Hec igitur sunt tempora quibus planeta mouetur in suo epiciclo a statione prima usque ad oppositionem augis uere uel ab oppositione augis 250 uere usque ad stationem secundam. Que quidem si duplaueris habebis tempora totarum retrogradationum in planetis predictis centro epicicli existente in predictis locis et erunt in uenere in loco quidem primo 43 dies una hora et 56 minuta; in loco secundo 41 dies 16 hore et 12 minuta; in loco tertio 40 dies et 8 hore. In marte autem erunt in loco primo 80 dies 255 2 hore et 48 minuta; in secundo loco 72 dies 23 hore et 54 minuta; in loco tertio 64 dies 13 hore et 10 minuta. In ioue autem erunt in loco primo 122 dies 21 hore et 16 minuta; in loco secundo 120 dies 9 hore et 46 minuta; in loco tertio 118 dies 3 hore et 10 minuta. In saturno autem erunt in loco primo 140 dies et 17 hore; in loco secundo 138 dies 8 hore et 14 260 minuta; in loco tertio 136 dies et 2 hore.

Attendere autem oportet diligenter qualiter istarum retrogradationum tempora contingat inuenire. Si enim centrum epicicli figeretur in auge usque dum planeta transiret arcum sue retrogradationis dimidie qui est a statione prima usque ad oppositionem augis uel ab oppositione augis 265 usque ad stationem secundam, sufficeret diuidere arcum dimidie retrogradationis, qui est centro epicicli existente in auge, per motum diurnum planete in epiciclo et exiret tempus retrogradationis dimidie. Idemque dico centro epicicli existente in opposito augis. At uero cum centrum epicicli mutetur dum planeta transit hunc arcum retrogradationis dimidie, centro 270 autem epicicli mutato mutantur loca ambarum stationum, diminuentia quidem arcum retrogradationis eo amplius quo centrum epicicli magis elongatur ab auge, et augentia ipsum quo magis elongatur ab oppositione augis: nam quanto amplius centrum epicicli accedit ad centrum orbis signorum tanto plus loca ambarum stationum accedunt ad oppositionem 275 uere augis epicicli: oportet centro epicicli existente in auge arcum dimidie retrogradationis esse minorem predicta quantitate descensus alterutrius stationis, et eo existente in opposito augis maiorem quantitate ascensus. Amplius autem cum loca ambarum stationum respiciant oppositionem

251–60 in planetis...2 hore *om. BCMmOP VΓ* (*in tabula habent BCMOPΓ*)
257 120: 129 φ*AIk*
261 diligenter *om.* β
262 contingat inuenire: potuerunt (poterunt *MqUV*) inueniri β
272 ab auge...elongatur *om. BV*
273 accedit: accedet θ
274 ambarum: aliarum δ/accedunt: accedent *MmOPY*ΓΨ
277 maiorem: minorem δ

at apogee, 70 days 8 hours 30 minutes; at mean distance, 69 days 4 hours 7 minutes; at perigee, 68 days 1 hour.[29]

These, then, are the times which the planet takes to move on the epicycle from first station to true perigee or from true perigee to second station. If you double them, you will get the periods of total retrogradation of the above planets for the above-mentioned positions of the center of the epicycle. These will be [as follows]: Venus, first position, 43 days 1 hour 56 minutes; second position, 41 days 16 hours 12 minutes; third position, 40 days 8 hours. Mars, first position, 80 days 2 hours 48 minutes; second position, 72 days 23 hours 54 minutes; third position, 64 days 13 hours 10 minutes. Jupiter, first position, 122 days 21 hours 16 minutes; second position, 120 days 9 hours 46 minutes; third position, 118 days 3 hours 10 minutes. Saturn, first position, 140 days 17 hours; second position, 138 days 8 hours 14 minutes; third position, 136 days 2 hours.

Careful attention must be paid to the method used to find the periods of those retrogradations.[30] For if the center of the epicycle remained fixed in the apogee the whole time that the planet was traversing the arc of half-retrogradation, namely, from first station to perigee [of the epicycle] or from perigee to second station, it would be sufficient to divide the arc of half-retrogradation for [the first position], when the center of the epicycle is in the apogee, by the daily motion of the planet on the epicycle; then the result would be the period of half-retrogradation. And I say the same for [the third position], when the center of the epicycle is in the perigee [of the deferent]. But, since the center of the epicycle changes position while the planet is traversing this arc of half-retrogradation, and [since,] once the center of the epicycle has changed position, the places of both stations change, [thus] decreasing the arc of retrogradation in proportion to the amount that the center of the epicycle moves away from the apogee and increasing that [arc] the more [the epicycle center] moves away from the perigee—for the nearer the center of the epicycle comes to the center of the ecliptic, the closer the positions of both stations approach the true perigee of the epicycle—[it follows that], when the center of the epicycle is in the apogee, the arc of half-retrogradation must be smaller [after the epicycle center has moved away from apogee] by the above-mentioned amount of the downward shift [toward epicyclic perigee] of each station, and, when [the epicycle] is in the perigee [of the deferent], [the arc of half-retrogradation] must be larger [after the epicycle center has moved away from perigee] by the amount of the upward shift [from epicyclic perigee]. Furthermore, since the positions of both

251 totarum retrogradationum: totalis retrogradationis β/retrogradationum: ut hic patet *add. BPV et* in auge in longitudine media in opposito augis hec sunt tempora totalis retrogradationis que secuntur *add. B* ut patet in tabula (secunda *add. C*) *add. CMΓ*

augis uere, hec autem uariatur a media, precedendo ipsam quantitate
280 equationis centri in medietate ecentrici descendente et sequendo eam eadem quantitate in reliqua, oportet centro epicicli existente in auge diminui arcum dimidie retrogradationis quantitate equationis eius centri ad quod statio uel retrogradationis dimidiatio peruenit, et eo existente in opposito augis augeri quantitate eadem.

285 Huius rei gratia describam deferentem cuiusuis predictorum 4 planetarum qui sit circulus *AB*. Et ponam ut punctus *C* sit centrum motus equalis, hoc est centrum equantis, et *D* centrum orbis signorum. Et protraham diametrum *ACDB* eritque punctus *A* aux deferentis, *B* uero oppositio augis. Volo itaque inuenire arcum retrogradationis planete quem dimi-
290 diabit centro epicicli existente in puncto *A* aut puncto *B*. Ponam igitur centrum epicicli in duobus punctis *E* et *F* equedistantibus augi, itemque in duobus punctis *G* et *H* equedistantibus oppositioni augis. Et ponam ut

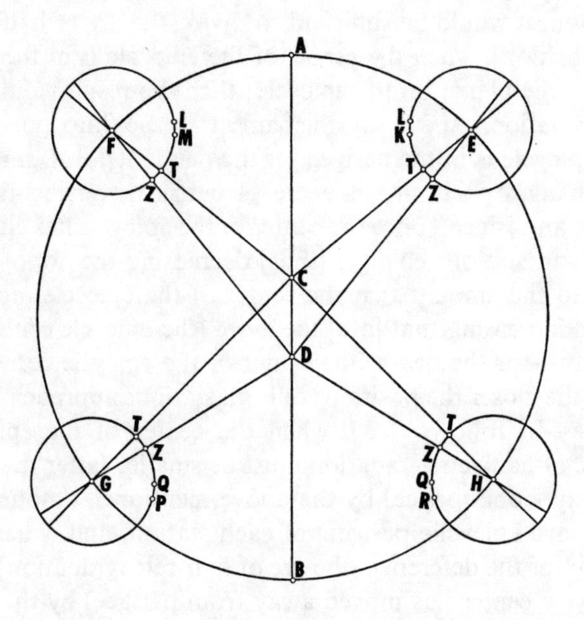

Fig. 19

281 reliqua: qualibet *MOPVY*Γ quolibet *B* 285 predictorum: supradictorum θ*HZ* supradicto *DX* 287 et D...signorum *om.* φ

stations vary with respect to the true perigee, and that [the true perigee] differs from the mean [perigee], preceding it by the amount of the equation of center on the descending half of the eccentric and lagging behind it by that amount on the other half,[31] the arc of half-retrogradation for [the first position], when the center of the epicycle is in the apogee, must be decreased by the amount of the equation of that centrum which pertains to the station or mid-point of retrogradation and for [the third position], when [the center of the epicycle] is in the perigee, [must be] increased by that amount.

To illustrate this [see Fig. 19, p. 316], I draw the deferent of any of the above four planets—let this be circle *AB*—and make point *C* the center of uniform motion, that is, the center of the equant, and *D* the center of the ecliptic. I draw the diameter *ACDB*: then point *A* will be the apogee of the deferent and *B* the perigee. I want, then, to find the arc of retrogradation in the middle of which the planet will be when the center of the epicycle is at point *A* and [the same for] point *B*. So I put the center of the epicycle at two points, *E* and *F*, equidistant from the apogee [of the deferent, and on either side of it], and also at two points, *G* and *H*, equidistant from the perigee. I posit that, when the planet is

Legend for Fig. 19

Illustration (after Campanus) of Campanus's procedure for correcting the length of the arc of half-retrogradation, derived from tabulated parameters, for the distance of station from apogee of the epicycle. *AFGBHE* represents the deferent, *C* the equant point, and *D* the earth. *A* is the apogee of the deferent, *B* its perigee. The epicycle is represented in four positions, *E, F, G,* and *H*. *E* and *F* are defined as follows: when the planet is in the middle of its retrograde arc (i.e., in the true perigee of the epicycle) as the epicycle is in the apogee of the deferent (*A*), then the planet is at first station as the epicycle is at *E* and at second station as the epicycle is at *F*. By symmetry, *E* and *F* are equidistant from *A*. The value tabulated is the position of the station for the epicycle at *A*, the apogee: namely, point *L*, which is marked as the first station for situation *E* and as the second station for situation *F*. However, we are interested in the position of the station for the epicycle, not at *A*, but at *E* and *F*. Since *E* and *F* are nearer to the earth *D* than *A* is, the station will be nearer the perigee of the epicycle in those situations. The true positions of the station at *E* and *F* are points *K* and *M*, respectively, and arc *LK* = arc *LM*, by symmetry. While the epicycle is moving from *E* to *A*, the true epicyclic perigee also moves from *Z* (the true perigee in all four positions drawn on the figure) to *T* (the mean epicyclic perigee in all four positions). Thus the arc of half-retrogradation as the epicycle goes from *E* (at which first station occurs) to *A* (at which the planet is in the middle of retrogradation) is given by arc *TK*, or $\widehat{LZ}-(\widehat{LK}+\widehat{TZ})$. By symmetry this is equal to arc *TM*, the arc of half-retrogradation from the middle of retrogradation (*A*) to second station (*F*). The analogous situation on either side of the perigee is represented in the lower part of the diagram. Here *Q* is the position of station for epicycle at perigee, *P* and *R* the positions of first station for epicycle at *G* and second station for epicycle at *H*, respectively, and the arc of half-retrogradation is arc *PT* (= arc *RT*), or $\widehat{QZ}+(\widehat{QP}+\widehat{TZ})$. Compare Figure 27 (p. 451).

---

281 auge: et planeta in puncto retrogradationis predicto scilicet arcu dimidie retrogradationis quem haberet si centrum epycicli figeretur in auge usque dum planeta transiret arcum sue retrogradationis dimidiam *add. fy*Λ

cum planeta fuerit in statione prima ad situm *E* quod mediet retrogradationem ad situm *A* et fiat in statione secunda ad situm *F*. Similiter quoque 295 ponam ut cum fuerit in statione prima ad situm *G* quod mediet retrogradationem ad situm *B* et fiat in statione secunda ad situm *H*. Et inquiram quantus sit arcus dimidie retrogradationis utrobique. Describam enim in quolibet istorum 4 situum signatorum super 4 puncta predicta epiciclum. Et protraham lineas motus equalis que sunt *CTE CTF CTG CTH* ut ubique 300 sit *T* oppositio augis medie. Protraham quoque a centro orbis signorum quod est punctum *D* lineas *DZE DZF DZG DZH* ut ubique sit *Z* oppositio augis uere. Et ponam ut ad situm *E* sit *K* statio prima; et ad situm *A* sit *L* prima et secunda; et ad situm *F* sit *M* secunda; item ad situm *G* sit *P* prima; et ad situm *B* sit *Q* prima et secunda; et ad situm *H* sit *R* secunda. 305 Quia igitur planeta est ad situm *E* in *K* et erit ad situm *A* in *T*: nam ad situm *A* fient idem *Z* et *T*: oportet ut arcus dimidie retrogradationis sit arcus *KT* qui est minor arcu *LZ* in duobus arcubus *LK* et *TZ*, quorum primus est differentia stationum situs *E* et *A* et secundus est equatio centri ad situm *E*. Arcus autem *LZ* est arcus dimidie retrogradationis ad situm 310 *A*. Quia similiter planeta erit ad situm *F* in *M* et erat ad situm *A* in *T*: tunc enim erant idem *T* et *Z*: oportet etiam ut secundus arcus dimidie retrogradationis sit *MT* qui est minor arcu *LZ* in duobus arcubus *LM* et *TZ*, quorum primus est differentia stationum situs *A* et *F* et secundus est equatio centri ad situm *F*. Contrarium autem erit in duobus sitibus *G* et *H*. 315 Quia enim planeta est ad situm *G* in *P* et erit ad situm *B* in *T*: tunc enim erunt idem *T* et *Z*: oportet ut arcus dimidie retrogradationis sit *PT* qui est maior arcu *QZ* in duobus arcubus *PQ* et *ZT*, quorum primus est differentia stationum situs *G* et *B* et secundus est equatio centri ad situm *G*. Arcus autem *QZ* est arcus dimidie retrogradationis ad situm *B*. Quia igitur 320 planeta erit ad situm *H* in *R* et erat ad situm *B* in *T*: nam tunc erant idem *T* et *Z*: oportet etiam ut secundus arcus dimidie retrogradationis sit *RT* qui est maior arcu *QZ* in duobus arcubus *RQ* et *ZT*, quorum primus est differentia stationum situs *B* et *H* et secundus est equatio centri ad situm *H*. Oportet ergo querere dato quod planeta mediet retrogradationem ad situm 325 *A* et *B* cum quo centro medio sit ante in statione prima uel post in secunda,

296 inquiram: queram βφ
299 ubique: utrobique βφ
301 quod est punctum: scilicet β similiter *V*/ DZH: et ponam *add.* θ/ut: et β/ubique: utrobique α
304 ad situm B...ad situm H sit: ad situm H *BMmV* ad situm H sit Q prima et Γ
306 A: hunc *BMmVy*

at first station for the position [of the epicycle at] *E*, it is in the middle of its retrogradation for position *A* and is at second station for position *F*. In like manner I posit that, when it is at first station for position *G*, it is in the middle of its retrogradation for position *B* and at second station for position *H*. I then consider the size of the arc of half-retrogradation for each of the two situations: I describe the epicycle at each of the four delineated positions about the four above-named points, and I draw the lines of uniform motion, *CTE*, *CTF*, *CTG*, and *CTH*, so that in each situation *T* is the mean perigee. I also draw lines *DZE*, *DZF*, *DZG*, and *DZH* from the center of the ecliptic, point *D*, so that in each situation *Z* is the true perigee. I put *K* as first station for position *E*, *L* as first and second [station] for position *A*, and *M* as second [station] for position *F*.[32] Similarly [I put] *P* as first [station] for position *G*, *Q* as first and second [station] for position *B*, and *R* as second [station] for position *H*. Since, then, the planet is at *K* for position *E* and will be at *T* for position *A*—for *Z* and *T* will become the same point for position *A*—the arc of half-retrogradation [at *E*] must be arc *KT*, which is less than arc *LZ* by the sum of the two arcs *LK* and *TZ*; the first of these arcs is the difference between [the places of] the station for positions *E* and *A*, while the second is the equation of center at position *E*. Arc *LZ* is the arc of half-retrogradation for position *A*. Similarly, since the planet will be at *M* for position *F* and was at *T* for position *A*—for then [at position *A*] *T* and *Z* were the same point—the second arc of half-retrogradation must be *MT*, which is less than arc *LZ* by the sum of the two arcs *LM* and *TZ*; the first of these is the difference between [the places of] the station for positions *A* and *F*, while the second is the equation of center at position *F*. The opposite will be true for the two positions *G* and *H*. For since the planet is at *P* for position *G* and will be at *T* for position *B*—for then *T* and *Z* will be the same point—the arc of half-retrogradation must be *PT*, which is greater than arc *QZ* by the sum of the two arcs *PQ* and *ZT*; the first of these is the difference between [the places of] the station for positions *G* and *B*, while the second is the equation of center at position *G*. Arc *QZ* is the arc of half-retrogradation for position *B*. Therefore, since the planet will be at *R* for position *H* and was at *T* for position *B*—for then *T* and *Z* were the same point—the second arc of half-retrogradation must be *RT*, which is greater than arc *QZ* by the sum of the two arcs *RQ* and *ZT*, the first of which is the difference between [the places of] the station for positions *B* and *H*, while the second is the equation of center at position *H*. Therefore, given that the planet is in the middle of its retrogradation at position *A* or at position *B*, we must find the mean centrum pertaining to it when it is at first station before [*A* or *B*][33] or at second [station] after

---

294 situm A: hoc est cum peruenerit centrum epycicli ad A *add.* φ*J* / situm F: hoc est cum peruenerit centrum epycicli ad F *add.* φ*J*

324 dato...planeta: supposito quod centrum epycicli figeretur (in auge *add. J*) usque dum planeta transiret arcum sue retrogradationis dimidium φ*J*

et illius centri medii sumere equationem eiusque uerificati stationem et accipere differentiam huius stationis ad stationem situs *A* aut *B* et coniungere differentiam stationum cum equatione centri totumque agregatum diminuere ex arcu dimidie retrogradationis situs *A* uel addere super arcum
330 dimidie retrogradationis situs *B* residuumque aut compositum diuidere per motum planete in una die, et quod exibit erit arcus dimidie retrogradationis ad situm *A* aut *B*. Si autem ad situm *A* aut *B* fuerit planeta in statione prima, solam equationem centri ad cuius situm mediabit retrogradationem ex arcu medie retrogradationis situs *A* minuemus aut super arcum medie
335 retrogradationis situs *B* addemus et residuum aut agregatum diuidemus ut prius. Nichil enim operatur hic differentia stationum. Cum autem in longitudine media contingit planetam esse in medio sue retrogradationis, non oportet aliquid addi uel minui quoniam tantum addit tunc aut minuit arcus dimidie retrogradationis sequens quantum addit aut minuit ante-
340 cedens, et si qua diuersitas fuerit non erit curanda. In mercurio autem licet ista contingant tamen ipsos eosdem arcus diuisimus per motum eius diurnum, de errore qui propter hoc accidit non curantes propter eius paruitatem. Nam et motus diurnus in eo est magnus et quantitas equationis centri est parua et differentia stationum non multa.

345 Quoniam autem uenus et mercurius quoad motum utriusque medium semper simul sunt cum sole, inpossibile est eos amplius elongari ab eo quam usque ad lineas egressas a centro orbis signorum contingentes epiciclum ex utraque parte. Cum uero uterque eorum fuerit super lineam contingentem epiciclum ex parte orientis, erit in ultima sui elongatione a sole
350 uespertina. Cum autem fuerit super lineam contingentem ipsum ex parte occidentis, erit in ultima sui elongatione ab eo matutina. Et huius quidem maxime elongationes uespertina et matutina diuersificantur secundum diuersitatem signorum. Vnde PTHOLOMEVS fecit tabulas ad hoc, ponens in eis maximam elongationem utriusque eorum a sole uespertinam et matuti-
355 nam in principiis omnium signorum. Et tu eandem tabulam facere poteris per instrumentum nostrum. Si enim hoc libuerit facies uerbi gratia in mercurio quod linea que egreditur a centro orbis signorum transeat per principium arietis ordinabisque centrum epicicli secundum talem situm quod hec linea contingat ipsum ex parte orientis sitque centrum deferentis in
360 situ correspondente huic situi epicicli. Facies quoque ut linea que egreditur

326 uerificari β*V*Π
336 hec β*V*
337 contingat β*V*
339 sequens: sequensque δ non *add.* *DN*/ quantum: tantum δ*OP*
354 maximam: ultimam β
356 uerbi gratia: ita β*Jo*
356–57 mercurio: fac *add.* β
357 linea: filum uel linea *Ff*Σ id est filum *add.* *Akvxy*Λ/linea que *om.* δ
359 hec linea: linea uel filum *F*
360 linea: uel filum *add.* *Ffkpy*Σ id est filum *add.* Λ

[*A* or *B*], and take the equation of that mean centrum and, having corrected it, take the station[34] pertaining to that corrected centrum, and find the difference between the latter station and the station for position *A* or *B*; then add the difference between the stations to the equation of center, and subtract the sum from the arc of half-retrogradation for position *A* or add it to the arc of half-retrogradation for position *B*, and then divide the result of this subtraction or addition by the motion of the planet [on the epicycle] in one day: the result will be the arc of half-retrogradation at position *A* or *B* [respectively]. However, if the planet is at first station in position *A* or *B*, all we do is subtract the equation of center for the position at which [the planet] is in the middle of its retrogradation from the arc of half-retrogradation for position *A*, or add it to the arc of half-retrogradation for position *B*, and divide the result of subtraction or addition as before. For in this case the difference between stations plays no part. But when it happens that the planet is in the middle of its retrogradation at mean distance [of the epicycle], there is no need to add or subtract anything, since then the arc of half-retrogradation [for the] succeeding [position] causes the same amount of addition or subtraction as the [arc of half-retrogradation for the] antecedent [position] causes subtraction or addition; and if there should be any difference,[35] it can be neglected. For Mercury, although these features are present, nevertheless we divided the [unmodified] arcs themselves by the daily motion [on the epicycle],[36] neglecting the resultant error on account of its smallness: for Mercury's daily motion is great, and the size of the equation of center is small, and the difference between stations not large.[37]

Now, since Venus and Mercury, insofar as the mean motion of each of them is concerned, are always at the same longitude as the sun, it is impossible for them to reach a greater elongation from it [the sun] than the tangents from the center of the ecliptic to the epicycle on either side. When either of them is on the line tangent to the epicycle on the eastern side, it will be in its greatest evening elongation from the sun, but, when it is on the line tangent to it on the western side, it will be in its greatest morning[38] elongation from it [the sun]. Now the greatest evening and morning elongations of the planet differ in the different signs. Hence PTOLEMY made tables for this purpose,[39] in which he put the maximum evening and morning elongation of both of the [planets] from the sun at the beginning of each sign. You, too, can compute the same table by our instrument.[40] If you want to do this, you must for Mercury, for instance [see Fig. 16, p. 246], make the line coming from the center of the ecliptic pass through the beginning of Aries, and arrange the center of the epicycle in such a position that this line touches it on the eastern side, while the center of the deferent is in the position corresponding to this position of the epicycle.[41] You

a centro equantis transeat per centrum epicicli cui facies equedistare eam que egreditur a centro orbis signorum, locusque in orbe signorum ubi ipsum secabit erit locus medii motus solis. In solaris igitur instrumenti figura facias transire per hunc locum lineam que egreditur a centro orbis
365 signorum eique facias equedistare eam que egreditur a centro sui ecentrici. Eritque sol in loco ubi ipsa secauerit ecentricum. Per hunc ergo locum facias transire denuo lineam que egreditur a centro orbis signorum. Eritque secundum eius uerum motum in loco in quo orbem signorum secauerit. Considera itaque quantus est arcus inter hunc locum et principium arietis.
370 Ipsa enim est maxima longitudo mercurii a sole uespertina in principio arietis. Si autem ordinaueris epiciclum secundum talem situm quod ipse contingat lineam predictam ex parte occidentis processerisque ad inueniendum uerum motum solis, sicud prius erat arcus orbis signorum inter uerum locum solis et principium arietis maxima longitudo mercurii a sole
375 uespertina, ita erit hic arcus orbis signorum inter uerum locum solis et principium arietis inuentus sicud prius maxima longitudo mercurii a sole matutina. Eodem modo inuenies has longitudines ad principia aliorum signorum nec solum ad principia sed ad quemcumque uoles situm. De uenere quoque idem.

380 Spera quoque cuiuslibet istorum 4 planetarum habet motum proprium equalem et similem motui sperarum trium predictorum de quo sufficienter dictum est. Habet itaque quilibet istorum 4 tres circulos: scilicet epiciclum, deferentem et equantem: tres quoque motus, sui in epiciclo, epicicli in deferente et motum spere sue. Motus igitur istorum quatuor planetarum
385 in longitudine eorumque circuli istis motibus necessarii sunt qui dicti sunt. De motibus autem eorum in latitudine circulisque eis necessariis post dicetur.

Nunc autem uolumus in hiis 4 planetis narrare magnitudines orbium trium predictorum qui sunt deferens, equans et epiciclus; distantiam quo-
390 que centrorum equantis, deferentis et orbis signorum et longitudinem cuiusque eorum longiorem et propiorem et magnitudinem sue spere et sui corporis sicud fecimus prius in mercurio et in luna. Promisimus enim supra quod illum ordinem seruaremus in omnibus. Narrabimus autem sicud ibi fecimus magnitudines premissas: primo quidem secundum partes
395 illas de quibus semidiameter deferentis que est linea *AC* habet 60 partes; deinde secundum partes illas de quibus semidiameter terre est pars una;

364 lineam: id est filum *add. fkxy*Λ hoc *F* ad hec Σ initium *f* ad unum Λ *om.* θ*H*/uolueris β
370 est: erit β*V*
373 uerum: medium β*V*/erat: erit αφ*H*
374–76 maxima...prius *om.* φ*FHMmOqrV*Σ
375–76 uespertina...sole *om.* α
378 ad principia: ibidem *hMUVYz*Γ idem *B*
382 4: planetarum *add.* β
392 prius: superius θ *om.* β
394 ibi *om.* α

must also make the line coming from the center of the equant pass through the center of the epicycle, and make the one coming from the center of the ecliptic parallel to the latter; then the place on the ecliptic where [the thread attached to the center of the ecliptic] cuts it [the ecliptic] will be the sun's mean longitude. Therefore, on the figure of the sun's instrument [see Fig. 10, p. 152], make the line coming from the center of the ecliptic pass through this place, and make [the line] coming from the center of the eccentric parallel to it; then the [mean] sun will be in the place where that [second line] cuts the eccentric. This, then, is the place through which you must make the line coming from the center of the ecliptic pass again: and the true [sun] will be in the place where it cuts the ecliptic.[42] So observe the size of the arc between this place and the beginning of Aries, for that is the greatest evening elongation of Mercury from the sun [when Mercury is] at the beginning of Aries. If, however, you arrange the epicycle in such a position that it touches the above-mentioned line on the western side and proceed to find the true position of the sun, then, just as previously the arc of the ecliptic between the true place of the sun and the beginning of Aries was the greatest evening elongation of Mercury from the sun, so too here the arc of the ecliptic between the true position of the sun and the beginning of Aries, found in the same way as before, will be the maximum morning elongation of Mercury from the sun. In the same way you can find these [two] elongations for the beginnings of the other signs, and not only for the beginnings, but for any position you want. The same is true for Venus.

Furthermore, the sphere of each of these four planets has its special motion, which is equal to and like the motion of the spheres of the three bodies previously described.[43] Enough has been said about that. Each of these four [planets], then, has three circles, namely, an epicycle, a deferent, and an equant, and also three motions, its own [motion] on the epicycle, that of the epicycle on the deferent, and the motion of its sphere. The motions of these four planets in longitude and the circles necessary for those motions are those that have been described. We will speak later of their motions in latitude and the circles necessary for them.[44]

Now, however, I wish to relate, for these four planets, the sizes of the three above-mentioned circles, namely, the deferent, equant, and epicycle, and also the distances between the centers of the equant, the deferent, and the ecliptic, and the greatest and least distances of each of them, and the size of the sphere and body [of each], just as we did before for Mercury and the moon. For we promised previously[45] that we would follow that [same] order for all. We will relate the above-mentioned sizes just as we did for them [Mercury and the moon], first of all in the units of which the radius of the deferent, which is line *AC*, contains 60, then in the units of which the radius of the earth is one,

ultimo uero secundum partes uulgariter notas que sunt miliaria. Primo quoque narrabo hoc in uenere; secundo in marte; tertio in ioue; quarto in saturno. Reiterabo igitur figuram premissam describendo circulum *AGB*
400 qui sit deferens ueneris circa centrum *C* et *LGH* qui sit eius equans circa centrum *K*. Sitque centrum orbis signorum punctum *D*. Et transeat per hec tria centra diameter *ADB*. Constat enim ex predictis quod ipsa sunt in linea una. Sitque *A* punctus longitudinis longioris; *B* uero propioris. Describam itaque epiciclum ueneris semel super punctum *A*, sitque tunc punc-
405 tus *E* longitudo longior ipsius, et semel super punctum *B* et sit tunc punctus *F* longitudo propior ipsius. Erit ergo *E* maxima elongatio ueneris a centro orbis signorum et *F* minima.

Dico igitur quod secundum quod linea *AC* que est semidiameter deferentis ueneris est 60 partes erit unaqueque 2 linearum que sunt *KC*, *CD* qua-
410 rum prima est distantia centri equantis a centro deferentis et secunda a centro deferentis ad centrum orbis signorum pars una et 15 minuta. Et linea *AE* que est semidiameter epicicli 43 partes et 10 minuta. Et linea *AD* que est distantia longitudinis longioris a centro orbis signorum 61 partes et 15 minuta. Et linea *DB* que est distantia longitudinis propioris 58 partes
415 et 45 minuta. Et linea *ED* que est maior longitudo centri ueneris que esse

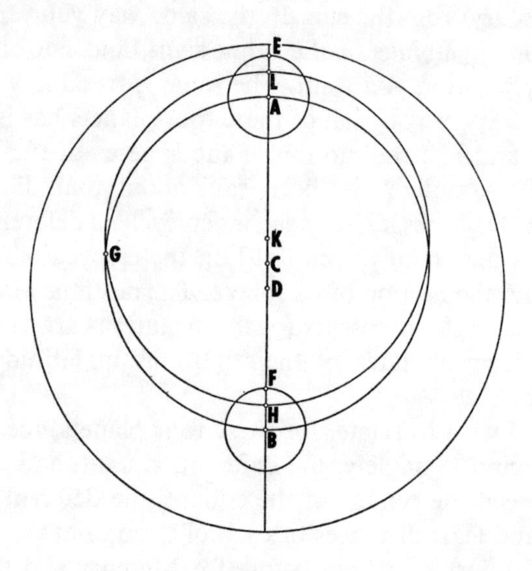

Fig. 20

403 uero *om.* αδ
406 maxima elongatio: in maxima elongatione θ
407 minima: in minima (elongatione *add.* *F*) θ
408 secundum quod *om.* β*V*
409 est: habet β*V*/60 partes: et secundum hoc *add.* β*V*
415 centri *om.* αφ*HV*

and finally in the well-known units, namely, miles. I will relate them first for Venus, second for Mars, third for Jupiter, and fourth for Saturn. Therefore I repeat the previous figure [i.e., Fig. 18, p. 308] and describe circle *AGB*, which is to be the deferent of Venus, about center *C*, and *LGH*, which is to be the equant, about center *K* [see Fig. 20, p. 324]. Let point *D* be the center of the ecliptic; and let the diameter *ADB* pass through these three centers. For it is established from what we said before that they lie on one straight line. Let *A* be the point of greatest distance and *B* [the point] of least distance. Thus I describe the epicycle of Venus once about point *A*—in this situation let point *E* be the [point of] greatest distance on it—and once about point *B*—in this situation let point *F* be the [point of] least distance on it. Then *E* will be the greatest possible distance of Venus from the center of the ecliptic and *F* the least possible.

I say, then, that in the units of which *AC*, the radius of Venus's deferent, is 60,[46] each of the two lines *KC* and *CD*, of which the first is the distance of the center of the equant from the center of the deferent and the second [the distance] from the center of the deferent to the center of the ecliptic, is 1 unit and 15 minutes;[47] line *AE*, which is the radius of the epicycle, is 43 units and 10 minutes;[48] line *AD*, which is the amount the greatest distance is removed from the center of the ecliptic, is 61 units and 15 minutes; line *DB*, which is the amount the least distance is removed [from the center of the ecliptic], is 58 units and 45 minutes; line *ED*, which is the greatest possible distance of the center of

Legend for Fig. 20

Schematic representation (after Campanus) of the model for Venus (or any of the three outer planets) to delineate the parts of its "sphere." The outer circle represents the ecliptic, with center *D*, the earth. *AGB* represents the deferent, about center *C*, and *LGH* the equant circle, about center *K* (*KC* = *CD*). Then *A* is the apogee and *B* the perigee of the deferent. The epicycle is drawn twice, once about center *A* and once about center *B*. *E* is the apogee of the epicycle in the first situation, and *F* the perigee of the epicycle in the second situation. Then *DE* represents the "greatest possible distance" of the planet, and *DF* its "least possible distance." These define the outer and inner boundaries, respectively, of the planet's "sphere."

399–417 Reiterabo...35 minuta: Ne autem prolixitas sermonis fastidium generet cuilibet planete suo loco tabulam subscribam in quibus facilime quis poterit dictas magnitudines contemplari. Sequitur tabula ueneris *f*

possit 104 partes et 25 minuta. Et linea *DF* que est minor centri corporis eius longitudo que esse possit 15 partes et 35 minuta.

At uero secundum quod semidiameter terre est pars una, possibile est nobis per ea que dicta sunt easdem magnitudines inuenire. Cum enim supremum mercurii sit infimum ueneris, si super lineam *ED* in mercurio addiderimus semidiametrum corporis mercurii que est 2 minuta et 8 secunda de partibus illis de quibus semidiameter terre est pars una, et semidiametrum corporis ueneris que est 26 minuta et 40 secunda de partibus eisdem: continet namque semidiameter corporis ueneris tertiam partem semidiametri terre et insuper tertiam tertie, hoc est quatuor nonas eius: habebimus lineam *DF* in uenere. Quapropter linea *DF* erit 178 partes 56 minuta et 48 secunda de partibus illis de quibus semidiameter terre est pars una. Et per hanc inuenientur omnes alie quemadmodum in mercurio dictum est. Eruntque ita linea *AC* 688 partes 59 minuta et 33 secunda. Et linea *DC* 14 partes 21 minuta et 14 secunda. Et linea *AE* 495 partes 41 minuta et 31 secunda. Et linea *AD* 703 partes 20 minuta et 47 secunda. Et linea *DB* 674 partes 38 minuta et 19 secunda. Et linea *ED* 1199 partes 2 minuta et 18 secunda.

De miliaribus autem erunt predicte magnitudines ita. Linea *AC* 2236093 miliaria et 555 sexagesime sexcentesime unius miliaris. Et linea *DC* 46584 miliaria et 590 sexagesime sexcentesime unius miliaris. Et linea *AE* 1608745 miliaria et 445 sexagesime sexcentesime unius miliaris. Et linea *AD* 2282678 miliaria et 485 sexagesime sexcentesime unius miliaris. Et linea *DB* 2189508 miliaria et 625 sexagesime sexcentesime unius miliaris. Et linea *ED* 3891424 miliaria et 270 sexagesime sexcentesime unius miliaris. Et linea *DF* 580763 miliaria et 180 sexagesime sexcentesime unius miliaris. Semidiameter autem corporis ueneris est 1442 miliaria et 280 sexagesime sexcentesime unius miliaris. Ambitus autem corporis eius erit 9095 miliaria et 63 sexagesime sexcentesime unius miliaris. Si autem semidiametrum ueneris subtraxerimus de linea sua *DF* eandemque addiderimus super lineam *ED*, fiet hec quidem distantia superficiei concaue spere ueneris a centro terre, illa uero distantia conuexe. Eritque hec 579320 miliaria et 560 sexagesime sexcentesime unius miliaris. Illa uero erit 3892866 miliaria et 550 sexagesime sexcentesime unius miliaris. Conuenit ergo precise infimum ueneris cum supremo mercurii. Fractiones enim utrobique precise equales sunt. Erit itaque spissitudo spere ueneris 3313545 miliaria et 650 sexagesime sexcentesime unius miliaris. Ambitus autem conuexe super-

416 centri corporis *om.* αφ*HV*
418 semidiameter: diameter αφ*V*
428 inueniuntur β
348 2282678: 2281618 *AdgiMmPStUuY*ΓΠ Ψ 2281628 *V* 2281678 *hvx* 228218 *y* 222-28161881618 Σ
446–47 a centro terre *om.* δ

Venus's [body], is 104 units and 25 minutes; line *DF*, which is the least possible distance of the center of its body, is 15 units and 35 minutes.[49]

Moreover, it is possible for us, by using [the method] stated [previously], to find the same magnitudes in the units of which the radius of the earth is one. For, since the uppermost point of Mercury's [sphere] is the lowest point of Venus's, if we add to line *ED* for Mercury the radius of Mercury's body, which is 2 minutes and 8 seconds[50] of the units of which the radius of the earth is one, and [also add] the radius of Venus's body, which is 26 minutes and 40 seconds[51] of the same units—for the radius of Venus's body contains a third plus a third of a third of the earth's radius, that is, four-ninths [in all]—we will have line *DF* for Venus. Hence line *DF* will be 178 units 56 minutes 48 seconds,[52] in those units of which the radius of the earth is one. From this [line *DF*] all the other [lines] can be found, as was explained for Mercury.[53] Thus line *AC* will be 688 units 59 minutes 33 seconds; line *DC* will be 14 units 21 minutes 14 seconds; line *AE* will be 495 units 41 minutes 31 seconds; line *AD* will be 703 units 20 minutes 47 seconds; line *DB* will be 674 units 38 minutes 19 seconds; and line *ED* will be 1,199 units 2 minutes 18 seconds.

In miles, however, the above magnitudes will be as follows:[54] line *AC* $2{,}236{,}093\frac{555}{660}$ miles; line *DC* $46{,}584\frac{590}{660}$ miles; line *AE* $1{,}608{,}745\frac{445}{660}$ miles; line *AD* $2{,}282{,}678\frac{485}{660}$ miles; line *DB* $2{,}189{,}508\frac{625}{660}$ miles; line *ED* $3{,}891{,}424\frac{270}{660}$ miles; and line *DF* $580{,}763\frac{180}{660}$ miles. The radius of Venus's body is $1{,}442\frac{280}{660}$ miles.[55] The circumference of its body will be $9{,}095\frac{63}{660}$ miles.[56] Now, if we subtract the radius of Venus's [body] from its line *DF* and add the same distance to line *ED*, the first [operation] will give the distance of the concave surface of Venus's sphere from the center of the earth and the second the distance of the convex [surface]. The first [distance] will be $579{,}320\frac{560}{660}$ miles, and the second will be $3{,}892{,}866\frac{550}{660}$ miles. Thus the lower surface of Venus's [sphere] coincides exactly with the upper surface of Mercury's. For in each case the [terminal] fraction is precisely the same. Thus the thickness of Venus's sphere will be $3{,}313{,}545\frac{650}{660}$ miles,[57] and the circumference of its convex surface will be

---

423 ueneris: tertiam partem diametri (semidiametri *M*Γ) terre secundum quod semidyameter (diameter *BCz*) terre (secundum...terre *om. KpV*Ψ) est pars una (secundum...una *om. M*Γ) et semidyameter corporis lune *add. BCKMOoPpVz*ΓΨ

426 uenere: et hoc modo inueniuntur omnes alie ut dictum est in mercurio *add. f*

ficiei spere ueneris erit 24469448 miliaria et 440 sexagesime sexcentesime unius miliaris. Ambitus autem sui deferentis erit 14055447 miliaria. Si ergo
455 ipsum diuiseris per 365 dies et quartam qui sunt tempus unius reuolutionis centri epicicli in ecentrico, exibit tibi motus centri epicicli in una die. Erit itaque dieta eius 38481 miliaria et 44 minuta unius miliaris. Ambitus itaque epicicli ueneris erit 10112115 miliaria et 40 minuta. Et si ipsum diuiseris per 584 dies qui sunt tempus unius reuolutionis ipsius in epiciclo suo, exi-
460 bit tibi dieta ipsius in suo epiciclo et est 17315 miliaria et 16 minuta unius miliaris. Corpus autem terre continet corpus ueneris 11 uicibus et insuper 25 quartas sexagesimas eius. Nam proportio diametrorum terre et corporis ueneris est sicud 9 ad 4 quorum cubi sunt 729 et 64. In eorum autem proportione necesse est ut se habeant corpora predicta. Quantitates autem
465 predicte secundum examinatam eorum numerationem per triplices modos predictos sunt sicud dictum est. Et ego posui eas per ordinem in subscriptis tabulis.

Conuenit autem ordini processus assumpti ut accedamus ad mensuras orbis solis et distantie centri sui ecentrici a centro orbis signorum et lon-
470 gitudinis longioris eius et propioris et etiam spere sue. PTHOLOMEVS enim nichil posuit de hiis nisi distantiam centri ecentrici a centro orbis signorum de partibus illis de quibus semidiameter ecentrici habet 60 partes: sic enim inuenit predictam distantiam duas partes et 30 minuta ut dictum est prius: et nisi distantiam centri corporis solis a centro terre de parti-
475 bus illis de quibus semidiameter terre est pars una. Sic enim inuenit ipsam 1210 partes. Sed non potuit certificare secundum partes istas distantiam longitudinis longioris eius aut propioris; quoniam ipse processit in inuentione istius distantie secundum partes istas per angulum qui subtenditur diametro corporis solis. Angulus autem ille quemadmodum ipse dixit fere
480 est idem in omni loco et non diuersificatur diuersitate sensibili propter longitudinem solis in suo ecentrico aut propter propinquitatem ipsius. Vnde distantia predicta quam ipse inuenit non potuit apropriari alicui certe parti ecentrici. Nec propter hoc potuit plus haberi quam quod in ecentrico solis est aliquis punctus qui distat a centro terre quantitate predicta.
485 Vtrum autem punctus ille esset in longitudine longiori aut propiori aut magis uicinius huic quam illi, non potuit haberi propter ambiguitatem medii assumpti ad hoc. In luna uero manifesta fuit diuersitas anguli qui

457 dieta: motus α*V*
461–62 insuper...sexagesimas: 2 quintas δ II 5 *H*
466 eas: eos βδφ ipsas *V*
466–67 subscriptis tabulis: subscripta tabula θ*H* tabula *BMmoPUVY* tabula que sequitur Γ subiecta figura *F*Σ subiecta tabula II
468 accedamus: ascendamus δ*A*ΠΨ
474 nisi: inuenit β*V*
481 longitudinem: diuersitatem β*V*/ecentrico: in longitudine *add.* β*V*/propter propinquitatem: propinquitate β*V*
486 uicinus *BFiMNPrSwZ*ΓΣΨ

24,469,448$\frac{440}{660}$ miles.[58] The circumference of its deferent will be 14,055,447 miles.[59] If, then, you divide the latter by 365$\frac{1}{4}$ days, which is the period of one revolution of the center of the epicycle on the eccentric, you will get the motion of the center of the epicycle in one day. Its daily motion will thus be 38,481 miles and 44 minutes of a mile.[60] The circumference of Venus's epicycle will be 10,112,115 miles and 40 minutes [of a mile]. If you divide this by 584 days,[61] which is the period of one revolution of Venus on its epicycle, the result will be its daily motion on its epicycle, and that is 17,315 miles and 16 minutes of a mile.[62] The body of the earth contains the body of Venus 11 times plus $\frac{25}{64}$ of it. For the proportion between the diameters of the earth and of Venus's body is 9 to 4; the cubes of these are 729 and 64,[63] and the bodies must be in proportion to the cubes. The above-mentioned quantities are, then, as we have given them, in properly calculated numbers, and in the three units named. I have set them down in order in the following tables [Table 3, p. 359].

It is appropriate to the order of procedure which we have chosen that we should go on to the measurement of the circle of the sun, of the distance of the center of its eccentric from the center of the ecliptic, of its greatest and least distances, and also of its sphere. For PTOLEMY set down nothing on these matters, excepting the distance of the center of the eccentric from the center of the ecliptic, expressed in the units of which the radius of the eccentric contains 60—in these terms he found the above distance to be 2 units and 30 minutes, as we stated before[64]—and also excepting the distance of the center of the sun's body from the center of the earth, in the units of which the earth's radius is one: he found it to be 1,210 of these units.[65] But he was unable to state with certainty the amount of the greatest distance or least [distance] in those units, since his method of procedure for finding that distance in those units was by the angle subtended by the diameter of the sun's body.[66] That angle, however, as he himself said,[67] is almost the same for all positions [of the sun] and does not change sensibly according to how far or how near the sun on its eccentric is. Hence the above distance which he found could not be assigned to any particular part of the eccentric.[68] And because of this [fact], the most that could be affirmed is that there is some position on the sun's eccentric which is the aforesaid distance from the center of the earth. It could not, however, be affirmed whether that point was the greatest distance or the least distance, or nearer the latter than the former, because of the ambiguity of the method adopted for [getting] this result. On the other hand, for the moon there was an

---

464 predicta: Proportio enim sperarum est sicut proportio duarum dyametrorum triplicata ut super (dictum est *add. J*) *add. Jo*

subtenditur diametro corporis eius in diuersis locis sui ecentrici quemadmodum ipse dixit et operatione probauit, unde possibile fuit ei certificare
490 in luna distantiam longitudinis longioris et propioris propter certitudinem medii assumpti ad hoc. Nos autem in omnibus distantiis quas posuimus processimus per illam distantiam lune quam ipse inuenit, que certa fuit et non ambigua, et per proportiones semidiametrorum deferentium ad distantias centrorum et ad semidiametros epiciclorum, quas etiam ipse posuit
495 in omnibus planetis certas et exquisitas. Aliter enim inpossibile erat ei inuenire aliquid certum eorum que ipse inquirebat. Et ista fuerunt nobis sufficientia ad inueniendum omnia que uolumus de distantiis et magnitudinibus predictis in omnibus planetis, supposito solum quod inter speras planetarum non sit uacuum et quod nulla earum sit maior quam requirat
500 diuersitas motus planete. Per hoc enim sequitur quod supremum inferioris sit infimum superioris sue. Per ista etiam inueniuntur magnitudines corporum planetarum et sperarum omnium et stellarum omnium; quoniam cum distantia alicuius eorum a centro terre fuerit nota et nos considerauerimus angulum qui subtenditur diametro corporis eius, cognoscemus per
505 hoc quantitatem diametri sui corporis; per hoc quoque corpus ipsius.

Vt igitur premissas magnitudines reperiamus in sole, describam ecentricum eius qui sit circulus *AB* circa centrum *C*. Sitque *A* punctus longitu-

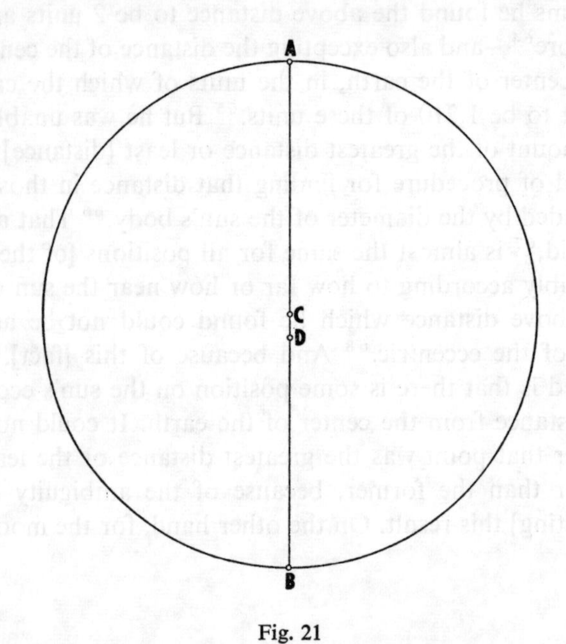

Fig. 21

489 operatione: experimento βV
501 sue: spere βV
502 sperarum omnium et *om.* δθφ
505 ipsius *om.* δθφ

obvious difference in the angle subtended by the diameter of its body in different positions on its eccentric, as he himself stated[69] and proved by experiment;[70] hence it was possible for him to calculate for the moon exactly the amount of greatest and least distance, because of the exactness of the method adopted for [getting] this [result]. Now, for all the distances which we have set down, we have proceeded by using that distance of the moon which [PTOLEMY] himself found, which was certain and contained no ambiguities, and also [by using] the ratios of the radii of the deferents to the distances between the centers and to the radii of the epicycles; these were also stated by [PTOLEMY] himself for all the planets as certain and exactly calculated. For otherwise it would have been impossible for him to find for certain any of the things which he was seeking. Those [elements] have been sufficient for us to find out everything we wish concerning the above-named distances and magnitudes for all the planets, provided only that we suppose that there is no empty space between the spheres of the planets and that none of them [the spheres] is greater than is required by the variation in the planet's motions.[71] For from this it follows that [for any two adjacent planetary spheres] the highest point of the lower coincides with the lowest point of the higher. From those [elements] can also be found the sizes of the bodies of the planets and of all the spheres and of all the stars.[72] For, when the distance of any of them from the center of the earth is known and we consider the angle subtended by the diameter of its body, we shall discover from this the actual size of the diameter of its body, and from the latter its actual volume.

In order, therefore, to find the previously named magnitudes for the sun, I describe its eccentric—let this be circle AB—about center C [see Fig. 21, p. 330].

Legend for Fig. 21

Schematic representation (after Campanus) of the sun's model to delineate the parts of its "sphere." AB represents the sun's eccentric, about center C; D represents the earth. Then CD is the eccentricity, AD is the sun's greatest distance, and DB is its least distance. These define the outer and inner boundaries, respectively, of the sun's "sphere."

dinis longioris, *B* uero propioris. Et protraham diametrum *ACB* et signabo in ea centrum orbis signorum quod sit punctum *D*. Dico ergo quod secun-
510 dum partes de quibus linea *AC* que est semidiameter deferentis est 60 partes est linea *CD* que est distantia centri ecentrici a centro orbis signorum 2 partes et 30 minuta. Et linea tota *AD* que est distantia longitudinis longioris solis a centro terre est 62 partes et 30 minuta. Et linea *DB* que est distantia longitudinis propioris eius a centro terre 57 partes et 30 minuta.
515 At uero secundum quod semidiameter terre est pars una, inueniuntur hee magnitudines per lineam *ED* in uenere. Illa enim est minor quam linea *DB* in sole quantitate semidiametrorum corporis ueneris et corporis solis. Semidiameter autem corporis ueneris est 26 minuta et 40 secunda prout semidiameter corporis terre est pars una, et semidiameter corporis solis
520 est 5 partes et 30 minuta secundum easdem partes. Erit ergo linea *DB* in sole 1204 partes 58 minuta et 58 secunda secundum partes illas de quibus semidiameter terre est pars una. Et linea *AC* erit 1257 partes 22 minuta et 24 secunda. Et linea *CD* 52 partes 23 minuta et 26 secunda. Et linea *AD* 1309 partes 45 minuta et 50 secunda. Manifestum est igitur quod distantia
525 centri corporis solis a centro terre quam posuit PTHOLOMEVS quam ipse quidem inuenit 1210 partes secundum quod semidiameter terre est pars una non est secundum ueritatem distantia puncti longitudinis longioris aut propioris sed cuiusdam puncti sui ecentrici qui non multum elongatur a longitudine propiori. Nam uera distantia puncti longitudinis propioris
530 quam paulo ante posuimus est minor illa distantia quam PTHOLOMEVS ponit 5 partibus fere, et uera distantia puncti longitudinis longioris quam etiam paulo ante posuimus est maior ea 100 partibus fere. Vicinum est igitur opus nostrum secundum nostram exquisitam numerationem operi PTHOLOMEI secundum suam communem numerationem. Nec debet censeri
535 differentia inter ipsa.

Ex quo manifestum erit attente consideranti quod uenus et mercurius sunt sub sole quemadmodum ipse posuit et non supra solem quemadmodum posuit GEBER secutus platonicos et egiptios. Mirabile enim uideretur si inter speras lune et solis esset tante magnitudinis spatium quod perueni-
540 ret usque ad 1138 partes fere secundum partes illas de quibus semidiameter terre est pars una et nullum corporum celestium ibi esset locatum. Adhuc

510 de *om.* θ*BHV*ΠΨ
515 semidiameter: dyameter α
525 centri *om.* βφ*V*
526 semidiameter: dyameter *ABmPStUw*Γ ΠΨ diametrum *di*
538 uidetur αφ*HV*
539–40 perueniret: prouenit δ*A* peruenit *H* proueniret *f* perueniat *F*Σ

Let *A* be the point of greatest distance, *B* that of least [distance]. I draw the diameter *ACB* and mark on it the center of the ecliptic—let this be point *D*. I say, then, that in the units of which *AC*, which is the radius of the deferent,[73] contains 60, line *CD*, which is the distance of the center of the eccentric from the center of the ecliptic, is 2 units and 30 minutes;[74] the whole line *AD*, which is the amount of the greatest distance of the sun from the center of the earth, is 62 units and 30 minutes;[75] line *DB*, which is the amount of its least distance from the center of the earth, is 57 units and 30 minutes.[76] But for the units of which the radius of the earth is one unit, these magnitudes are derived from line *ED* for Venus. For that [line *ED*] is less than line *DB* for the sun by the amount of [the sum of] the radii of Venus's body and the sun's body. Now the radius of the body of Venus is 26 minutes and 40 seconds[77] where the radius of the earth's body is one unit, and the radius of the sun's body is 5 units and 30 minutes in the same units.[78] Thus line *DB* for the sun will be 1,204 units 58 minutes 58 seconds[79] in those units of which the radius of the earth is one; line *AC* will be 1,257 units 22 minutes 24 seconds; line *CD* will be 52 units 23 minutes 26 seconds; and line *AD* will be 1,309 units 45 minutes 50 seconds.[80] It is clear, then, that the distance between the center of the sun's body and the center of the earth stated by PTOLEMY—he found that it was 1,210 units where the radius of the earth is one unit—is, strictly speaking, not the distance of the point of greatest distance or of the point of least distance, but of a point on [the sun's] eccentric not far removed from the least distance. For the true distance of the point of least distance, as we stated it just above, is approximately 5 units less than the distance stated by PTOLEMY,[81] and the true distance of the point of greatest distance, also stated by us just above, is approximately 100 units greater.[82] Therefore our working, carried out with our very exact calculations, comes close to the working of PTOLEMY, carried out with his rough calculations; and the difference between them should be neglected.

From this it will be clear to anyone who gives careful thought to the matter that Venus and Mercury are below the sun, as he [PTOLEMY] assumed,[83] and not above the sun, as was assumed by JĀBIR,[84] following the disciples of Plato and the Egyptians.[85] For it would seem amazing if there were between the spheres of sun and moon a space of such size that it reached about 1,138 of the units[86] of which the radius of the earth is one, and yet none of the heavenly

---

516 DB: et linea DE in uenere terminatur (causatur *f*) in centro corporis ueneris scilicet in E et in D usque ad conuexum spere ueneris et est distantia tanta quanta est semidyameter corporis ueneris. Item linea DB in sole terminatur (causatur *f*) in centro corporis solis scilicet in B et in descendendo uersus centrum orbis signorum quod est D est tanta distantia usque ad concauum spere solis quanta est semidyameter corporis solis (conuexum autem spere ueneris contiguatur concauo spere solis *add. Jo*) *add. fJo*

autem mirabilius esset spatium illud precise continere proportiones orbium et corporum mercurii et ueneris et istos planetas alibi esse locatos. Amplius autem si alius ordo planetarum ponatur, inpossibile erit planetas dominari 545 in primis horis dierum denominatorum ab ipsis uel oportebit alium ordinem dierum septimane ponere. Conueniens est itaque confiteri ordinem illum sperarum celestium quem posuit PTHOLOMEVS et quem nos ipsum secuti posuimus in hiis que prius dicta sunt.

De miliaribus autem erunt predicte magnitudines ita. Linea *AC* 4080748 550 miliaria. Et linea *CD* 170031 miliaria et 110 sexagesime sexcentesime unius miliaris. Et linea *AD* 4250779 miliaria et 110 sexagesime sexcentesime unius miliaris. Et linea *DB* 3910716 miliaria et 550 sexagesime sexcentesime unius miliaris. Semidiameter autem corporis solis 17850 miliaria. Et ambitus corporis eius 112200 miliaria. Longitudo autem concaue superficiei 555 spere ipsius a centro terre 3892866 miliaria et 550 sexagesime sexcentesime unius miliaris. Longitudo uero conuexe superficiei spere eius ab eodem centro 4268629 miliaria et 110 sexagesime sexcentesime unius miliaris. Spissitudo itaque spere sue erit 375762 miliaria et 220 sexagesime sexcentesime unius miliaris. Ambitus autem conuexe superficiei spere sue 560 26831383 miliaria et 220 sexagesime sexcentesime unius miliaris. Ambitus quoque sui ecentrici erit 25650416 miliaria. Et si diuiseris ipsum per 365 dies et quartam qui sunt tempus unius reuolutionis eius in suo ecentrico, exibit tibi motus eius in die una de partibus sui ecentrici. Erit itaque dieta ipsius 70227 miliaria. Corpus autem solis continet corpus terre 166 uicibus 565 et insuper 3 octauas eius. Proportio enim diametri corporis solis ad diametrum terre est sicud proportio 11 ad duo. Cubi autem horum numerorum sunt 1331 et 8; hii uero se habent in proportione predictorum corporum et in proportione predicta. Et hec omnia patent in tabula subscripta.

Postquam autem hee magnitudines determinate sunt in uenere et in 570 sole, conueniens est ut determinemus eas in marte. Ad hoc uero reiterabo figuram premissam. Dico itaque quod secundum partes de quibus linea *AC* que est semidiameter deferentis martis est 60 partes erit utraque duarum linearum *KC* et *CD* quarum prima est distantia centri equantis a centro deferentis et secunda centri deferentis a centro orbis signorum 6 partes 575 et 30 minuta. Et linea *AE* que est semidiameter epicicli 39 partes et 30 minuta. Et linea *AD* que est distantia longitudinis longioris 66 partes et 30 minuta. Et linea *DB* que est distantia longitudinis propioris 53 partes et 30 minuta. Et linea *ED* que est maior longitudo martis que esse possit

542 esset *om.* α*H*
549–64 predicte...miliaria *om. f*
549–50 AC 4080748 miliaria: et 110 sexagesime sexcentesime unius miliaris *add.* β*V*
559–60 Ambitus...miliaris *om. BbGKmNWw*
568 subscripta: subiecta θ*HNZ*ΠΥ hac *BOoVY* una *M*Γ
570 conueniens: consequens θφ*H*

bodies should be located in that space. However, it would be even more amazing that that space should exactly fit the dimensions of the circles and bodies of Mercury and Venus, and yet those planets should be located somewhere else.[87] Moreover, if we suppose a different order of the planets, it will not be possible for [each] planet to be regent in the first hour of the weekday named after it, or alternatively we will have to assume a different order for the days of the week.[88] Thus it is proper to admit the order of the celestial spheres which PTOLEMY assumed and which we, following him, assumed in what was said before.

In miles the magnitudes named will be as follows: line *AC* 4,080,748 miles; line *CD* 170,031$\frac{110}{660}$ miles; line *AD* 4,250,779$\frac{110}{660}$ miles; and line *DB* 3,910,716$\frac{550}{660}$ miles.[89] The radius of the sun's body [will be] 17,850 miles, and the circumference of its body 112,200 miles.[90] The distance of the concave surface of its sphere from the center of the earth [will be] 3,892,866$\frac{550}{660}$ miles, and the distance of the convex surface of its sphere from that center 4,268,629$\frac{110}{660}$ miles. Therefore the thickness of its sphere will be 375,762$\frac{220}{660}$ miles.[91] The circumference of the convex surface of its sphere [will be] 26,831,383$\frac{220}{660}$ miles.[92] The circumference of its eccentric will be 25,650,416 miles. If you divide the latter by 365$\frac{1}{4}$ days, which is the period of one revolution [of the sun] on its eccentric, the result will be its daily motion on its eccentric. Thus its daily motion will be 70,227 miles.[93] The body of the sun contains the body of the earth 166 times plus $\frac{3}{8}$.[94] For the ratio between the diameter of the sun's body and the diameter of the earth is as 11 to 2, and the cubes of these numbers are 1,331 and 8; these are in the ratio of the above-named bodies, and the ratio is as we have said. All of this is clearly displayed in the following table [Table 4, p. 360].

Now that these magnitudes have been calculated for Venus and the sun, it is appropriate to calculate them for Mars. To this end I repeat the previous figure [Fig. 20, p. 324]. I say, then, that in the units of which *AC*, which is the radius of the deferent of Mars, contains 60, each of the two lines *KC* and *CD*, of which the first is the distance of the center of the equant from the center of the deferent and the second [the distance] of the center of the deferent from the center of the ecliptic, will be 6 units and 30 minutes;[95] line *AE*, which is the radius of the epicycle, [will be] 39 units and 30 minutes;[96] line *AD*, which is the amount of the greatest distance, [will be] 66 units and 30 minutes; line *DB*, which is the amount of the least distance, [will be] 53 units and 30 minutes; line *ED*, which is the greatest possible distance of Mars, [will be] 106 units;

---

546 due prime rationes sunt necessarie, ultima uero persuasiua *mg. p*

568 subscripta: Nota si diuiseris ambitum ecentrici solis per 365 dies qui sunt tempus reuolutionis eius in suo ecentrico et quartam unius diei, exibit tibi motus eius in una die de partibus sui ecentrici *add. f*

106 partes. Et linea *DF* que est minor longitudo martis que esse possit
580 14 partes.

At uero secundum partes illas de quibus semidiameter terre est pars una, inueniemus eas per lineam *AD* in sole. Nam linea *DF* in marte addit super lineam *AD* in sole semidiametrum corporis solis et semidiametrum corporis martis. Semidiameter autem corporis solis est 5 partes et 30 minuta
585 secundum partes illas de quibus semidiameter terre est pars una. Semidiameter autem corporis martis est pars una et 10 minuta secundum easdem partes. Ergo linea *DF* in marte est 1316 partes 25 minuta et 50 secunda secundum partes illas de quibus semidiameter terre est pars una. Per hanc autem inuenientur omnes alie sicud superius dictum est. Eruntque ita
590 linea *AC* 5641 partes 50 minuta et 43 secunda. Et linea *DC* 611 partes et 12 minuta. Et linea *AE* 3714 partes 12 minuta et 53 secunda. Et linea *AD* 6253 partes 2 minuta et 43 secunda. Et linea *DB* 5030 partes 38 minuta et 43 secunda. Et linea *ED* 9967 partes 15 minuta et 36 secunda.

De miliaribus autem erunt predicte magnitudines ita. Linea quidem *AC*
595 18310352 miliaria et 265 sexagesime sexcentesime unius miliaris. Et linea *CD* 1983621 miliaria et 540 sexagesime sexcentesime unius miliaris. Et linea *AE* 12054315 miliaria et 35 sexagesime sexcentesime unius miliaris. Et linea *AD* 20293974 miliaria et 145 sexagesime sexcentesime unius miliaris. Et linea *DB* 16326730 miliaria et 385 sexagesime sexcentesime unius
600 miliaris. Et linea *ED* 32348289 miliaria et 180 sexagesime sexcentesime unius miliaris. Et linea *DF* 4272415 miliaria et 350 sexagesime sexcentesime unius miliaris. Semidiameter autem corporis martis est 3786 miliaria et 240 sexagesime sexcentesime unius miliaris. Ambitus uero corporis martis 23800 miliaria. Longitudo autem concaue superficiei spere ipsius a centro
605 terre 4268629 miliaria et 110 sexagesime sexcentesime unius miliaris. Et longitudo conuexe superficiei spere ipsius a centro terre 32352075 miliaria et 420 sexagesime sexcentesime unius miliaris. Spissitudo itaque spere sue erit 28083446 miliaria et 310 sexagesime sexcentesime unius miliaris. Ambitus autem conuexe superficiei spere ipsius erit 203355904 miliaria. Ambi-
610 tus quoque sui deferentis erit 115093643 miliaria et 440 sexagesime sexcentesime unius miliaris. Et si diuiseris ipsum per 687 dies qui sunt tempus unius reuolutionis centri epicicli ipsius, exibit tibi motus centri epicicli in una die. Erit itaque dieta ipsius 167530 miliaria et 47 minuta. Ambitus autem epicicli eius erit 75769980 miliaria et 220 sexagesime sexcentesime
615 unius miliaris. Et si diuiseris ipsum per 780 dies qui sunt tempus unius

582 eas: eadem β*V*
589 inueniuntur β*V*/ita: itaque θφ*DX*
594–602 Linea quidem AC...miliaris *om. BmPOV* (*habent hic tabulam BmO*)
600 32348289: 32344289 *ABbcDGHisWwXZ*
606 superficiei...terre *om.* δ/ipsius: martis *add.* β

and line *DF*, which is the least possible distance of Mars, [will be] 14 units.[97]

In those units of which the radius of the earth is one, we will find those [magnitudes] from line *AD* for the sun. For line *DF* for Mars exceeds line *AD* for the sun by [the sum of] the radius of the sun's body and the radius of Mars's body; now the radius of the sun's body is 5 units and 30 minutes, expressed in those units of which the radius of the earth is one, and the radius of Mars's body is 1 unit and 10 minutes,[98] expressed in the same units. Thus line *DF* for Mars is 1,316 units 25 minutes 50 seconds in the units of which the radius of the earth is one.[99] From this [magnitude] all the others can be found, as was stated above. Thus line *AC* will be 5,641 units 50 minutes 43 seconds; line *DC* [will be] 611 units 12 minutes; line *AE* [will be] 3,714 units 12 minutes 53 seconds; line *AD* [will be] 6,253 units 2 minutes 43 seconds; line *DB* [will be] 5,030 units 38 minutes 43 seconds; and line *ED* [will be] 9,967 units 15 minutes 36 seconds.[100]

In miles the above-named magnitudes will be as follows: line *AC* $18,310,352\frac{265}{660}$ miles; line *CD* $1,983,621\frac{540}{660}$ miles; line *AE* $12,054,315\frac{35}{660}$ miles; line *AD* $20,293,974\frac{145}{660}$ miles; line *DB* $16,326,730\frac{385}{660}$ miles; line *ED* $32,348,289\frac{180}{660}$ miles; and line *DF* $4,272,415\frac{350}{660}$ miles.[101] The radius of Mars's body is $3,786\frac{240}{660}$ miles; the circumference of Mars's body 23,800 miles;[102] the distance of the concave surface of its sphere from the center of the earth $4,268,629\frac{110}{660}$ miles; and the distance of the convex surface of its sphere from the center of the earth $32,352,075\frac{420}{660}$ miles. Thus the thickness of its sphere will be 28,083,446 $\frac{310}{660}$ miles.[103] The circumference of the convex surface of its sphere will be 203,355,904 miles.[104] The circumference of its deferent will be $115,093,643\frac{440}{660}$ miles; if you divide this by 687 days,[105] which is the period of one revolution of the center of its epicycle, the result will be the motion of the center of the epicycle in one day. Thus its day's journey will be 167,530 miles and 47 minutes [of a mile].[106] The circumference of its epicycle will be $75,769,980\frac{220}{660}$ miles; if you divide this by 780 days,[107] which is the period of one revolution of it

reuolutionis eius in suo epiciclo, exibit tibi dieta ipsius in suo epiciclo et est 97141 miliaria. Corpus autem martis continet corpus terre semel et insuper 127 sedecimas duocentesimas. Nam proportio dictorum corporum est sicud proportio 7 ad 6 quorum cubi sunt 343 et 216. Ipsi uero se habent
620 in proportione ipsorum et in proportione predicta. Et hec omnia patent in subiecta tabula.

In ioue autem sunt predicte magnitudines ita. Secundum quod linea *AC* que est semidiameter deferentis est 60 partes erit utraque duarum linearum que sunt *KC* et *CD* quarum prima est distantia centri equantis a centro
625 deferentis et secunda centri deferentis a centro orbis signorum 2 partes et 45 minuta. Et linea *AE* que est semidiameter epicicli 11 partes et 30 minuta. Et linea *AD* que est distantia longitudinis longioris 62 partes et 45 minuta. Et linea *DB* que est distantia longitudinis propioris 57 partes et 15 minuta. Et linea *ED* que est maior longitudo iouis que esse possit 74
630 partes et 15 minuta. Et linea *DF* que est minor longitudo iouis que esse possit 45 partes et 45 minuta.

At uero secundum partes illas de quibus semidiameter terre est pars una, inueniemus eas per lineam *ED* in marte. Nam linea *DF* in ioue addit super lineam *ED* in marte semidiametrum corporis iouis et semidiametrum cor-
635 poris martis. Semidiameter autem corporis martis est pars una et 10 minuta secundum partes illas de quibus semidiameter terre est pars una ut dictum est prius. Semidiameter autem corporis iouis est 4 partes et 34 minuta secundum easdem partes. Ergo linea *DF* in ioue erit 9972 partes 59 minuta et 36 secunda secundum partes illas de quibus semidiameter terre est pars
640 una. Et linea *AC* 13079 partes 20 minuta et 8 secunda. Et linea *CD* 599 partes 28 minuta et 10 secunda. Et linea *AE* 2506 partes 52 minuta et 22 secunda. Et linea *AD* 13678 partes 48 minuta et 18 secunda. Et linea *DB* 12479 partes 51 minuta et 58 secunda. Et linea *ED* 16185 partes 40 minuta et 40 secunda.

645 De miliaribus autem predicte magnitudines erunt ita. Linea quidem *AC* 42448389 miliaria et 20 sexagesime sexcentesime unius miliaris. Et linea *CD* 1945550 miliaria et 550 sexagesime sexcentesime unius miliaris. Et linea *AE* 8135941 miliaria et 430 sexagesime sexcentesime unius miliaris. Et linea *AD* 44393939 miliaria et 570 sexagesime sexcentesime unius milia-
650 ris. Et linea *DB* 40502838 miliaria et 130 sexagesime sexcentesime unius miliaris. Et linea *ED* 52529881 miliaria et 340 sexagesime sexcentesime unius miliaris. Et linea *DF* 32366896 miliaria et 360 sexagesime sexcentesime unius miliaris. Semidiameter autem corporis iouis erit 14820 miliaria

619 proportio *om.* δθ
622 In ioue autem *om.* β*V*
641 10 secunda: 18 secunda *BdFhMoPrV*Γ ΣΨ *mg. v*
645–53 Linea quidem AC...miliaris *om. BoPV* (*hic tabulam habet B*)

[Mars] on its epicycle, the result will be its day's journey on the epicycle, and that is 97,141 miles.[108] The body of Mars contains the body of the earth once plus $\frac{127}{216}$. For the ratio between the above-named bodies is as 7 to 6, the cubes of which are 343 and 216;[109] these are in the ratio of [the bodies], and the ratio is as we have stated. All this is set out clearly in the following table [Table 5, p. 361].

For Jupiter the above-named magnitudes are as follows: in units where *AC*, which is the radius of the deferent, is 60, each of the two lines *KC* and *CD*, of which the first is the distance of the center of the equant from the center of the deferent and the second [the distance] of the center of the deferent from the center of the ecliptic, will be 2 units and 45 minutes;[110] line *AE*, which is the radius of the epicycle, [will be] 11 units and 30 minutes;[111] line *AD*, which is the amount of the greatest distance, [will be] 62 units and 45 minutes; line *DB*, which is the amount of the least distance, [will be] 57 units and 15 minutes; line *ED*, which is the greatest possible distance of Jupiter, [will be] 74 units and 15 minutes; and line *DF*, which is the least possible distance of Jupiter, [will be] 45 units and 45 minutes.[112]

For the units of which the radius of the earth is one, we shall find these [magnitudes] from line *ED* for Mars. For line *DF* for Jupiter exceeds line *ED* for Mars by [the sum of] the radius of Jupiter's body and the radius of Mars's body. Now, in the units of which the radius of the earth is one, the radius of Mars's body is 1 unit and 10 minutes, as was stated above. The radius of Jupiter's body is 4 units and 34 minutes[113] in the same units. Therefore line *DF* for Jupiter will be 9,972 units 59 minutes 36 seconds,[114] expressed in the units of which the radius of the earth is one; line *AC* [will be] 13,079 units 20 minutes 8 seconds; line *CD* [will be] 599 units 28 minutes 10 seconds; line *AE* [will be] 2,506 units 52 minutes 22 seconds; line *AD* [will be] 13,678 units 48 minutes 18 seconds; line *DB* [will be] 12,479 units 51 minutes 58 seconds; and line *ED* [will be] 16,185 units 40 minutes 40 seconds.[115]

In miles the above-named magnitudes will be as follows: line *AC* 42,448,389$\frac{20}{660}$ miles; line *CD* 1,945,550$\frac{550}{660}$ miles; line *AE* 8,135,941$\frac{430}{660}$ miles; line *AD* 44,393,939$\frac{570}{660}$ miles; line *DB* 40,502,838$\frac{130}{660}$ miles; line *ED* 52,529,881$\frac{340}{660}$ miles; and line *DF* 32,366,896$\frac{360}{660}$ miles.[116] The radius of Jupiter's body will

---

621 tabula: Adinplentes processum ordinis quem polliciti sumus, consequens (conueniens *B*) est ut ascendamus (accedamus *B*) ad magnitudines iouis. Dicimus igitur quod *add.* β

644 40 secunda: et linea DF 9972 partes 59 minuta et 36 secunda *add.* β

et 600 sexagesime sexcentesime unius miliaris. Ambitus uero corporis
655 iouis erit 93160 miliaria. Longitudo autem concaue superficiei spere ipsius a centro terre erit 32352075 miliaria et 420 sexagesime sexcentesime unius miliaris, longitudo autem superficiei conuexe 52544702 miliaria et 280 sexagesime sexcentesime unius miliaris. Spissitudo itaque spere ipsius est 20192626 miliaria et 520 sexagesime sexcentesime unius miliaris. Ambitus
660 autem conuexe superficiei spere ipsius est 330280986 miliaria et 440 sexagesime sexcentesime unius miliaris. Ambitus quoque sui deferentis est 266818445 miliaria et 220 sexagesime sexcentesime unius miliaris. Et si diuiseris ipsum per 4334 dies qui sunt tempus unius reuolutionis centri epicicli ipsius in suo deferente, exibit tibi motus centri epicicli eius in una
665 die. Erit itaque dieta ipsius 61564 miliaria. Ambitus autem epicicli eius est 51140204 miliaria et 440 sexagesime sexcentesime unius miliaris. Et si diuiseris ipsum per 399 dies qui sunt tempus unius reuolutionis eius in epiciclo suo, exibit tibi dieta ipsius in epiciclo et est 128171 miliaria. Corpus autem iouis continet corpus terre 95 uicibus et insuper 6353 septem
670 uiginti millesimas ipsius. Nam proportio diametrorum dictorum corporum est sicud 137 ad 30 quorum cubi sunt 2571353 et 27000. Ipsi uero se habent in proportione dictorum corporum et in proportione predicta. Et hec omnia patent in subiecta tabula.

In saturno autem sunt predicte magnitudines ita. Secundum quod linea
675 *AC* que est semidiameter deferentis est 60 partes, erit utraque duarum linearum *KC* et *CD* tres partes et 25 minuta. Et linea *AE* 6 partes et 30 minuta. Et linea *AD* 63 partes et 25 minuta. Et linea *DB* 56 partes et 35 minuta. Et linea *ED* 69 partes et 55 minuta. Et linea *DF* 50 partes et 5 minuta.

680 At uero secundum partes illas de quibus semidiameter terre est pars una inueniemus eas per lineam *ED* in ioue. Nam linea *DF* in saturno addit super lineam *ED* in ioue semidiametrum corporis iouis et semidiametrum corporis saturni. Semidiameter autem corporis saturni est 4 partes et 30 minuta secundum partes illas de quibus semidiameter terre est pars una,
685 et semidiameter corporis iouis 4 partes et 34 minuta secundum easdem partes. Erit igitur linea *DF* in saturno 16194 partes 44 minuta et 40 secunda secundum partes illas de quibus semidiameter terre est pars una. Et linea *AC* 19401 partes 21 minuta et 28 secunda; et linea *CD* 1104 partes 47 minuta et 58 secunda. Et linea *AE* 2101 partes 48 minuta et 50 secunda. Et

655 ipsius: iouis *add.* β*V*
657 conuexe: spere iouis *add.* β*V*
668 suo...epiciclo *om.* β*V*
672–73 Et...tabula *om.* βφ
674 In saturno autem *om.* β*V*
685–86 et semidiameter...easdem partes *om.* αφ*H*
686 40 secunda: 4 secunda *ABdImoPUuVYy* ΠΨ 42 secunda Γ

be 14,820$\frac{600}{660}$ miles; the circumference of Jupiter's body will be 93,160 miles.[117] The distance of the concave surface of its sphere from the center of the earth will be 32,352,075$\frac{420}{660}$ miles; the distance of the convex surface [will be] 52,544,702$\frac{280}{660}$ miles. Thus the thickness of its sphere is 20,192,626$\frac{520}{660}$ miles.[118] The circumference of the convex surface of its sphere is 330,280,986$\frac{440}{660}$ miles.[119] The circumference of its deferent is 266,818,445$\frac{220}{660}$ miles. If you divide the latter by 4,334 days,[120] which is the period of one revolution of the center of its [Jupiter's] epicycle on its deferent, the result will be the motion of the center of its epicycle in one day. Thus its day's journey will be 61,564 miles.[121] The circumference of its epicycle is 51,140,204$\frac{440}{660}$ miles; if you divide this by 399 days,[122] which is the period of one revolution of it [Jupiter] on its epicycle, the result will be its day's journey on the epicycle, and that is 128,171 miles.[123] The body of Jupiter contains the body of the earth 95 times plus $\frac{6353}{27000}$. For the ratio of the diameters of the above-named bodies is as 137 to 30, the cubes of which are 2,571,353 and 27,000.[124] These are in the ratio of the above-named bodies, and the ratio is as we have stated. All this is clearly laid out in the following table [Table 6, p. 362].

For Saturn the above-named magnitudes are as follows: in the units in which line *AC*, which is the radius of the deferent, is 60, each of the two lines *KC* and *CD* will be 3 units and 25 minutes;[125] line *AE* [will be] 6 units and 30 minutes;[126] line *AD* [will be] 63 units and 25 minutes; line *DB* [will be] 56 units and 35 minutes; line *ED* [will be] 69 units and 55 minutes; and line *DF* [will be] 50 units and 5 minutes.[127]

For those units of which the radius of the earth is one, we shall find these [magnitudes] from line *ED* for Jupiter. For line *DF* for Saturn exceeds line *ED* for Jupiter by [the sum of] the radius of Jupiter's body and the radius of Saturn's body. Now the radius of Saturn's body is 4 units and 30 minutes[128] in those units of which the radius of the earth is one, and the radius of Jupiter's body is 4 units and 34 minutes in the same units. Therefore line *DF* for Saturn will be 16,194 units 44 minutes 40 seconds in those units of which the radius of the earth is one;[129] line *AC* [will be] 19,401 units 21 minutes 28 seconds; line *CD* [will be] 1,104 units 47 minutes 58 seconds; line *AE* [will be] 2,101 units 48 minutes 50 seconds; line *AD* [will be] 20,506 units 9 minutes 26 seconds;

---

673 tabula: Postulat nunc demum ordo (denum ordinem *B*) nostri tractatus ut postquam singule magnitudines singulorum planetarum a saturno premisse sunt ut (ire *M*) et nunc altius ascendentes (ascendere *BM*) ipsius saturni magnitudines enodemus (enarrando *M*). Dicimus igitur quoniam *add.* β*JV*

690 linea *AD* 20506 partes 9 minuta et 26 secunda. Et linea *DB* 18296 partes 33 minuta et 30 secunda. Et linea *ED* 22607 partes 58 minuta et 16 secunda.

De miliaribus autem erunt dicte magnitudines ita. Linea quidem *AC* 62966224 miliaria et 520 sexagesime sexcentesime unius miliaris. Et linea *CD* 3585576 miliaria et 250 sexagesime sexcentesime unius miliaris. Et 695 linea *AE* 6821341 miliaria et 290 sexagesime sexcentesime unius miliaris. Et linea *AD* 66551801 miliaria et 110 sexagesime sexcentesime unius miliaris. Et linea *DB* 59380648 miliaria et 270 sexagesime sexcentesime unius miliaris. Et linea *ED* 73373142 miliaria et 400 sexagesime sexcentesime unius miliaris. Et linea *DF* 52559306 miliaria et 640 sexagesime sex-700 centesime unius miliaris. Semidiameter autem corporis saturni erit 14604 miliaria et 360 sexagesime sexcentesime unius miliaris. Ambitus uero corporis saturni erit 91800 miliaria. Longitudo autem concaue superficiei spere ipsius a centro terre est 52544702 miliaria et 280 sexagesime sexcentesime unius miliaris. Longitudo uero conuexe est 73387747 miliaria et 705 100 sexagesime sexcentesime unius miliaris. Spissitudo itaque spere eius erit 20843044 miliaria et 480 sexagesime sexcentesime unius miliaris. Ambitus autem conuexe superficiei spere ipsius et ipse est ambitus concaue superficiei spere stellarum fixarum erit 461294410 miliaria et 440 sexagesime sexcentesime unius miliaris. Ambitus quoque sui deferentis erit 395787698 miliaria 710 et 440 sexagesime sexcentesime unius miliaris. Et si diuiseris ipsum per 10800 dies qui sunt tempus unius reuolutionis centri epicicli in suo deferente, exibit tibi dieta centri epicicli et est 36647 miliaria. Ambitus autem sui epicicli est 42877003 miliaria et 220 sexagesime sexcentesime unius miliaris. Et si diuiseris ipsum per 378 dies qui sunt tempus unius reuolu-715 tionis eius in suo epiciclo, exibit tibi dieta eius in suo epiciclo et est 113431 miliaria. Corpus autem saturni continet corpus terre 91 uicibus et insuper octauam eius. Nam proportio diametrorum predictorum corporum est sicud 9 ad 2 quorum cubi sunt 729 et 8. Ipsi uero se habent in proportione dictorum corporum et in proportione predicta. Et hec omnia patent in 720 subiecta tabula.

Et quia conuexa superficies spere saturni est concaua superficies spere stellarum fixarum ut prius dictum est, prouenit autem spere superficies ex multiplicatione sue diametri in circumferentiam maioris circuli descripti in superficie ipsius qui est idem quod ambitus eius: nam ex multiplicatione 725 medietatis diametri in medietatem circumferentie prouenit superficies circuli, superficies uero spere quadrupla est ad superficiem maioris circuli in eius superficie descripti: si duplum longitudinis conuexe superficiei spere

692–700 Linea quidem AC...miliaris *om.* *BoPV* (*habet hic tabulam B*)
700–701 Semidiameter...miliaris *om.* δ
711 in suo: ecentrico *add.* β*V*
720 subiecta *om.* β*V*
721 spere²: huius *V om.* β

line *DB* [will be] 18,296 units 33 minutes 30 seconds; and line *ED* [will be] 22,607 units 58 minutes 16 seconds.[130]

In miles the named magnitudes will be as follows: line *AC* $62{,}966{,}224\frac{520}{660}$ miles, line *CD* $3{,}585{,}576\frac{250}{660}$ miles; line *AE* $6{,}821{,}341\frac{290}{660}$ miles; line *AD* $66{,}551{,}801\frac{110}{660}$ miles; line *DB* $59{,}380{,}648\frac{270}{660}$ miles; line *ED* $73{,}373{,}142\frac{400}{660}$ miles; and line *DF* $52{,}559{,}306\frac{640}{660}$ miles.[131] The radius of Saturn's body will be $14{,}604\frac{360}{660}$ miles; the circumference of Saturn's body will be 91,800 miles.[132] The distance of the concave surface of its sphere from the center of the earth is $52{,}544{,}702\frac{280}{660}$ miles, while the distance of its convex [surface] is $73{,}387{,}747\frac{100}{660}$ miles. Thus, the thickness of its sphere will be $20{,}843{,}044\frac{480}{660}$ miles.[133] The circumference of the convex surface of its sphere, which is the circumference of the concave surface of the sphere of the fixed stars, will be $461{,}294{,}410\frac{440}{660}$ miles.[134] Furthermore the circumference of its deferent will be $395{,}787{,}698\frac{440}{660}$ miles. If you divide the latter by 10,800 days,[135] which is the period of one revolution of the center of the epicycle on its deferent, the result will be the day's journey of the center of the epicycle, and that is 36,647 miles.[136] The circumference of its epicycle is $42{,}877{,}003\frac{220}{660}$ miles. If you divide the latter by 378 days,[137] which is the period of one revolution of it [Saturn] on its epicycle, the result will be its day's journey on the epicycle, and that is 113,431 miles.[138] The body of Saturn contains the body of the earth 91 times plus $\frac{1}{8}$. For the ratio between the diameters of the above-named bodies is as 9 to 2, the cubes of which are 729 and 8.[139] These are in the ratio of the above-named bodies, and the ratio is as we have stated. All of this is clearly laid out in the following table [Table 7, p. 363].

Now, since the convex surface of Saturn's sphere is [i.e., coincides with] the concave surface of the sphere of the fixed stars, as was stated before,[140] and the [area of the] surface of a sphere is found by multiplying the diameter by the circumference of a great circle described on its surface, which [great circle] is the same as its circumference—for the area of a circle is found by multiplying half the diameter by half the circumference, and the surface area of a sphere is four times the area of a great circle described on its surface[141]—[then] if you

saturni a centro terre multiplicaueris in ambitum ipsius superficiei, proueniet tibi superficies concaua spere stellarum fixarum et ipsa est 730 67706715144825054 miliaria quadrata et 38 none nonagesime unius miliaris. Per hunc quoque modum inuenies superficies quarumlibet sperarum si tu cognoueris diametrum et ambitum ipsarum. Ideoque possibile erit tibi si uolueris inuenire superficies concauas et conuexas sperarum omnium planetarum, superficies quoque circulorum ipsorum.

735 Dicte sunt itaque uere magnitudines distantiarum orbium sperarum et corporum in planetis omnibus secundum quod conuenit probationibus PTHOLOMEI. Quidam autem eorum qui has magnitudines uoluerunt inuenire obmiserunt corpora omnium planetarum quasi due superficies sperice sibi equedistantes contingentes corpus uniuscuiusque planete in sua longi- 740 tudine longiori et propiori essent una superficies. Ideoque accidit eis error non modicus in numeratione tali.

Quod si uera est positio PTHOLOMEI in motu stellarum fixarum quod uidelicet moueantur ad orientem secundum successionem signorum in 100 annis uno gradu super polos orbis signorum, non erit dieta earum que 745 sunt in circulo orbis signorum nisi 35 miliaria et duodecima pars unius miliaris fere qui motus ualde paruus est et pene insensibilis.

Hiis itaque determinatis conueniens est demonstrare qualiter faciamus instrumentum nostrum per quod equemus 4 planetas predictos qui sunt uenus mars iupiter et saturnus. Demonstrabimus autem hoc in omnibus 750 eis simul quoniam idem modus est operandi in eis, excepto quod distantie centrorum et magnitudines epiciclorum erunt in eis diuerse, et quod in saturno et ioue oportebit nos uti epiciclo oportune circulationis quemadmodum fecimus in luna. Paruitas enim epicicli uere circulationis in hiis duobus non sustineret diuisionem in 360 partes nisi magnitudo instrumen- 755 ti multum excederet magnitudinem in ceteris sufficientem. Sumam igitur alias duas tabulas rotundas de quibus precessit mentio in principio istius operis quas oportet esse ualde planas et ualde pollitas. Et in una earum ponam uenerem et martem et in alia iouem et saturnum. Scribam enim in medio utriusque earum et ex utraque parte punctum *D* supra quem descri- 760 bam unum circulum maiorem quem potero describere in ea, ut sit iste circulus orbis signorum. Et faciam 4 circulos concentricos ut possim inter eos ponere nomina signorum et diuisiones graduum singulares et collectas per 5 et 5 et numerum ipsorum graduum quemadmodum feci in planetis tribus

737–38 inuenire: numerare δ uel numerare *add.* *F*Σ

739 sibi: ibi *ABFhiJKkmPqSUuVv*$_1$ *Y*ΠΣΨ

747 conueniens: consequens θ*fH*

750 est: erit δ

multiply twice the distance of the convex surface of Saturn's sphere from the center of the earth by the circumference of that surface, you will get [the area of] the concave surface of the fixed stars: this is 67,706,715,144,825,054$\frac{38}{99}$ square miles.[142] By the same method you can find the surface area of any of the spheres, if you know their diameters and circumferences. Therefore it will be possible for you to find [the areas of] the concave and convex surfaces of all the planets if you want, as well as the areas of their [various] circles.

Thus we have stated the true magnitudes of the distances, circles, spheres, and bodies for all the planets, according to the determinations of PTOLEMY. However, some of those who tried to find these magnitudes left out the bodies of all the planets, as if the two concentric spherical surfaces which touch the body of each planet at its greatest and least distances [i.e., on its outer and inner sides][143] were a single surface. The result is that a not inconsiderable error accrues during such a process of numerical calculation [as is involved here].

Now, if the view of PTOLEMY about the motion of the fixed stars is correct, namely, that they move about the poles of the ecliptic toward the east, [i.e.,] following the order of the signs [of the zodiac], one degree in 100 years, the day's journey of those [fixed stars] which lie in the ecliptic[144] will be only about 35$\frac{1}{12}$ miles,[145] which is a very small and almost undetectable motion.

Now that we have determined these [quantities], it is appropriate to describe how to make our instrument for equating the four above-mentioned planets, namely, Venus, Mars, Jupiter, and Saturn. We shall describe it for all of them at once, since the mode of operation is the same for all, except that the distances between the centers and the sizes of the epicycles will differ from one to the other and that for Saturn and Jupiter we will have to use an epicycle with convenient circular representation, as we did for the moon.[146] For the small size of the epicycle in its true representation for these two [planets] would not allow us to divide it into 360 parts, unless the size of the instrument greatly exceeded that which suffices for the other [planets]. Therefore I take two more of the circular boards which were described at the beginning of this work:[147] these must be very flat and very smooth. On one of them I put Venus and Mars and on the other Jupiter and Saturn [see Fig. 22, p. 346].[148] I mark in the middle of each, on both faces, point *D*, about which I describe the largest circle that I can fit onto it [the board]; this [circle] is to be the circle of the ecliptic. I draw four concentric circles[149] such that I can put between them the names of the signs, the divisions into single degrees and groups of five degrees, and the numbers of the degrees, as I did for the three planets previously described.[150]

---

746 insensibilis: licet secundum THEBIT (Tebiz *k*) et alios (quamplures *add. g*) sit ab hoc (adhuc *Kk*) diuersus *add. dgKkRSt*

747 *tit.* Capitulum de fabricatione instrumentorum equationum pro uenere marte ioue et saturno Γ

-·-·-·-·- **Tabula motus epicicli**

═══ **Tabula motus centri epicicli**

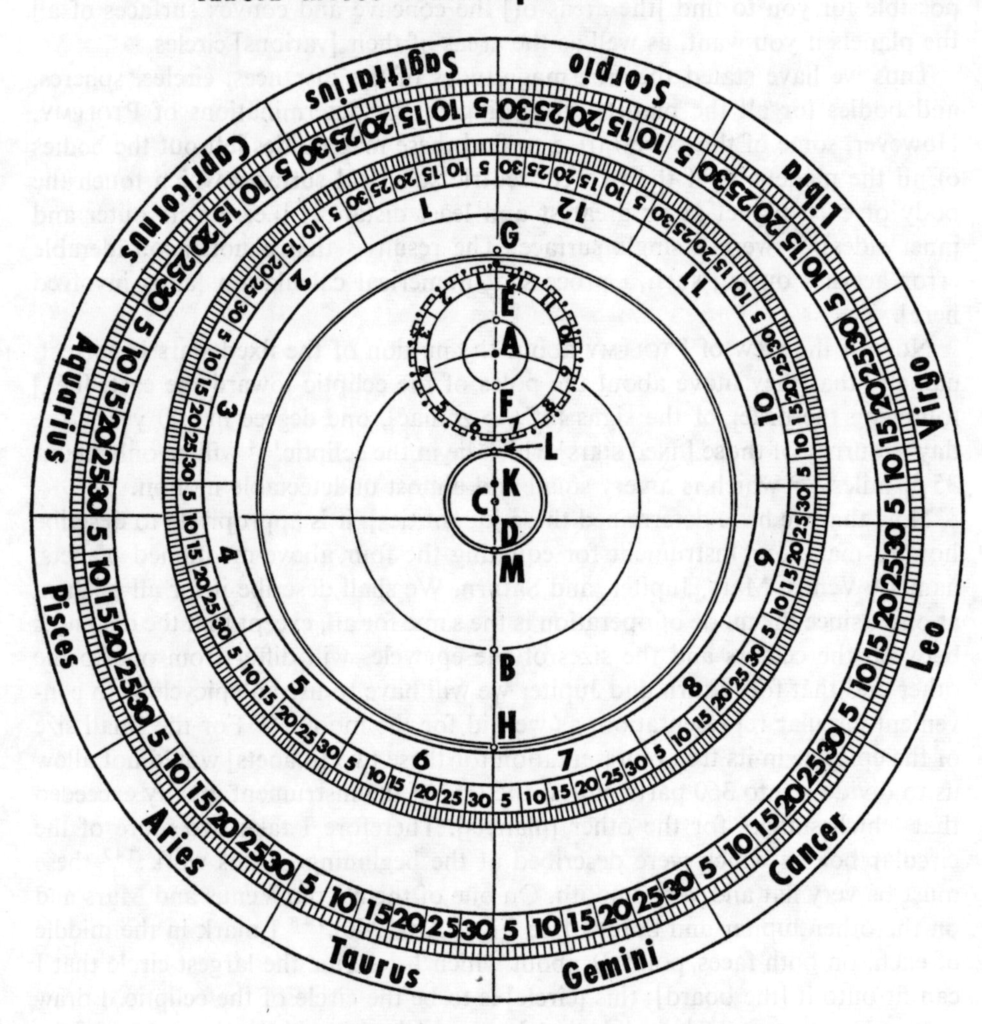

INSTRVMENTVM EQVATIONIS
SATVRNI IOVIS MARTIS ET VENERIS

Fig. 22

Legend for Fig. 22

Instrument for Venus or one of the three outer planets (after Campanus). The particular instrument represented is Saturn's, but, except for the dimensions and the position of the apogee, the plan is the same for all four. The outer graduated system of circles (with center *D*, the earth) represents the ecliptic. The graduated system of circles immediately within that (with center *K*) represents the equant. The zero point of the equant is aligned with the planet's apogee (in this case Saturn's, in Sagittarius 0;5°). *AB* represents the deferent, with center *C*. *EF* represents the epicycle, with center *A* (here, since Saturn's epicycle is so small in comparison with the deferent, the graduation of the epicycle is marked on the "epicycle in its convenient representation," drawn on the same center outside the true epicycle). The double and dotted and dashed lines represent the edges of the moving parts. The smallest of these (indicated by the dotted and dashed line) is the "disk of the motion of the epicycle"; it revolves about center *A* within the "disk of the motion of the center of the epicycle" (indicated by the double lines), which in turn revolves about center *C*. Threads are attached by one end to points *K*, *D*, and *A* (the last is necessary only if there is an "epicycle in its convenient representation"), and a mark is made at the apogee of the epicycle (or the "epicycle in its convenient representation"). The true position of the planet is found as follows. Stretch the thread at *K* through the point on the equant marking the mean centrum, and revolve the "disk of the motion of the center of the epicycle" until point *A* coincides with the thread. Keeping that disk fixed, revolve the "disk of the motion of the epicycle" until the mark at the epicyclic apogee also coincides with the thread. Stretch the thread at *A* through the point on the epicycle marking the mean anomaly, and note where it intersects circle *EF*. Stretch the thread at *D* through that intersection on circle *EF* (or, if the epicycle is big enough so that *EF* itself can be graduated, as is true for Venus and Mars, omit the step with the thread from *A* and simply stretch the thread at *D* through the point on the epicycle marking the mean anomaly). Where the thread at *D* cuts the ecliptic is the true position of the planet. Compare Figure 7 (p. 48) and Plate 3.

predictis. Et quadrabo quemlibet eorum duabus diametris ortogonaliter
765 se secantibus supra centrum *D*. Et transeat una istarum diametrorum in uenere quidem per 17 gradus et 50 minuta geminorum; in marte autem per unum gradum et 50 minuta leonis; in ioue uero per 14 gradus et 30 minuta uirginis; in saturno uero per 5 minuta sagittarii. Eruntque auges istorum 4 planetarum in diametris predictis sub predictis partibus orbis
770 signorum. Oppositiones autem augium erunt in eadem diametro sub partibus orbis signorum condiametralibus partibus predictis. Relinquam autem in istis diametris que transeunt per auges dictorum planetarum tantum spatium infra interiorem 4 circulorum qui sunt propter orbem signorum quantum uolam distare circulum unum cum suis diuisionibus descriptum
775 supra centrum equantis ab eo. Faciam enim in istis 4 planetis equantem maiorem deferente ut in eius circumferentia possit numerari medium centrum cuiuslibet eorum, ita quod totus equans cum suis diuisionibus sit extra deferentem quemadmodum feci in mercurio. Residuum uero spatium quod erit in predicta diametro usque ad *D* diuidam in uenere et in marte in
780 60 partes. Et sumam ex eis in uenere quidem a puncto *D* unam partem et quartam partis. Et ubi terminabitur ponam notam *K* eritque *K* centrum equantis. Et diuidam lineam *KD* per medium super punctum *C* eritque *C* centrum deferentis. Sumamque ex ea a puncto *C* uersus circumferentiam 30 partes de predictis partibus. Et ubi terminabuntur ponam notam *A*
785 eritque punctus *A* aux deferentis et linea *AC* erit semidiameter ipsius. Ipsum ergo describam et sit circulus *AB*. Item sumam ex predicta diametro a puncto *A* uersus circumferentiam 21 partes et tertiam et quartam unius partis de predictis partibus. Et ubi terminabuntur ponam notam *E* eritque linea *AE* semidiameter epicicli. Ipsum ergo describam et sit *EF*. In marte
790 autem sumam ex predictis 60 partibus a puncto *D* 6 partes et medietatem partis que sit linea *DK* quam diuidam per medium in *C*. Eritque *K* centrum equantis martis et *C* centrum deferentis. Et ponam lineam *CA* 30 partes et lineam *AE* 19 partes et tres quartas unius partis de partibus predictis. Eritque linea *CA* semidiameter deferentis et linea *AE* semidiameter
795 epicicli, quos etiam describam. Sitque hic circulus *AB*; ille uero circulus *EF*. Remanebitque ex predictis 60 partibus in uenere quidem 7 partes fere. In marte autem 4 partes minus quarta parte unius partis. In spatio igitur illo describam 4 circulos supra centrum *K* qui erunt loco equantis. Et ipsos diuidam et intitulabo sicud feci superius in equante mercurii.

764 diametris: lineis α*V*
768 5 minuta: 5 prima minuta β*V* primi gradus *add.* θφ
770 eadem diametro: eisdem dyametris β*V*
773 infra: intra β*V*
784 de: ex β*V*
789 EF: circulus EF φ
796 Remanebuntque θφ/fere: et sexta pars unius partis δ

I divide each of the [circles] into four [sections] by two diameters intersecting one another at right angles on center *D*. Let one of these diameters pass through Gemini 17 degrees 50 minutes for Venus; through Leo 1 degree 50 minutes for Mars; through Virgo 14 degrees 30 minutes for Jupiter; and through Sagittarius [0 degrees] 5 minutes for Saturn:[151] the apogees of the four planets will lie on those diameters in the above-mentioned places on the ecliptic, and the perigees will lie on the same diameters in the places on the ecliptic diametrically opposite to the above-mentioned. On those diameters which pass through the apogees of the above-mentioned planets I leave vacant, below the inmost of the four circles representing the ecliptic, just as much space as I want to separate it [the inmost circle] from a circle described about the center of the equant and divided in the appropriate way.[152] For, in the case of these four planets, I make the equant, along the circumference of which the mean centrum is to be counted, greater than the deferent, so that the whole of the equant, with its divisions, lies outside the deferent, just as I did for Mercury.[153] I divide the interval remaining on the above diameter up to point *D* into 60 units[154] for Venus and Mars. From this I cut off, for Venus, $1\frac{1}{4}$ units, beginning from point *D*. At their end I put point *K*, which will be the center of the equant. I bisect line *KD* at point *C*, which will be the center of the deferent. I cut off from it [line *KD* produced] 30 of the above-mentioned parts, from point *C* toward the circumference, and at their end I mark point *A*: point *A* will be the apogee of the deferent, and line *AC* will be its radius. Then I draw the [deferent] itself—let this be circle *AB*. Next I cut off on the same diameter 21 plus $\frac{1}{3}$ and $\frac{1}{4}$[155] of the above-mentioned units from point *A* toward the circumference. At their end I mark point *E*: line *AE* will be the radius of the epicycle. Then I draw [the epicycle] itself: let this be *EF*. For Mars I cut off $6\frac{1}{2}$ of the above-mentioned 60 units from point *D* to make line *DK* and bisect *DK* at *C*. Then *K* will be the center of Mars's equant and *C* the center of the deferent. I make line *CA* 30 units and line *AE* $19\frac{3}{4}$ units in the above-mentioned units; line *CA* will be the radius of the deferent and line *AE* the radius of the epicycle, [both of] which I then draw: let the first be circle *AB* and the second circle *EF*. Thus there will remain [between the epicycle and the ecliptic] from the above-mentioned 60 units about 7 units for Venus and 4 units minus $\frac{1}{4}$ of a unit for Mars.[156] In that [remaining] space, therefore, I describe about center *K* four circles to represent the equant, and I divide and label them as I did previously for the equant of Mercury.[157]

800 In ioue autem et saturno diuidam ipsum residuum usque ad punctum *D* in 96 partes. Et ponam lineam *DK* in ioue quidem 5 partes et medietatem partis, in saturno autem 6 partes et 5 sextas unius partis ex ipsis 96 partibus. Et diuidam ipsam in duo equalia super punctum *C*. Et ponam lineam *CA* 60 partes et lineam *AE* ponam in ioue quidem 11 partes et medietatem 805 partis, in saturno autem 6 partes et medietatem partis. Eritque punctum *K* centrum equantis et punctum *C* centrum deferentis et linea *AC* semidiameter ipsius deferentis et linea *EA* semidiameter epicicli. Et describam in utroque eorum deferentem quidem circulum *AB* et epiciclum circulum *EF*. Et erit in utroque istorum planetarum epiciclus iste uere circulationis. 810 Residuum autem ex predictis partibus 96 erit in ioue quidem 19 partes; in saturno uero 22 partes et due tertie unius partis. In extremitate igitur istarum partium uersus circumferentiam circuli signorum describam circulos 4 super centrum *K* qui erunt loco equantis. Et ipsos diuidam et intitulabo sicud feci superius in equante mercurii. In reliquo autem spatio quod 815 est infra describam in utroque eorum super punctum *A* quod est centrum epicicli circulum maiorem quem potero infra equantem qui erit epiciclus oportune circulationis.

Et diuidam epiciclos ueneris et martis et istos duos epiciclos oportune circulationis in 360 partes faciendo 4 circulos concentricos, et intitulabo 820 eos prorsus sicud feci in epiciclo mercurii. Describam postmodum super centrum *C* duos circulos quorum unus parum excedat epiciclum uere circulationis in uenere et in marte et epiciclum oportune circulationis in ioue et saturno: et sit circulus *GH*. Et alius parum subsistat eisdem et sit circulus *LM*. Et concauabo totum spatium quod est inter 2 circulos *GH* et 825 *LM*, ita quod in concauitatem illam ingrediatur tabula una plana ita spissa quod ipsa possit etiam concauari. Fiat quoque sic ista concauitas quod tabula sibi inposita non possit inde egredi. Postea faciam isti concauitati tabulam similem quam uocabo tabulam motus centri epicicli. Et lineabo in ea deferentem et epiciclum cum diuisionibus et inscriptionibus epicicli uere 830 quidem lineationis tantum in uenere et marte, uere tamen et oportune in ioue et saturno. Postea concauabo iterum in ista tabula totum illud spatium quod continet diuisiones et inscriptiones epicicli. Et faciam tabulam aliam isti concauitati similem quam uocabo tabulam motus epicicli. Et diuidam eam et inscribam sicud diuiditur et inscribitur epiciclus. Postea ordinabo 835 quamlibet earum in concauitate sua, ita quod interior contingat exteriorem ex omni parte et ex omni modo et quod moueatur in ea motu leui et equali

800–801 usque...D *om.* β*V*
821 epiciclum: epiciclos δ*A* circulum β*V*
822 epiciclum: epiciclos δ*A* uere et *add.* δ
824–25 Et concauabo...LM *om.* β*V*
828 epicicli: ecentrici β*V*
830 lineationis: circulationis δ/oportune: lineationis *add.* *F* circulationis *add.* δ

For Jupiter and Saturn, however, I divide the [interval] remaining between *D* [and the point chosen for the outer edge of the equant] into 96 units.[158] I make line *DK* $5\frac{1}{2}$ of the 96 units for Jupiter and $6\frac{5}{6}$ of the 96 units for Saturn. I bisect it at point *C* and make line *CA* 60 units, and I make line *AE* $11\frac{1}{2}$ units for Jupiter and $6\frac{1}{2}$ units for Saturn. Then point *K* will be the center of the equant, point *C* the center of the deferent, line *AC* the radius of the deferent, and line *EA* the radius of the epicycle. For each of them I draw the deferent, circle *AB*, and the epicycle, circle *EF*. That will be the epicycle in its true representation for both those planets. There will be a remainder, from the above-mentioned 96 units, of 19 units for Jupiter and of $22\frac{2}{3}$ units for Saturn.[159] Therefore, at the end of those [remaining] units toward the circumference of the ecliptic, I describe about center *K* four circles to represent the equant, and I divide and label them as I did previously for the equant of Mercury. In the space remaining below [i.e., toward the center] I describe for both [planets] about center *A*, which is the center of the epicycle, the biggest circle I can fit inside the equant: this [circle] will be the epicycle in its convenient representation.

I divide the epicycles of Venus and Mars and these two epicycles in convenient representation into 360 parts, using four concentric circles and labeling them exactly as I did for Mercury's epicycle.[160] Next I describe about center *C* two circles, of which one should pass just outside the epicycle, in its true representation for Venus and Mars and in its convenient representation for Jupiter and Saturn: let this be circle *GH*. The other should pass just inside it [the epicycle]: let this be circle *LM*. I hollow out the whole space between the two circles *GH* and *LM*, in such a way that the cavity will contain a flat disk thick enough to be hollowed out itself.[161] That cavity should, furthermore, be made in such a way that a disk fitted into it cannot slip out of it. Next I make a disk similar [in size and shape] to that cavity and call it the "disk of the motion of the center of the epicycle." I draw on it the deferent and the epicycle, with the divisions and labeling of the epicycle, in its true representation alone for Venus and Mars, but in both true and convenient [representations] for Jupiter and Saturn. Then I once more hollow out in that disk the whole of the area containing the divisions and labeling of the epicycle and make another disk similar [in size and shape] to the latter cavity; I call this the "disk of the motion of the epicycle." I divide and label it as the epicycle is divided and labeled. Next I fit each disk into its cavity, so that [each] inner [disk] touches its outer [container] on every side and in every way and moves in it with a smooth and even

---

807–9 Si leuaueris rotam apparebit tibi ecentricus deferens literaliter descriptus per autorem *mg. M*

et non possit egredi ab ea. Sitque omnium istarum tabularum continentium et contentarum superficies una quemadmodum dictum est in mercurio et in luna.

840 Faciamque unum signum notabile in auge epicicli uere circulationis in uenere et marte aut oportune circulationis in ioue et saturno. Figam quoque unum clauum in puncto *K* et alium in puncto *D* itemque alium in centro epicicli iouis et saturni. Et ligabo ad pedem cuiuslibet istorum clauorum unum filum de seta tenuissimum et equale sicud feci in mercurio et in 845 luna.

Cum ergo uoluero equare quemlibet istorum 4 planetarum per istud instrumentum, sumam medium centrum et medium argumentum cuiuscumque ipsorum uoluero ad tempus datum et numerum medii centri queram in equante, numerum autem medii argumenti in epiciclo. Et faciam 850 notam unam uel materialem uel solum in intellectu in termino utriusque, scilicet in equante et in epiciclo. Et extendam filum ligatum ad punctum *K* super notam equantis. Et circumuoluam tabulam motus centri epicicli quousque centrum epicicli cadat sub eodem filo. Tabulam quoque motus epicicli circumuoluam etiam quousque aux epicicli cadat sub eodem filo. 855 Et si fuerit opus ad uenerem uel ad martem, filum ligatum ad punctum *D* faciam transire per signum epicicli ubi terminatum fuerit medium argumentum. Et locus in orbe signorum ubi filum illud secabit orbem signorum erit uerus locus eius. Si autem fuerit opus ad iouem uel ad saturnum, filum ligatum ad centrum epicicli faciam transire per signum epicicli ubi termi- 860 natum fuerit medium argumentum. Et per punctum epicicli uere circulationis ubi filum illud secat ipsum, faciam transire filum ligatum ad punctum *D*. Et locus in orbe signorum ubi filum istud ipsum secabit erit uerus

844 seta: serico φ*AhM*Γ serica Λ
847–48 cuiuscumque: cuiusuis α cuius *fH* cuiusque Π
852 super...equantis *om.* β*V*
861 secat: secabit β*V*

motion and cannot slip out of it. The surface of all these disks, whether containing or contained, should be one and the same, as we prescribed for Mercury and the moon.

I make a clearly visible mark at the apogee of the epicycle, in its true representation for Venus and Mars or in its convenient representation for Jupiter and Saturn. I also affix a nail at point *K* and another at point *D*, and yet another at the center of the epicycle of Jupiter and Saturn. I tie to the base of each of these nails a silk thread, very fine and of uniform thickness, just as I did for Mercury and the moon.

Then, when I want to equate any of these four planets by this instrument, I take the mean centrum and the mean anomaly of whichever [planet] I want for the datum time, and I look on the equant for the number constituting the mean centrum and on the epicycle for the number constituting the mean anomaly. At the place where each brings me, that is, on the equant and on the epicycle, I make a mark either on the actual instrument or just in my memory. I stretch the thread attached to point *K* through the mark on the equant and revolve the disk of the motion of the center of the epicycle until the center of the epicycle falls beneath that thread. Furthermore, I revolve the disk of the motion of the epicycle until the apogee of the epicycle [also] falls beneath the same thread. If the operation concerns Venus or Mars, I make the thread attached to point *D* pass through the mark on the epicycle where the mean anomaly brings me. Then the point on the ecliptic where that thread cuts the ecliptic will be the true longitude of that [planet]. If, however, the operation concerns Jupiter or Saturn, I make the thread attached to the center of the epicycle pass through the mark on the epicycle where the mean anomaly brings me. I make the thread attached to point *D* pass through the point on the epicycle in its true representation where that thread [attached to the center of the epicycle] cuts it. Then the place on the ecliptic where this thread [attached to point *D*] cuts it will be the true longitude of that [planet]. The mean centrum of Venus

locus eius. Medium autem centrum ueneris est medium argumentum solis. Ad cetera autem sunt tabule. Directionem uero stationem et retrogradatio-
865 nem istorum planetarum per istud instrumentum facile inuenies quemadmodum de mercurio supra docuimus.

is the mean anomaly of the sun.[162] For the other [mean motions][163] there are tables. As for the forward motions, stations, and retrogradations of these planets, you can easily find them by this instrument in the way which we explained above for Mercury.[164]

---

866 docuimus: *add.* Explicit theorica planetarum CAMPANI quod xx xviii *b*; in auxilio dei omnipotentis et sic est finis anno 1415 in crastino Scolastice. Scriptum colonie anno domini 1415° incompleto in crastino Scolastice uigilie per manus M. Petri de umbre prope hassel *C*; Explicit theorica CAMPANI de ciuitate NOVARIA Super motibus et magnitudinibus septem planetarum *D*; Amen. Explicit tractatus magistri CAMPANI (CAPANI Σ) de equatione planetarum. Deo gratias *F*Σ; Deo gratias. Scriptum per me dominum Cominum de Pontenigo et expletum XXV Julii 1477, hore 18 *g*; Laus deo amen *i*; Et in hoc finitur theorica planetarum CAMPANI NOVARIENSIS *J*; Et sic est finis de equatoriis CAMPANI uerorum motuum septem planetarum cum aliquibus communibus extractis ex theorica sua tradita in presenti libro quando de theorica sua faciliter non possunt intelligi relata sunt... Finita sunt hic per Godefridum de Molendino Treuerensi in medicinis practicantem protunc anno M442° 13 die octobris Deo gratias nunc et in perpetuum *j*; Explicit deo gratias. Hic liber est finitus qui finit sit benedictus *K*; Finis *k*; cum auxilio Dei omnipotentis qui est benedictus in secula amen *m*; Explicit *O*; Nota quod modus operandi sequentium rotarum debet fieri in epyciclo uere circulationis et non in epyciclo opportune circulationis ut patuit supra. Explicit theorica planetarum CAMPANI per me NICOLAVM GRODVNE Carpen[te] 22 octobris 1454 *o*; Explicit theorica motuum planetarum *P*; Explicit liber CAMPANI de equationibus planetarum et cetera *p*; Explicit theorica CAMPANI *Qx*; Explicit theorica CAMPANI NOVARIENSIS. Explicit de instrumentis equationis planetarum *R*; Ffinitus est liber *rz*; Expliciunt theorice magistri Campani nauariensis *S*; Per me Johannem ?Curanomsbs finitus est CAMPANVS iste feria tertia proxima ante palmarum anno domini mi° cccc° 78 *s*; Deo gratias. Amen. Explicit theorica CAMPANI de motibus planetarum cum multis aliis exquisite elaboratis *T*; Explicit liber CAMPANI de equationibus planetarum per instrumenta *U*; Hec scriptor: Finis adest libri CAMPANVS sit suus auctor instrumenta sua si quis et acta uelit arguat hec pridem me culpa labor inertis scribe sopitus deuiet astra poli *u*; Explicit in Dei nomine hic liber finitus anno Dei 1356 ydus Septembris scilicet 13 die in locis multis incorrectus *V*; cum auxilio dei cunctipotentis. Explicit theorica magistri CAMPANI planetarum *Y*; *post descriptionem machinae cuiusdam motus planetarum per sphaeras dentatas imitantis* Explicit theorica planetarum CAMPANI *y*; *post eandem descriptionem* Per Ihesum Christum Filium scripsi... Anno 1466 3° ydus septembris Λ; Explicit magister Campanus cum theorica sua super instrumentum suum Π; Deo gratias amen Ψ

[Table 1]
[Summary of lunar distances and dimensions; see lines IV.403–69]

1

Prout linea que est a centro terre ad longitudinem longiorem est 60 partes erunt hee magnitudines ita

Magnitudines	Partes	Minuta	Totum in minutis
AC	49	41	2981
CD	10	19	619
AE	5	15	315
AD	60	0	3600
DB	39	22	2362
ED	65	15	3915
DF	34	7	2047
AB	99	22	5962

2

Prout semidiameter terre est pars una erunt predicte magnitudines ita

Magnitudines	Partes	Minuta	Totum in minutis
AC	48	51	2931
CD	10	9	609
AE	5	10	310
AD	59	0	3540
DB	38	43	2323
ED	64	10	3850
DF	33	33	2013
AB	97	43	5863

Ad magnitudines inueniendas secundum partes in tabula prima et tertia positas ponit PTHOLOMEVS principia et radices et tu ex eis elicias easdem secundum partes positas in secunda et quarta tabula

5
[Magnitudes of the moon and its sphere]

	Miliaria	Tricesime tertie
Semidiameter corporis lune	948	13
Tota diameter corporis lune	1896	26
Totus ambitus lunaris corporis	5958	32
Ambitus superficiei conuexe	1314961	11
Ambitus superficiei concaue	678458	22
Ambitus deferentis aut equantis	996540	0
Dieta centri epicicli lune de partibus equantis	36458	47 minuta
Ambitus epicicli lune	105400	0
Dieta lune in suo epiciclo	3826	56 minuta

[Table 1—*Continued*]

3				*4*		
Prout linea *AC* que est semidiameter ecentrici est 60 partes erunt predicte magnitudines ita				Prout semidiameter terre est 3245 miliaria erunt predicte magnitudines ita		
Magnitudines	Partes	Minuta	Totum in minutis	Magnitudines	Miliaria	Undecime unius miliaris
AC	60	0	3600	*AC*	158540	5
CD	12	28	748	*CD*	32941	4
AE	6	20	380	*AE*	16768	2
AD	72	28	4348	*AD*	191481	9
DB	47	32	2852	*DB*	125653	2
ED	78	48	4728	*ED*	208250	0
DF	41	12	2472	*DF*	108885	0
AB	120	0	7200	*AB*	317135	0

Isti duo numeri sunt cubi illorum duorum qui secuntur et sunt illi scilicet qui sequentur in quibus reperitur proportio diametrorum terre et lune secundum fractiones positas et si diuideris maiorem cubum per minorem exibunt proportiones corporis terre ad corpus lune

Cubi	Latera cubitorum
1,635,108,092,541,000,000	1,178,100
40,802,444,725,282,163	344,267

[Table 2]
[Summary of distances and dimensions for Mercury; see lines V.380–473]

	Magnitudines quantitatum mercurii secundum triplicem modum predictum					
	1 Prout semidiameter ecentrici est 60 partes		*2* Prout semidiameter terre est pars una		*3* Prout 4000 cubiti faciunt unum miliare	
Magnitudines	Partes	Minuta	Partes	Minuta	Miliaria	Undecime unius miliaris
AC	60	0	117	2	379826	4
CD	3	0	5	51	18985	10
AE	22	30	43	53	142421	4
AD	69	0	134	35	436784	1
DB	55	34	108	23	351753	2
ED	91	30	178	28	579205	5
DF	33	4	64	30	209331	9
DG	51	0	99	29	322868	7
DM	57	0	111	11	360840	5

4
[Magnitudes of Mercury and its sphere]

	Miliaria	Tricesime tertie
Semidiameter corporis mercurii	115	13
Diameter corporis mercurii	230	26
Ambitus corporis mercurii	725	11
Longitudo concaue spere mercurii a centro terre	209198	13
Longitudo conuexe spere mercurii a centro terre	579320	28
Spissitudo spere mercurii	370122	5 undecime
Ambitus superficiei conuexe spere mercurii	3641446	2 septime
Ambitus deferentis mercurii	2387480	0
Dieta centri epicicli mercurii	6536	34 minuta
Ambitus epicicli mercurii	895220	0
Dieta mercurii in epiciclo suo	7717	24 minuta
Ambitus superficiei concaue spere mercurii	1314961	11

[Table 3]
[Summary of distances and dimensions for Venus; see lines VI.408–61]

	Magnitudines predictarum quantitatum ueneris secundum triplicem modum						
	1 Prout semidiameter ecentrici est 60 partes		*2* Prout semidiameter terre est pars una			*3* Prout 4000 cubiti sunt unum miliare	
Magnitudines	Partes	Minuta	Partes	Minuta	Secunda	Miliaria	660 unius miliaris
AC	60	0	688	59	33	2236093	555
CD	1	15	14	21	14	46584	590
AE	43	10	495	41	31	1608745	445
AD	61	15	703	20	47	2282678	485
DB	58	45	674	38	19	2189508	625
ED	104	25	1199	2	18	3891424	270
DF	15	35	178	56	48	580763	180

4
[Magnitudes of Venus and its sphere]

	Miliaria	660 unius miliaris
Semidiameter corporis ueneris	1442	280
Ambitus corporis ueneris	9095	63
Distantia superficiei concaue spere ueneris a centro terre	579320	560
Distantia superficiei conuexe spere ueneris a centro terre	3892866	550
Spissitudo spere ueneris	3313545	650
Ambitus totius conuexe superficiei spere ueneris	24469448	440
Dieta centri epicicli ueneris	38481	44 minuta
Ambitus epicicli ueneris	10112115	40 minuta
Ambitus deferentis ueneris	1405547	0
Dieta ueneris in suo epiciclo	17315	16 minuta

[Table 4]

[Summary of distances and dimensions for the sun; see lines VI.509–64]

	Magnitudines solis predicte secundum triplicem modum						
	1 Prout semidiameter ecentrici est 60 partes		*2* Prout semidiameter terre est pars una			*3* Prout 4000 cubiti faciunt unum miliare	
Magnitudines	Partes	Minuta	Partes	Minuta	Secunda	Miliaria	660 unius miliaris
AC	60	0	1257	22	24	4080748	0
CD	2	30	52	23	26	170031	110
AD	62	30	1309	45	50	4250779	110
DB	57	30	1204	58	58	3910716	550

4

[Magnitudes of the sun and its sphere]

	Miliaria	660 unius miliaris
Semidiameter corporis solis	17850	0
Ambitus corporis solis	112200	0
Longitudo concaue superficiei spere solis a centro terre	3892866	550
Longitudo conuexe superficiei spere solis a centro terre	4268629	110
Spissitudo spere solis	375762	220
Ambitus superficiei spere solis	26831383	220
Ambitus solaris circuli qui est ecentricus eius	25650416	0
Dieta solis	70227	0 minuta

[Table 5]
[Summary of distances and dimensions for Mars; see lines VI.571–617]

	Magnitudines martis predicte secundum triplicem modum						
	1 Prout semidiameter ecentrici est 60 partes		*2* Prout semidiameter terre est pars una			*3* Prout 4000 cubiti faciunt unum miliare	
Magnitudines	Partes	Minuta	Partes	Minuta	Secunda	Miliaria	660 unius miliaris
AC	60	0	5641	50	43	18310352	265
CD	6	30	611	12	0	1983621	540
AE	39	30	3714	12	53	12054315	35
AD	66	30	6253	2	43	20293974	145
DB	53	30	5030	38	43	16326730	385
ED	106	0	9967	15	36	32348289	180
DF	14	0	1316	25	50	4272415	350

4
[Magnitudes of Mars and its sphere]

	Miliaria	660 unius miliaris
Semidiameter corporis martis	3786	240
Ambitus corporis martis	23800	0
Longitudo concaue superficiei spere martis a centro terre	4268629	110
Longitudo conuexe superficiei spere martis a centro terre	32352075	420
Spissitudo spere martis	28083446	310
Ambitus superficiei conuexe spere martis	203355904	0
Ambitus deferentis martis	115093643	440
Dieta centri epicicli martis	167530	47 minuta
Ambitus epicicli martis	75769980	220
Dieta martis in suo epiciclo	97141	0 minuta

[Table 6]
[Summary of distances and dimensions for Jupiter; see lines VI.622–68]

	Magnitudines iouis predicte secundum triplicem modum						
	1 Prout semidiameter ecentrici est 60 partes		*2* Prout semidiameter terre est pars una			*3* Prout 4000 cubiti faciunt unum miliare	
Magnitudines	Partes	Minuta	Partes	Minuta	Secunda	Miliaria	660 unius miliaris
AC	60	0	13079	20	8	42448389	20
CD	2	45	599	28	10	1945550	550
AE	11	30	2506	52	22	8135941	430
AD	62	45	13678	48	18	44393939	570
DB	57	15	12479	51	58	40502838	130
ED	74	15	16185	40	40	52529881	340
DF	45	45	9972	59	36	32366896	360

4
[Magnitudes of Jupiter and its sphere]

	Miliaria	660 unius miliaris
Semidiameter corporis iouis	14820	600
Ambitus corporis iouis	93160	0
Longitudo concaue superficiei spere iouis a centro terre	32352075	420
Longitudo conuexe superficiei spere iouis a centro terre	52544702	280
Spissitudo spere iouis	20192626	520
Ambitus conuexe superficiei spere iouis	330280986	440
Ambitus deferentis iouis	266818445	220
Dieta centri epicicli iouis	61564	
Ambitus epicicli iouis	51140204	440
Dieta iouis in suo epiciclo	128171	—

[Table 7]
[Summary of distances and dimensions for Saturn; see lines VI.674–716]

	Magnitudines saturni predicte secundum triplicem modum						
	1 Prout semidiameter ecentrici est 60 partes		*2* Prout semidiameter terre est pars una			*3* Prout 4000 cubiti faciunt unum miliare	
Magnitudines	Partes	Minuta	Partes	Minuta	Secunda	Miliaria	660 unius miliaris
AC	60	0	19401	21	28	62966224	520
CD	3	25	1104	47	58	3585576	250
AE	6	30	2101	48	50	6821341	290
AD	63	25	20506	9	26	66551801	110
DB	56	35	18296	33	30	59380648	270
ED	69	55	22607	58	16	73373142	400
DF	50	5	16194	44	40	52559306	640

4
[Magnitudes of Saturn and its sphere]

	Miliaria	660 unius miliaris
Semidiameter corporis saturni	14604	360
Ambitus corporis saturni	91800	0
Longitudo concaue superficiei spere saturni a centro terre	52544702	280
Longitudo conuexe superficiei spere saturni a centro terre	73387747	100
Spissitudo spere saturni	20843044	480
Ambitus conuexe superficiei spere saturni	461294410	440
Ambitus deferentis saturni	395787698	440
Dieta epicicli saturni	36647	—
Ambitus epicicli saturni	42877003	220
Dieta saturni in suo epiciclo	113431	—

Commentary

Commentary

Section I

"Clementissimo...dicendum" (I.4–110). A defective version of some sections of this dedication was printed from an Ambrosian MS (?*H*) by Tiraboschi, *Storia della letteratura*, vol. 4, bk. 2.2, chap. 5, pp. 244–46.

The influence on Campanus's address here to Urban IV of the famous *Eulogy* of the Emperor Frederick II by the emperor's secretary Petrus de Vineis is considerable. It consists not so much in verbal borrowings (though there are echoes: e.g., line I.19, "faminis"—"amplo famine"; line I.27, "Exultet...totus mundus" —"totus mundus exultet"; line I.40, "O dulcissima mirande misericordie"—"O miranda diuina clementia"; lines I. 8–9, "quadam prerogatiua mirabili"—"quanti priuilegii prerogatiua") as in the bombastic style. To facilitate comparison, a text of the *Eulogy* is reproduced on pp. 448–49 as Appendix B. For information on editions of the *Eulogy*, see E. H. Kantorowicz, "The Prologue to *Fleta* and the School of Petrus de Vinea," *Speculum*, vol. 32 (1957), p. 233, note 10. Kantorowicz not only points out the influence it had on the prologue to *Fleta*, but also gives other examples of its popularity as a model in the late thirteenth century. Many comparable stylistic exercises will also be found in Rockinger's collection of later medieval letter-writing handbooks (Ludwig Rockinger, *Briefsteller und Formelbücher des eilften bis vierzehnten Jahrhunderts*. Quellen und Erörterungen zur bayerischen und deutschen Geschichte [Munich, 1863–64], vol. 9, pts. 1–2).

1 "Vrbano Qvarto" (I.5). Jacques Pantaléon, pope from August 29, 1261, to October 2, 1264. See pp. 4–7, above.

"Qvarto" (I.5). According to Artaud de Montor (*The Lives and Times of the Popes* [New York, 1910], vol. 7, p. 163), Urban was the first pope to differentiate himself from his predecessors of the same name by adopting a number. However, the practice of referring to *previous* popes by number is much older.

2 "beatorum pedum osculum" (I.6–7). According to the thirteenth-century German writer of a treatise entitled *Summa prosarum dictaminis* (Rockinger, *Briefsteller*, pt. 1, p. 262), it was the Italian custom to use such phrases as "se totum ad sanctorum oscula pedum" in letters to the pope. He comments, "...que nos ob adola-

cionis similitudinem non probamus." Cf. ibid., p. 454 (Conradi *Summa de arte prosandi*): "deuota pedum oscula beatorum", and pt. 2, p. 957.

3 "noctes insompnes ducitis" (I.12). Cf., e.g., Vergil *Aeneid* 9. 166–67: "noctem custodia ducit insomnem ludo"; Statius *Thebaid* 2. 74; and Urban himself (Guiraud, *Les Régistres d'Urbain IV*, vol. 2, p. 305, col. 1), in a letter dated Orvieto, May 20, 1264: "noctes ducentes insompnes nec diebus otia indulgentes, studiosam demus operam...ad eandem ecclesiam."

4 "Petri nauiculam" (I.15). The figure was evidently a trite one, as it occurs more than once in the section on papal letters in the Baumgartenberg formulary of about 1300 (Rockinger, *Briefsteller*, pt. 2, p. 811). See also the examples from the twelfth century cited in *Novum glossarium mediae latinitatis* (Copenhagen, 1957——), cols. 1118–19, s.v. "navicula."

5 "fermentauit" (I.19–20). The use of "fermentare" to mean "corrupt" is derived from the biblical use of "fermentum." Cf. Matt. 16:6: "cauete a fermento pharisaeorum et sadducaeorum," also Mark 8:15, Luke 12:1, 1 Cor. 5:6–8 (an expansion of the metaphor). St. Augustine uses "fermentare" as a synonym of "corrumpere": see *Thesaurus linguae latinae, editus auctoritate et consilio Academiarum Quinque Germanicarum* (Leipzig, 1900——), s.v. "fermentare II." At Gal. 5:9, where the Vulgate reads "Modicum fermentum totam massam corrumpit," the Itala have "fermentat."

6 "Ad istius...fermentauit" (I.17–20). The phrases used here seem to indicate that there was something especially providential about the mode of Urban's election. The known facts are these: on the death of Alexander IV a conclave of eight cardinals assembled at Viterbo to elect a new pope. They were unable to gather the requisite six votes for any of their own number, and the conclave dragged on inconclusively for three months. They then unanimously elected Jacques Pantaléon, who was at that time Patriarch of Jerusalem, but happened to be in Viterbo on quite other business (not being a cardinal). It is hard to see anything in this account to justify the strong insistence of Campanus that it was God alone, and no human being, who was responsible for the election. The only other evidence which would seem to bear on this election comes from the fourteenth-century Florentine historian Giovanni Villani (Lodovico Antonio Muratori, *Rerum italicarum scriptores* [Milan, 1723–51], vol. 13, col. 219, note a). He says that the cardinals, being unable to reach accord, locked themselves into the conclave and made a secret agreement to choose as pope the first cleric who knocked on the door. Urban "come piacque a Dio" was the first. This account not only is intrinsically implausible but is combined with other details that are certainly false (e.g., that Urban was a poor priest who was trying to get back a wretched living that had been taken from him). Nonetheless it may conceivably be a distorted reminiscence of some arrangement made to give an impression that the choice of God had fallen on Pantaléon.

7 "ruenti...erroribus" (I.20). For the wretched state of the papacy at the death of

Alexander IV, see Johannes Haller, *Das Papsttum* (Basel, 1951–53; 2d ed.), vol. 4, pp. 291–96.

8 "per strati...equoris" (I.25). Cf. Anicius Manlius Severinus Boethius *Philosophiae consolatio* 2. 2. 8 (ed. Ludovicus Bieler, Corpus Christianorum series latina 94, pt. 1 [Turnhout, 1957]): "...ius est mari nunc strato aequore blandiri, nunc procellis ac fluctibus inhorrescere."

9 "umbra...sopori" (I.31–33). Several phrases in this sentence are reminiscent of Vergil *Eclogues* 5. 46–47: "Quale sopor fessis in gramine, quale per aestum/dulcis aquae saliente sitim restinguere riuo."

10 "cesset galea...ensis" (I.33–34). Cf. Ovid *Metamorphoses* 1. 99: "...non galeae, non ensis erat" (in a description of the Golden Age).

11 "Is quippe...sitienti" (I.36–39). The whole passage is a reminiscence of Isa. 58:7: "Frange esurienti panem tuum, et egenos uagosque induc in domum tuam; cum uideris nudum operi eum, et carnem tuum ne despexeris," with echoes from Matt. 25:35–38.

12 "necessaria" (I.40). This usage of the word is strange: could Campanus have supposed that it can mean "unceasing," by a false etymology (*ne+cessare*)?

13 "Philosophiam" (I.45). The idea of personifying Philosophy in the vivid way here and in the following sentences is taken straight from Boethius's *Philosophiae consolatio*.

14 "rei...pudica" (I.47–48). Cf. Boethius *Philosophiae consolatio* 2. pr. 4. 13: "...angustia rei familiaris inclusus esse mallet ignotus."

15 "in uere...ridiculum" (I.49–50). The meaning of this phrase, if the text is right, is quite uncertain. A possible alternative translation would be, "...she has always, among those who are truly her familiars, considered it ridiculous that she whose...." For the two interpretations of "domesticis," see note 17, below.

16 "uenerabile...collegium" (I.55–56). One may conjecture that this group included both Thomas Aquinas and William of Moerbeke; for, though neither was strictly speaking a chaplain, both were attached to the papal court at the time, the former as *lector curiae*, the latter as *poenitentiarius minor*. Cf. Martin Grabmann, *Forschungen über die lateinischen Aristoteles-Übersetzungen des XIII Jahrhunderts*, Beiträge zur Geschichte der Philosophie des Mittelalters, vol. 17 (Münster i.W., 1916), p. 161; Clagett, *Archimedes*, vol. 1, p. 443 and note 16; and p. 11, above.

17 "ubi...proponitis" (I.61–62). The connotation of "domesticus" here is uncertain, and hence the translation is only a suggestion. I have taken it as equivalent to οἰκεῖος ("belonging to"), as in the well-known expression "domestici fidei" (Gal. 6:10). Cf. also Boethius, *Philosophiae consolatio* 3. pr. 12. 35: "Atque haec nullis extrinsecus sumptis, sed ex altero ⟨altero⟩ fidem trahente insitis domesticisque probationibus explicabas" (the subject of the sentence is Philosophy).

The "domestica problemata" may, however, refer to a particular category of "household" problems which fall into the sphere of "practical philosophy." Cf.

Chalcidius, commentary 296 (J. H. Waszink, ed., *Timaeus a Calcidio translatus commentarioque instructus*..., Plato Latinus, vol. 4 [London, 1962], p. 270): "...ex his quippe constat alterum philosophiae genus, quod actiuum uocatur. Id porro diuiditur trifariam, in moralem domesticam [= οἰκονομικήν] publicam."

18 "illa saturnalia...uacauisse" (I.65–66). Campanus is probably thinking of Macrobius *Saturnalia* 7. 1. 13, where details are given of philosophers' banquets: "...sic Agathonis conuiuium, quia Socratas Phaedros Pausanias et Erysimachos habuit, sic ea cena quam Callias doctissimis dedit, Charmadam dico, Antisthenen et Hermogenen ceterosque his similes, uerbum nullum nisi philosophum sensit."

19 "Iste uero sunt epule...eisdem" (I.66–68). Campanus has derived this from the opening of Plato's *Timaeus*. In Chalcidius's translation (Waszink, *Timaeus*, p. 7) this reads, "Socrates: Unus duo tres; quartus e numero, Timaee, uestro requiro, ut, qui hesterni quidem epuli conuiuae fueritis, hodierni praebitores inuitatoresque ex condicto resideatis." However, Chalcidius misrepresented the sense of the Greek, which implies only that Socrates gave dinner to his companions yesterday and is being feasted by them today, and not that he *demanded* the return banquet.

20 "duplicis sancte" (I.70). "Doubly sacred" because attended by both religion and philosophy.

21 "que datorum...habundanter" (I.79). Cf. James 1:5: "Deo qui dat omnibus affluenter et non improperat."

22 "in...nouum" (I.81). Campanus is presumably here claiming originality for his equatorium, since he could hardly be referring to his description of the planetary system, which he himself (lines I.98–100) ascribes to Ptolemy, as "a novelty." Moreover, the reference to "pleasure plus utility" here is echoed at lines II.64–65, which refer explicitly to the instrument.

23 "Qui pauperculam...dragmam meam" (I.89–91). Luke 21:1–3: "Respiciens autem uidit eos qui mittebant *munera* sua in *gazophylacium* diuites. uidit autem et quandam *uiduam pauperculam* mittentem aera minuta duo et dixit uere dico uobis quia uidua haec pauper plus quam omnes misit." (Italics added to emphasize parallel language.) Perhaps we should read "uiduam" for "mulierem" here (the manuscripts vary between "mulierem" and "minorem"), but Campanus, as always, paraphrases rather than quotes the Vulgate.

24 "Quin imo...reperiat demonstratum" (I.96–101). Here Campanus admits that his work consists of "ymaginationes" (which I have translated "models," but which might be paraphrased as "picturings of the arrangements of things") rather than proofs, but claims correctly that he is merely stating a system for which Ptolemy has furnished the groundwork of observation and proof.

25 "retunditur" (I.105). The subjunctive "retundatur" would give a better sense. Cf. Solinus, *Collectanea rerum memorabilium*, chap. 27, sec. 26 (ed. Theodor Mommsen [Berlin, 1864], p. 136), speaking of the hyena: "dens unus atque perpetuus, qui ut numquam retundatur, naturaliter capsularum modo clauditur."

26 "cernentibus" (I.109). If correct, this must presumably be translated "sifted," i.e., "examined thoroughly." MS *L*'s "comedentibus" ("who eat it") would make sense if construed as picking up the metaphor of lines I.94–96.

Section II

1 "Primus...naturale" (II.2–4). The reference is to Aristotle *Metaphysics* E. 1. $1026^{a}18$: "...so it follows that there are three theoretical types of philosophy, the mathematical, the physical, and the theological." But the citation of Aristotle is taken from the *Almagest* 1. 1 (Ptolemy, Manitius, vol. 1, p. 1, line 24–p. 2, line 1), and the following philosophical excursus also owes much to Ptolemy's treatment there, as my quotations will show.

2 "medium...extremorum" (II.4–5). Cf. the *Almagest* (Ptolemy, Manitius, vol. 1, p. 2, lines 25–30: "...such a subject [mathematics] falls as it were in the middle of the other two...because it is an attribute of all things both mortal and immortal." Gerard of Cremona translates: "hec quidem natura quasi medium tenet inter illas duas naturas..." (Ptolemy, *Almagesti Cl. Ptolemei Pheludiensis Alexandrini Astronomorum principis Opus ingens ac nobile*...[Venice: Petrus Liechtenstein, 1515], fol. 2r). But Campanus is not merely echoing the Latin *Almagest* here; he is expressing himself in Aristotelian terminology: "medium" = μέσον and "extrema" = ἀκρά. These are most familiar from Aristotle's logical works (e.g., *Analytica priora* 1. $25^{b}32$). However, the technical meaning, "middle and extreme *terms*," is inapplicable here, and Aristotle himself uses the disjunction constantly in a nontechnical sense. Very like Campanus's expression here is that in *De partibus animalium* 3. $661^{b}9$–11: ὁρίζουσι δ'ἑκατέρας οἱ κυνόδοντες, μέσοι τὴν φύσιν ἀμφοτέρων ὄντες· τό τε γὰρ μέσον ἀμφοτέρων μετέχει τῶν ἄκρων.

3 "doctrinalis modi certitudine" (II.7). Ptolemy, too (Manitius, vol. 1, p. 2, line 36–p. 3, line 12), says that mathematics alone of the three gives certain knowledge. Gerard's translation of this passage reads, in part, "dico quod duo reliqua genera diuisionis theorice sola estimatione cognoscuntur: et non scientie veritate comprehenduntur....Genus vero doctrinale ipsum solum replet eum qui ipsum studiose reponit" (Ptolemy, *Almagesti*, ed. Liechtenstein, fol. 2r).

4 "anthonomasice" (II.8). Similar misspellings (for the correct "antonomastice") are also found in most manuscripts of Sacrobosco (Thorndike, "*Sphere*" *of Sacrobosco*, chap. 3, p. 95) and in John of Salisbury's *Metalogicon libri IIII* (ed. C. C. J. Webb [Oxford, 1929], p. 873c), bk. 2, chap. 16: "...omnes se Aristotilis adorare uestigia gloriantur. Nam et antonomasice, id est excellenter, Philosophus appellatur." The spelling therefore probably is that of the author. The term "antonomasia" is defined by Flavius Sosipatrus Charisius (*Artis grammaticae libri V*, ed. K. Barwick [Leipzig, 1925], bk. 4, sec. 273, p. 360, line 21) as "dictio per accidens proprium significans, ut cum domitor maris dicitur et intelligitur

Neptunus." Cf. Aelius Donatus, *Ars maior*, pt. 3, chap. 6 (Heinrich Keil, *Grammatici latini ex recensione Henrici Keilii* [Leipzig, 1857–80], vol. 4, p. 400, lines 15–19) and *Anecdota Helvetica*, ed. H. Hagen (Keil, *Grammatici*, suppl., p. XLVIII, lines 13–18).

5 "doctrinale...contraire" (II.8–9). Campanus is giving and justifying the etymology of "mathematicus." Strictly speaking, the Greek μαθηματικός means "to do with *learning*," but Gerard, in his translation of the *Almagest*, regularly renders μαθηματικός by "doctrinalis" (see passage quoted in note 3, above), and furthermore Campanus is merely repeating the formulation of Cassiodorus and others (see note 6, below, on "quadriuium").

6 "quadriuium" (II.14). The *quadrivium* of arithmetic, music, geometry, and astronomy was the more advanced part of the curriculum of the seven liberal arts (the preliminary part, the *trivium*, consisted of grammar, rhetoric, and dialectic). The "seven liberal arts" was a classification made in antiquity, perhaps going back as far as Varro (see A. Cornelius Celsus, *Quae supersunt*, ed. F. Marx, Corpus medicorum latinorum, vol. 1 [Leipzig, 1915], pp. ix ff., and, in general, P. Merlan, *From Platonism to Neoplatonism*, 2d ed. [The Hague, 1960], pp. 88–95). But it was almost certainly Boethius who applied the name "quadrivium" to the four "scientific" arts, as appears from the following passages of his *De institutione arithmetica* 1. 1 (ed. G. Friedlein [Leipzig, 1867], p. 7): "...haud quemquam in philosophiae disciplinis ad cumulum perfectionis evadere, nisi cui talis prudentiae nobilitas quodam quasi *quadruvio* vestigatur." And (ed. Friedlein, p. 9): "...horum ergo illam multitudinem, quae per se est, arithmetica speculatur integritas, illam vero, quae ad aliquid, musici modulaminis temperamenta pernoscunt, immobilis vero magnitudinis geometria notitiam pollicetur, mobilis vero scientiam astronomicae disciplinae peritia vendicat....hoc igitur illud quadruvium est, quo his viandum sit."

Campanus's classification in the passage of the present text clearly goes back to Boethius. He also owes much, directly or indirectly, to Cassiodorus *Institutions* 2. 21 (itself dependent on the above passage of Boethius): "Mathematica, quam Latine possumus dicere 'doctrinalem,' scientia est quae abstractam considerat quantitatem....haec ita dividitur—divisio mathematicae: arithmetica—musica—geometria—astronomia. arithmetica est disciplina quantitatis numerabilis secundum se. musica est disciplina quae de numeris loquitur, qui ad aliquid sunt, his qui inveniuntur in sonis. geometria est disciplina magnitudinis immobilis et formarum. astronomia est disciplina cursus caelestium siderum." Cassiodorus's classification is repeated almost word for word in Isidore *Etymologies* 3, Introduction.

However, the grouping of the four subjects, arithmetic, music, geometry, and astronomy, as the subdivisions of mathematics is much earlier, going back at least to Archytas in the early fourth century B.C. See Hermann Diels, *Die Fragmente der Vorsokratiker*, 10th ed., ed. Walther Kranz (Berlin, 1961), no. 47, fragment B.1, lines 4–8. Cf. also Plato *Republic* 525[a]–530[d]. The Archytas passage is quoted by Nicomachus of Gerasa (between A.D. 50 and 150), who seems to have been the first to associate the four with the pairs of opposites, discrete/continuous, absolute/relative, rest/motion (*Introductionis arithmeticae libri II*, ed. R. Hoche [Leipzig, 1866], bk. 1, chap. 3, secs. 1–2, pp. 5–6; cf. translation by M. L. D'Ooge, *Introduction to Arithmetic*, Univ. of Mich. Studies, Humanistic Series, vol. 16 [New York, 1926], p. 184): "It is clear that two scientific disciplines will be concerned with explaining the whole field of investigation of number (τοῦ ποσοῦ), arithmetic number in itself, music number in relation to other things. Furthermore, since one part of extension (τοῦ πηλίκου) is at rest and stable, and another in motion and revolution, there will be two further sciences concerned with extension, geometry with the motionless part, and astronomy (σφαιρική) with the part that moves and revolves." Proclus (*In primum Euclidis Elementorum librum commentarii*, ed. G. Friedlein [Leipzig, 1873], p. 35, lines 21ff.) attributes this same classification to "the Pythagoreans." Boethius's *Arithmetic* is a mere adaptation of Nicomachus's work.

7 "que consonantiarum... melorum" (II.18–19). This is a reminiscence of two chapters in Boethius *De institutione musica* 4. 14 (ed. Friedlein, p. 337), beginning, "Nunc de speciebus primarum consonantiarum tractandum est," and 1. 21 (ed. Friedlein, p. 212), beginning, "His igitur expeditis dicendum de generibus melorum."

8 "concepta" (II.24). For this meaning ("concise"), cf. Chalcidius (Waszink, *Timaeus*, p. 332, line 11): "ut conceptim dicatur," and *Thesaurus linguae latinae*, vol. 4, col. 62, s.v. "concepte."

9 "duo capita" (II.28). The "two headings" are, in modern parlance, astronomy and astrology.

10 "irradiant influunt" (II.29). These are technical terms in astrology, as is clear from, for example, Campanus, *Tractatus de sphera* (in *Sphera mundi nouiter recognita cum commentariis 7 authoribus* [Venice, 1518], chap. 14, fol. 154r[a]): "propter quod oportuit ea [*scilicet* corpora celestia] circulariter moueri vt orbiculariter inferant [influant: MS Venice, Biblioteca Marciana, VIII, 69, fol. 34r] hos effectus in omnem mundi distantiam. Hec autem agunt irradiando in ista inferiora... irradiatio autem est influentia secundum latitudinem [rectitudinem: MS]. Unde constat quod orbiculariter mouentur corpora celestia et elementa [circa elementa et elementata: MS] vt per eorum motum circularem continua successione influant in ipsa irradiando directe in singula eorum successiue."

11 "iudicantis" (II.30). "Judgment" implies that one is going to predict the effects on earthly things of the rays of the heavenly bodies. Hence the term "judicial astrology." For the contrast between "demonstrantis" and "iudicantis," cf. Campanus, *Tractatus de sphera*, ed. Venice, 1518, chap. 29, fol. 156r[a]: "Idcirco [sic] autem de tot circulis fecimus mentionem: quia omnes veniunt in magnam vtilitatem astronomi siue considerantis siue iudicantis."

12 "in suam... diuisa" (II.31–32). The division of astronomy into theoretical and practical parts would imply, for a modern, something like the distinction between the theory of motion (celestial mechanics) and the observation of heavenly phenomena. The difference between this and Campanus's division is significant: for him "theoretical" implies qualitative description, "practical" quantitative description. His distinction is exemplified in the arrangement of his descriptions of the systems for the individual bodies, where first the layout and relationships of the various circles are given, and then the relative and absolute parameters. Ptolemy too often gives first a qualitative description (with geometrical proofs) of the model which will produce the *kind* of effects exemplified by the phenomena in question (themselves described qualitatively), before going on to the determination of the actual parameters; see, e.g., *Almagest* 3. 3 (Ptolemy, Manitius, vol. 1, pp. 148–65). But the establishment of his system as a whole is achieved by a constant interplay of theory and observation, and Campanus's classification here is ludicrous if it is meant as a description of the *methodology* of astronomy.

13 "prima principia geometrie" (II.35). Euclid's exposition of geometry is entitled στοιχεῖα ("elements").

14 "per duplicem introitum" (II.48–49). This does not mean "double-entry" in the modern sense, i.e., entering a table which is a function of two variables with both vertical and horizontal arguments (examples of such a table do occur in the *Almagest*, but they are not used for computing planetary positions and are never mentioned by Campanus). Instead Campanus is referring to the practice of entering the *same* column *twice* with arguments less than and greater than the desired one and interpolating between them. E.g., if one wants to find the equation corresponding to an argument of 30;42°, since the functions are (normally) tabulated only for integer degrees, one enters first with 30°, then again with 31°, and interpolates (linearly) between the two corresponding values of the equation.

15 "in hoc opere... corrigere" (II.45–50). Campanus describes in order the various operations needed to compute a planetary position from Ptolemaic or similar tables. For a detailed account of the latter and of their mode of operation, see pp. 41–51, above, where the various technical terms are explained (for the latter, see also the index of technical terms, below).

16 "indebiles" (II.53). Though this form is not found in classical Latin or in any of the standard dictionaries of medieval Latin, it has been posited by Romance philologists to explain the Old French (thirteenth century) and Catalan form "endeble" ("feeble"). See *Tobler-Lommatzsch, Altfranzösisches Wörterbuch* (Wiesbaden, 1954——), vol. 3, s.v. "endeble."

17 "almanach" (II.54). By this word Campanus probably means a work giving the true places of the planets for a given year or period of years every *n* days (*n* usually equals 5 or 10). Such works were very popular in the later Middle Ages and thereafter (in fact, they were a standard source of income for the professional astronomer in the sixteenth century). There is a description of the construction

of this kind of almanac, supposedly by Theon of Alexandria (fl. A.D. 364), but more probably from the later Byzantine era, printed in N. Halma, ed., *Commentaire de Théon sur les Tables Manuelles de Ptolemée; Tables Manuelles Astronomiques de Ptolemée et de Théon* (Paris, 1822–25), pt. 3, pp. 38–42, and translated with commentary in J. B. J. Delambre, *Histoire de l'astronomie ancienne* (Paris, 1817), vol. 2, p. 635. Almanacs are mentioned as a common thing before A.D. 1000 by al-Bīrūnī (*The Chronology of Ancient Nations*, trans. C. E. Sachau [London, 1879], p. 6; however, he does not use the word "manāk").

Another meaning of "almanach" (less likely here) is "perpetual almanac"—true positions of the planets worked out for every *n* days of a planetary "period" (an integer number of years in which the planet returns to about the same position), with instructions for adaptation from one such period to the next. For an example, see the "Almanac of az-Zarqāl" (printed in Millás Vallicrosa, *Estudios sobre Azarquiel*, chaps. 3 and 4). There is a good deal of information on subsequent medieval almanacs of that type in chapters 6 to 9 of the same work. It is the type mentioned in a contemporary reference by Roger Bacon, *Opus tertium* (in *Opera quaedam hactenus inedita*, ed. Brewer, vol. 1, p. 36): "Sed hae tabulae vocantur *Almanach* vel *Tallignum*, in quibus semel sunt omnes motus coelorum certificati a principio mundi usque in finem."

The origin of the word "almanach" is still unknown: it is not from an Arabic root, but it probably came into European languages via the Arabic of North Africa and Spain. No certain instance of its occurrence has been found in Arabic texts earlier than the fourteenth century, but it is found in an Arabic-Spanish glossary of the late thirteenth century (see R. P. A. Dozy, *Supplément aux dictionnaires arabes*, 2d ed. [Leiden, 1927], vol. 2, p. 734). For details see G. Levi della Vida, "Appunti e questi di storia litteraria araba 5. 'Almanacco,'" *Revista degli studi orientali*, vol. 14 (1934), pp. 265–70, to which H. J. P. Renaud, "L'Origine du mot 'Almanach,'" *Isis*, vol. 37 (1947), pp. 44–46, adds little of significance. The word is found in medieval Latin texts from the twelfth century onward; for a good collection of examples, see Moritz Steinschneider, "Über das Wort Almanach," *Bibliotheca mathematica*, n.s., vol. 2 (1888), pp. 13–16; see also *Oxford English Dictionary* (Oxford, 1933), s.v. "almanac," and P. F., "Bulletin critique et chronique bibliographique," *Bulletin du Cange*, vol. 14 (1939), p. 77. An earlier example than any cited by these is found in R. Abraham ibn Ezra, *De rationibus tabularum* (ca. 1148) (in *El libro de los fundamentos de las Tablas astronómicas de R. Abraham ibn 'Ezra*, ed. J. M. Millás Vallicrosa [Madrid, 1947], p. 119): "Sed vere sunt tabule que singulis diebus docent coequare planetas vel a tempore determinato dant rationes componendi almanac, id est tabulas per quas semel factas per totum annum planetas coequatos habebis."

18 "instrumentum" (II.62). On Campanus's equatorium, see pp. 30–33, above.

19 "tabule" (II.74). This was the technical term for the removable disks, containing the stereographic projection of various circles for a given latitude, which fitted

into a hollow in the main body of the astrolabe, known as the "mother." These disks were often inscribed on both sides, each side for a different horizon. See William H. Morley, *Description of a Planispheric Astrolabe Constructed for Sháh Sultán Husain Safawí* (London, 1856) (reprinted in R. T. Gunther, *The Astrolabes of the World* [Oxford, 1932], vol. 1, p. 9), and the treatise of Severus Sabokt (ibid., p. 83). There are several examples of astrolabes with several interchangeable disks described in Gunther's work, e.g., no. 3 (ibid., p. 114). Campanus proposes that his three disks, for moon and Mercury, Venus and Mars, and Jupiter and Saturn, respectively, should likewise all be of the same size and fit into the same hollow and that both sides should be utilized. This proposed construction might cause practical difficulties, for which see pp. 32–33, above.

Section III

For the Ptolemaic theory of the sun, which Campanus is paraphrasing, see pp. 41–42, above.

1 "59 minuta et 8 secunda" (III.3). Ptolemy, in the *Almagest* 3. 1 (Ptolemy, Manitius, vol. 1, p. 147), gives 0;59,8,17,...° per day. But 0;59,8° would be the first approximation from any parameter, and Campanus probably derived it from the Toledan tables, where it is the entry for day 1 in the solar mean motion table (Toomer, "Survey of the Toledan Tables," p. 47, Table 28). In chapter 13, sec. 3, al-Farġānī (Alfraganus, Carmody, p. 24) says, "Soli autem sunt duo motus... quorum unus est...in omni die et nocte 59 minutorum fere."

2 "die naturali" (III.3). I.e., a νυχθήμερον of 24 hours. The adjective is added to distinguish it from the "day" in popular parlance, the period from sunrise to sunset. Cf. Isidore *Etymologies* 5. 30: "...dies legitimus uiginti quattuor horarum, usque dum dies et nox spatia sui cursus ab oriente usque ad alium orientalem solem caeli uolubilitate concludit. abusiue autem dies unus est spatium ab oriente sole usque ad occidentem." The earliest example of this expression quoted by du Cange (*Glossarium mediae et infimae latinitatis*, new ed. [Niort, 1883–87], vol. 3, p. 109, s.v. "dies naturalis") is from 1375, but it is found in Sacrobosco's *Spera* (e.g., Thorndike, "*Sphere*" *of Sacrobosco*, p. 101), and in the *De sphaera* of Grosseteste (*Die philosophischen Werke*, ed. Baur, p. 22), both from the first half of the thirteenth century.

3 "uersus finem geminorum" (III.9). The solar apogee is placed by Campanus in Gemini 17;50°, as appears from lines III.124–25 (see also note 29, below). Though not "the end" of the sign, it is nearer the end than the beginning.

4 "duabus...60 partes" (III.9–11). Ptolemy (*Almagest* 3. 4: Ptolemy, Manitius, vol. 1, p. 170) finds 2;29,30 parts, which he immediately takes as $\frac{1}{24}$ (= $2\frac{1}{2}$ parts exactly). It was standard practice in Greek trigonometry and astronomy to express the diameter of any circle as 120 parts (no doubt so that one degree should

subtend at the circumference an amount roughly corresponding to one part). Al-Farġānī (16. 1: Alfraganus, Carmody, p. 30) also gives the sun's eccentricity as $2\frac{1}{2}$ parts out of 60.

5 "sonat" (III.16). The use of this word to express the "etymological meaning" appears strange, but it is found not infrequently in medieval works, e.g., MS Paris, Bibliothèque Nationale, fonds latin, 16208, fol. 94r[a], line 14: "Kardaga nomen siriacum est et in latino sonat abscissio"; and MS Oxford, Bodleian Library, Arch. Seld. B.34, fol. 24v, line 5: "de tabulis in suo azig ad alcardeiet at alieb atque almaiel dispositis et quid sonet alieb et almaiel." A somewhat similar use of "sonare" is occasionally found in classical writers, e.g., Cicero *De finibus* 2. 6: "dico...Epicurum...non intellegere interdum, quid sonet haec uox uoluptatis, id est, quae res huic uoci subiciatur." Solinus, *Collectanea*, chap. 11, sec. 8 (ed. Mommsen, p. 81): "Britomartem gentiliter nominantes, quod sermone nostro sonat virginem dulcem." Modern Italian "s(u)onare" can also bear this meaning.

6 "aux...eleuatio" (III.16). "Aux" is the Latinized form of the Arabic "awj," itself derived from the Sanskrit "ucca," which was used by the Hindus for "apogee" but literally means "upper point," so Campanus's etymology is correct.

7 "oppositio augis" (III.19). "Perigee," or, literally, "opposite point to the apogee" (cf. Thorndike, "*Sphere*" *of Sacrobosco*, chap. 4, p. 113; and Grosseteste, *De sphaera*, chap. 4, in *Die philosophischen Werke*, ed. Baur, p. 22). But the etymology given by Campanus points to the usual Arabic word for perigee, "ḥaḍīḍ," which does indeed mean "lowest point."

8 "Si...solis" (III.20–26). For the system of spheres, see pp. 53–56, above. Here Campanus says that, if you draw around the center of the universe two circles, the inner one touching the sun at its least distance from the earth and the outer one touching it at its greatest distance, and then rotate each circle about its own diameter, the space enclosed between the two spheres so formed will be the "sphere" of the sun. The rotation is described as the movement of a semicircular arc chosen at random about the diameter forming its base, a description derived from the definition of a sphere in Euclid's *Elements*, bk. 11, definition 14 (ed. J. L. Heiberg [Leipzig, 1883–88], vol. 4, p. 4). Cf. Thorndike, "*Sphere*" *of Sacrobosco*, chap. 1, p. 76: "Spera igitur ab Euclide sic describitur: spera est transitus circumferentie dimidii circuli quotiens fixa diametro quousque ad locum suum redeat circumducitur." Cf. also Grosseteste, *De sphaera* (*Die philosophischen Werke*, ed. Baur, p. 11), and Campanus, *Tractatus de sphera*, ed. Venice, 1518, chap. 2, fol. 153r[a].

9 "uestigia" (III.27). Literally "tracks," i.e., the paths traced through space by the moving semicircles. Cf. Campanus, *Tractatus de sphera*, ed. Venice, 1518, chap. 1, fol. 153r[a]: "Punctus motus describit lineam. Cum enim habet situm in magnitudine et nullam habet partem: vestigium motus eius non potest esse aliud quam longitudo."

10 "trahens secum augem" (III.29–30). I.e., the apogees of the planets (including

the sun, but not of course the moon) are fixed in relation to the fixed stars and share their motion of precession (see note 11, below). Though Campanus does not point it out, the idea that the sun is included in the precession differs from Ptolemaic doctrine: Ptolemy thought that the apogee of the sun, unlike that of the planets, was fixed in relation to the tropical points (and therefore moved through the fixed stars), for he had calculated its tropical longitude to be the same as that determined by Hipparchus about 270 years before (*Almagest* 3. 4: Ptolemy, Manitius, vol. 1, p. 167). The ninth-century Islamic astronomer al-Battānī showed on the basis of his observations that the sun's apogee, too, was subject to the motion of the fixed stars (see C. A. Nallino, *Al-Battānī sive Albatenii "Opus astronomicum,"* Pubblicazioni del Reale Osservatorio di Brera in Milano, no. 40 [Milan, 1899–1907], pt. 1, p. 216). This became the standard medieval view, and is found in al-Farġānī 13. 2 (Alfraganus, Carmody, p. 24): "...mutantur auges 7 planetarum [i.e., sun, moon, and five planets]...per successionem signorum in omnibus 100 annis...uno gradu."

11 "omnibus...gradu" (III.30–31). Cf. chap. 13 of al-Farġānī (Alfraganus, Carmody, p. 24): "...alius est motus tardus qui est spericus super axes circuli signorum qui est aequalis motui spere stellarum fixarum, id est in omnibus 100 annis uno gradu."

The "motion of the fixed stars" is the motion of precession (the slow increase in the longitudes of the fixed stars with respect to the point of the vernal equinox). It was discovered by Hipparchus (about 130 B.C.) and tentatively estimated by him to be about 1° in 100 years. Ptolemy (*Almagest* 7. 2: Ptolemy, Manitius, vol. 2, pp. 12–15) confirmed Hipparchus's figure about 270 years afterward (and hence Campanus uses the same figure). This speed is, however, too low (because of a small error in Ptolemy's parameter for the sun's mean motion: see Augustinus Ricius, *De motu octauae Sphaerae* [Paris, 1521], fols. 38v–40r; Pierre Simon de Laplace, *Exposition du système du monde* [Paris, 1813], bk. 5, p. 383; and J. L. E. Dreyer, "On the Origin of Ptolemy's Catalogue of Stars. II," *Monthly Notices of the Royal Astronomical Society*, vol. 78 [1918], p. 346); it was only on the basis of more exact determination of the sun's motion that later astronomers, notably Islamic ones, were able to correct this figure. (But the earliest known examples of the value of 1° in 66 years, or rather 1½° in 100 years, come from Hindu astronomical works [see H. T. Colebrooke, "On the Notion of the Hindu Astronomers concerning the Precession of the Equinoxes and Motions of the Planets," *Asiatick Researches*, vol. 12 (1816), pp. 209–50].) Estimates of 1° in 70 or 66 years, much nearer the truth, were usually accepted in medieval times.

However, because of discrepancies in the observations, it was thought by some that precession was not constant, and eventually the curious theory of "trepidation" was propounded, according to which the equinoctial points moved only over a limited range, and when they reached the bounds of this range, the direction of precession was reversed. Though the trepidation theory originated in antiquity (see Theon's *Introduction*, in Halma, *Commentaire de Théon*, pt. 1, p. 53) and is also found in Hindu astronomy, it is particularly associated with the ninth-century astronomer Ṯābit ibn Qurra of Harran, who wrote in Arabic and is in the Islamic tradition though he was himself a pagan ("Sabaean"). There exists a Latin translation of a work ascribed to him, entitled *De motu octaue spere* (printed in Millás Vallicrosa, *Estudios sobre Azarquiel*; Millás Vallicrosa, "El *Liber de motu octave sphere* de Ṯābit ibn Qurra," *Al-Andalus*, vol. 10 [1945], pp. 89–108; Millás Vallicrosa, *Nuevos estudios sobre historia de la ciencia española* [Barcelona, 1960]; and Carmody, *Astronomical Works of Thabit b. Qurra*, pp. 102–13; and translation and commentary in O. Neugebauer, "Thabit ben Qurra 'On the Solar Year' and 'On the Motion of the Eighth Sphere,'" *Proceedings of the American Philosophical Society*, vol. 106 [1962], pp. 264–99). This work is clearly the source of Campanus's remarks on the theory here. In the *De motu* we are asked to imagine an "ecliptic" set at a fixed angle to a celestial "equator." Set at right angles to the planes of this ecliptic and equator are two small circles, having their centers at the intersections of the circumferences of those two. Another ecliptic (the one in which the sun actually moves) is attached to two points which move around the circumference of these small circles, their vectors always differing by 180°. The two points by which the second ecliptic is attached are known as "caput arietis" and "caput libre" and each must be some fixed star of the appropriate constellation. Since the two points on the second ecliptic 90° distant from the "caput arietis" and the "caput libre" always touch the fixed ecliptic, the second ecliptic moves back and forth relative to the "equator." The sphere of the fixed stars is immovable relative to the second ecliptic. (For a more detailed account see Neugebauer, "Thabit ben Qurra," p. 290, or Toomer, "Survey of the Toledan Tables," pp. 118–22.) Campanus's description fits the above account except that he seems to identify the "capita arietis et libre" with the centers of the small circles instead of the points moving around their circumferences. This is merely an error, not a deliberate change.

12 "diuidens...equalia" (III.34–35). For Campanus the zodiac is a belt 12° wide, through the center of which runs the ecliptic. See note 14, below.

13 "nodus capitis...nodus caude" (III.41). The origin of this "head" and "tail" nomenclature for the nodes is to be found in Hindu mythology. The primitive notion that eclipses were caused by a monster swallowing the sun or moon was retained in the names applied in Sanskrit astronomy to the lunar nodes (in or near which eclipses must occur), Rāhu and Ketu. In the mythology Rāhu was a demon whose head had been set in the sky by Vishnu to devour the two luminaries. According to some, he was in the form of a snake with only head and tail (see Varāhamihīra, *Brihat Samhita*, ed. and trans. V. Subrahmany Sastri [Bangalore, 1947], p. 45). This version was taken over by the Islamic astronomers, who therefore called the lunar nodes "ra's at-tinnīn" and "ḏanab at-tinnīn" (head and

tail of the serpent); cf. al-Farġānī. 12. 15 (Alfraganus, Carmody, p. 24): "nominaturque figura que accidit ex abscisione circuli planetarum et circuli signorum atannin i.e. draco, et punctus quo incipit planeta progredere [?] uersus septentrionem a circulo signorum nominatur rasatannin i.e. caput draconis quod est genzahar [read 'geuzahar']; punctus uero ei oppositus nominatur adeneb id est cauda." In medieval Latin these terms are translated as "caput draconis" and "cauda draconis" or, as in Arabic also, "caput" and "cauda" alone. (See, e.g., al-Farġānī 18. 6: Alfraganus, Carmody, p. 32. Cf. also Jābir, in Petrus Apianus, *Instrumentum Primi Mobilis.... Accedunt iis Gebri filii Affla Hispalensis astronomi vetustissimi pariter et peritissimi libri IX de Astronomia, ante aliquot secula Arabice scripti et per Giriardum Cremonensem latinitate donati* [Nuremberg, 1534], p. 74: "...quando est in coniunctione prima, recedens a nodo caudae, et est in coniunctione secunda uadens ad nodum capitis.") Here the terminology is extended to cover the nodes of all the planets.

14 "latitudinis...graduum" (III.47). The zodiac was originally just the group of 12 constellations through which the sun passes on its annual course. But it came to be thought of early in antiquity as a belt of a definite width on either side of the ecliptic. No doubt this view arose from the practice of constructing celestial globes, on which such a conventional representation would be convenient. (The Farnese globe has just such a band, about 12° wide, representing the zodiac. See Georg Thiele, *Antike Himmelsbilder* [Berlin, 1898], plates 2–6.) The germ of the idea can be seen in Aristotle *Metaphysics* 12. 8. 1073^{b}20, where one of Eudoxus's spheres is said to move κατὰ τὸν λελοξωμένον ἐν τῷ πλάτει τῶν ζῳδίων (κύκλον). The idea is clearly expressed by Theon of Smyrna (2d century A.D.), *Expositio rerum mathematicarum ad legendum Platonem utilium*, ed. Eduard Hiller (Leipzig, 1878), p. 133, line 17: ἔτι τῶν μὲν ἄλλων κύκλων ἕκαστος ὄντως ἐστὶ κύκλος ὑπὸ μιᾶς γραμμῆς περιεχόμενος. ὁ δὲ λεγόμενος ζῳδιακὸς ἐν πλάτει τινὶ φαίνεται καθάπερ τυμπάνου κύκλος. (Cf. also Hyginus, *Astronomica*, ed. Émile Chatelain and Paul Legendre [Paris, 1909], bk. 1, sec. 6, p. 6; Attalus, as quoted by Hipparchus [*In Arati et Eudoxi Phaenomena Libri Tres*, ed. K. Manitius (Leipzig, 1894), p. 88, line 23–p. 90, line 1].) The figure of 12° is first mentioned in Geminus (1st century B.C.), if the passage is not a later interpolation (Geminus, *Elementa astronomiae*, ed. Karl Manitius [Leipzig, 1898], chap. 5, sec. 53, p. 62, line 8): τὸ δὲ πλάτος ἐστὶ τοῦ ζῳδιακοῦ κύκλου μοιρῶν ιβ′. However, the connection of the width of the zodiac with the latitudinal limits of the planets' orbits appears first in Pliny the Elder *Naturalis historia* 2. 66: "...huic conexa latitudinum signiferi obliquitatisque causa est. per hunc stellae quas diximus feruntur...Veneris tantum stella excedit eum binis partibus...luna quoque per totam latitudinem eius uagatur, sed omnino non excedens eum. ab his Mercurii stella laxissime, ut tamen e duodenis partibus (tot enim sunt latitudinis) non amplius octonas pererret, neque has aequaliter, sed duas medio eius et supra quattuor, infra duas. sol deinde medio fertur inter duas partes, flexuoso draconum meatu inaequalis,

Martis stella quattuor mediis, Iouis media et super eam duabus, Saturni duabus ut sol." Ignoring the peculiar doctrine about a latitudinal motion of the sun (for which cf. the opinion of Attalus quoted by Hipparchus, above), we find here a zodiac 12° wide within the bounds of which all the planets move, except Venus, which goes as much as 2° outside it. This is almost certainly the source of Martianus Capella's remarks (*De nuptiis Philologiae et Mercurii*, ed. Adolf Dick [Leipzig, 1925], bk. 8, sec. 867, pp. 456–57): "nam in zodiaco XII esse latitudinis partes superius intimaui [cf. ibid., sec. 834, p. 438] per quas diuersis modis sidera spatiantur. nam alia per tres partes, alia per quattuor, alia per octo, quaedam per omnes XII deferuntur." He omits the exception of Venus (thus misrepresenting the situation, for Venus can achieve a southerly latitude of up to 8°). (See also Cleomedes, *De motu circulari corporum caelestium libri duo*, ed. Hermann Ziegler [Leipzig, 1891], bk. 1, chap. 4, p. 34, where, however, no values are given.) Campanus very probably derived his statement here directly or indirectly from Martianus. Cf. Grosseteste, *De sphaera* (*Die philosophischen Werke*, ed. Baur, p. 14), and Campanus, *Tractatus de sphera*, ed. Venice, 1518, chap. 23, fol. 155r[b].

15 "1210 uicibus" (III.49). Ptolemy *Almagest* 5. 15 (Ptolemy, Manitius, vol. 1, p. 312).

16 "quinquies...et medietatem" (III.49–50). Ptolemy *Almagest* 5. 16 (Ptolemy, Manitius, vol. 1, p. 313). The ratio of the volumes, 166, is merely $5\frac{1}{2}$ cubed to the nearest whole number (manuscripts *D*, *J*, and *K* give the exact cube, $166\frac{3}{8}$, for which see also lines VI.564–65, below). Ptolemy (*Almagest* 5.16: see above) more sensibly says "about 170."

17 "spatium...cubitis" (III.51–53). This figure for the length of one degree in latitude comes from chap. 8, sec. 2, of al-Farġānī (Alfraganus, Carmody, pp. 13–14). The story about the measurement by the wise men in the time of al-Ma'mūn comes from the same source. "Cumque abierimus in terra uersus meridiem et septentrionem super lineam medii diei, augebitur altitudo axis septentrionalis a circulo emisperii et minuetur ex eo secundum quantitatem ambulationis nostre in terra, inueniemus quoque post hoc quod portio unius gradus circuli ex rotunditate terre sit 56 miliariorum et 2 tertiarum unius miliarii per miliarium quod est 4000 cubitorum per gradus equales secundum quod solicite probatum est in diebus Almenon [*al.* Almehon] et conuenerunt super probationem eius sapientes plures numero." Al-Farġānī was a contemporary of the Caliph al-Ma'mūn (reigned 813 to 833), and is the earliest witness. But the story is often repeated in Islamic sources. See C. A. Nallino, "Il valore metrico del grado di meridiano secondo i geografi arabi," in *Raccolta di scritti editi e inediti*, vol. 5 (Rome, 1944), pp. 420–26, for details, and pp. 435–52 of the same article for a discussion of the metrical value of the units employed and the accuracy of the measurement. If Nallino is right, the Arabic "mile" ("mīl") employed by al-Ma'mūn's measuring teams was considerably longer than the "mile" ("miliare") of the West in Campanus's time. However, except for his definition of "miliare" here, Campanus does not seem

to consider any possible discrepancy, and hereafter he expresses absolute distances in "miles"; all such absolute distances are ultimately derived from this one measurement of the terrestrial degree of latitude.

18 "sumpta...ambitus" (III.54–59). The method suggested for measuring a degree of latitude, namely, finding the absolute distance between two points on the same meridian whose latitudinal difference is known, is essentially the same as that of Eratosthenes, who was the first to carry out such a measurement to determine the size of the earth, and of most subsequent attempts in antiquity (see M. R. Cohen and I. E. Drabkin, *Source Book in Greek Science* [Cambridge, Mass., 1948], pp. 149–53) and the Middle Ages. However, the particular feature of actually measuring exactly one degree to north or south strongly recalls the account given by ibn Khallikān of the measurement in the time of al-Ma'mūn (see Nallino, "Il valore metrico," p. 424).

19 "ALMEON" (III.60). This spelling is proof that Campanus used John of Seville's translation of al-Farġānī. Gerard of Cremona's version (Alfraganus, *Il 'libro dell'aggregazione,'* ed. Campani, p. 89) renders al-Ma'mūn's name as "Maimonis," "Maymonis," or the like. Cf. p. 34, above.

20 "Si...tertiam" "(III.61–63). The value of π implied in this procedure, $\frac{22}{7}$, is that invariably employed by Campanus. Its origin is the treatise of Archimedes, *Measurement of the Circle* (see *Opera omnia cum commentariis Eutocii*, ed. J. L. Heiberg, 2d ed. [Leipzig, 1910–15], vol. 1, p. 236), in which he established that π lies between the limits $3\frac{1}{7}$ and $3\frac{10}{71}$. The value $3\frac{1}{7}$ became standard in antiquity: it is used, for instance, by Heron *Metrica* 1. 26 (*Heronis Alexandrini Opera quae supersunt omnia*, vol. 3, *Rationes dimetiendi et commentatio dioptrica*, ed. H. Schoene [Leipzig, 1903], p. 64) and the Gromatici (Karl Lachmann, ed., *Gromatici Veteres*, in *Die Schriften der Römischen Feldmesser* [Berlin, 1848], vol. 1, p. 355, line 14). It is found in the *Geometry* attributed to Boethius (*De institutione arithmetica*, ed. Friedlein, p. 423) and hence, probably, occasionally in the early Middle Ages. Examples are Adelbold's letter to Gerbert (ca. 1000) (see Maximilian Curtze, "Die Handschrift No. 14836 der Königl. Hof- und Staatsbibliothek zu München," *Abhandlungen zur Geschichte der Mathematik*, vol. 7 [1895], p. 133) and Franco of Liège (ca. 1040) (see C. Winterberg, "Der Traktat Franco's von Lüttich *De quadratura circuli*," ibid., vol. 4 [1882], no. 2). In general see Johannes Tropfke, *Geschichte der Elementar-Mathematik*, vol. 4, *Ebene Geometrie*, 3d ed. by Kurt Vogel [Berlin, 1940], pp. 280–83). By the time of Campanus, however, several Latin versions of the above treatise of Archimedes were in circulation: the texts are printed in Clagett, *Archimedes*, vol. 1, pp. 20–432. (A text on the quadrature of the circle falsely attributed to Campanus is in the same work, pp. 588–604.) The value $\frac{22}{7}$ was undoubtedly standard long before Campanus's time.

21 "6490...miliaris" (III.64). $6490\frac{10}{11} = 20{,}400 \cdot \frac{7}{22}$.

22 "miliaria 3923754...miliaris" (III.65). $3{,}923{,}754\frac{6}{11} = (6490\frac{10}{11} \div 2) \cdot 1209$, that

is, the earth's radius multiplied by the distance earth–sun in earth-radii (see lines III.49–50, above) less one earth-radius.

"35700 miliaria" (III.66). $35{,}700 = 6490\frac{10}{11} \cdot 5\frac{1}{2}$ (earth's diameter multiplied by linear ratio sun–earth (see lines III.49–50, above).

"miliaria 112200" (III.67). $112{,}200 = 35{,}700 \cdot \frac{22}{7}$.

23 "piramidalem" (III.73–74). Strictly speaking, a "piramis" as used in medieval optics is not a cone, but a translation of the Arabic maḫrūṭ, meaning any body formed by the meeting in a single point of straight lines proceeding from the perimeter of its base, which can be an arbitrary surface. (See Matthias Schramm, *Ibn al-Hay<u>th</u>ams Weg zur Physik*, Boethius, vol. 1 [Wiesbaden, 1963], p. 112 with note 1, for this use in the Latin translation of ibn al-Hayṯam's [Alhazen's] *Optics*. See also Albert Lejeune, *Euclide et Ptolémée: Deux stades de l'optique géometrique grecque*, Univ. Louvain, Recueil de travaux d'histoire et de philologie, ser. 3, fasc. 31 [Louvain, 1948], p. 34, note 8. Leonardo of Pisa, too, understands by "piramis" the genus including cone and pyramid [Baldassarre Boncompagni, *Scritti de Leonardo Pisano*, vol. 2 (Rome, 1862), dist. 6, p. 169]: "Cum itaque piramidem aliquem metiri desideras, embadum sue basis, *cuiuscumque sit forme*, per tertiam altitudinis ipsius multiplica." Witelo [Vitellio; ca. 1270] uses "pyramis rotunda" specifically for cone [*Opticae libri decem*, in *Opticae thesaurus*, ed. Friedrich Risner (Basel, 1572), propositions 89–91, 95–98, etc.].) Since "piramis" includes both pyramid and cone in the modern sense, it is understandable that Campanus, like Witelo, uses "piramis rotunda" (line III.74) when he means "cone" in the strict sense.

The expressions "umbra piramidalis" and "conus" (= "apex," line III.76) also occur in Grosseteste's *De sphaera* (*Die philosophischen Werke*, ed. Baur, p. 29). Cf. also Jābir (Apianus, *Instrumentum Primi Mobilis*, p. 46): "...causa faciens eclipses lunares non sunt nisi introitus lunae in pyramidem umbrae terrae, et est pyramis quae accidit ex casu radii solis super illud." For a much earlier use of "pyramis" in the sense of "cone" (also in the context of the earth's shadow-cone), see the Epistle of Sisebut (contemporary of Isidore of Seville, early seventh century), in Jacques Fontaine, ed., *Isidore de Seville*, *"Traité de la nature," suivi de l'épitre en vers du roi Sisebut à Isidore*, Bibliothèque de l'école des hautes études hispaniques, fasc. 28 (Bordeaux, 1960), p. 333, lines 38–41: "solis lumina... tendunt per inania uasta/Donec pyramidis peragat uicta umbra cacumen."

24 "267 uicibus" (III.86). Ptolemy (*Almagest* 5. 15: Ptolemy, Manitius, vol. 1, p. 312) gives 268 earth-radii as the distance from the earth's *center* to the tip of the shadow-cone; Campanus, who is giving the distance from the earth's *surface*, has subtracted one earth-radius from that figure.

25 "866536...miliaris" (III.88). $866{,}536\frac{4}{11} = (6{,}490\frac{10}{11} \div 2) \cdot 267$.

26 "sensibili" (III.89). This word, which I have translated "material," means literally "available to the senses" (in contradistinction to the parts of the actual "solar sphere," which are invisible and untouchable).

27 "instrumento...necessaria" (III.89–90). Most of Campanus's description of the construction of the part of the equatorium for the sun should be sufficiently illustrated by Figure 10, p. 152.

28 "inter quamlibet" (III.117). If Campanus had written "sub qualibet" (cf. line III.121), the Latin would be less confused, and for clarity I have here translated what he means rather than what he has written.

29 "17 gradus...geminorum" (III.124–25). Like all of Campanus's apogee parameters, this value is taken from the Toledan tables. See p. 35, above.

30 "numerus signorum" (III.142). Here Campanus uses "signum" to mean a 30-degree division of a circle which is not that of the ecliptic. Hence these signs, not being signs of the zodiac, have numbers, not names.

31 "seta" (III.156). That this should be translated "silk," and not "animal hair," which is a theoretically possible alternative, is shown by the requirement that it be very fine, and also by "Chaucer's" expression (Price, *Equatorie of the Planetis*, p. 26, line 11), "in euery equant mot be a silk thred." This meaning ("silk thread") of "seta" is not found in classical Latin, but is attested from the early twelfth century (hence French "soie"). See du Cange, *Glossarium*, s.v. "seta," and Walther v. Wartburg, *Französisches etymologisches Wörterbuch* (Bonn, Leipzig, Tübingen, and Basel, 1928——), vol. 11, p. 49, s.v. "saeta."

32 "equare solem...instrumentum" (III.157). The process laboriously described by Campanus for finding the place of the sun amounts to this: one finds from tables the mean motion of the sun from apogee for the given time, marks this off on the eccentric (the inner band, circle *BE*, on the equatory in Fig. 10, p. 152), and stretches the thread through this point; the point on the ecliptic (the outer band in Fig. 10) through which the thread then passes will give the true position of the sun.

33 "intrabo...solis" (III.157–59). Campanus's account of the process of finding the "argument of the sun" (its mean motion counted from the apogee) is best illustrated by a worked example. Examples of mean motion tables of the general type which he presupposes here may be found in the "Handy Tables" (Halma, *Commentaire de Théon*, vol. 2, pp. 66–77), in al-Battānī (Nallino, *Al-Battānī*, pt. 2, pp. 19–23), and in al-Khwārizmī (H. Suter, "Die astronomischen Tafeln des Muḥammed ibn Mūsā al-Khwārizmī in der Bearbeitung des Maslama ibn Aḥmed al-Madjrīṭī und der latein. Übersetzung des Athelhard von Bath," *Skrifter, Kongelige Danske Videnskabernes Selskab*, ser. 7, historical and philosophical sec., vol. 3 [1914], no. 1, pp. 115–16, tables 4 and 5). However, we will use the tables for Novara in MS Dublin, Trinity College, D.4.30, which may have been composed by Campanus himself. The relevant tables are on fols. 43r–44r. Our "datum time" is December 23, 1266, 6 P.M. In the table of the Dublin manuscript, "anni collecti" are given in groups of 28 years; thus the "anni expansi" table gives the mean motion for each of 28 consecutive years. The result is the distance of the sun from its position in mean motion at the vernal equinox.

		Signa	Gradus	Minuta	Secunda
[Result]		8	29	24	27
Anni collecti	1260	11	5	43	42
Anni expansi	5	11	29	43	17
Menses	Nouember	9	1	2	33
Dies	23	0	22	40	8
Hore	6	0	0	14	47
[Excess carried from preceding column]		1 (= 1·30°)	2 (= 2·60′)	2 (= 2·60″)	

The tables described in the text differ from those in the Dublin MS in two respects: (1) they do not contain seconds; (2) they tabulate the mean distance of the sun, not from the vernal equinox, but from apogee (i.e., the longitude of the apogee would be embedded in the figures for "anni collecti"). In both these respects the tables described in the text are unlike other medieval tables known to me, but are like the "Handy Tables." The second respect is the more important: it means that the result extracted from the mean motion table can be used directly as the argument for finding the sun's equation, whereas the usual procedure is first to subtract the longitude of the sun's apogee. That his departure from the usual procedure is not an oversight of Campanus is shown by lines V.601–3, where, to determine the mean motion of Mercury, he subtracts from the mean motion of the sun, not the longitude of Mercury's apogee, but the *difference* between the longitudes of the sun's apogee and Mercury's apogee.

34 "distantia" (III.159). Here and often elsewhere throughout the work, the "distance" is to be understood as *angular* distance. The intended meaning should always be clear from the context.

35 "deferentis" (III.160). The word "deferent" is used loosely here for the sun's eccentric: strictly speaking, it should be used only for a circle which "carries" an epicycle. Since, however, the deferent of any of the planets is also an eccentric, it is applied to the sun's nondeferent eccentric by extension. It can be justified by the expression at line III.128, "circulum deferentem solem." Cf. lines V.1345 and VI.510, and Campanus, *Tractatus de sphera*, ed. Venice, 1518, chap. 53, fol. 158v[a]: "deferens lune secat deferentem solis in duas partes."

36 "annis residuis" (III.166). These "remaining years" are the number which results from subtracting the number of "collected years" (the highest multiple of 28, or whatever is the unit of the collected years, which is less than the number of years of the datum time) from the number of years of the datum time (in the above-worked example, from the number of *completed* years of the datum time).

37 "super" (III.175). For this way of writing the sum *above* the summands, see D. E. Smith, *History of Mathematics* (Boston, 1923–25), vol. 2, p. 92, with references.

38 "patet...sole" (III.203). See pp. 41–42, above.

39 "firmamentum" (III.205). This is a biblical way of saying "the outermost sphere" (i.e., the sphere of the fixed stars). Cf. Sacrobosco, *Spera*, chap. 1 (Thorndike,

"*Sphere*" *of Sacrobosco,* p. 77): "speram stellarum fixarum, que firmamentum nuncupatur." See also Grosseteste, *De sphaera,* chap. 1 (*Die philosophischen Werke,* ed. Baur, p. 15, line 24). The usage is derived from such passages as Gen. 1:6 and is first found in the Church Fathers, but less specifically than here. See, e.g., Isidore *De natura rerum* 13. 1 (Fontaine, *Isidore de Seville,* p. 225): "...circulum inferioris caeli...solidauit nuncupans eum firmamentum." Campanus gives an imaginative though incorrect explanation for the usage in his *Tractatus de sphera,* ed. Venice, 1518, chap. 10, fol. 153v[b]: "...stellis quas fixas nominauimus [nominamus: MS Venice, Marc. VIII, 69, fol. 32r]: quarum sphera dicitur octaua....Et dicitur firmamentum quoniam ipsius motus semper videtur esse firmus et vniformis et quia in eo stelle fixe videntur firmari."

Section IV

For the theory of the moon, see pp. 42–47, above.

1 "5...circulos" (IV.2). See lines IV.278–89, with note 46, below, for details of the five circles, and compare the passage in al-Farġānī, chap. 13, secs. 5–7 (Alfraganus, Carmody, pp. 24–25): "Lune uero sunt 5 motus uolubiles: motus scilicet corporis lune, quo mouetur in circulo breui...."

2 "13 gradus...fere" (IV.4). Ptolemy's parameter for the moon's mean daily motion in anomaly is 13;3,53,56,17,51,59°: *Almagest* 4. 4 (Ptolemy, Manitius, vol. 1, p. 210).

3 "capud draconis et cauda" (IV.9). See section III, note 13, above.

4 "declinatio" (IV.12). Here and in what follows, "declination" is used, not in the technical sense of measurement perpendicular to the equator, but more generally of any angular measurement perpendicular to the axis of reference; here the axis of reference is the ecliptic, so the measurement is one of latitude.

5 "uersus orientem" (IV.18). The natural way to translate this would be "toward the east." However, we cannot suppose Campanus made the error of saying that the apogee of the deferent moves from west to east (at line IV.63 he correctly states that it moves toward the west). MSS *f, J,* and *y* have corrected to "occidentem," but the text here is confirmed by "uersus occidentem" at line IV.22, where the movement of the ecliptic is involved (here again "orientem" of some manuscripts is a later correction). Campanus must mean here something like "one part succeeds another for an observer *looking toward* the east," i.e., the parts come up in succession *from* the east.

6 "uersus occidentem" (IV.22). See note 5, above.

7 "apparebit...dicentur" (IV.22–23). The reference is to lines IV.83–87.

8 "5 graduum" (IV.24). So Ptolemy states, *Almagest* 5. 7 (Ptolemy, Manitius, vol. 1, p. 285).

9 "24 gradus...fere" (IV.49). Actually it is twice the daily elongation of the moon

from the sun, which, according to Ptolemy, would be 24;22,53,22,...° (see *Almagest* 4. 4: Ptolemy, Manitius, vol. 1, p. 211). At line IV.65 Campanus gives the more accurate figure of 24;22,52°. The rounded figure here is probably derived from *Almagest* 5. 2 (Ptolemy, Manitius, vol. 1, p. 261), or from al-Farġānī, chap. 13, sec. 10 (Alfraganus, Carmody, p. 25).

10 "ad meridiem" (IV.54). This could equally well be translated "at the meridian of the city of Novara." But the translation "noon" emphasizes that there is an interval of exactly one day between the two situations, and in the similar passage at lines IV.129–30, the alternative translation, "meridian," is excluded.

11 "13 gradibus...34 secundis" (IV.62). Ptolemy gives the mean daily motion of the moon as 13;10,34,58,33,30,30° (*Almagest* 4. 4: Ptolemy, Manitius, vol. 1, p. 210). Campanus's figure here is probably derived from that by dropping, rather than rounding off, the terminal fraction. This procedure introduces an error of 1 in the seconds, but the practice was a common one.

12 "circuli...deferentis" (IV.60–62). This circle which is concentric with the center of the earth and in the plane of the deferent is one of Campanus's "five circles of the moon" (see lines IV.278–89 and note 46, below), but is used purely as a reference circle on which to measure the moon's mean motion.

13 "11 gradibus...18 secundis" (IV.64). This figure for the motion of the apogee of the eccentric relative to the ecliptic was derived as follows: Ptolemy gives (*Almagest* 4. 4: Ptolemy, Manitius, vol. 1, pp. 210, 211) the moon's mean motion per diem as 13;10,34,...° and the moon's mean elongation per diem as 12;11, 26,...°; Campanus doubled the latter, getting 24;22,52° (see line IV.65), and subtracted from the result the former, getting 11;12,18°. Using Ptolemy's full parameters, one would have 11;12,18,24,...°.

14 "respicit" (IV.67). Literally "looks at." This word and the cognate "respectus" were used in medieval astronomy in the situation where a point *A* moves with uniform angular velocity (but not necessarily uniform distance) about point *B*. Then point *A* is said to "look at" point *B* in order to regulate its motion. The usage is extended at line IV.105, where point *A*, the mean apogee of the epicycle, is said to "look at" point *B* because the place of the mean apogee is determined by the straight line joining point *B* to the center of the epicycle; i.e., in this case point *A* is always at the opposite end of a given straight line to the *moving* point *B*.

15 "longitudinem longiorem" (IV.80). This technical term is used for the planets as well as the moon (see the index of technical terms, below, s.v.) and is borrowed from planetary tables (see p. 398, note 62, below). It does not refer, as one might suppose, to the maximum distance of the body from the earth (Campanus expresses the latter by "longior longitudo que unquam esse potest" or similar expressions: see, e.g., lines IV.411–12), for that would require the body to be in the apogee of the epicycle, a situation which is not relevant here; rather it refers to the position of the epicycle in the apogee of the deferent.

16 "11 graduum...7 secundorum" (IV.81–82). This figure is obtained by subtracting from the westward motion of the apogee of the deferent, which is 11;12,18° per day (line IV.64), the westward motion of the nodes, 0;3,11° per day (lines IV. 89–90). It really represents (as stated at lines IV.96–98) only the daily elongation of the apogee of the deferent from the nodes, a quantity totally devoid of astronomical interest. However, the formulation here is taken from *Almagest* 5. 2 (Ptolemy, Manitius, vol. 1, p. 260). Ptolemy gives the motion of the apogee with respect to the nodes because apogee, epicycle, and node are all conceived as moving in the same plane, which is inclined to the ecliptic, and he has chosen the node as the point in this plane from which to measure the other motions.

17 "trium...11 secundorum" (IV.89–90). This is the westward daily motion of the nodes with respect to the ecliptic. Ptolemy tabulates the westward motion of the nodes with respect to the mean longitude of the moon (or vice versa) as 13;13,45, 39,48,56,37° per day (*Almagest* 4. 4: Ptolemy, Manitius, vol. 1, p. 211), the same figure Campanus gives (cut off, as usual, after the seconds) at line IV.100. If one subtracts from the latter the mean daily motion in longitude, one gets 13;13,45° − 13;10,34° = 0;3,11°, as here. The exact Ptolemaic figure would be 0;3,10,41, 15,26,7° per day.

18 "13 gradus...45 secunda" (IV.100). See note 17, above.

19 "respicit" (IV.105). See note 14, above. For the usage here we may compare Jābir (Apianus, *Instrumentum Primi Mobilis,* p. 56): "...diameter orbis reuolutionis transiens per longitudinem longiorem et propiorem, non semper recte *respicit* per motum centri orbis reuolutionis centrum orbis signorum, imo semper recte *respicit* punctum, cuius elongatio...."

20 "Eritque...augem (IV.117–23). This is a remarkable muddle of false and inaccurate statements. (1) Campanus says that the maximum difference between mean and true apogee (that is, when MN in Figure 23 is a maximum) occurs "at about" the mean distance. The "mean distance" can be only the mean between greatest and least distances, i.e., in Figure 23, $DK = CK = CL$, the radius of the deferent. Calculation shows that the double mean elongation (2η) is then 84;2°. However MN is a maximum at a 2η of 114°, as Campanus himself states at lines IV.247–48. So it is 30° away from the position of mean distance. (2) He states that the mean distance occurs when the center of the epicycle is approximately halfway between apogee and perigee of the deferent. This is perhaps not too inaccurate a way of describing a 2η of 84;2°. But he then elucidates this by saying it is a little *more* than 90° and a little *less* than 270°, which is the reverse of the truth. He seems, then, to have been led astray by supposing the condition to be like that of a simple eccentric, where the maximum equation does occur at a mean longitude of slightly more than 90°; compare the false statement at lines IV.264–67, with note 44. In the condition discussed here, neither the mean distance nor the maximum equation occurs at a 2η of slightly more than 90°: the mean distance is before 90°, the maximum equation well past it.

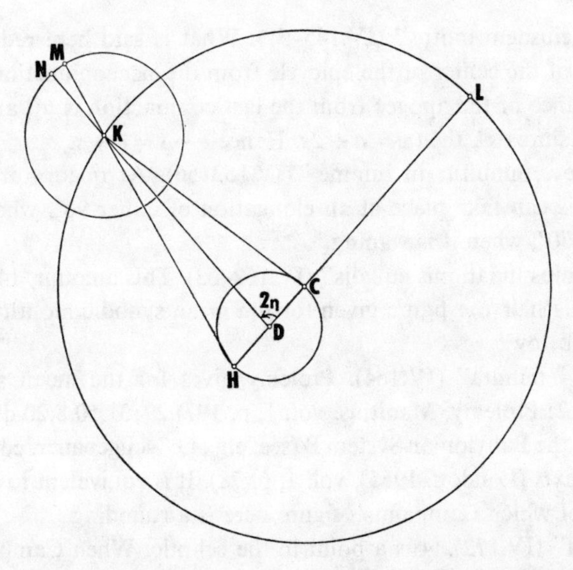

Fig. 23

Illustration of the variation of the distance between mean and true apogees of the moon as a function of the double mean elongation (2η). D represents the earth, C the (momentary) center of the moon's deferent. L is the (momentary) apogee of the deferent, and H the point on the "small circle" opposite C. Then, by Ptolemy's definition, the true apogee of the epicycle is N, on the extension of line DK, and the mean apogee of the epicycle is M, on the extension of line HK. The particular situation depicted is at "mean distance," i.e., when DK, the distance of the epicycle center from the earth, equals CK or CL, the radius of the deferent. Obviously this equality occurs when the double mean elongation ($\angle CDK$) is slightly less than 90°. Campanus asserts that "at about" a 90° double elongation arc MN is a maximum. Inspection of the figure will show that Campanus's assertion is not so, but that angle CDK must be considerably more than 90° for arc MN to be a maximum.

21 "Fuit... deferentis" (IV.124–26). The fact that the center of the epicycle is always in the apogee at mean conjunction is due not, as Campanus imagines, to a peculiar bounty of the Creator, but to the construction of Ptolemy's lunar theory. See pp. 44–45, above.

22 "dictum est prius" (IV.127). See lines III.204–17 and lines IV.58–62.

23 "Ponamus... meridies" (IV.129–30). The idea of assembling all the points in Aries 0° and then examining their positions 24 hours later is directly borrowed from *Almagest* 5. 2 (Ptolemy, Manitius, vol. 1, pp. 261–62).

24 "ariete 13... 34 secunda" (IV.132). See line IV.62, with note 11, above.

25 "piscibus 18... 42 secunda" (IV.133). I.e., the apogee has moved 11;12,18° against the succession of the signs. See line IV.64, with note 13, above.

26 "piscibus 29... 49 secunda" (IV.134). I.e., the node has moved 0;3,11° against the succession of the signs. See lines IV.89–90, with note 17, above.

27 "prefati tres motus" (IV.142). The motions of the center of the epicycle, of the apogee, and of the mean sun.

28 "In omni...eiusdem motus" (IV.145–51). What is said here reduces to this: if the distance of the center of the epicycle from the last conjunction at time t is $e°$, and the distance of the apogee from the last conjunction is $a°$, and the distance of the mean sun is $s°$, then $e = a + 2s$. Hence $e - s = a + s$.

29 "quadratione...minuitur in lumine" (IV.153–56). At quadrature the moon is half full. This can take place at an elongation of either 90°, when the moon is waxing, or 270°, when it is waning.

30 "14 dies...unius lunationis equalis" (IV.162–63). This amount, 14 days 18 hours 22 minutes, is half the figure given for the mean synodic month at line IV.164, see note 31, below.

31 "29 dies...44 minuta" (IV.164). Ptolemy gives for the mean synodic month (*Almagest* 4. 2: Ptolemy, Manitius, vol. 1, p. 197) 29;31,50,8,20 days. This is the parameter of the Babylonian System B (see, e.g., O. Neugebauer, ed., *Astronomical Cuneiform Texts* [London, 1955], vol. 1, p. 78). It is equivalent to 29 days 12;44, 3,20 hours, of which Campanus's figure here is a rounding.

32 "in puncto *A*" (IV.172). *A* is a point in the ecliptic. When Campanus says that the various points are "in" point *A*, he means that *A* is the projection of their actual positions from *D*, the center of the universe, onto the ecliptic (and the same is true of their positions *G*, *E*, etc., on the following day), because the ecliptic is the circle of reference for measuring angular distances.

33 "proprius motus augis" (IV.184). By this term is meant the motion of the apogee with respect to the node. See p. 388, note 16, above.

34 "1 gradus...45 secunda" (IV.193–94). This value for the moon's equation in the particular position described by Campanus and the 1;35,27° for the moon near the perigee of the epicycle (lines IV.198–99) are derived by computation from the tables (on which see pp. 46–47, above).

The data are double mean elongation (2η), 24;22,52° (line IV.216); mean anomaly ($\bar{\alpha}$), 0^s13;3,54° near apogee (line IV.236) and 6^s13;3,54° near perigee. Using the lunar tables from the Toledan tables, which are essentially the same for our purposes here as the "Handy Tables" (Halma, *Commentaire de Théon*, vol. 2, pp. 78–89), we find, with argument 2η, the *equatio centri* (δ) to be 3;34,27°. Since $\bar{\alpha}+\delta=\alpha$, then α equals 0^s16;38,21° (apogee) and 6^s16;38,21° (perigee). With argument α we find the *equatio argumenti* (c_1) to be 1;19,27° and 1;33,58°, respectively. With argument α we find the *equatio circuli breuis* (c_2) to be 0;39° and 0;57°, respectively. With argument 2η we find the *minuta proportionalia* (c_3) to be 0;2. Since $c_2 \cdot c_3$ equals 0;1,18° and 0;1,54°, respectively, and the corrected equation $c = c_1 + c_2 \cdot c_3$, then c equals 1;20,45° and 1;35,52°, respectively.

Campanus gives 1;20,45° and 1;35,27°. The discrepancy in the second figure is very probably due to a small calculating error. For Campanus's own explanation of the use of the tables for the moon, see lines IV.656–749.

35 "equatio...diminuens motum" (IV.195–96). "Subtractive" and, in lines IV.200–

201, below ("equatio...augens motum"), "additive": literally "decreasing its motion" and "increasing its motion," respectively.

36 "Quia...stationaria" (IV.201–6). It is indeed true that the moon moves slowest (with respect to the earth) when it is in both apogees: since the sense of the moon's rotation on the epicycle is opposite to that of the epicycle on the deferent, the motion of the moon on the epicycle will have its strongest negative effect when it is in the apogee of the epicycle. The epicycle itself will have its slowest apparent motion when it is in the apogee of the deferent (cf. Ptolemy's remarks on this type of model in *Almagest* 3. 3 [Ptolemy, Manitius, vol. 1, pp. 153–54]). However, Campanus does not adopt the above reasoning, but instead draws his proof from the tables: he envisages comparing the motion of the moon in one day when it is near the perigee of the deferent with the motion in one day when it is near the apogee. In note 34, above, we have dealt with the situation when the epicycle is near the apogee of the deferent. Let us now take the situation near perigee. For $2\eta = 6^s\,0°$ and $\bar{\alpha} = 0°$, the equation is obviously zero. On the following day, with $2\eta = 6^s\,24;22,52°$ and $\bar{\alpha} = 0^s 13;3,54°$, $\delta = -7;54°$; so α is only $0^s 5;10°$ and c, the equation, is consequently only $0;36,19°$, much smaller than the $1;20,45°$ at the corresponding position near the apogee of the deferent. Since c is negative in both cases, the moon near the perigee of the deferent travels farther in one day—i.e., its speed is greater—than it does near the apogee. When Campanus says that the greatest subtraction from the moon's anomaly for one day's motion occurs when the center of the epicycle is in the perigee of the deferent and the moon is in the apogee of the epicycle, he is quite right, for the change in δ is greatest in the neighborhood of $2\eta = 180°$, and for $2\eta > 180°$, δ is negative.

The conclusion at lines IV.205–6, "...therefore the moon can never become retrograde or even stationary," is a logical one only if one adds the further premise, "...and since the moon cannot be retrograde or stationary when it is in both apogees...." That premise is only contingently true: the model described for the moon *could* produce retrogradation, but with the parameters assumed it *does not*. It is doubtful whether Campanus realized this; he probably thought that the fact that the moon's rotation on the epicycle is contrary to that of the planets (which are retrograde in the perigee of the epicycle) was sufficient to guarantee that it could not be retrograde or stationary.

37 "super...ipsius" (IV.209). That is, at the two points where the vector of the moon's motion on the epicycle is along the line joining moon and center of universe.

38 "in canonibus tabularum" (IV.215). The term "canones" was generally applied in medieval Latin to the rules for the use of astronomical tables which are often found annexed to the tables. One of the commonest sets of rules is the "Canones Arzachelis" which frequently accompanies the Toledan tables.

39 "centrum" (IV.215). This arc or angle (the angle between the apogee and the center of the epicycle) was called κέντρον by Ptolemy, though not in the *Almagest*.

See, e.g., Προχείρων κανόνων ὑποτύπωσις (Ptolemy, *Opera quae exstant omnia*, ed. J. L. Heiberg, vol. 2, *Opera astronomica minora* [Leipzig, 1907], p. 166, line 24). The word became standard in this sense in Greek astronomy and was translated into Arabic as "markaz": see, e.g., Nallino, *Al-Battānī*, pt. 2, p. 335. From Arabic it was translated into Latin as "centrum."

40 "equatio...uariatur" (IV.238–39). The reason why this function is called the equation of center is not the one Campanus gives. See p. 51, above.

41 "inter...oppositionem" (IV.241). I.e., for $0° < 2\eta < 180°$ δ is positive while for $180° < 2\eta < 360°$ δ is negative.

42 "si...circumferentiam" (IV.243–44). This situation is depicted on the right side of Figure 11, p. 172.

43 "3 signa et 24 gradus" (IV.248). The maximum equation of center occurs at this argument in the lunar table of the *Almagest* 5. 8 (Ptolemy, Manitius, vol. 1, p. 286, cols. 1–3). At this point the function is tabulated every 3°. However, in the lunar table of the Toledan tables, which tabulates the function for every degree, the maximum also occurs in this place.

44 "ipsarum...*MN* maximus" (IV.264–67). It is completely untrue that $\angle DKH$ is a maximum when $\angle DHK$ is a right angle. Campanus is presumably confusing the situation with that of a simple eccentric, such as that of the sun, in which the maximum equation occurs at a true longitude of 90° from the apogee (see *Almagest* 3. 3: Ptolemy, Manitius, vol. 1, pp. 155–57). However that means that in the present case $\angle CKH$, *not* $\angle DKH$, is at a maximum when $\angle DHK = 90°$. In fact we can see directly from the lunar table (*Almagest* 5. 8: Ptolemy, Manitius, vol. 1, p. 286) that the statement is false: the greatest difference between the two apogees, 13;9°, occurs at a double elongation of 114°, i.e., $\angle DKH = 13;9°$ and $\angle CDK = 114°$; therefore $\angle DHK = \angle CDK - \angle DKH = 100;51°$. Compare the false statement at lines IV.117–23 (see note 20, above).

45 "dictum...sole" (IV.277). See note 11 of section III, above.

46 "circulum descriptum...lune" (IV.280–81). This circle, as was mentioned in note 12, above, is purely a reference circle against which to measure the motion of the apogee of the deferent and of the center of the epicycle; both of the latter move around a center which is different from the center of the ecliptic. They cannot be measured with reference to the primary carrying circle, that of the nodes (the "fifth circle" at line IV.286), because that primary circle is not in the plane of the ecliptic (to which all motions in longitude must ultimately be referred). Therefore this auxiliary circle, which is immovable with respect to the deferent, but eccentric to it, was introduced by Ptolemy (*Almagest* 4. 6: Ptolemy, Manitius, vol. 1, p. 218).

47 "Habet...11 secundis" (IV.286–89). See note 16, above.

48 "spera lune" (IV.295). With reference to the spheres, see pp. 53–56, above.

49 "cuius...spera" (IV.299–302). For this formulation, see lines III.20–26, with note 8 of section III, above.

50 "speras omnium elementorum" (IV.302–3). For the spheres of the elements, see lines IV.342–65, below.

51 "locus" (IV.317). For this usage, see lines IV.335–36, with note 55, below.

52 "celum esse...mansio" (IV.328). The notion that the highest heaven is "empyrean" and that it is inhabited by angels arose from Christian commentaries on the opening verses of Genesis (cf. "celum cristallinum" in note 54, below). Particular problems were (1) the absence of any mention of the creation of angels in the scriptural account of Creation; (2) in Gen. 1:1 God is said to have created the "heaven" ("In principio creauit Deus caelum et terram"), and in Gen. 1:7 to have created the "firmament" ("Et fecit Deus firmamentum"), which must therefore be different from the heaven of verse 1. A popular solution of both difficulties was to interpret "caelum" of verse 1 as meaning, not the "heaven" which we see (this is the "firmament"), but a superior sphere, and to suppose that its creation involved the creation of the angels, whose natural abode is there. The oldest version of this known to me (and perhaps the original one) is that of Walafrid Strabo (d. 849), in his *Glossa ordinaria* (Gen. I, in J. P. Migne, *Patrologiae cursus completus*, *Series latina* [Paris, 1844–64], vol. 113, col. 68): "Coelum, non visibile firmamentum, sed empyreum, id est, igneum vel intellectuale, quod non ab ardore, sed a splendore dicitur, quod statim repletum est angelis." Walafrid's description is quoted verbatim by Petrus Lombardus, *Sententiarum libri quatuor* (bk. 2, dist. 2, in Migne, *PL*, vol. 192, col. 656), and is the source of Hugh of St. Victor, *Summa sententiarum* (tract. 3, chap. 1, in Migne, *PL*, vol. 176, col. 81). The application of the term "empyreum" to this "extramundane" area is derived by Walafrid Strabo from Martianus Capella, *De nuptiis Philologiae et Mercurii libri VIIII* (bk. 2, line 200: ed. Dick, p. 76): "Philologia... cum...conspiceret...ipsam...quae ambitum coercet ultimum sphaeram miris raptibus incitatam...tanti operis...patrem deumque non ⟨ne⟩sciens ab ipsa etiam deorum notitia secessisse, quoniam extramundanas beatitudines eum transscendisse cognouerat, empyrio quodam intellectualique mundo gaudentem, iuxta ipsum extimi ambitus murum annixa etc." The thirteenth-century *Summa philosophiae* ascribed to Grosseteste has a whole chapter "De caelo empyreo" (*Die philosophischen Werke*, ed. Baur, pp. 545–50). For a convenient collection of medieval references to the subject, see Thomas Litt, *Les Corps célestes dans l'univers de saint Thomas d'Aquin* (Louvain, 1963), pp. 255–61.

53 "nonum" (IV.329). The necessity for a "ninth sphere" was astronomical: precession was regarded as the motion of the sphere of the fixed stars ("the eighth sphere") eastward with respect to a stationary ninth sphere. The problem of whether this ninth sphere was to be identified with the "crystal heaven" was, of course, purely theological. By the time Campanus came to write his *Tractatus de sphera*, he seems to have decided, hesitantly, in favor of the identification: see the Venice, 1518, edition, chap. 12, fol. 154r[a].

54 "celum cristallinum" (IV.329). This concept, like that of the empyrean heaven,

arose out of biblical exegesis. A perpetual difficulty for Christian commentators was to explain the existence of the "waters above the firmament" (Gen. 1:7), a notion irreconcilable with the common ancient view that all water was confined below the sphere of the moon. One theory held that these waters are solidified into "crystal" ("crystallum" in Latin normally means "crystal"; however, the original Greek meaning, "ice," helped to gloss over the difficulty; cf. Solinus, *Collectanea*, chap. 15, sec. 31 [ed. Mommsen, p. 99]: "putant glaciem coire et in crystallum corporari"). Support was found in another passage of scripture, Ezek. 1:22: "Et similitudo super capita animalium firmamenti, quasi aspectus crystalli horribilis, et extenti super capita eorum desuper." The theory is expounded, e.g., by Bede, *Hexaemeron*, bk. 1 (in Migne, *PL*, vol. 91, col. 18): "qui enim cristallini lapidis quanta firmitas, quae sit perspicuitas ac puritas novimus, quem de aquarum concretione certum est esse procreatum, quid obstat credi quod idem dispositor naturarum in firmamento coeli substantiam solidarit aquarum" (referred to by Hugh of St. Victor, *Summa sententiarum*, tract. 3, chap. 1, in Migne, *PL*, vol. 176, col. 89). Its origin lies in antiquity, however, as is shown by Jerome's comment on the above Ezekiel passage, *Comm. in Ezech.* 1. 1. 20 (in Migne, *PL*, vol. 25, col. 29): "similitudo firmamenti, quod nos appellamus coelum, habens speciem crystalli quod est purissimum, et ex aquis mundis atque lucentibus, nimio frigore concrescere dicitur"; cf. ibid., col. 23: "firmamenti etiam similitudinem crystallo comparatam, coelum hoc quod suspicimus intellegi volunt." For references to other ancient authors and modern literature, see J. K. Wright, *The Geographical Lore of the Time of the Crusades*, American Geographical Society Research Series, no. 15 (Washington, D.C., 1925), pp. 58–59, with notes; for medieval views, see ibid., pp. 182–84.

55 "locus...locatis" (IV.335–36). Cf. Campanus, *Tractatus de sphera*, ed. Venice, 1518, chap. 12, fol. 154r[a]: "Eritque conuexa superficies summe omnium locus vniuersalia [*om.* MS Venice, Marc. VIII, 69, fol. 33r; read "uniuersalis"] omnium inferiorum et vniuersaliter omnium rerum"; ibid., chap. 15, fol. 154r[b]: "Et accidunt ista propter naturale desiderium cuiuslibet locati [quod est se habere secundum similitudinem situs *add.* MS, fol. 34v] ad suum locum. Ex premissis enim constat quod primus locans est sphericum et omnino [omnia: MS] locata ab ipsa [ipso: MS] et ad [a: MS] se inuicem vsque ad mixta sunt etiam spherice locantia et locata."

56 "quinta essentia" (IV.339). The idea that the heavenly bodies are made of some substance different from all earthly elements is at least as old as Aristotle. See *De caelo* 269[a]20: ἐκ δὴ τούτων φανερὸν ὅτι πέφυκέ τις οὐσία σώματος ἄλλη παρὰ τὰς ἐνταῦθα συστάσεις, θειοτέρα καὶ προτέρα τούτων ἁπάντων. Aristotle himself referred to it by expressions like ἡ πρώτη οὐσία τῶν σωμάτων. But later writers, building on the theory of the four elements (see next note) and influenced by Aristotle's own enumeration (cf. *De caelo* 270[b]20: διόπερ ὡς ἑτέρου τινὸς ὄντος τοῦ πρώτου παρὰ γῆν καὶ πῦρ καὶ ἀέρα καὶ ὕδωρ, αἰθέρα προσωνόμασαν τὸν ἀνωτάτω

τόπον), called it πεμπτὴ οὐσία—"fifth substance." The earliest certain occurrence seems to be the lost work of Xenarchus (1st cent. B.C.)— πρὸς τὴν πεμπτὴν οὐσίαν. Cf. also Lydus *De mensibus* 2. 8, p. 28, lines 12–14.

The Greek term was translated by Cicero as "quintum genus" or "quinta natura" (e.g., *Tusculanae disputationes* 1. 26. 45: "sin autem est quinta quaedam natura, ab Aristotele inducta primum, haec et deorum est et animorum"; cf. *Academica priora* 1. 7. 26). The translation of οὐσία by "essentia" is also due to Cicero, but the expression "quinta essentia" is not found in his extant works. It appears, for example, in the fourth century Latin translation of Philo *Quaestiones in Genesim* 3. 6. In medieval times it would be familiar, for example, from Macrobius, *Commentary on the Somnium Scipionis*, bk. 1, chap. 14, sec. 20 (*Opera*, ed. J. Willis [Leipzig, 1963], p. 59). Since for Aristotle the universe was a continuum, he (like Campanus here) supposed the fifth substance to constitute everything above the lower limit of the moon's sphere, and not merely the visible heavenly bodies. Hence the popularity of the alternative title of αἰθήρ (medieval Latin "ether"), which was, however, also applied at times to the fiery element. The motion natural to the fifth substance was supposed to be circular, in contradistinction to the four elements, which have only vertical motion. For an exhaustive treatment of the ancient evidence, see Paul Moraux, "Quinta essentia," *Paulys Realencyclopädie der classischen Altertumswissenschaft* (Stuttgart, 1894——), vol. 24, cols. 1171–1263.

57 "quatuor elementa" (IV.339). The theory of the four elements, at least in the canonical form of earth, water, air, and fire, is the invention of Empedocles (see, e.g., Diels, *Die Fragmente*, ed. Kranz, fragment B.17, line 18). Its adoption by others, including Aristotle (in a modified form), insured almost universal acceptance in antiquity and the Middle Ages. The innumerable versions all agree with what Campanus states at lines IV.366–67, that all "corruptible" bodies are formed by the mixture of two or more of the elements. The elaboration of the theory in lines IV.342–64, according to which the elements are separated into the four concentric "spheres," in order of heaviness from the center, of earth, water, air, and fire, was not universally accepted but had become commonplace by the early twelfth century. See Wright, *Geographical Lore*, pp. 151, 226–27, who quotes among others Abelard, *Expositio in hexaemeron*, cols. 735–36 (see also note 59, below), and William of Conches, *De philosophia mundi libri quatuor*, bk. 4, chap. 1. An account similar to that of Campanus will be found in Grosseteste, *De sphaera* (*Die philosophischen Werke*, ed. Baur, pp. 11–12). Campanus himself gives a much more elaborate account of his views on the nature of the four elements in his *Tractatus de sphera*, ed. Venice, 1518, chaps. 3–6, fols. 153r[a]–v[a]. The essence of the theory is expounded in Macrobius's *Commentary on the Somnium Scipionis*, bk. 1, chap. 22, secs. 5–6 (ed. Willis, p. 92): "quidquid ex omni materia, de qua facta sunt omnia, purissimum ac liquidissimum fuit, id tenuit summitatem et aether uocatus est; pars illa, cui minor puritas et inerat aliquid leuis ponderis,

aer extitit et in secunda delapsus est, post haec quod adhuc quidem liquidum sed iam usque ad tactus offensam corpulentum erat, in aquae fluxum coagulatum est, iam uero quod de omni siluestri tumultu uastum impenetrabile densetum, ex defaecatis abrasum resedit elementis, haesit in imo...terrae nomen accepit."

58 "Tertia" (IV.347). The passage beginning here and running through line IV.377 is repeated almost word for word in Campanus's *Computus maior*, MS *R*, fol. 254r. Cf. Campanus, *Tractatus de sphera*, ed. Venice, 1518, chap. 5, fol. 153r[b].

59 "congregentur...arida" (IV.349–50). Gen. 1:9. This text was frequently used as an authority for various cosmological theories in the Middle Ages. See Wright, *Geographical Lore*, pp. 184–87, especially the passage from Abelard translated on p. 186. Grosseteste, *De sphaera* (*Die philosophischen Werke*, ed. Baur, p. 12), gives essentially the same account as Campanus of the reason for the violation of the perfectly spherical order: "Verumtamen ut animalia terrena habitaculum et receptaculum haberent, aqua in concavitates terrae recessit et apparuit superficies terrae arida et separata," also echoing the Genesis passage.

60 "Quia...aquis" (IV.353–57). The concept of the οἰκουμένη, the inhabited or habitable part of the world, has a long history. Campanus probably derived his formulation here from al-Farġānī 6. 4 (Alfraganus, Carmody, pp. 9–10): "si ergo cogitatione rationali considerauerimus in planitie terre esse circulum magnum abscindentem circulum equinoctialem per medium super rectos angulos, absideritque eum in ultimis locis habitabilibus orientis et occidentis, necesse erit ut hii circuli abscindant planitiem terre per 4 quartas; eritque una quartarum septentrionalis circumdans uniuersa loca terre habitabilia, eritque longitudo eius ab oriente in occidentem dimidium circuli id est 180 gradus," and from Boethius *Philosophiae consolatio* 2. 7. 4: "huius igitur tam exiguae in mundo regionis quarta fere portio est, sicut Ptolomaeo probante didicisti, quae nobis cognitis animantibus incolatur." He may also have had in mind *Almagest* 2. 1 (Ptolemy, Manitius, vol. 1, p. 58), where Ptolemy says that the οἰκουμένη lies approximately within the bounds of the quarter of the earth which is defined as in the al-Farġānī passage cited above (which is directly derived from this passage of the *Almagest*). Campanus's words here might lead us to suppose that he thought the whole of that quarter inhabited; but it is clear from both Ptolemy and al-Farġānī that even within the habitable quarter there are considerable uninhabited sections, and in the more detailed account of the habitable parts of the earth which Campanus gives in his *Tractatus de sphera* (ed. Venice, 1518, chaps. 46–49, fols. 157v[a]–58r[a]) he makes it clear that he believes only a small part of the habitable quarter to be actually inhabited. See especially chap. 47, fol. 157v[b]: "vnde concluditur quod tota longitudo [*scilicet* quarte habitabilis] est habitata: latitudo non est tota habitata. Nam partes vicine equatori non inveniuntur habitate: non quidem propter intemperiem loci...sed propter alias causas incertas, videlicet aut propter maria: aut propter montes inaccessibiles aut feras aut serpentes aut similia. Similiter etiam ex parte poli arctici multum est declaratum [de eo: MS Venice,

Marc. VIII, 69, fol. 56v] quod non est habitatum....Intelligantur igitur duo circuli paralelli equidistantes equatori: quorum vnus separet partes habitabiles ab illis inhabitabilibus que sunt versus equatorem: et alius separet easdem partes habitabiles ab illis inhabitabilibus que sunt versus polum. Eritque sola illa pars terre que inter istos duos parallelos in predicta quarta includitur habitabilis: sed nec ipsa tota cum sint in ea multa maria: multi paludes: multi magni et inaccessibiles montes. multe arene: multa deserta et alia quamplura habitationem impedientia."

In this passage of the *Theorica* Campanus seems to be deliberately opposing the not uncommon view (stemming from Macrobius's *Commentary on the Somnium Scipionis*, bk. 2, chap. 9, secs. 4–7 [ed. Willis, p. 123]) that there are *four* habitable "islands" on the earth. Belief in the existence of *inhabited* antipodes was heretical, and, as a good Aristotelian, Campanus was unwilling to believe in habitable but uninhabited areas. Contrast Grosseteste, *De sphaera* (*Die philosophischen Werke*, ed. Baur, p. 24): "Haec duo maria dividunt terram in quattuor partes, quarum una sola inhabitatur."

The use of the expression οἰκουμένη goes back far before Ptolemy, and the conceptions of its nature and extent varied greatly. The word was probably used initially by the early Ionian geographers, but it is first found in extant literature in Herodotus (e.g., 3. 106). Crates of Mallos (2d cent. B.C.) said that the shape of the οἰκουμένη was semicircular (though the location of the semicircle is unknown): see H. J. Mette, *Sphairopoiia, Untersuchungen zur Kosmologie des Krates von Pergamon* (Munich, 1936), fragments 8a and 8b. The notion that the whole of the rest of the world is covered by water was also widespread in antiquity; see, e.g., Strabo, bk. 2, chap. 5, sec. 5: ἐν θατέρῳ δὴ τῶν τετραπλεύρων... ἱδρῦσθαί φαμεν τὴν καθ' ἡμᾶς οἰκουμένην περίκλυστον θαλάττῃ καὶ ἐοικυῖαν νησῷ. In general see the article "Oikumene" by Friedrich Gisinger in *Paulys Realencyclopädie*, vol. 17, pt. 2, especially cols. 2130 ff. and 2141 ff. Much of interest will also be found in the article, "Die Grenzen der Menschheit: I. Die antike Oikumene," by J. Partsch (*Berichte über die Verhandlungen der K. sächsischen Gesellschaft der Wissenschaften*, phil.-hist. ser., vol. 68 [1916], no. 2).

61 "In aliis...letentur" (IV.372–75). Campanus seems to be implicitly denying here the theory that each of the elements is the habitation of living creatures most like it in makeup. According to this theory, air is the natural home of birds, as water of fishes (and fire of the salamander). The earliest surviving example of such a theory is the passage in Plato's *Timaeus*, 39^{e}–40^{a}, in which the demiurge is said to have created four kinds of living creatures, each to live in the appropriate element—stars in fire, birds in air, etc. In Chalcidius's translation (Waszink, *Timaeus*, p. 33) this reads in part: "diuersa animalium genera statuit esse debere constituitque quattuor, primum caeleste plenum diuinitatis, aliud deinde praepes aeriuagum, tertium aquae liquoribus accommodatum, quartum quod terrena soliditas sustineret." Campanus must have been familiar with this and with

Chalcidius's commentary on it (ibid., p. 164): "caelestium quidem stellas, terrenorum uero uolatilia et item nantia quaeque per terram feruntur." Later in time than the *Timaeus*, but probably derived from an earlier source, is Diodorus Siculus, bk. 1, chap. 7, sec. 5 (*Diodori Bibliotheca historica*, ed. Fr. Vogel [Leipzig, 1888], vol. 1, p. 13). Campanus, however, adopts the view that the air is not the "natural element" of birds, and by implication rejects the whole theory.

62 "longitudinemque...propiorem" (IV.382–83). "Longitudo longior" and "longitudo propior" are technical terms for the greatest and least distance, respectively, from the center of the system, not of the body itself, but of the center of its epicycle. They are applied in the theory of the five planets. Their origin lies in the tables for the equation. The equation of anomaly (see pp. 46 and 50, above) for any given anomaly varies according to the distance of the epicycle from the observer, being greatest at least distance of the epicycle and least at greatest distance. In order to achieve a simple tabulation, Ptolemy in the *Almagest* (see, e.g., 11. 11: Ptolemy, Manitius, vol. 2, p. 261) computed the equation of anomaly for the epicycle at greatest, least, and mean distances, tabulated the anomaly at mean distance and the differences to be subtracted for greatest distance and added for least distance, and gave a simple interpolation rule for intermediate positions (compare lines V.876–1275 of present treatise). This procedure became standard, and is found in the majority of Islamic planetary tables. There the columns for the differences at greatest and least distance are headed "al-buʿd al-abʿad" and "al-buʿd al-aqrab," respectively (meaning "farthest distance" and "nearest distance": see, e.g., Nallino, *Al-Battānī*, pt. 2, p. 108). Through misunderstanding of the Arabic usage of the comparative plus the definite article to express the superlative (for this, compare the remark of Curtze on the practice of Gerard of Cremona, in Anaritius, *In decem libros priores Elementorum Euclidis commentarii*, ed. Maximilian Curtze as Supplementum to Euclid, *Opera omnia* [Leipzig, 1899], pp. xiii–xiv), these expressions were literally translated into Latin as "longitudo longior" and "longitudo propior" (instead of "longissima" and "proxima"). See, e.g., the Toledan tables (Toomer, "Survey of the Toledan Tables," p. 60). Campanus applies the expression to the moon as well, though the lunar tables are somewhat differently arranged.

63 "primo...cubitis" (IV.399–403). Ptolemy uses none of these units, but instead uses a system in which $AD = 60$ (cf. lines IV.485–89). He does so because, in his simple theory of the moon (see pp. 42–44, above), the distance corresponding to AD is the radius of the unit circle. Accordingly, $AD = 60$, $AE = 5;15$ (radius of epicycle: *Almagest* 4. 9: Ptolemy, Manitius, vol. 1, p. 244), and $CD = 10;19$ (radius of small circle: *Almagest* 5. 4: Ptolemy, Manitius, vol. 1, p. 269). From these all the other distances are derived by simple addition and subtraction; they are tabulated in section 1 of Table 1 (p. 356, above: see also Appendix A, p. 447, below). In particular $AC = AD - CD = 49;41$. Campanus first reduced this to a norm such that AC, the radius of the deferent, is 60. The other distances are derived

by proportion sums; they are tabulated in section 3 of Table 1. Next the distances are given in earth-radii; the basic parameter is AD, the mean distance of the moon from the earth. That distance was calculated by Ptolemy (*Almagest* 5. 13: Ptolemy, Manitius, vol. 1, p. 304) as 59 earth-radii; the other distances are derived from it and are tabulated in section 2 of Table 1 (Ptolemy [ibid.] also gives $DB = 38;43$ and $AE = 5;10$). Finally the distances are given in "miles" (note 17 of section III, above). Campanus has already established (lines III.63–64) that the diameter of the earth is $6{,}490\frac{10}{11}$ miles: this immediately gives him the transformation from section 2 of Table 1. The resulting figures are given in section 4 of Table 1. In all the measurements a "minute" means $\frac{1}{60}$ of the unit in question.

The figures in sections 1–3 of Table 1 are all correctly computed to the nearest minute; the figures in section 4 are exactly those of section 2 multiplied by $3{,}245\frac{5}{11}$.

64 "minuta. At" (IV.414–15). For the passage interpolated by some manuscripts between these two sentences, see Appendix A, p. 447, and its introduction.

65 "17 minuta et 32 secunda" (IV.422). Ptolemy found that $r_{☾} = 0;17{,}33 r_{\oplus}$ (*Almagest* 5. 16: Ptolemy, Manitius, vol. 1, p. 313). The figure here probably reflects a scribal error in Campanus's text of the *Almagest* (however, the printed text of Gerard's translation gives the correct 0;17,33 [Ptolemy, *Almagesti*, ed. Liechtenstein, fol. 56r]). In the same place Ptolemy gives the approximation $r_{\oplus} \approx 3\frac{2}{5} r_{☾}$. This approximation is also found in al-Farġānī 22. 2 (Alfraganus, Carmody, p. 39).

66 "948 miliaria...miliaris" (IV.431). The diameter of the earth was calculated as $6{,}490\frac{10}{11}$ miles (lines III.63–64, above). This divided by 2 and multiplied by 0;17,32 (cf. line IV.422, above) gives exactly $948\frac{13}{33}$ miles.

67 "5958 miliaria...miliaris" (IV.433–34). Multiplying the lunar diameter, $1{,}896\frac{26}{33}$ miles, by $3\frac{1}{7}$ gives $5{,}961\frac{11}{33}$ miles. The erroneous figure in the text probably resulted from a miscalculation by Campanus: if you subtract $3 \cdot \frac{26}{33}$ from $5{,}961\frac{11}{33}$ miles, you get $5{,}958\frac{32}{33}$ miles. Thus it seems that, when multiplying by 3, Campanus forgot to include the terminal fraction, and $\frac{32}{33}$ was later changed to $\frac{22}{33}$ by a scribal error. The other figures for circumferences below (lines IV.441–44 and IV.448) are derived by multiplication of the radii by $2 \cdot 3\frac{1}{7}$ exactly.

68 "101261 miliaria...miliaris" (IV.440–41). $ED + r_{☾} = 208{,}250$ miles $+ 948\frac{13}{33}$ miles; hence, far edge of moon at greatest distance $= 209{,}198\frac{13}{33}$ miles. $DF - r_{☾} = 108{,}885$ miles $- 948\frac{13}{33}$ miles; hence, near edge of moon at least distance $= 107{,}936\frac{20}{33}$ miles. The thickness of the moon's sphere, then, is $209{,}198\frac{13}{33}$ miles less $107{,}936\frac{20}{33}$ miles, or $101{,}261\frac{26}{33}$ miles.

69 "27 dies et 8 horas" (IV.445). This is a mean sidereal month. If we take the crude value given by Campanus (line IV.62) for the mean lunar motion in longitude in one day, 13;10,34°, and divide it into 360°, we get 27;19,19,... days, or 27 days 8 hours to the nearest hour.

There is an apparent contradiction between this passage and lines IV.161–63, where Campanus says that the epicycle traverses the deferent in 14 days 18 hours 22 minutes. The contradiction could be resolved as follows: the earlier passage is considering the motion of the epicycle with respect to the apogee of the deferent (which itself moves around the ecliptic in the opposite direction to the sense of motion of the epicycle), whereas the passage here is considering the motion of the epicycle with respect to a fixed point of the ecliptic (e.g., Aries 0°). Although both passages are correct, the use of the same wording for different motions is clumsy, and this further supports the view that the "dieta" passages, of which this is one, are spurious additions (see pp. 28–29, above).

70 "equante" (IV.446). The use of the term "equant" for the moon's deferent is strange, and the only parallel elsewhere in the work is in Table 1, section 5, where we read in most manuscripts, except the δ group, "ambitus deferentis aut equantis" (for the proper meaning of "equant," see p. 49, note 5, above, and p. 405, section V, note 3, below). The text is here discussing the uniform motion of the center of the epicycle: this motion takes place about the earth, which is not the center of the deferent and so is analogous to an equant point (see p. 45, above). Hence one might justify the description of this motion as being "on the moon's equant," But the usage is curious, and it is significant that this, too, occurs in a passage on the "dieta" (cf. preceding note).

71 "dieta" (IV.446). The meaning of this word here and elsewhere in the work is "distance traveled in one day," and the connection with "dies" is obvious. This is, however, a false etymology: the real derivation is from the Greek δίαιτα, and the basic meaning is "way of life." But the confusion with "dies" is an early medieval one, and "dieta" in the sense of a day's march is found early in the twelfth century. See du Cange, *Glossarium*, s.v., sec. 2, quoting Alanus ab Insulis (d. 1203); W. B. Sedgwick, "Some Poetical Words of the Twelfth Century," *Bulletin du Cange*, vol. 7 (1932), p. 224, quoting numerous passages from twelfth-century poets; and Einar Löfstedt, *Coniectanea, Untersuchungen auf dem Gebiete der antiken und mittelalterlichen Latinität* (Uppsala, 1950), pp. 130–33.

72 "36458 miliaria...miliaris" (IV.447–48). 996,540 miles $\div$ $27\frac{1}{3}$ = $36{,}458\frac{32}{41}$ miles, which is indeed $36{,}458\frac{47}{60}$ miles to the nearest sixtieth.

73 "27 dies et 13 horas" (IV.449). This is an "anomalistic month." The value given here is derived by taking the crude figure given by Campanus (line IV.4) for the daily lunar motion in anomaly, 13;4°, and dividing it into 360°. The result is 27;3,3,... days, i.e., about 27 days 13;13 hours, or 27 days 13 hours to the nearest hour. Theoretically, it could also have been derived from a more accurate figure for the mean daily anomaly, or from al-Farġānī, who says (chap. 17: Alfraganus, Carmody, p. 31, line 2), "luna ambulat circulum breuem in 27 diebus et 13 horis et 3ª unius hore fere." But it seems certain that in all the passages on the "dieta" the underlying mean motion figures are simply the round numbers for daily mean motion given by Campanus. The critical cases are the

figures for the sidereal periods of Jupiter and Saturn (lines VI.663 and VI.711: see section VI, notes 120 and 135), where the errors are explicable only by the assumption that the crude parameters for daily mean motion were used. Since Campanus usually computes to far greater accuracy (though he may round his final results), this is yet another indication that the "dieta" passages are later interpolations (see pp. 28–29, above).

74 "3826 miliaria...miliaris" (IV.450–51). 105,400 miles $\div$ $27\frac{13}{24} = 3{,}826\frac{614}{661}$ miles, which is indeed $3{,}826\frac{56}{60}$ miles to the nearest sixtieth.

75 "PTHOLOMEVS...ueritati" (IV.455–57). The reference is to *Almagest* 5.16 (Ptolemy, Manitius, vol. 1, p. 313), where Ptolemy uses the approximation $r_{\oplus} = 3\frac{2}{5}r_{\leftmoon}$, and then says that the cube of $3\frac{2}{5}$ is $39\frac{1}{4}$. Campanus's procedure is as follows: he takes the figures derived for the radii of the two bodies in miles, which are $r_{\oplus} = 3{,}245\frac{5}{11}$ miles (diameter $= 6{,}490\frac{10}{11}$ miles, lines III.63–64) and $r_{\leftmoon} = 948\frac{13}{33}$ (line IV.431). He then multiplies both numbers by $11 \cdot 33$ to remove the fractions, getting 1,178,100 and 344,267, respectively (cf. lines IV.468–69). The highest common factor of the latter two numbers is 1,309. Dividing by 1,309, he gets 900 and 263, respectively. So the latter are relatively prime numbers, or, as Campanus says, they are the smallest integers expressing the ratio between the diameters of the two bodies. He then cubes each, getting 729,000,000 and 18,191,447; dividing the greater by the lesser gives 40;4 to the nearest minute, or $40\frac{1}{15}$. This is a very roundabout way of finding the result: Campanus's basic parameter is $r_{\leftmoon} = 0;17{,}32r_{\oplus}$ (line IV.422). Cubing that and taking the reciprocal gives 40;4,25,... immediately.

76 "tres tabulas...quartam tabulam" (IV.484–86). As stated in the text, the table is in four parts; however, for convenience I have separated the last part of the fourth section, which contains quantities not found in the first three, and numbered it 5. Section 1 (the numbering of the sections has been added by me) gives the various lengths in the units of which *AD* (distance from earth to apogee) is 60 (derived from Ptolemy; see note 63, above). Section 2 gives them in earth-radii, and section 3 in the units of which *AC*, the radius of the deferent, is 60. Section 4 gives them in "miles." In sections 1–3 the third column gives the numbers of the second column reduced to minutes.

There is considerable variation in the wording and numbers of the table among different manuscripts. The wording adopted in the text follows that of MS *F* for the most part. The numbers printed are the correct ones.

The table, like Tables 2–7, is little more than a consolidation of the various quantities stated in the body of the text. It adds the following additional information: (1) the length of *AB*, the diameter of the deferent. This should be twice the radius *AC*; the discrepancies between 2*AC* and *AB* which can be derived from sections 2 and 4 are explicable on the supposition that *AB* was found by adding the values for *AD* and *DB*. (2) In the last part of section 5 the sides of the cubes found at lines IV.468–69 are (unnecessarily) cubed. It seems likely to me that both of these are later interpolations. It is significant that the only other place where the length of *AB* is mentioned is in the passage interpolated in some manuscripts after line IV.414 (see Appendix A, with introduction there).

77 "in principio" (IV.495). Line II.74, above.

78 "sicud in sole dictum est" (IV.504). Lines III.89–153, above.

79 "circulationem maiorem epicicli" (IV.507–8). This is the "epicycle in its convenient representation," the meaning and purpose of which are explained below at lines IV.526–44. The word "circulatio" here probably means "system of circles," though at lines V.64–65 and V.77 it certainly means "revolution."

80 "96 partes" (IV.509). The number 96 is chosen for convenience: the number of parts has to be somewhat greater than 78;48 (the sum of the radius of the small circle, the radius of the deferent, and the radius of the epicycle); because the factors of 96 are all 2 or 3, it is easy to divide a length into 96 parts (Campanus suggests dividing by 12 and then again by 8).

81 "12 partes et 28 minuta" (IV.511). Cf. line IV.406, above.

82 "de predictis...60 partes" (IV.516). Cf. line IV.405, above.

83 "6 partes et...20 minuta" (IV.521). Cf. line IV.407, above.

84 "circulum alium" (IV.530). The figure described here is what is later (e.g., lines IV.570–71) called "epiciclus oportune circulationis": "the epicycle in its convenient representation."

85 "23 partes...partibus" (IV.531–32). In fact 23½ parts, since the 12;28 parts were taken as 12½ parts (line IV.511).

86 "limbo" (IV.556). "Limb" is the name given to the graduated edge of an astrolabe, quadrant, or other measuring instrument. It was often in the form of a raised section fastened onto the main body of the instrument. See Price, *Equatorie of the Planetis*, p. 47, lines 12–16, for an example on an equatorium. Campanus uses the word here with an extended meaning, for a raised and removable outer section in general. It is worth noting that even in antiquity, when "limbus" meant any "edge" or "border," it is also found especially applied to the zodiac, thus foreshadowing the later usage: e.g., Varro *De re rustica* 2. 3. 7 uses the expression "limbus duodecim signorum."

87 "concauitas ista...collocari" (IV.559–62). In fact the series of cavities containing disks must consist, not of two of each, as is implied here, but of three of each.

88 "Et fiat...tabule" (IV.562–66). Campanus's meaning is best illustrated by a drawing of the cross-section of the disk ($X'C'K'Y'$) fitted into the cavity ($XCKY$), as in Figure 24.

89 "lunaris forme" (IV.579). Strictly speaking, these shapes are not lunes (a lune is the crescent-shaped figure formed between the arcs of two intersecting circles); but in appearance they are very similar, for each is the area left when from one circle a smaller circle which is completely inside the first but near one edge has been removed. As can be seen from Figure 13 (p. 196), the limbs would be shaped very like a crescent.

Fig. 24

Diagram to illustrate Campanus's instructions for "beveling" the parts of the instrument so that an inserted disk shall not slip out of its container. Shown is a vertical cross-section of a ring-shaped disk *X′C′K′Y′* (with outer circumference *X′Y′* and inner circumference *C′K′*) inserted in its container in the cavity *XCKY* (outer circumference *XY*, inner circumference *CK*). The disk and the cavity, as Campanus says, slope from top to bottom toward the center on the side of *CK*, and from top to bottom toward the circumference on the side of *XY*. While it is true that such beveling would achieve its object, Campanus does not explain how one could get the disk *into* the cavity after the sides of each had been so shaped.

90 "in puncto *K*" (IV.608). It is essential for the proper working of the instrument, though Campanus does not mention it, that the nail on point *K*, which is on the boundary between the movable "disk of the motion of the center of the eccentric" and the stationary circle *CK*, be actually located on the movable disk.

91 "tabulam...sole" (IV.613–14). Like the *Almagest*, and unlike the Toledan tables, the tables for Novara in MS Dublin, Trinity College, D.4.30 list the mean motion in elongation (53v–54v). However, see note 93, below.

92 "superius" (IV.616). Lines III.160–93.

93 "Considerabo...motum" (IV.619–21). To find the position of the mean sun, given the mean elongation, one would count *backward* from the position of the mean moon. That may be what Campanus means here, and the expression "beginning [to count] from Aries" may refer merely to how one describes the sun's position in the ecliptic. But it looks rather as if he has confused two different procedures, and that what he is describing here is how to find the position of the mean sun from a table for mean solar motion. One would have to perform such an operation if one did not have the mean elongation tabulated (as would be true if one were using the Toledan tables). With a mean elongation table, the most natural procedure would be to double the elongation and subtract it from the moon's position. Campanus's roundabout procedure here, by way of the mean sun, may perhaps be a further indication that he is thinking of the operation with a solar table. Moreover, such operation is what is described at lines IV.674–79.

94 "cursu tarda aut cursu uelox" (IV.646–47). That is, whether its velocity is less or greater than the mean. The expressions "uelox cursu" and "cursu tardior" are applied to the moon in al-Farġānī 25. 5–6 (Alfraganus, Carmody, p. 43). Cf. Jābir (Apianus, *Instrumentum Primi Mobilis*, p. 46).

95 "quemadmodum...sole" (IV.657–58). Lines III.203–34.

96 "ut...prius" (IV.660). Not explicitly, but implied at, e.g., lines IV.49–54 and IV. 187–88.

97 "fuit" (IV.666). The past tense is used here, presumably, because Campanus is analyzing the process of reasoning adopted by those who first tackled the problem.

98 "de quibus...sermo" (IV.671). The table for mean anomaly was mentioned at lines IV.614–15.

99 "sumunt...centrum" (IV.674–79). Compare the procedure described here for finding the double elongation with note 93 at lines IV.619–21, above.

100 "sicud...ostensum est" (IV.681). Lines IV.237–61.

101 "ut prius demonstrauimus" (IV.685–86). Lines IV.112–15 and IV.240–61.

102 "Supposuerunt...compositores" (IV.692). For what follows, cf. pp. 46–47, above.

103 "termini" (IV.699). I.e., the points where the lines meet the ecliptic.

104 "equationem circuli breuis" (IV.724). This is the name given to this function in the tables for Novara, MS Dublin, Trinity College, D.4.30, fols. 49r–51v. It is probably derived from the heading in the Toledan tables, "diuersitas diametri circuli breuis." The natural way to interpret the latter would be the "difference resulting from the diameter of the small circle," i.e., the difference in the equation of anomaly that results from the epicycle center's being at perigee instead of apogee. (See Figure 12, p. 186: the difference in the distance of the epicycle from D is $AD - BD = 2CD$, which is equal to the diameter of the small circle CK.) But it should be noted that in the *Almagest* 5. 8 (Ptolemy, *Syntaxis mathematica*, vol. 1, p. 390 [*Opera quae exstant*, ed. Heiberg]) it is called ἐπικύκλου διαφορά (i.e., the difference resulting from the position of the epicycle), and it is possible that "circulus breuis" originally meant the epicycle; indeed, it is the term used to translate "epicycle" in Johannes Hispalensis's translation of al-Farġānī 12. 10 (Alfraganus, Carmody, p. 23). It is also used for the moon's epicycle by Grosseteste in *De sphaera* (*Die philosophischen Werke*, ed. Baur, p. 27) and is mentioned as one of its names in *Theorica planetarum Gerardi* 8 (Gerardus, *Theorica*, p. 17).

105 "inuestigauerunt...in auge deferentis" (IV.726–29). Campanus talks as if the proportion between the amount of the equation of anomaly when the epicycle is at apogee and its amount for another given position of the epicycle were the same for all values of the anomaly; and that is indeed the approximation on which this function in the table is based, but it *is* only an approximation: when the epicycle is small compared with the deferent, the proportion between the sines of the equations of anomaly is very nearly constant, and, since the maximum equation of the moon is 7;40° (*Almagest* 5. 7: Ptolemy, Manitius, vol. 1, p. 282), the error is never great. The function is actually calculated from the *maximum* equation of center at each degree of anomaly (see *Almagest* 5. 7: Ptolemy, Manitius, vol. 1, pp. 283–84; see also pp. 46–47, above).

Section V

For the following description, refer to Figure 14, p. 228, and cf. p. 50, above.

1 "4...circulos" (V.2). Cf. al-Farġānī 14. 7 (Alfraganus, Carmody, p. 27): "Motus autem mercurii sunt 4...."

2 "in longitudine" (V.2). The qualification is made because Mercury has a latitudinal motion as well, which Campanus mentions but does not describe because it cannot be represented on his instrument. See lines V.340–42 and note 5, below.

3 "equans" (V.14). Ptolemy has no fixed technical term for the equant circle or point. He sometimes describes the circle as ὁ τὴν ἀνωμαλίαν ποιῶν ἔκκεντρος (e.g., *Almagest* 9. 5: Ptolemy, *Syntaxis*, ed. Heiberg, vol. 2, p. 252, line 19); cf. the similar description of the equant point at *Almagest* 10. 6 (ibid., p. 317, line 1). The definition of the term given here by Campanus is a free translation of the Arabic "al-falak al-mumaṯṯil li'l-masīr" (Nallino, *Al-Battānī*, pt. 2, p. 238) or "al-falak al-muʿaddil li'l-masīr (e.g., al-Bīrūnī, *Tafhīm*, *The Book of Instruction in the Elements of the Art of Astrology*..., trans. R. R. Wright [London, 1934], sec. 180, p. 93), both meaning "the circle which gives uniformity to the motion." In Johannes Hispalensis's translation of al-Farġānī 14. 2 (Alfraganus, Carmody, p. 27) we find the expression "circulus equans motum," which is an exact translation of the Arabic phrases; "circulus equans," or plain "equans," is an abbreviation of this.

4 "in superficie una" (V.15). In fact, in the Ptolemaic system the plane of Mercury's deferent is not always in the ecliptic (i.e., in the same plane as the equant; cf. also lines V.20–21, below), but for the purpose of longitude computations it can be treated as if it were (*Almagest* 9. 6: Ptolemy, Manitius, vol. 2, p. 123). Campanus ignores this aspect of Ptolemaic theory (but see note 16, below, and section VI, note 4).

5 "Superficies...dicetur" (V.15–17). For an account of the Ptolemaic latitude theory for Mercury, see pp. 52–53, above. Campanus does not keep this promise to describe Mercury's latitudinal motions, except for a brief dismissal at lines V.340–42. Also note lines VI.386–87, below, and p. 29, above.

6 "Equans...inconueniens" (V.17–18). Since the sole purpose of the equant is to provide a *point* from which the angular velocity of the center of the epicycle appears constant, the size of the circle of which it is the center is, as Campanus says, irrelevant.

7 "aux" (V.19). In all circles referred to in this passage, the apogee is the point farthest from the center of the ecliptic, the perigee the point nearest to it.

8 "17 gradus...libre" (V.24). Like all the other apogee values of Campanus, this is taken from the Toledan tables. See Toomer, "Survey of the Toledan Tables," p. 45.

9 "solui" (V.57). The meaning "to be replaced by" is dictated by the context, but I have been unable to find an example of a comparable use of "soluere" from any period. The possibility of corruption must be admitted, but the reading of the δ group of manuscripts seems to be a stopgap correction. Perhaps Campanus has extended the meaning "pay for" which "soluere" commonly has.

10 "ad quarum...inpossibile" (V.54–58). Campanus says that the apogee of the deferent (point *K* in Figure 14, p. 228) moves back and forth over a limited path

(bounded by the tangents *MPR* and *LNQ* in Figure 14) and that, furthermore, its motion is *continuous*; i.e., there is no "rest" at its turning-points, since the motion of the center of the deferent (point *C* in Figure 14), which governs the motion of the deferent apogee, is itself continuous (around the small circle *CNEP* in Figure 14). Now Aristotle, in Book 8 of the *Physics*, where he "proves" that only circular motion can be continuous and infinite, argues that infinite motion on a straight line is impossible, because something moving along a straight line would have to turn back when it reached the end of the line (since, according to Aristotle, a straight line cannot be infinite in length), and where it turned back it would pause. On the last point he says (262^a12–14), "But the chief reason why it is clear that it is impossible for motion along a straight line to be continuous is that, when it turns back, it must necessarily pause [στῆναι], [and that is true] not only [if the motion is] along a straight line, but also if it is along a circle." I.e., if a motion, whatever its path, reverses itself, it cannot be continuous, for at the point of reversal there is a pause.

It is hardly possible to doubt that Campanus had in mind here that passage of the *Physics* and that he is implicitly refuting it. This conviction is reinforced by comparison of his wording here with that of medieval commentators on the above passage of Aristotle, e.g., Aquinas, *In octo libros Physicorum Aristotelis expositio*, chap. 8, lect. 16, line 5 (Sancti Thomae Aquinatis Opera Omnia, iussu Leonis XIII P.M. edita, vol. 2 [Rome, 1884], p. 425): "Dicit ergo quod maxime ex hoc manifestum est quod impossibile est motum rectum esse continuum in infinitum, quia necesse est id quod reflectitur quiescere inter duos motus." Campanus has in fact produced a counterexample, a motion which does turn back but does not undergo pause ("quies") at the point of reversal. (However, the validity of the counterexample is questionable, for the motion is neither along a straight line nor along a circle, nor even along the same path in opposite directions, but rather along a sausage-shaped curve.)

It was probably already customary to dispute this thesis of Aristotle in the schools in Campanus's time. Roger Bacon, in his *Questiones supra libros octo Physicorum Aristotelis* (ed. F. Delorme [Oxford, 1935], in *Opera hactenus inedita Rogeri Baconi*, ed. R. Steele, fasc. 13, pp. 423–24), writes, "Queritur tunc an solus motus localis sit continuus," etc. (The passage is translated from an Amiens manuscript by Duhem, *Système*, vol. 8, pp. 256–57.) Bacon expounds (and then rejects) an attempt to refute the thesis by means of what became the standard counterexample for the following three centuries: a small object projected upward meets a large object falling downward. Denial of the thesis was common in the Parisian school of physics in the fourteenth century (see Duhem, ibid., pp. 258–93). For fifteenth-century refutations of it, see Proposition 13, part 2, of the *Or Adonai* of the Jewish philosopher Ḥasdai Crescas (H. A. Wolfson, *Crescas' Critique of Aristotle* [Cambridge, Mass., 1929], pp. 279–81 and note on p. 626), and Johannes Versor, *Quaestiones Physicae*, bk. 8, quest. 11

(quoted by Wolfson, ibid.). The question was still being discussed in the sixteenth century and later (see, e.g., G. B. Benedetti, *Diversarum speculationum mathematicarum*, chap. 23, in *Mechanics in Sixteenth-Century Italy*, trans. and annotated Stillman Drake and I. E. Drabkin [Madison, 1969], pp. 215–16, and note 157 on p. 215). Though refutations of Aristotle's thesis were no doubt familiar to Campanus from scholastic disputation, his use of this particular counter-example is surely original. It is not repeated in any of the later refutations known to me, though Benedetti's refutation, which is also drawn from planetary motion, provides an interesting parallel.

There is evidence that Aristotle's thesis of the "pause" was disputed much earlier in the Middle Ages, by Islamic scholars. In Arabic bibliographical works we read that Abū'l-Aḥmad, known as ibn Karnīb, wrote a reply to Ṯābit ibn Qurra's "denial of the necessity of the existence of a pause between every two contrary motions." See ibn an-Nadīm, *Kitâb al-Fihrist*, ed. Gustav Flügel (Leipzig, 1871), vol. 1, p. 263, lines 5–6 (read "sukūn" for "sukūnayn," as in ibn al-Qifṭī); ibn al-Qifṭī, *Ta'rīkh al-Ḥukamā'*, ed. Julius Lippert (Leipzig, 1903), p. 169, and ibn Abī Uṣaybi'a, *Kitāb 'Uyūn al-Anbā' fī Ṭabaqāt al-Aṭibbā'*, ed. A. Müller (Königsberg, 1884), vol. 1, p. 234, give more corrupt versions of the same notice.

11 "accessionis et recessionis" (V.58). The terminology is that of the theory of "trepidation" associated with Ṯābit ibn Qurra (see section III, note 11, above), and properly so, since here, too, there is motion back and forth along a limited arc of a curve.

12 "Erit. . . parte sue oppositionis" (V.89–91). It is not clear, when Campanus says that the motion of the apogee will be "greater" than that of the perigee, whether he means only "over a greater arc of the equant," as is suggested by the second half of the sentence. This statement is true but has no application to real motion and is thus pointless. Or does he mean "faster"? That is false, for obviously, since the perigee is always 180° away from the apogee, both must traverse an equal arc of the ecliptic in the same time. If "faster" is what Campanus means, he has made the mistake of thinking that, because the center of the epicycle moves uniformly about the equant, the apogee of the deferent does as well. But in fact the uniform movement of the apogee is about the center of the small circle (*F*, not *E*, in Figure 14).

13 "3 gradus…24 secunda" (V.96). Ptolemy (*Almagest* 9. 3: Ptolemy, Manitius, vol. 2, p. 101) gives 3;6,24,6,59,35,50° per day. Al-Fargānī 14. 8 (Alfraganus, Carmody, p. 27) gives 3;6°.

14 "a quo mouetur corpus mercurii equaliter" (V.102). So we must read, against all manuscripts, to make sense. The variant readings recorded in the critical apparatus seem to be attempts to mend the nonsense of the majority reading. But they still give only a feeble sense. That the sense required here is what I have suggested is indicated by the cross-reference at lines V.768–72 to this pas-

sage. What Campanus means is that only if the anomaly is measured from the mean apogee is it uniform; if measured from the true apogee, it is nonuniform. Cf. lines VI.43–53.

15 "secundum...medii" (V.110–11). See p. 48, above.

16 "Semper...terminat" (V.112–13). This is a way of saying that the center of Mercury's epicycle and the mean sun have the same longitude. It is a necessary periphrasis only if the orbit of Mercury is supposed not to be in the plane of the ecliptic, a refinement of the Ptolemaic system which Campanus otherwise ignores (see note 4, above, and section VI, note 4, below).

17 "quasi duplum" (V.120). Only "*about* double," because the motion of the center of the epicycle is uniform with respect to the equant (*E* in Figure 14), while the motion of the deferent is uniform with respect to the center of the small circle (*F* in Figure 14). Therefore, strictly speaking, these motions cannot simply be added to get the motion of the epicycle with respect to the apogee of the deferent at any given time. However, Campanus compares these motions (e.g., at lines V.121–22 or V.129–30) only when both have completed a whole or half circle, at which points the vectors lie in the same straight line, and the comparison is legitimate.

18 "ex creatoris beneplacito" (V.124). See section IV, note 21.

19 "equalitatem motuum" (V.128–29). The fact that the motion of the epicycle on the deferent is equal to the motion of the deferent about the center of the small circle (both being equal to the sun's mean motion).

20 "Possibile...potest" (V.140–44). The statement that the closest approach of the center of the epicycle to the earth occurs at the tangential points, i.e., at a distance from apogee of 120°, is derived from Ptolemy, who says (*Almagest* 9. 9: Ptolemy, Manitius, vol. 2, p. 144) that the closest approach is found at a distance from apogee of 120° on either side. Ptolemy, of course, realized that this neat position was a consequence of the particular parameters chosen for the distances between the centers, and attached no significance to it. Campanus, however, seems to think that it is a qualitative consequence of the model (it is he who introduces the tangential points here, a confusing concept; for, whereas in the description of the limits of the motion of the apogee they are really relevant, here it is only contingently true that the tangential points are 120° from apogee [because, in Figure 14, $DE = EF$]). In fact, the minimum does not occur at exactly 120°, as has been shown by Willy Hartner ("The Mercury Horoscope of Marcantonio Michiel of Venice," in *Vistas in Astronomy*, ed. A. Beer, vol. 1 [London, 1955], pp. 109–17), by means of an exhaustive numerical computation with Ptolemy's parameters. Unfortunately on page 115 Hartner made a calculating error in a crucial place (corrected in Hartner, "Mediaeval Views on Cosmic Dimensions and Ptolemy's Kitāb al-Manshūrāt," in *Mélanges Alexandre Koyré*, vol. 2 [Paris, 1964], p. 267, note 25). Hartner's value of s for $\varphi = 121°$ should read 18.5203 instead of 18.5176; thus the true value is the same as that for $\varphi =$

120°, and so Hartner's assertion ("The Mercury Horoscope," p. 117) that the minimum occurs at $\varphi = 121°$ is not quite correct. However, it is clear from examination of the symmetry of the function *s* in Hartner's Table I (ibid., p. 115) that the minimum occurs, not at exactly 120°, but at about $120\frac{1}{2}°$.

21 "In tali...augis duo" (V.146–49). For (in Figure 14) $\angle FND = 90°$ (*DNL* is a tangent to the circle). Therefore $\cos \angle DFN = FN/FD = \frac{1}{2}$. Therefore $\angle DFN = 60°$, and $\angle CFN = 120°$.

22 "Erunt...4 signa" (V.175–77). Ptolemy, too, notes that the epicycle comes to a position nearest to the earth twice in every revolution of the epicycle about the equant (*Almagest* 9. 5: Ptolemy, Manitius, vol. 2, p. 122) and proves it from observations in *Almagest* 9. 8 (Ptolemy, Manitius, vol. 2, pp. 136–40).

23 "in quo quamdiu...stationis secunde" (V.192–206). In describing the phenomena of station and retrogradation, we should say that, *at the point* in which the westward angular velocity of Mercury becomes equal to the eastward angular velocity of the epicycle center (both as seen from the center of the universe), the planet becomes stationary, and, as soon as the first angular velocity exceeds the second, it becomes retrograde. Campanus, however, has no conception of momentary velocity or of speed at a point, but only of speed over an interval of time or space. Hence the clumsiness and inaccuracy of his explanation here, which has to be in terms of the actual arc of the ecliptic traversed by the two moving points in a definite interval of time (e.g., one day). For practical purposes, however, the method described here of comparing the two components of motion over one day at a time would work perfectly adequately for determining the stations, as the motions involved are comparatively slow. Cf. lines V.629–42 and note 87, below.

24 "Hec...uera epicicli" (V.207–8). The figures tabulated below are given for the distance of the stations from the apogee for the corresponding "equated centra" (for the concept of the equated centrum, see p. 51, note 10, above, and note 25, below). They are taken from the Toledan tables (Toomer, "Survey of the Toledan Tables," Table 44), which in their turn are taken from al-Khwārizmī's table (H. Suter, "Die astronomischen Tafeln des Muḥammed ibn Mūsā al-Khwārizmī

Position	Equated centrum	Distance, station–apogee	Location of epicycle center
(1)	0°	$4^s27;14°$	Apogee (maximum distance)
(2)	120°	$4^s24;29°$	Perigee
(3)	180°	$4^s24;42°$	Minimum distance
(4)	64°	$4^s25;9°$	Mean distance

in der Bearbeitung des Maslama ibn Aḥmed al-Madjrīṭī und der latein. Übersetzung des Athelhard von Bath," *Skrifter, Kongelige Danske Videnskabernes Selskab*, ser. 7, historical and philosophical sec., vol. 3 [1914], no. 1, Tables 51–56,

pp. 162–67); in both for each planet a column entitled "Statio prima" gives the above information for every degree of centrum ($\varkappa$). Al-Khwārizmī's tables are merely an expansion of those in the "Handy Tables" (Halma, *Commentaire de Théon*, pt. 3, pp. 11–15), which are for every three degrees. The whole problem is dealt with by Ptolemy (*Almagest* 12. 1–8), who also constructs a table (Ptolemy, Manitius, vol. 2, pp. 314–15), which is essentially the same as those mentioned above, except that the argument is the uncorrected centrum instead of the equated centrum (see note 25, below).

25 "Centro...equato" (V.210). In the *Almagest* 12. 8 (Ptolemy, Manitius, vol. 2, pp. 314–15), Ptolemy's table has as argument the uncorrected centrum. But it is clear that for his "Handy Tables" he changed to the equated centrum—that is, the centrum modified by the equation of center (see Ptolemy, *Opera astronomica minora*, p. 173, lines 16–18 [*Opera quae exstant*, ed. Heiberg]). The equated centrum is found in the Theonic "Handy Tables," too, and its use became the general rule: see, e.g., al-Khwārizmī, chap. 13 (Suter, "Die astronomischen Tafeln," p. 12), where the term used is "centrum examinatum" ($\approx \varkappa$).

26 "sicud...supra" (V.212). Lines V.149–51.

27 "2 signis...media" (V.217). The place of the mean distance, that is, the place where the center of the epicycle is exactly the distance of the radius of the deferent from the center of the universe, can be read off from tables of the Ptolemaic type for finding the true place of the planet by observing where the column of "minutes of proportion" reaches zero (cf. p. 51, above). In the Toledan tables for Mercury (Toomer, "Survey of the Toledan Tables," Table 44) the "minutes" do indeed reach zero between $2^s4°$ and $2^s5°$.

The reason for selecting this particular position of Mercury here is that it is one of the basic positions used by Ptolemy for establishing his table of stations (for Mercury, see *Almagest* 12. 6: Ptolemy, Manitius, vol. 2, pp. 297–98).

28 "tempus...18 minuta" (V.219–23). The times of retrogradation, i.e., of the passage from first to second station, corresponding to the four positions above (see note 24, above) are given as (1) 21 days 2;16 hours, (2) 22 days 20;45 hours, (3) 22 days 17;24 hours, and (4) 22 days 5;18 hours.

With position (1) as example, these are derived as follows: The distance of each station from the apogee is $4^s27;14°$. Therefore the distance of each from the perigee is $6^s - 4^s27;14° = 32;46°$. So the distance between the two stations is $2 \cdot 32;46 = 2^s5;32°$ $(= 1,5;32°)$. This distance is divided by 3;6,24,7° per day, the daily motion of Mercury in anomaly, and the result is 21;5,39 days or 21 days 2 hours 16 minutes to the nearest minute. Campanus gives 3;6,24° for the daily motion of Mercury in anomaly (e.g., line V.96), but at least one more significant place is required to achieve the accuracy of the above figures for retrogradation times. He must have taken the more accurate figure from Ptolemy, who gives (*Almagest* 9. 3: Ptolemy, Manitius, vol. 2, p. 101) 3;6,24,6,59,...° for Mercury's motion in anomaly per day.

All the above figures for retrogradation times are correctly computed to the nearest minute, except the fourth, which has a large error. The correct period is 22 days 10;27 hours. The error is to be explained as follows: when subtracting $4^s25;9°$ from 6^s, Campanus got, instead of 34;51°, 34;31° (i.e., he probably misread his own 5 as 3); 34;31° leads to the above incorrect result for position (4).

The foregoing procedure is confirmed by Campanus's own statement at lines VI.340–42. See further section VI, note 29, below.

29 "115 diebus...5 minutis" (V.223–24). This number is derived by dividing 360° (the whole circumference of the epicycle) by 3;6,24,7° (the daily motion of Mercury on the epicycle). Cf. preceding note.

30 "tempus...47 minuta" (V.226–28). These figures for the periods of direct motion in the four above positions of the epicycle (see note 24, above) are as follows: (1) 94 days 18;49 hours, (2) 93 days 0;20 hours, (3) 93 days 3;41 hours, and (4) 93 days 15;47 hours. All are correctly derived by subtraction of the period of retrogradation (given in note 28, above) from the period of one revolution on the epicycle (115 days 21;5 hours). Consequently the fourth shares the error of the corresponding retrogradation period. It should be 93 days 10;38 hours.

31 "in opposito augis epicicli" (V.277). This is not the mean but the true perigee of the epicycle, i.e., the point on the epicycle nearest to the earth at the given moment.

32 "uicinius" (V.288). See note 17, above.

33 "*Z*...augis" (V.292). I.e., the perigee of the deferent in that position (which is not, however, the point of the orbit nearest to the observer *D*).

34 "est centrum epicicli ex...*AB*" (V.311–12). Since the centers of the epicycle and the deferent rotate at the same speed but in different directions, they must always be on opposite sides of the axis of symmetry *AB*, except when they both lie on that axis.

35 "*EK* et *DK*" (V.317). *K* has now to be taken, not as the apogee of the deferent, as depicted in Figure 14 (p. 228), but as the center of the epicycle wherever it happens to be on the deferent (as is explained by the interpolation found in some manuscripts).

36 "in una medietate" (V.327). I.e., on the first half (from *K* to *Z*).

37 "similem...lune" (V.331). For this motion, see section III, notes 10 and 11. See also lines IV.274–77.

38 "quod...dictum est" (V.332–33). Lines IV.295–377.

39 "motus in latitudine" (V.339). For these, see pp. 52–53, above.

40 "quemadmodum...luna" (V.348–49). Lines IV.378–483 and Table 1, p. 356.

41 "*A* locus augis" (V.352–53). Here *A* is the position of the apogee of the *equant*. Later (lines V.361–62) *A* is the apogee of the *deferent*, which in the first position shown in Figure 15, p. 236, has the same longitude as the apogee of the equant. So here we must take it to indicate the direction only, and not the precise position, of the apogee of the equant.

42 "ea...motuum" (V.369–70). Lines V.108–37. See especially lines V.128–30.

43 "Eruntque...recta (V.370–71). For $\angle KHN = 60°$. Therefore $\angle HKN = \angle HNK = 60°$. And $\angle HKB = 120°$. Therefore $\angle HKN + \angle HKB = 180°$; i.e., *N*, *K*, and *B* lie on a straight line.

44 "Erit...mercurius" (V.373). See note 20, above.

45 "secundum...57 partes (V.380–90). The basic parameters here are $AC = 60$ (assumed), $CH = HK = KD = 3$, $AE = 22;30$. The latter two are taken from Ptolemy *Almagest* 9. 9 (Ptolemy, Manitius, vol. 2, p. 144; same figures in al-Farġānī 16. 4–5 [Alfraganus, Carmody, p. 30]). There is also the figure for *DB*, 55;34, which, though it could theoretically be derived from the above by trigonometry, is in fact also taken from the *Almagest* 9. 9 (Ptolemy, Manitius, vol. 2, p. 146, line 12). All the other measurements are then correctly derived from the above by addition and subtraction, as follows: $AD = AC + 3CH = 69$. $ED = AD + AE = 91;30$. $DF = DB - AE = 33;4$. $DG = AC - 3CH = 51$. $DM = AC - CH = 57$. Cf. section 1 of Table 2, p. 358.

46 "sicud...precedentibus" (V.394). Lines IV.317–19.

47 "Nisi...uidetur" (V.394–97). There are four possibilities: (1) The spheres are contiguous and as small as possible. This Campanus accepts. (2) The spheres pass through one another, so that, e.g., the lowest point reached by Mercury would be lower than the highest point reached by the moon. This possibility is expressed by the rather ill-chosen phrase "the planets go outside their own spheres": better would have been "the planets enter one another's spheres." That is impossible for Campanus; he supposes the spheres to be solid, and to allow one solid to pass through another would be to suppose that two bodies occupy the same space, which is contrary to Aristotelian thinking. (3) The spheres are not contiguous. This possibility would mean that there is empty space between them. That, too, is rejected as impossible, for Aristotelian thinkers rejected the existence of void. (See, e.g., Aristotle *Physics* 214^a16–217^b28, summed up at 217^b20: φανερὸν ὡς οὔτ᾽ ἀποκεκριμένον κενὸν ἔστιν, οὐθ᾽ ἁπλῶς οὔτ᾽ ἐν τῷ μανῷ, οὔτε δυνάμει. For a twelfth-century example, see, e.g., Daniel of Morley [in Karl Sudhoff, "Daniels von Morley *Liber de naturis inferiorum et superiorum*, nach der Handschrift Cod. Arundel 377 des Britischen Museums zum Abdruck gebracht," *Archiv für die Geschichte der Naturwissenschaften und der Technik*, vol. 8 (1918), p. 28]: "cum nichil sit uacuum in mundo.") (4) The spheres are contiguous, but not as small as possible, so that, e.g., there would be an interval between the highest point reached by the moon and the lowest point reached by Mercury in which no movement would take place, but which would be filled by the "quinta essentia" only, whether belonging to Mercury's sphere or the moon's or both. Campanus admits the possibility of this, but objects to it on the ground that it involves superfluity (the supposition is that the Creator always does things in the most economical way possible).

48 "hee...esse" (V.398–99). The same point is made by Ptolemy in the *Planetary*

Hypotheses (Goldstein, "The Arabic Version," p. 8): "But if there is space or emptiness between the [spheres], then it is clear that the distances can not be smaller, at any rate, than those mentioned." It is not found in al-Farġānī.

49 "64 partes et 10 minuta" (V.403). Line IV.419.

50 "17 minuta et 32 secunda" (V.404). Line IV.422 and section IV, note 65.

51 "semidiameter quoque...terre" (V.405–7). Al-Farġānī 22. 4 (Alfraganus, Carmody, p. 39) gives $d_{☿} = \frac{1}{28}d_{\oplus}$ (see p. 56, note 16). Campanus takes $\frac{1}{28}$ as equivalent to 0;2,8 (0;2,9 would be slightly more accurate). At line V.476, 0;2,8 is converted to the fraction $\frac{8}{225}$.

52 "multiplicabo...60 partes," (V.414–19). I.e., each parameter is multiplied by $\frac{64;30}{33;4}$. In fact it is necessary only to multiply the "basic parameters" AC, CH, and AE thus, and then the others can be obtained by addition and subtraction as before ($DB = DF + AE$). The figures given by Campanus are all consistent with that procedure, and are correctly computed to the nearest minute. They are listed in section 2 of Table 2, p. 358.

53 "De miliaribus...ita" (V.427). The following measurements are obtained by multiplying the preceding figures for the distances in earth-radii by $3{,}245\frac{5}{11}$, Campanus's parameter for the earth's radius in "miles" (see lines III.63–64; cf. section IV, note 63). The figures are all computed exactly. They are listed in section 3 of Table 2, p. 358.

54 "Si igitur...lune a centro terre" (V.447–56). $64;29,40 \cdot 3{,}245\frac{5}{11}$ is indeed $209{,}313\frac{26}{33}$ miles, and $0;2,8 \cdot 3{,}245\frac{5}{11}$ is indeed $115\frac{13}{33}$ miles. Then $209{,}313\frac{26}{33}$ miles $-$ $115\frac{13}{33}$ miles $= 209{,}198\frac{13}{33}$ miles, which is, of course, exactly the same as the distance found for the moon's convex surface (lines IV.436–37).

55 "spissitudo...5 undecime unius miliaris" (V.459–60). Distance to convex surface of ☿'s sphere − distance to concave surface of ☿'s sphere = ($579{,}320\frac{5}{11}$ miles + $\frac{13}{33}$miles) − (209,198 miles + $\frac{13}{33}$miles); hence, thickness of ☿'s sphere = $370{,}122\frac{5}{11}$ miles.

56 "ambitus...2 septime unius miliaris" (V.460–62). 579,321 miles $\cdot 2 \cdot \frac{22}{7} =$ $3{,}641{,}446\frac{2}{7}$ miles.

57 "ipsumque...trigesimas unius miliaris" (V.464–65). Lines IV.442–43.

58 "Ambitus uero...trigesime unius miliaris" (V.465–66). $115\frac{13}{33}$ miles $\cdot 2 \cdot \frac{22}{7} =$ $725\frac{1}{3}$ miles.

59 "Ambitus autem...2387480 miliaria" (V.466–67). The length of AC is $379{,}826\frac{4}{11}$ miles; this multiplied by $2 \cdot \frac{22}{7} = 2{,}387{,}480$ miles.

60 "dieta...34 minuta unius miliaris" (V.469–70). 2,387,480 miles $\div\ 365\frac{1}{4} =$ $6{,}536\frac{34}{60}$ miles to the nearest sixtieth.

61 "Ambitus...895220 miliaria" (V.470–71). The length of AE is $142{,}421\frac{4}{11}$ miles; this multiplied by $2 \cdot \frac{22}{7} = 895{,}220$ miles.

62 "116 dies" (V.471). If we divide Campanus's rounded figure (line V.96) for the mean daily motion of Mercury on the epicycle, 3;6,24°, into 360°, we get 1,55;52,... days, or 116 days to the nearest day (cf. section IV, note 73). Al-

Farġānī, too, in chap. 17. 1 (Alfraganus, Carmody, p. 31), gives "3 mensibus et 26^d fere"; i.e., $(3 \cdot 30+16)$ days = 116 days. Contrast lines V.223–24, where Campanus gives the more accurate figure of 115 days 21 hours 5 minutes (see note 29, above).

63 "dieta...24 minuta unius miliaris" (V.472–73). 895,220 miles ÷ 116 equals, not $7{,}717\frac{24}{60}$ miles, but $7{,}717\frac{25}{60}$ miles to the nearest sixtieth.

64 "Corpus...eius" (V.473–75). $225^3/8^3 = 22{,}247\frac{161}{512}$. Al-Farġānī (22. 5: Alfraganus, Carmody, p. 39), who also starts from $d_{☿} = \frac{1}{28}d_{\oplus}$, more sensibly gives $v_{☿} \approx \frac{1}{22{,}000}\, v_{\oplus}$.

65 "225 ad 8" (V.476). Campanus takes the approximate figure $r_{☿} \approx 0;2,8r_{\oplus}$ of lines V.405–6 (see note 51, above) as exact: thus, $1:0;2,8 = 225:8$.

66 "17 gradus...libre" (V.493–94). See line V.24 and note 8, above.

67 "sicud in sole" (V.496). Lines III.89–153.

68 "quantum...diuisionibus" (V.499–500). As the sequel (lines V.535–43) makes clear, this means that one should leave enough space for the four circles which are used for the graduation of the equant, since it is from the *inmost* of these four circles that the 32 parts of line V.510 are to be counted.

69 "medium centrum" (V.502). I.e., $\bar{\lambda}_{\odot} - \lambda_{A☿}$ (see p. 51, above). For "centrum" see section IV, note 39. The "mean centrum" is opposed to the "equated centrum," which is derived from it by application of the equation of center.

70 "Posset...idem" (V.503–4). As was explained in note 6, above, it is only the position of the center of the equant which is significant, not that of its circumference.

71 "faciam ipsum...intra ipsum" (V.504–9). What Campanus is saying here is that since he has to use the space inside the deferent for a circle (*TZ* in Figure 16, p. 246) on which to measure the movement of the center of the deferent, there is no room there for the graduation of the equant, which consequently has to be put outside the deferent.

72 "in 32 partes" (V.510). 32 (or 96, see next note) is chosen as a number into which a line can conveniently be divided. See section IV, note 80.

73 "diuisionem primam" (V.512). I.e., each of the 32 parts represents 3 units in the system in which the radius of the deferent contains 60 units. But the actual division of each into 3 is not necessary, as all the lengths concerned are multiples of 3 ($DK = KH = HC = 3, AC = 60$), except AE $(=22;30)$ which is exactly $7\frac{1}{2}$ times 3.

74 "in epiciclo lune" (V.530–31). Lines IV.532–42.

75 "eo ingenio...luna" (V.553–54). Lines IV.551–69.

76 "unus...*TZ*" (V.559–60). I.e., $HR - HM = \frac{3}{4}$ part; $HQ - HZ = \frac{3}{4}$ part. So the radius HR is 30 parts, and the radius HZ is 10 parts.

77 "Diuidam...deferentis" (V.562–63). Circle *TZ* fulfills the same function here as the "epicycle in its convenient representation" for the moon (see lines IV.526–42). Since, on any reasonably-sized instrument, the "small circle" *CK* would be too

small to be graduated, the larger circle *TZ* is drawn concentric with it, and the graduations inscribed within that.

78 "ita spissa...concauari" (V.572–73). This first disk has to be thick enough to hold not just one but two layers: the "disk of the motion of the center of the epicycle," which in its turn contains the "disk of the motion of the epicycle."

79 "quod...egredi" (V.573–74). The method is explained at lines IV.562–66 (see section IV, note 88).

80 "faciam...epicicli" (V.591–93). The purpose of the first mark is made clear at lines V.615–17. That of the second is not explicitly stated: in fact it is used in the operation at lines V.621–23, where the apogee of the epicycle is lined up with one of the threads.

81 "tabulam...solis" (V.599). This is a table of the sun's mean motion counted from its apogee. See section III, note 33.

82 "sicud...sunt" (V.600–601). E.g., lines V.111–12.

83 "tria...quantitate" (V.601–3). $\lambda_{A☿}$ (line V.24) minus $\lambda_{A\odot}$ (lines III.124–25) gives $6^s 17;30° - 2^s 17;50° = 3^s 29;40°$.

84 "ut...est" (V.609). Not said explicitly, but derivable from lines V.97–101.

85 "circulo motus centri deferentis" (V.611). This is circle *TZ* in Figure 16, p. 246. Cf. line V.570.

86 "tabulam motus centri deferentis" (V.615–16). This is the same as what was defined as the "tabulam motus centri ecentrici" at line V.575.

87 "Si autem per...retrogradus est" (V.629–42). The theory underlying this procedure is correct, though in practice it would produce only approximately correct results. If the planet is on the upper arc between the two tangents to the epicycle from the center of observation *D*, then the component of its motion on the epicycle as seen from *D* is from west to east (or nil if it is actually in one of the points of tangency), and, since the motion of the center of the epicycle is always from west to east, the combined motion must be in a west-to-east direction, i.e., direct. So Mercury can be retrograde only if it is on the lower arc, and then only if its east-to-west motion on the epicycle, as seen from the center *D*, is greater than the west-to-east motion of the center of the epicycle, as seen from the same point. To determine this, Campanus compares the two motions over a period of one day (cf. note 23, above). He uses the round figures of three degrees a day (instead of 3;6,24°, as in line V.96) for the motion of Mercury on the epicycle, and one degree a day (instead of 0;59,8°, as in line V.110) for the motion of the center of the epicycle. (With this crudity of approximation, it does not matter that the motion of the epicycle is uniform, not with respect to *D*, but with respect to the equant point *K*.) According as the 3° arc of the epicycle appears under an angle of less than 1°, of 1°, or of greater than 1° as seen from *D*, the planet is considered to be in direct motion, stationary, or retrograde, respectively.

88 "ascendens uel descendens" (V.645–46). "Ascending" means "between perigee and apogee"; "descending," "between apogee and perigee." Cf. lines IV.121–23.

89 "quemadmodum...luna" (V.652). Lines III.203–34 and IV.656–749.

90 "linea medii...perueniat" (V.670–73). Refer to Figure 17, p. 256, which is adapted from a figure found in some manuscripts (e.g., MS *A*, fol. 131r), though not referred to in the text of any of them. In this figure, *D* is the center of the ecliptic, *K* the center of the equant, *B* the center of the epicycle at the time given (by subscript). Then *DS*, parallel to *KB*, is the "line of Mercury's mean motion in the ecliptic" (line V.670).

91 "anguli...signorum" (V.675–77). The two angles are *AKB* and *ADS* in Figure 17.

92 "quemadmodum geometre probant" (V.679). E.g., Euclid 1. 29, as noted by the φ MSS.

93 "sicud...est" (V.684). Lines III.204–16.

94 "linea...epicicli" (V.691–92). This is line *DB* in Figure 17.

95 "3 signis et 13 aut 14 gradibus" (V.716). This seems to be a plain mistake of Campanus (repeated at line V.725), but I can find no explanation for it. The figure is much too high for the position of the maximum equation of center. The table in the *Almagest* 11.11 (Ptolemy, Manitius, vol. 2, p. 265) gives 90°–96°, the "Handy Tables" (Halma, *Commentaire de Théon*, pt. 2, pp. 188–89) 93°–100°, the Toledan tables (Toomer, "Survey of the Toledan Tables," Table 44) 93°–97°. It is conceivable that we have in both places a scribal error for "iii aut iiii" or "viii aut viiii."

96 "in medietate altera...minus" (V.740–46). I.e., if $\bar{\varkappa}$ is the mean centrum and δ the equation of center, then $\delta(\bar{\varkappa}) = -\delta(360° - \bar{\varkappa})$.

97 "Processerunt...modo" (V.752). The procedure can be illustrated by any standard ancient or medieval planetary table. For instance, the first three columns of the Toledan tables for Mercury begin and end as shown in the accompanying table.

Linee numeri				Equatio centri	
Signa	gradus	Signa	gradus	Gradus	minuta
0	1	11	29	0	3
0	2	11	28	0	6
0	3	11	27	0	9
......	......	......	......	......	......
5	29	6	1	0	8
6	0	6	0	0	0

98 "linee numeri" (V.760). This appellation for the argument to a planetary table is indeed frequently found in medieval Latin astronomical tables (e.g., the Toledan tables, as in the preceding note, and in the tables for Novara, MS Dublin, Trinity College, D.4.30, fol. 58r), though other headings, such as "tabule numeri" also occur. The expression "linee numeri" is a literal translation of the Arabic "suṭūru'l-ʿadad," ("lines of number": see Nallino, *Al-Battānī*, pt. 2, p. 78) and

preserves the Arabic idiom of putting the dependent descriptive genitive in the singular.

99 "uias...arismetricis (V.760–61). The calculation is explained in *Almagest* 11. 9 (Ptolemy, Manitius, vol. 2, pp. 252–53), and demands the use of trigonometry, which is what is really meant by Campanus's expression.

100 "Dictum est supra" (V.768). E.g., lines V.94–106.

101 "prout autem...difformis" (V.771–72). The true apogee of the epicycle coincides with the mean at apogee and perigee of the equant; in between it differs from the mean. Thus it is obvious that, if the motion of the planet's body is uniform with respect to the mean apogee of the epicycle while the epicycle is traveling from apogee to perigee of the equant, it cannot at the same time be uniform with respect to the true apogee.

102 "prius diximus" (V.775). Lines V.666–73.

103 "Aux...centri" (V.776–77). The true apogee is, of course, that intersection of line and epicycle farthest from the center of the ecliptic (point *M* in Figure 17).

104 "similem" (V.780). "Similar" is here a technical term, as Campanus goes on to explain. Here and in what follows (e.g., line V.793), it must be remembered that Campanus thinks of the equation of center as an *arc of the ecliptic* rather than as an angle at the center of the ecliptic.

105 "arcus...proportio" (V.783–84). The equivalent of this definition will be found in Euclid, bk. 3, Definition 11 (cf. *mg.* MS *F*), where, however, Euclid speaks of sectors rather than arcs.

106 "linea ueri...in centro orbis signorum" (V.784–90). I.e., in Figure 17, line *MD* cuts the two parallel lines *KN* and *DS* at *B* and *D*, respectively. Therefore $\angle MBN = \angle BDS$.

107 "quemadmodum...est" (V.792). Lines V.677–79, and see note 92, above.

108 "equationi centri" (V.793). See note 104, above.

109 "quod...equales" (V.796–97). This is equivalent to the statement at lines V.783–84.

110 "uerificationem...argumenti" (V.811–12). I.e., finding the true centrum and true anomaly from their respective means, as Campanus explains.

111 "Prcipiunt" (V.812). Here and in much of this section Campanus is referring to the explicit rules for operating with astronomical tables given in the "canones" (see section IV, note 38). For a printed example of such rules for the planets, see the tables of al-Khwārizmī, chap. 10 (Suter, "Die astronomischen Tafeln," pp. 10–11).

112 "Dictum...supra" (V.816). E.g., lines V.111–13.

113 "lineas numeri" (V.831). Literally, "lines of numbers": cf. line V.760 and note 98, above.

114 "si centrum est...habemus argumentum uerum" (V.834–47). This is best illustrated by Figure 17, p. 256: there $\angle AKB$ is the mean centrum. Clearly $\angle ADB$, the true centrum, equals $\angle AKB_1$ *minus* $\angle KB_1D$, which equals the equation of center. On the other hand, if *P* is the position of the planet on the epicycle, then

$\angle NBP$ is the mean anomaly, and $\angle M_1B_1N_1$, which equals $\angle KB_1D$, must be *added* to it to get $\angle MBP$, the true anomaly. The converse is true when the epicycle is on the other half of the equant, for $\angle ADB_2$ (>180°) = $\angle AKB_2$ (>180°) + $\angle KB_2D$, and $\angle M_2B_2P_2 = \angle N_2B_2P_2 - \angle M_2B_2N_2$.

115 "in fine operis" (V.849). See lines V.870–75 and V.1303–10.

116 "ambe distantie predicte" (V.870). I.e., both the equation of center $\delta(\bar{\kappa})$ and the equation of anomaly $c(\alpha)$. The following rules (lines V.870–75) for finding the true position λ can then be summarized by the formula $\lambda = \bar{\lambda} + \delta + c$ if we remember that δ is negative for $0° < \bar{\kappa} < 180°$ and c is negative for $180° < \alpha < 360°$. The rules are repeated below, lines V.1303–10.

117 "sicud indicabunt sequentia" (V.884–85). The promise is unfulfilled in the work we have. It is (contingently) true that for all planets the size of the epicycle has a bigger effect on the equation of anomaly than does the variation in the distance of the epicycle from the observer. This could have been verified either by examples from the tables or by suitable manipulation of the instrument, but it is not a legitimate conclusion from Campanus's subsequent account.

118 "premisse distantie" (V.920). I.e., the equations of anomaly for an epicycle position other than at the apogee.

119 "sue relatiue" (V.920). Literally "the ones corresponding to them." But here and throughout I have used the paraphrase "at the same anomalies," or the like, to avoid confusion with the use of "corresponding" to translate "compares" (see lines V.901 and V.926).

120 "4 signa...superius" (V.923). See lines V.141–46 and note 20, above.

121 "32580 equationes" (V.936). That is, $180 \cdot 181$; the equations for anomalies of 1° to 180° are calculated 181 times, for true centra of 0° to 180°, inclusive. Since the equation of anomaly for $\alpha = 180°$ is always zero, a better figure would have been $179 \cdot 181$, or 32,399.

122 "secundum...5 gradus" (V.946–47). This figure is chosen because it is where the center of the epicycle is at mean distance from the observer (cf. lines V.977–78, V.985–86). In the Toledan tables for Mercury (Toomer, "Survey of the Toledan Tables," Table 44), the column "minuta proportionalia" reaches a minimum (i.e., the distance of the center of the epicycle from the center of the ecliptic reaches the mean) at an argument of $2^s 4°$ and $2^s 5°$. The argument is indeed the true centrum (cf. line V.1278), and that fact explains the slight discrepancy from the table in the *Almagest* 11. 11 (Ptolemy, Manitius, vol. 2, p. 265, col. 8), where the minimum occurs at about 67°: for, as is explained by Ptolemy at Manitius, vol. 2, p. 266, line 26, the argument there is the *mean* centrum. It was changed to the true centrum by Ptolemy himself for his "Handy Tables" (see the introduction to the latter in *Opera minora*, ed. Heiberg, p. 169, line 24–p. 170, line 3), and remained so for all subsequent tables of this type known to me.

123 "equatio argumenti" (V.956–57). This is indeed the heading of column (5) of the planetary tables in the Toledan tables (see Toomer, "Survey of the Toledan

Tables," p. 60); there we find, too, the "longitudo longior" of line V.965 and the "longitudo propior" of line V.975 (columns (4) and (6), respectively). The same headings are found in MS Dublin, Trinity College, D.4.30.

124 "longitudo longior" (V.965.). See note 123, above.

125 "longitudo propior" (V.975). See note 123, above.

126 "cum centrum...4 gradus" (V.985–86). Cf. note 122. The mean distance, according to the Toledan tables, which Campanus is following here, is not at exactly $2^s 4°$ for the true centrum, but between $2^s 4°$, where the value of the minuta proportionalia is -1, and $2^s 5°$, where it is $+1$. Hence at $2^s 4°$ a small subtraction has to be made from the equation of anomaly. This is clear from Campanus's own narrative (lines V.1078–79).

127 "sine...diminutione" (V.988). It is true that there is no increase or decrease at the ideal mean distance. In the actual Toledan tables, however, it is not true for any tabulated value of Mercury, such as $2^s 4°$. See preceding note.

128 "minuende...addende" (V.1035). The feminines "minuenda" and "addenda" are here used as nouns. The reason for the gender is that the expression was originally "pars minuenda," etc.

129 "Quamlibet...remotior" (V.1044–50). If taken in the most obvious way, the meaning of these lines would be that they divided the 60 representing the value of each "greatest distance" by the arc of the equant between apogee and maximum distance, i.e., $2^s 4°$ for Mercury, and assigned to each degree of that arc the fraction of 60 accruing from the resultant proportion sum. That is, if d is the number of degrees between the apogee of the equant and the position of the center of the epicycle, then m, the number of minutes at that point, is given by $m = 60 - \frac{60}{64} \cdot d$.

However, this equation is *not* the theoretical basis of the column of "minutes of proportion" in the tables. Ptolemy explains how he computed it at *Almagest* 11. 10 (Ptolemy, Manitius, vol. 2, pp. 256–60): having obtained the equations of anomaly for the epicycle at maximum, mean, and minimum distances, he computed the maximum equation of anomaly trigonometrically for every 3 or 6 degrees of centrum between these positions, normed the difference between the maximum equations of anomaly at mean distance and at maximum distance as 60, then expressed the differences between the maximum equations at mean distance and at the distances between mean and maximum in terms of this normed difference. He repeated the process for minimum distance and the distances between the minimum and the mean distance. He correctly states that the proportion between the maximum equations of anomaly for any two centra can be assumed to hold between the equations of any same anomaly for the same two centra without significant error (see pp. 46–47 and 50–51, above).

Since Campanus at least hints at Ptolemy's procedure (ll.1052–54), it seems unlikely that he held the above-described gross misconception of the basis of computation. I therefore suggest the following alternative explanation of this passage: "secundum uiam proportionis inuente" (ll.1047–48) refers, not to the simple ratio $\frac{60}{64} \cdot d$, but rather to the final result of Ptolemy's computation; i.e., Campanus is making the trivial point that the "minutes of proportion," arrived at by some unspecified method, can be applied to any ("quamlibet," line V.1044) greatest distance (i.e., for any degree of anomaly), and that the argument for the minutes of proportion is the arc between the maximum and mean distances (the true centrum). Then lines V.1052–54 are a vague explanation of "secundum uiam proportionis inuente" in lines V.1047–48. This interpretation must be what Campanus means, but his expression is very ambiguous.

130 "titulum...minuatur" (V.1057–58). The headings "minuta proportionalia" and "minuatur" are found in the Toledan tables (where many manuscripts, however, have "proportionalia minuta") and in MS Dublin, Trinity College, D.4.30. Cf. note 123, above.

131 "In hac...unum minutum" (V.1062–78). The details of the occurrence of the numbers in the column "minuta proportionalia" tally exactly with the Mercury table of the Toledan tables.

132 "quod quasi...diminueretur" (V.1080). Campanus betrays his uneasiness at the fact that the values do not reach zero, as they theoretically should. The reason is that the actual mean position occurs between $\varkappa = 2^s 4^\circ$ and $\varkappa = 2^s 5^\circ$. Cf. notes 126 and 127, above.

133 "3 signa et 28 gradus" (V.1089). The minimum distance is supposed to occur at a mean centrum of 4 signs (lines V.141–51). To a mean centrum of 4 signs in the Toledan tables corresponds an equation of center of 2;41°. So the actual figure for the true centrum here should be ($4^s - 2;41^\circ$), or $3^s 27^\circ$ to the nearest degree. Campanus may have been influenced in his choice of $3^s 28^\circ$ by the fact that it is the middle one of the 11 degrees to which corresponds a value of 60 for the minutes of proportion (see lines V.1132–36).

134 "Quamlibet...propinquior" (V.1096–103). Cf. lines V.1044–50 and note 129, above.

135 "Posuerunt...consimilem" (V.1115–27). The column of minutes of proportion for Mercury in this area, as described by Campanus, would appear as shown in the accompanying table.

Linee numeri		
Signa	minuta	Minuta proportionalia
2	4	1
		addatur
2	5	1
2	6	2
2	7	4
2	8	6

This is exactly as it appears in the Toledan tables.

136 "Non solum...equationem argumenti" (V.1132–47). Here, too, the appearance

of this part of the column in the Toledan tables is exactly as Campanus describes.

137 "Vnde... minimam" (V.1159–64). It is, of course, true that the same distances, and hence the same values for the minutes of proportion, occur on either side of the minimum distance. However, it is not true, as one might incautiously conclude from Campanus's expressions here, that the function of the minutes of proportion is symmetric about the centrum for the minimum distance. In fact the tables show that the function decreases from its maximum at $3^s27°$ much more slowly than it had increased up to it, so that, e.g., it does not reach +40 until $\varkappa = 5^s23°$, or 56° after the minimum distance, whereas its previous value of +40 came at $\varkappa = 2^s27°$, or only 30° before the minimum.

138 "centrum... 25 gradus" (V.1176–77). Indeed, in the Toledan tables, for $\varkappa = 3^s5°$ and $\varkappa = 4^s25°$ the minutes of proportion are +50.

139 "In mercurio... argumenti" (V.1178–80). This inference is, however, valid for the other planets, which therefore dispense with the headings "add" and "subtract" in the column of minutes of proportion. See lines V.1220–25.

140 "ista diuersitas" (V.1200). This refers to "diuersificantur" in lines V.1153–54; i.e., Campanus is saying that in the tables of Venus, Mars, Jupiter, and Saturn the minutes of proportion for the least distance increase monotonically, and thus do not differ from the minutes of proportion for the greatest distance, which decrease monotonically, whereas in the Mercury table the minutes of proportion for the least distance increase to a maximum and then decrease (see lines V.1113–47).

141 "2 signa et 28 gradus" (V.1202). Indeed, in the Toledan tables (see Toomer, "Survey of the Toledan Tables," Tables 40–43), the minutes of proportion reach zero at the following arguments: Saturn, between $2^s27°$ and $2^s28°$; Jupiter, between $2^s28°$ and $2^s29°$; Mars, between $2^s27°$ and $2^s28°$; and Venus, between $2^s28°$ and $2^s29°$. The same is true of the tables for Novara in MS Dublin, Trinity College, D.4.30 (fols. 57v, 61r, 65r, 69r).

142 "diuiserunt... mediam" (V.1214–16). See note 129, above, for an explanation of the expressions used here.

143 "ut dictum est" (V.1237). Campanus here refers to the account at lines IV.692–719, from which this statement can easily be derived, though it is not explicit.

144 "In luna... oppositionem" (V.1237–38). What Campanus is emphasizing here is that the moon's minutes of proportion take the whole arc from apogee to perigee to go from 0 to 60, while for Venus, etc., they go from (negative) maximum to zero and again to (positive) maximum in the same space.

145 "circulus breuis" (V.1245). See section IV, note 104, for this term.

146 "inceperunt a cifris" (V.1250). In contradistinction to the table for the planets, where the column begins with 60 and decreases, for the moon it begins with 0 and increases.

The word "cifra," derived from the Arabic "ṣifr" (from the root "ṣfr," "be empty, vacant"), is a standard medieval term for the symbol 0. See, e.g., the

Algorismus ascribed to Sacrobosco (J. O. Halliwell, ed., *Rara mathematica*, vol. 1, *Johannes de Sacro-Bosco: Tractatus de arte numerandi* [London, 1839], p. 3): "Decima figura dicitur theta, vel circulus, vel cifra"; and cf. Vogel, *Mohammed ibn Musa Alchwarizmi's "Algorismus,"* p. 45, note 3. For information on this and other names for zero, see D. E. Smith and L. C. Karpinski, *The Hindu-Arabic Numerals* (Boston, 1911), pp. 57–62.

147 "cifram... 60 minuta proportionalia" (V.1250–60). The appearance of the column of "minuta proportionalia" for the moon as described here corresponds exactly to the columns in both the tables for Novara (MS Dublin, Trinity College, D.4.30, fols. 49r, 51v) and the Toledan tables (Toomer, "Survey of the Toledan Tables," Table 39).

148 "omnibus planetis" (V.1263). Here "all the planets" includes the moon.

149 "simpliciter" (V.1264). The meaning is that the minutes of proportion in the lunar table form a single "stretch," increasing monotonically from 0 to 60. For Venus, Mars, Jupiter, and Saturn, they form one stretch decreasing from 60 to 0, and a second increasing from 0 to 60; while Mercury has, in addition to that, a third stretch decreasing from 60 to 40. In modern terminology, we should say that the function has respectively 0, 1, and 2 extrema between the arguments 0° and 180° (noninclusive: Campanus considers only the stretch between apogee and perigee. If we consider it over a complete revolution, then the figures become 2, 4, and 6, respectively. But Campanus's description is very apt when applied to the actual appearance of the tables).

150 "conuenientis longitudinis" (V.1281). I.e., either the greatest distance or the least distance, as appropriate. See what follows.

151 "titulum minutorum proportionalium" (V.1282). I.e., "addatur" or "subtrahatur." See lines V.1058–62, V.1111–13, above.

152 "utrum crescant aut decrescant" (V.1283). See lines V.1220–24.

153 "super... argumenti" (V.1303–4). The equation of anomaly has been dealt with immediately above. The rules for writing "add" or "subtract" over the equation of center were given at lines V.847–49. See lines V.870–75 (with note 116, above) for the rules Campanus gives in the subsequent lines.

154 "si sermo... prolixior" (V.1317–18). Cf. lines II.24–25: "ne fiat sermo concepta mensuratione prolixior." See section II, note 8.

155 "instrumentum premissum" (V.1323–24). As the sequel shows, this refers to the separate equatorial plates described earlier, and indeed to those (for Venus and the three superior planets) which have yet to be described. For Campanus it is an instrument (in the singular) because it all fits into one "mother," though in fact the plate for each body works independently from the rest. See the initial description at lines II.74–80.

The point of using the instrument to find these elements of the tables is not to indulge in the useless exercise of finding the planetary positions by a tabular method when one could get them directly from the instrument, but rather to check existing tables, as Campanus says at lines V.1381–82. He there rather optimistically adds that one could correct and even compose tables in this way. In fact it is doubtful whether tables so derived, even from the most carefully constructed instrument possible at the time, could have compared in accuracy with those derived from trigonometrical computation.

It is tempting to think that Roger Bacon had this passage of the *Theorica* in mind when he wrote in his *Opus tertium*, composed in 1267 (in *Opera quaedam hactenus inedita*, ed. Brewer, vol. 1, pp. 35–36): "nam sine instrumentis mathematicis nihil potest sciri, et instrumenta haec non sunt facta apud Latinos, et non fierent pro ducentis libris, nec trecentis. Adhuc autem sunt tabulae meliores; nam licet *certificatio tabularum sit per instrumenta* [my italics], tamen instrumenta, nisi sint immensae quantitatis, nihil valent: et haec mali usus sint, et difficilis conservationis propter rubiginem nec possint portari de loco ad locum, sine periculo fractionis; et homo non potest habere ubique et semper nova instrumenta, cum tamen oporteat hoc, nisi habeat tabulas semel certificatas." However, Bacon probably meant observational rather than calculating instruments.

156 "illi loco epicicli conueniat" (V.1328). If the position of the epicycle is given with respect to the apogee of the equant, then the position of the center of the deferent on the small circle is immediately known, since the two angles concerned are equal. See p. 50, above.

157 "si filum...propositum" (V.1328–33). See Figure 16, p. 246. The first thread is that at point *K*, the second that at *D*.

158 "immobile...dicetur" (V.1335). This is not explicitly stated, but is obviously implicit in the description at lines VI.27–32.

159 "extende filum...propositum" (V.1336–43). See Figure 13, p. 196. The first thread is that at point *K*, the second that at *D*.

160 "deferentis" (V.1345). For this inaccurate use of "deferens" to mean the sun's eccentric (by analogy with the other planets), see section III, note 35.

161 "In sole...propositum" (V.1345–48). See Figure 10, p. 152. The threads mentioned here should be at points *C* and *D*, respectively. However, in the description of the making of the plate for the sun (lines III.89–156) there is no mention of any thread at point *C*, and indeed it is unnecessary if the instrument is used only for finding the true position of the sun. This slip is perhaps an indication that Campanus never actually manufactured the instrument. See pp. 32–33, above.

162 "longitudine media" (V.1350). Campanus does not explain how one determines the position of mean distance: in fact all one has to do is to measure off from *D* along the thread attached to *D* (see Figure 16, p. 246, and Figure 22, p. 346) an amount equal to the radius of the deferent (*AC*), and move the epicycle according to Campanus's instructions until the point thus marked on the thread coincides with the epicycle center.

163 "loca predicta" (V.1367). Apogee, perigee, and position of mean distance.

164 "cuiusuis argumenti" (V.1367–68). In computing the minutes, Ptolemy took

the maximum equations rather than equations of equal anomalies (see note 129, above). Theoretically, different anomalies would give different minutes of proportion for the same centrum, but in fact the variation would be so slight that it would not be detectable by the rather crude method Campanus is advocating here.

165 "parte aut partibus" (V.1378–79). Campanus is distinguishing between "pars," a "unit fraction" (e.g., "quarta pars," $\frac{1}{4}$, or "quinta pars," $\frac{1}{5}$), and "partes," a "compound fraction" (e.g., "duae partes," $\frac{2}{3}$, or "quinque undecimae partes," $\frac{5}{11}$). The distinction is quite pointless here, but was felt to be an important one in general in ancient and medieval times. See Kurt Sethe, *Von Zahlen und Zahlworten bei den alten Ägyptern* (Strasbourg, 1916), pp. 60–69; O. Neugebauer, *Vorlesungen über Geschichte der antiken mathematischen Wissenschaften*, vol. 1, *Vorgriechische Mathematik* (Berlin, 1934), pp. 86–93.

166 "tabulas...componere" (V.1381). This is unduly optimistic. See note 155, above.

Section VI

1 "licet...necessaria" (VI.4–6). For an account of the Ptolemaic model for these planets, see pp. 47–49, above. As Campanus says, the models have the same structure; only the parameters are different.

2 "distantia centrorum" (VI.8). I.e., the eccentricities (the "centers" are the centers of the ecliptic, the deferent, and the equant: hence, there are two "distances between the centers" for each planet).

3 "differunt a luna" (VI.17–18). Because the moon moves on its epicycle in the opposite sense to the motion of the five planets on their epicycles. See lines IV.4–5.

4 "sicud postea dicetur" (VI.27). The same promise to give an account of the planetary motions in latitude is made at lines VI.386–87, but it is unfulfilled. The latitudes could not be represented on Campanus's instrument, and hence it might be considered reasonable for him to omit discussion of them; and indeed at lines V.340–42 (theory of Mercury) he appears to disclaim any intention of going into the latitude theory. However, he does discuss the moon's latitudinal motion (lines IV.6–36), and the statement here is quite unambiguous: this is what leads us to suspect that the work as we have it is not complete. See pp. 29–30, above.

For the Ptolemaic latitude theory, see pp. 51–53, above.

5 "Sunt...signorum" (VI.27–29). For a better understanding of this description, refer to Figure 18, p. 308.

6 "locus augis...unus" (VI.37–38). I.e., they will have the same longitude, though they are separate points: e.g., as represented on Figure 18.

7 "in uenere...sagittarii" (VI.38–41). The following positions of the apogees of the planets are given: Venus, ♊ 17;50° (the same as that for the sun); Mars,

♌ 1;50°; Jupiter, ♍ 14;30°; and Saturn, ♐ 0;5°. They agree exactly with the list given in the Toledan tables (see Toomer, "Survey of the Toledan Tables," p. 45), as do those for the sun (lines III.124–25) and Mercury (line V.24). The tables for Novara have the same values.

8 "illa diametro...equantis" (VI.45–46). For the same rather odd expression, see lines V.98–99.

9 "Demonstrauimus...supra" (VI.55–56). Lines V.780–99 and see section V, note 104.

10 "effectiue et finaliter...non" (VI.59–64). For the Aristotelian doctrine of the four "causes" (material, formal, efficient, and final), see, e.g., Aristotle *Physics* 2. 3 (194^b16–195^b30), and for a good short account, see W. D. Ross, *Aristotle* (London, 1923), pp. 71–75. Campanus's point here is the following: the name "equation of center" is given to two arcs equal in angular size but different in position, one an arc of the ecliptic, the other an arc of the epicycle. Only the first can be called "equatio centri" in the full sense of the term—i.e., "that which equates the centrum"—for the second is applied as equation not to the centrum but to the anomaly. For the moon, that is the only use of the arc of the epicycle. Campanus's way of expressing the distinction is not a happy one, but it is intelligible.

The "efficient cause," according to Aristotle (*Physics* 194^b29–31), is "that from which comes the immediate origin of the change...the producer of change [is the cause] of the thing changed." Hence, the equation of center can be regarded as the efficient cause of the corrected centrum and of the corrected anomaly. The "final cause" (194^b32–35) is the οὗ ἕνεκα, the end for which something is done; for instance, "health" is the final cause of "walking." Now the "equatio centri," in the sense of "the equating of the center," can properly be regarded as the final cause of the application of the equation of center to the mean centrum, but not, of course, as the final cause of its application to the mean anomaly. However, this justification depends on an ambiguity in the word "equatio," which can mean both "that which equates" and "the act of equating." An arc may be identified with the first, but not with the second. We may suspect that Campanus's command of Aristotelian logic was superficial.

11 "Quantitates...sunt ita" (VI.65–66). Campanus gives the mean daily motions shown in the accompanying table.

	(1) Planet on epicycle (anomaly)	(2) Epicycle in longitude
Venus	0;36,59°	0;59,8°
Mars	0;27,42°	0;31,26°
Jupiter	0;54,9°	0;4,59°
Saturn	0;57,8°	0;2,0°

As he states at lines VI.81–83, for the superior planets the sum of these two motions is equal to the mean daily motion of the sun (0;59,8°). The parameters for the superior planets in column (2) are the same as those in the Toledan tables (Toomer, "Survey of the Toledan Tables," Tables 34, 33, and 32), and those in column (1) are derived by subtracting the values in column (2) from 0;59,8°. The parameters in column (1) for Venus should agree with what is in Table 35 of the Toledan tables. There, however, stands 0;37,0°, while in the tables for Novara (MS Dublin, Trinity College, D.4.30, fol. 37v), which give parameters identical with those of the Toledan tables in all other cases discussed here, we find 0;36,59°. So it looks as if Campanus took his figures not directly from the Toledan tables, but from that intermediary work.

12 "Vnus...parte" (VI.76–78). For the terminology used here, cf. lines V.112–13, and section V, note 16.

13 "si in planetis...medio" (VI.81–83). See p. 49, above. What is stated here is that $\overline{\lambda}_P + \overline{\alpha}_P = \overline{\lambda}_\odot$; i.e., the sum of the movement on the eccentric and the move-

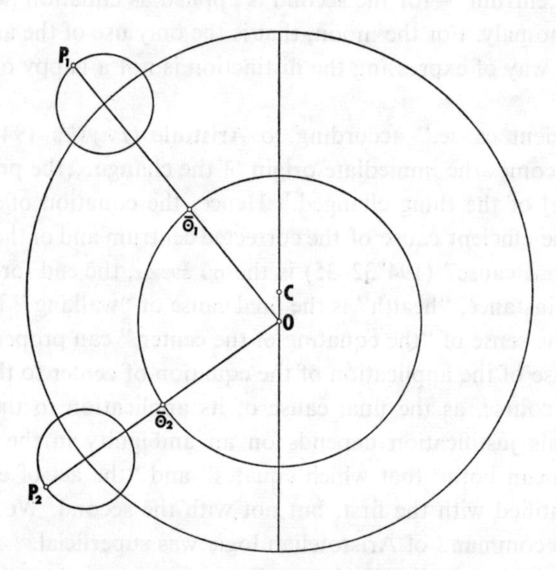

Fig. 25

Schematic illustration of two successive mean conjunctions between the sun and an outer planet in the Ptolemaic system. O is the earth, C the center of the planet's deferent. At the first conjunction the mean sun (traveling on an imaginary circle with O as center) is at $\overline{\odot}_1$. The center of the planet's epicycle lies on the extension of $O\overline{\odot}_1$, and the planet is at the apogee of the epicycle (as it must be at every mean conjunction according to Ptolemaic theory), at P_1. At the next mean conjunction the earth, the mean sun, and the planet again lie on one straight line, $O\overline{\odot}_2P_2$. Since the first conjunction, the mean sun has completed a whole revolution of its circle plus the arc $\overline{\odot}_1\overline{\odot}_2$; the epicycle center has traversed the arc of the deferent corresponding to arc $\overline{\odot}_1\overline{\odot}_2$, and the planet has completed one revolution on the epicycle, which has brought it back to the same position with respect to O, namely, to the apogee of the epicycle. Thus the angular sum of the epicycle's motion plus the planet's motion equals the mean sun's motion from conjunction to conjunction.

ment on the epicycle is constant for the three outer planets, and equal to the mean motion of the sun.

14 "in eadem habitudine epicicli" (VI.87). I.e., the true anomaly will always be the same (namely, zero) at conjunction.

15 "Sit...medium" (VI.89–90). Illustrated by Figure 25, where two successive mean conjunctions of the sun and an outer planet are represented (*O* is the earth, *C* the center of the planet's deferent).

16 "arcum...secunde" (VI.94–95). Angle P_1OP_2 in Figure 25.

17 "ex creatoris beneplacito" (VI.122). Cf. lines IV.124 and V.124. The presence of the planet in the apogee of the epicycle at mean conjunction is, of course, determined by the condition that the vector of the epicycle radius carrying the planet shall always be parallel to the direction from earth to mean sun. See pp. 47–48, above.

18 "mutuantur...motus" (VI.131–32). This acute observation (elaborated in one chapter of Campanus's *Tractatus de sphera*, ed. Venice, 1518, chap. 52, fols. $158r^b$–v^a) singles out a feature of the Ptolemaic system which is inexplicable in its own (geocentric) terms, but which immediately makes sense when one substitutes a heliocentric model. For then the epicycles of Venus and Mercury become their orbits around the sun, and the epicycles of the three outer planets are seen to be reflections of the earth's orbit around the sun. Only the moon's connection with the sun requires further explanation; it becomes clear when we remark that it is only Ptolemy's "*second* lunar anomaly" which is a function of the moon's elongation from the sun. This is roughly equivalent, in modern terms, to the effect of the gravitational pull of the sun on the moon. Cf. p. 44, above.

19 "in quibus...solis" (VI.136–37). This is nonsense, and it is hard to see how it has arisen. The epicycle centers of Venus and Mercury coincide in longitude with the mean sun, and so the planets are "on one side or the other" of the mean sun except when in the apogee or perigee of their epicycle, i.e., in the two points where the line from the observer to the mean sun cuts the epicycle. This is what Campanus should have said, but the only way in which the text could be reconciled with that would be to translate, "the *lines to them [the planets] from the center* cut [i.e., coincide with] the *line to it [the sun] from the center*." Such a meaning ("from the center") for "ecentricus" would be unique, and it is perhaps preferable to suppose mere error on Campanus's part. Corruption and interpolation seem equally unlikely.

20 "ut dictum est prius" (VI.148). Lines VI.38–41, and see note 7, above. All these values are taken from the Toledan tables.

21 "Eritque...linea *DA*" (VI.156–57). As depicted in Figure 18, *E* is, of course, the true apogee as well as the mean. But, as Campanus says below, the identity occurs only contingently ("secundum accidens," line VI.176), because the epicycle happens to be in the apogee of the eccentric, where *DA* coincides with *KA*. To make the point clearer I append Figure 26, showing the position after the epicycle has

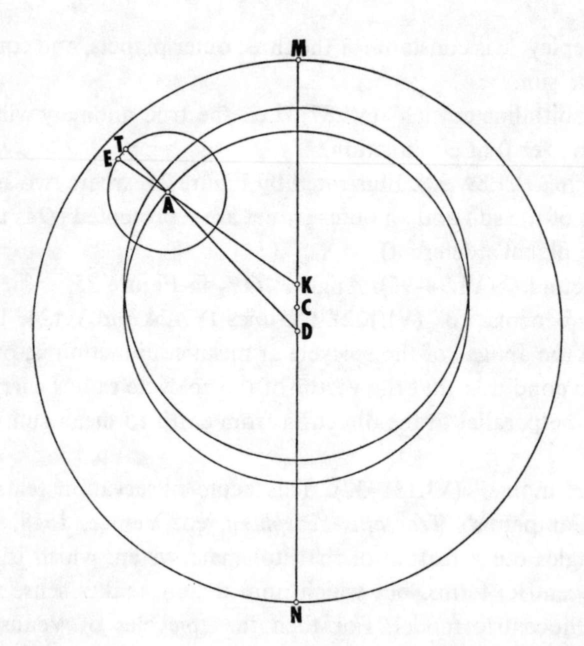

Fig. 26

Illustration of the difference between mean and true apogees of the epicycle for a planet whenever the epicycle is not in the apogee or perigee of the deferent (compare Figure 18, p. 308). *D* represents the earth, *C* the center of the deferent, *K* the equant point, and *M* and *N* the points in the ecliptic which are the projections of the apogee and perigee of the deferent. The epicycle is drawn at point *A* in the deferent, away from its apogee and perigee. Then the mean apogee of the epicycle is *E*, on the extension of line *KA*, while the true apogee of the epicycle is *T*, on the extension of line *DA* (these two points coincide when *A* lies on line *MN*, as in Figure 18). Arc *ET*, the difference between mean and true epicyclic apogees, is obviously equal to angle *DAK*, the equation of center.

moved away from the apogee. The true apogee is now *T*, and the distance between true and mean apogee, arc *ET*, is equal to the equation of center, $\angle DAK$.

22 "equalitas... lineam *AK*" (VI.167–68). See p. 49, above.

23 "Ibique... maxima" (VI.180). It is not true that the equation of center will be a maximum when $\angle MDA$ in Figure 26 is a right angle (i.e., when the true longitude of the epicycle is 90°): it is a maximum when $\angle MCA$ is 90°; i.e., if the equation of center is θ, the maximum occurs when the true longitude of the epicycle is $(90° - \theta/2)$. Campanus has forgotten that this is not a simple eccentric (where the maximum equation does indeed occur at a true longitude of 90°), but an equant model. The error is repeated at lines VI.184–87.

24 "modo contrario" (VI.190). I.e., whereas from apogee to perigee of the eccentric the true epicyclic apogee first moved westward away from *E* and then eastward toward it, from perigee to apogee it first moves eastward away from *E* and then westward toward it.

25 "dicte sunt prius" (VI.194). Lines VI.66-80 and see note 11, above.

26 "estque...maior eo" (VI.198–201). For this curious way of expressing the comparative speeds and the reason for it, see section V, note 23.

27 "sicud dictum...mercurio" (VI.216–17). Lines V.179–206.

28 "In uenere...30 minuta" (VI.221–33). Like the figures for Mercury (see section V, note 24), these are taken from the Toledan tables (Toomer, "Survey of the Toledan Tables," Tables 40–45), with interpolation of the values for mean distance. The tables for Novara (MS Dublin, Trinity College, D.4.30, fols. 56v–70v) give exactly the same values as the Toledan tables here. What is given in each case is the distance of the station from true apogee of the epicycle; to get the arc of half-retrogradation from each position, subtract the appropriate distance from 6 signs.

29 "Tempora...una hora" (VI.234–47). The times of half-retrogradation are derived from the preceding figures for distance of station from apogee (which is equivalent to the supplement of the arc of half-retrogradation). For Mercury (see section V, note 28) the corresponding figures were derived by dividing the distance of station from perigee by the mean daily motion in anomaly. For these four planets, Campanus follows a somewhat different procedure, which is explained at lines VI.261–340: for the position of epicycle at mean distance, the procedure is the same as that for Mercury; i.e., if p is the distance of first station from true epicyclic perigee and a the daily motion of the planet in anomaly, t, the time of half-retrogradation, is given by

$$t = \frac{p}{a} \text{ days.} \tag{1}$$

But for the position of the epicycle at apogee, we consider not just p, but p_A for the epicycle at apogee and p' for the epicycle in such a position before apogee that, if the planet is at first station, it will be in the middle of its arc of retrogradation when the epicycle is at apogee. We find $\delta(p')$, the equation of center for position p', and form $p_A-(p_A-p')-\delta$, i.e., $p'-\delta$. Then

$$t = \frac{(p'-\delta)}{a} \text{ days.} \tag{2}$$

Similarly for the position of the epicycle at perigee

$$t = \frac{(p'+\delta)}{a} \text{ days.} \tag{3}$$

Campanus is correct in not using formula (1), since p gives the distance of station from epicyclic perigee at a given *point*, whereas he now requires the length of time the planet takes to traverse an arc (from first station to perigee), during which time the epicycle too is *in motion*. It is clear from lines VI.289–90 that the "periods [corresponding to] the three...positions" (lines VI.234–35) mean "such that the planet has traversed half the arc of retrogradation (i.e., is at mean opposition or, for Venus, at mean inferior conjunction) when the center of the epicycle is at apogee, mean distance, or perigee." This is as expected, for it corresponds to the explicit definition of Ptolemy (e.g., *Almagest* 12. 2: Ptolemy, Manitius, vol. 2, p. 281).

Ptolemy's procedure (*Almagest* 12. 2–6) is quite different from that described above, since it starts from basic geometrical and kinematical principles. Nevertheless, his results (which are rounded) agree quite well with those of Campanus, as the accompanying comparative table of retrogradation times shows. Campanus's figures in the table are all correctly computed according to formulas (1) to (3). For an example of such a computation and a discussion of the problem of finding p', see note 33, below. (Incidentally, some manuscripts of the β group [e.g., *M*, fol. 181r[b], and *P*, fol. 184v[a]] exhibit the times of half-retrogradation and total retrogradation for Venus, Mars, Jupiter, and Saturn in a table; the table, however, merely replaces the text [see critical apparatus at lines VI.235–47 and VI.251–60] and clearly is not due to Campanus himself.)

Time of total retrogradation in days

	Epicycle at apogee	Epicycle at mean distance	Epicycle at perigee
Mercury			
Almagest	21^d	$22\frac{1}{2}^d$	23^d
Campanus	$21^d 2;16^h$	$22^d 5;18^h$	$22^d 20;45^h$
Venus			
Almagest	43^d	$41\frac{2}{3}^d$	$40\frac{2}{3}^d$
Campanus	$43^d 1;56^h$	$41^d 16;12^h$	$40^d 8^h$
Mars			
Almagest	80^d	73^d	$64\frac{1}{2}^d$
Campanus	$80^d 2;48^h$	$72^d 23;54^h$	$64^d 13;10^h$
Jupiter			
Almagest	123^d	121^d	118^d
Campanus	$122^d 21;16^h$	$120^d 9;46^h$	$118^d 3;10^h$
Saturn			
Almagest	$140\frac{2}{3}^d$	138^d	136^d
Campanus	$140^d 17^h$	$138^d 8;14^h$	$136^d 2^h$

30 "Attendere...inuenire" (VI.261–62). This rather elegant procedure (described in the previous note and by Campanus here through line VI.340) is known to me from no other source, and seems to be Campanus's own work. It is true that Ptolemy, on finding the durations of retrogradation at *Almagest* 12. 2–6, does explicitly take into account the fact that the epicycle is in motion during the retrogradation, and makes adjustments accordingly. But his procedure is so different that it would have been of little help to Campanus here. Much more important is the influence of a passage in Book 8 of Jābir (printed at Appendix C, pp. 450–52, below). Jābir does not give explicitly the rules that Campanus

develops here, but they can be deduced from his argumentation, and the two rough numerical results he gives agree well with Campanus's figures (see introductory paragraph preceding Appendix C).

31 "precedendo...reliqua" (VI.279–81). I.e., in Figure 19, p. 316, the true perigee of the epicycle is to the left (east) of the mean (as seen from the earth *D*) when the epicycle is traveling from *A* to *B* ("the descending half"), but to the right (west) of it from *B* to *A*. Thus we can take "precedere" to mean "be farther advanced in the succession of the signs," and "sequi" the opposite.

32 "ponam...*R* secunda" (VI.302–4). In Figure 19 the position of *L* as first station is marked on the situation *E*, and its position as second station on the situation *F* (*Q* is used analogously in situations *G* and *H*, respectively). The reason for choosing the same letter to mark the two different stations is that the arc *LZ* is equal for both situations.

33 "Oportet...prima" (VI.324–25). Campanus's analysis up to this point has been impeccable. But here we find a serious difficulty: we *ought* indeed to find the mean centrum for which the planet is at first station, but Campanus does not tell us how to do it with the available data. That he was in fact able to solve the problem can be shown from the figures he gives at lines VI.234–47 for the periods of retrogradation (see example below). The only mathematically correct way open to Campanus would be the iteration process described in the following example. It might seem surprising to find Campanus using such a refined technique. However, examples of iteration are found in the *Almagest* (e.g., 10. 7), though not in the same context (cf. note 29, above). Furthermore, the whole of this discussion, which seems to be Campanus's own work, shows a surprising insight. The only other conceivable way for him to have reached correct results would be by trial and error, using the tables.

The following is an example of the iteration method of finding the period of half-retrogradation from the position of first station for the apogee position of Mars. The first station (s_1) is at an anomaly of $5^s7;28°$ for epicycle at apogee (line VI.225); i.e., p, the distance of the first station from perigee, is 6^s (or 180°) minus $5^s7;28°$, or 22;32°. The mean daily motion of Mars in longitude (l) is 0;31,26°, and the daily motion of Mars in anomaly (a) is 0;27,42°.

We make the false assumption that p', the distance of the station from epicyclic perigee in this position (which is before the apogee, since at apogee the planet is at mean opposition), is equal to p. (Compare Jābir in Appendix C, p. 452, below: "arcus *GM*, qui est superfluitas diuersitatis arcus quem abscindit centrum orbis reuolutionis in tempore in quo abscindit stella arcum *BG secundum propinquitatem*, sed secundum ueritatem arcum *BM*, uerum *ipse est quesitus*" [italics added].) Then the mean centrum, $\bar{\varkappa}_1$, equals $(p \cdot l)/a$. With $\bar{\varkappa}_1$ as argument, we find from the tables (I use the Toledan tables) the equation of center $\delta_1(\bar{\varkappa}_1)$, and hence the true centrum $\varkappa_1$ by $\varkappa_1 = \bar{\varkappa}_1 - \delta$. With $\varkappa_1$ as argument, we find from the tables the first station $s_1(\varkappa_1)$. We then form $180° - s_1 - \delta_1$ to get p'_1, the amended distance of the first station from perigee. In figures: $\bar{\varkappa}_1 = 22;32 \cdot 0;31,26/0;27,42 = 25;34$. Since $\delta(\bar{\varkappa}_1) = 4;31$, then $\varkappa_1 = 25;34 - 4;31 = 21;3$. And since $s_1(\varkappa_1) = 5^s7;52°$, then $p'_1 = 180° - 5^s7;52° - 4;31° = 17;37°$. We now iterate the whole process, using p'_1 instead of p. Thus, $\bar{\varkappa}_2 = 17;37 \cdot 0;31,26/0;27,42 = 19;59$. Since $\delta_2(\bar{\varkappa}_2) = 3;35$, then $\varkappa_2 = 19;59 - 3;35 = 16;24$. And since $s_2(\varkappa_2) = 5^s7;44°$, then $p'_2 = 180° - 5^s7;44° - 3;35 = 18;41°$. Further iteration gives $p'_3 = 18;27°$, $p'_4 = 18;30°$, and $p'_5 = 18;30°$; when $p'_{n+1} = p'_n$, we have reached the true value of p'.

We then divide p' by a, the daily motion in anomaly, to obtain the period of half-retrogradation: hence, $18;30/0;27,42 = 40;4^d = 40^d 1\frac{3}{5}^h$ (Campanus, line VI.239, gives $40^d 1;24^h$, or $40^d 1\frac{2}{5}^h$, probably through a small slip).

It is only for Mars, which has a large eccentricity, that so many iterations are necessary. For Venus there are fewer, while for Jupiter and Saturn the eccentricities are so small that the first trial with p gives the correct answer.

34 "stationem" (VI.326). Brachyology for "the distance of the station from (true) apogee." Cf. *Theorica planetarum Gerardi* 80 (Gerardus, *Theorica*, p. 38).

35 "et si qua diuersitas fuerit" (VI.340). Campanus is right to make this qualification, since there is not the exact symmetry on either side of the position of mean distance that there is about apogee and perigee.

36 "ipsos...diurnum" (VI.341–42). This statement is confirmed by my calculations: see section V, note 28.

37 "differentia...non multa" (VI.344). It is in fact greater than that for Venus, but compared with the speed of Mercury on the epicycle it is indeed small, and Campanus is justified in neglecting the errors involved. For the epicycle at apogee, for instance, the more accurate procedure would make a difference of only about $6\frac{1}{2}$ hours in the 21 days of retrogradation.

38 "uespertina...matutina" (VI.350–51). The names "uespertina" and "matutina" are appropriate because, when the planet is to the east of the sun, it is visible in the evening just after sunset; when it is to the west of it, it is visible in the morning just before sunrise. Cf. Ptolemy's expression μέγισται ἀποστάσεις ἑῷοί τε καὶ ἑσπέριοι (e.g., *Almagest* 12. 9: Ptolemy, *Syntaxis*, ed. Heiberg, vol. 2, pp. 508–9). The expressions "uespertina" and "matutina" are applied to Mercury in al-Farġānī 26. 4 (Alfraganus, Carmody, p. 44).

39 "PTHOLOMEVS...hoc" (VI.353). *Almagest* 12. 10 (Ptolemy, Manitius, vol. 2, p. 324).

40 "tu...nostrum" (VI.355–56). This section (through line VI.379) on the construction of tables for the greatest elongation of Venus and Mercury by means of the instrument, though technically correct, contains features which lead me to suspect that it may be an interpolation. Firstly, the word "line" ("linea") is used consistently (e.g., line VI.357) in circumstances where Campanus elsewhere uses "thread" ("filum") (see, e.g., line III.196). Secondly, use is made of a "line" from the center of the sun's eccentric (line VI.365) of which there is no mention

in Campanus's description of that part of the instrument at lines III.89–156 (see, however, section V, note 161). Thirdly, there is the odd expression "on the *figure* of the sun's instrument" (lines VI.363–64). This, and the strange use of "linea," suggests that the writer was working entirely from drawn *figures* and has naively expressed himself accordingly.

41 "in situ...epicicli" (VI.359–60). See Figure 15, p. 236. There N is the center of the deferent. The rule is that $\angle CHN = \angle HDB$. See p. 49, above.

42 "facias transire...signorum secauerit" (VI.364–68). This is a neat method of going from mean to true longitude of the sun on the instrument. Strictly speaking, mean longitude of the sun is measured on the eccentric about center C of Figure 10 (p. 152). But it can equally well be measured on the ecliptic about center D, provided the vector from D to the ecliptic is always parallel to the vector from C to the eccentric. The procedure described here is a direct consequence of that parallelism.

43 "motum...predictorum" (VI.380–81). This is the motion of precession, common to all the spheres (lines III.28–31), of 1° in 100 years according to Ptolemy. The "three bodies previously described" are the sun, the moon, and Mercury.

44 "De motibus...dicetur" (VI.386–87). This promise to describe the motions in latitude is not fulfilled. See also lines V.15–17 with note 5 of section V, lines VI. 26–27 with note 4, above, and p. 29, above.

45 "Promisimus...supra" (VI.392–93). Line V.379.

46 "secundum...60 partes" (VI.408–9). As elsewhere, the basic parameters (here AC = radius of deferent, $KC = CD$ = half-eccentricity, AE = radius of epicycle) are taken from the *Almagest*, and the others are obtained from them by addition or subtraction.

47 "pars una et 15 minuta" (VI.411). *Almagest* 10. 3 (Ptolemy, Manitius, vol. 2, p. 163).

48 "43 partes et 10 minuta" (VI.412). *Almagest* 10. 2 (Ptolemy, Manitius, vol. 2, p. 161).

49 "linea AD...35 minuta" (VI.412–17). $AD = AC+CD = 60+1;15 = 61;15$. $DB = AC-CD = 60-1;15 = 58;45$. $ED = AD+AE = 61;15+43;10 = 104;25$. $DF = DB-AE = 58;45-43;10 = 15;35$.

50 "2 minuta et 8 secunda" (VI.421–22). Cf. section V, note 51. The procedure here is exactly analogous to that used for Mercury, lines V.399–409. (Indeed, that passage is the reference of "per ea que dicta sunt," line VI.419.)

51 "26 minuta et 40 secunda" (VI.423). This probably comes from al-Farġānī 22. 4 (Alfraganus, Carmody, p. 39), who gives $r_{\oplus} : r_{♀} = 3\frac{1}{3} : 1$ ("diametrum ueneris est 1 pars ex 3 partibus et 3[a] unius partis"). Campanus seems to have misunderstood, however, and taken him to say that $r_{♀} : r_{\oplus} = \frac{1}{3}+(\frac{1}{3}\cdot\frac{1}{3})$ (cf. lines VI.424–25). The latter is exactly equivalent to 0;26,40.

52 "linea DF...48 secunda" (VI.426–27). With line ED for Mercury (lines V.424–25) and the radius of Venus as derived above (see preceding note), line DF for Venus is

found from *ED* for Mercury $+r_{☿}+r_{♀} = 178;28r_{\oplus}+0;2,8r_{\oplus}+0;26,40r_{\oplus}$, or $178;56,48r_{\oplus}$.

53 "per...dictum est" (VI.428–29). I.e., each quantity is multiplied by $\frac{178;56,48}{15;35}$. In fact it is necessary to multiply thus only the "basic parameters" *AC*, *DC*, and *AE*, and then the others can be derived as above (see note 49, above). All are correctly computed to the nearest second. They are listed in Table 3, section 2, p. 359.

54 "De miliaribus...ita" (VI.434). Cf. section V, note 53. Here, for the planets, as was done for the moon, the figures just given in earth-radii are multiplied by $3{,}245\frac{5}{11}$, and the results are all correctly computed. They are listed in Table 3, section 3, p. 359.

55 "Semidiameter...280...miliaris" (VI.442–43). 0;26,40 (radius of Venus in earth-radii) multiplied by $3{,}245\frac{5}{11}$ equals $1{,}442\frac{280}{660}$ miles.

56 "Ambitus...63...miliaris" (VI.443–44). Campanus has made a calculating error here: $1{,}442\frac{280}{660}\cdot 2\cdot\frac{22}{7}$ ($\frac{22}{7}$ is invariably used for π in thiswork; see section III, note 20) gives not $9{,}095\frac{63}{660}$ but $9{,}066\frac{440}{660}$. I cannot explain the error.

57 "Eritque...560...miliaris" (VI.447–52). Least distance of Venus (*DF*: see line VI.441) less $r_{♀}$ gives the near edge of Venus's sphere: thus $580{,}763\frac{180}{660}$ miles $-$ $1{,}442\frac{280}{660}$ miles $= 579{,}320\frac{560}{660}$ miles. Greatest distance of Venus (*ED*: see line VI.440) plus $r_{♀}$ gives the far edge of Venus's sphere: $3{,}891{,}424\frac{270}{660}$ miles $+\,1{,}442\frac{280}{660}$ miles $= 3{,}892{,}866\frac{550}{660}$ miles. The far edge of Venus's sphere less the near edge equals $3{,}892{,}866\frac{550}{660}$ miles less $579{,}320\frac{560}{660}$ miles; therefore the thickness of Venus's sphere is $3{,}313{,}545\frac{650}{660}$ miles.

58 "Ambitus...conuexe...440...miliaris" (VI.452–54). The radius of the far edge of Venus's sphere (lines VI.448–49) is $3{,}892{,}866\frac{550}{660}$ miles. This multiplied by $2\cdot\frac{22}{7}$ gives $24{,}469{,}448\frac{440}{660}$ miles.

59 "Ambitus...deferentis...14055447 miliaria" (VI.454). The radius of Venus's deferent (*AC*: lines VI.434–35) is $2{,}236{,}093\frac{555}{660}$ miles. This multiplied by $2\cdot\frac{22}{7}$ gives 14,055,447 miles.

60 "dieta...44...miliaris" (VI.457). 14,055,447 miles divided by $365\frac{1}{4}$ is in fact 38,481 plus *43* sixtieths of a mile to the nearest sixtieth.

61 "584 dies" (VI.459). If we divide the crude figure given by Campanus (line VI.67) for Venus's mean daily motion on the epicycle, 0;36,59°, into 360°, we get 9,44;2,... days, or 584 days to the nearest day (cf. section IV, note 73). Al-Farġānī 17. 1 (Alfraganus, Carmody, p. 31), too, says, "ambulat circulum breuem...uenus in anno persico et 7[m] et 9[d] fere," i.e., in $(365+7\cdot 30+9)$ days $=$ 584 days.

62 "Ambitus...16...miliaris" (VI.457–61). The radius of Venus's epicycle (*AE*: lines VI.436–37) is $1{,}608{,}745\frac{445}{660}$ miles. This multiplied by $2\cdot\frac{22}{7}$ gives $10{,}112{,}115\frac{440}{660}$ miles. The latter divided by 584 is indeed $17{,}315\frac{16}{60}$ miles to the nearest sixtieth.

63 "Corpus...et 64" (VI.461–63). $r_{\oplus} : r_{♀} = 9:4$ (see line VI.423, with note 51, above), and $9^3 : 4^3 = \frac{729}{64} = 11\frac{25}{64}$.

64 "duas...prius" (VI.473–74). E.g., lines III.9–11 and see section III, note 4.
65 "1210 partes" (VI.476). Cf. line III.49, with note 15 of section III.
66 "Sed...solis" (VI.476–79). This statement is true. Ptolemy's calculation is found at *Almagest* 5. 14–15, where the distance of the sun is derived from the observed diameter and the previously calculated distance of the moon.
67 "quemadmodum ipse dixit" (VI.479). E.g., *Almagest* 5. 14 (Ptolemy, Manitius, vol. 1, p. 305).
68 "distantia...ecentrici" (VI.482–83). This comment is true as far as concerns the *Almagest*. But in the *Planetary Hypotheses* (which Campanus could not have known directly), it is clear that Ptolemy regards the distance of $1210r_{\oplus}$ as the *mean* distance of the sun. See Goldstein, "The Arabic Version," p. 7.
69 "quemadmodum ipse dixit" (VI.488–89). *Almagest* 5. 11 (Ptolemy, Manitius, vol. 1, p. 294).
70 "operatione probauit" (VI.489). Ptolemy's observations, with the consequent calculations, are detailed in *Almagest* 5. 13.
71 "supposito...motus planete" (VI.498–500). This supposition is indeed the theoretical basis for the absolute distances in Campanus's universe. Cf. lines V.394–97, and see p. 54, above.
72 "stellarum omnium" (VI.502). It is not immediately clear what this phrase refers to: since it is not the planets, the only two possibilities are (1) sun and moon or (2) the fixed stars. Alternative (1) seems impossible on linguistic grounds. The objection to (2) is that Campanus says nothing about the sizes of the fixed stars. However, al-Farġānī 22. 5–6 (Alfraganus, Carmody, pp. 39–40) does give sizes for stars of the first six magnitudes, and Campanus is probably thinking of that information. (Though al-Farġānī does not say so, such sizes could theoretically be derived from the apparent magnitudes, which he gives, and the actual distance.)
73 "deferentis" (VI.510). The eccentric of the sun is not, of course, a deferent, but Campanus's slip is a natural one, as the line AC has, in the analogous data for the planets (e.g., Venus, lines VI.408–9), been a deferent. See also line III.160, and section III, note 35.
74 "2 partes et 30 minuta" (VI.512). *Almagest* 3. 4; cf. section III, note 4.
75 "linea tota...62 partes et 30 minuta" (VI.512–13). $AD = AC+CD$, which is equivalent to $R+e$; therefore $AD = 60+2;30$, or $62;30$.
76 "linea DB...57 partes et 30 minuta" (VI.513–14). $DB = R-e$; therefore $DB = 60-2;30$, or $57;30$.
77 "26 minuta et 40 secunda" (VI.518). See note 51, above.
78 "5 partes et 30 minuta" (VI.520). See lines III.49–50 and section III, note 16. This value comes from *Almagest* 5. 16 (Ptolemy, Manitius, vol. 1, p. 313).
79 "linea DB...58 secunda" (VI.520–21). With line ED for Venus (lines VI.432–33) and the radius of Venus as derived above (see note 51, above), line DB for the sun is found as ED for Venus$+r_{♀}+r_{\odot} = 1199;2,18r_{\oplus} + 0;26,40r_{\oplus} + 5;30r_{\oplus}$, or $1204;58,58r_{\oplus}$.

80 "linea *AC*...50 secunda" (VI.522–24). As before (see note 53, above), the transformation to earth-radius units is achieved by multiplication of the Ptolemaic distances by a constant factor, in this case $\frac{1204;58,58}{57;30}$. Doing so leads to the exact figures given by Campanus.

81 "minor...5 partibus fere" (VI.530–31). 1204;58,58 (*DB*: line VI.521) subtracted from 1210 gives 5;1,2 earth-radius units.

82 "maior...100 partibus fere" (VI.532). 1309;45,50 (*AD*: line VI.524) less 1210 gives 99;45,50 earth-radius units.

83 "uenus...ipse posuit" (VI.536–37). Ptolemy gives a short discussion of the planetary order at *Almagest* 9. 1 (Ptolemy, Manitius, vol. 2, p. 93). He there says that the older astronomers put Venus and Mercury below the sun, but some of the later ones put them above, because no transits over the disk of the sun had been observed. He says that the latter fact could be explained by the latitudinal motion of the bodies, and prefers the traditional order, though he seems to admit that there is no decisive argument, since no planetary parallax has been detected.

84 "GEBER" (VI.538). Jābir ibn Aflaḥ, of Seville (fl. early twelfth century: see Suter, "Mathematiker und Astronomen," no. 284, p. 119. He is dated by the statement of Maimonides [b. 1135] that Maimonides himself knew ibn Aflaḥ's son: see Moritz Steinschneider, *Die hebräischen Übersetzungen des Mittelalters und die Juden als Dolmetscher* [Berlin, 1893], p. 543; and Moses Maimonides, *The Guide for the Perplexed*, trans. M. Friedländer, 2d ed. [London, 1904], p. 164). Jābir's work criticizing Ptolemy (of which the Arabic text has never been printed) was translated into Latin by Gerard of Cremona, and that translation was printed by Petrus Apianus (see Apianus, *Instrumentum Primi Mobilis*). For Campanus's knowledge of it, see p. 35, above. For Jābir's criticism of Ptolemy on the planetary order, see Apianus, *Instrumentum Primi Mobilis*, p. 3 and, in detail, ibid., bk. 7, pp. 103–6. Jābir's Book 7 is well summarized and criticized by J. B. J. Delambre, *Histoire de l'astronomie du Moyen Age* (Paris, 1819), p. 184.

85 "platonicos et egiptios" (VI.538). Campanus gets this information from Macrobius, *Commentary on the Somnium Scipionis*, bk. 1, chap. 19, secs. 1–2 (ed. Willis, p. 73): "in quo [scilicet sphaerarum ordine] dissentire a Platone Cicero uideri potest, cum hic solis sphaeram quartam de septem, id est in medio locatam dicat, Plato a luna sursum secundam, hoc est inter septem a summo locum sextum tenere commemoret. Ciceroni Archimedes et Chaldaeorum ratio consentit, Plato Aegyptios omnium philosophiae disciplinarum parentes secutus est" and so on. As far as it concerns Plato, Macrobius's statement is correct (see Plato *Timaeus* 38^{c}–38^{d}; and cf. Plato *Republic* 616^{d}–617^{b}, on which see Proclus Diadochus, *In Platonis Rem Publicam commentarii*, ed. W. Kroll [Leipzig, 1899], vol. 2, p. 218, line 1, with Kroll's note on p. 413). The "Egyptians," however, must be some Greek work apocryphally attributed to an Egyptian author.

86 "1138 partes fere" (VI.540). The nearest point of the solar eccentric is 1204;58,

$58r_\oplus$ (line VI.521); the farthest point of the lunar eccentric is $64;10r_\oplus$ (line IV. 419); and therefore their difference is $1140;48,58r_\oplus$. If we take the actual edges of the *spheres*, then we must subtract from the difference the radius of the sun's body ($5;30r_\oplus$: line VI.520) plus the radius of the moon's body ($0;17,32r_\oplus$: line IV.422). This gives 1135;1,26. It remains mysterious where Campanus got 1138 from, though the error is small. Scribal corruption (1138 for 1135) is distinctly possible, though there is no manuscript authority for 1135.

87 "Adhuc...locatos" (VI.541–43). The same argument is used by Ptolemy in *Hypotheseis* (*Opera minora*, ed. Heiberg, p. 118, lines 10–20). See p. 55, above. Campanus probably got the idea from al-Fargānī 21. 2 (Alfraganus, Carmody, p. 38): "Cumque posuissemus longitudinem longiorem utriusque circuli lune, circuli uidelicet egresse cuspidis et circuli breuis, longitudinem mercurii propiorem, et usi fuissemus hac affinitate quam premisimus, fecissemusque similiter in mercurio et uenere, inuenimus longitudinem longiorem utriusque circuli ueneris esse propiorem longitudinem solis quam patefecit Tholomeus. Et hoc indicio precepimus quod nulla uacuitas sit inter circulos."

88 "si alius...ponere" (VI.544–46). This phraseology refers to the astrological practice of assigning to each hour of the day a planet as "regent" (χρονοκράτωρ, "gubernator," "dominus") in the "Ptolemaic" order Saturn, Jupiter, Mars, Sun, Venus, Mercury, Moon. The regent of the first hour of the day was considered the regent of the whole day. This resulted in the planetary week as we now know it: Saturday ("dies Saturni"), Sunday ("dies solis"), etc. See F. K. Ginzel, *Handbuch der mathematischen und technischen Chronologie* (Leipzig, 1906–14), vol. 3, pp. 97–105; A. Bouché-Leclercq, *L'Astrologie grecque* (Paris, 1899), pp. 479–86. Campanus is, of course, correct in stating that a change in the planetary order would result in a change in the weekday order, but to use that eventuality as an argument for the correctness of the Ptolemaic planetary order was an absurdity even in the thirteenth century.

89 "De miliaribus...550...miliaris" (VI.549–53). As before, these figures are obtained by multiplying the preceding figures in earth-radius units by $3{,}245\frac{5}{11}$. All are correctly computed. They are tabulated in Table 4, section 3, p. 360.

90 "Semidiameter...112200 miliaria" (VI.553–54). Radius: $5\frac{1}{2} \cdot 3{,}245\frac{5}{11} = 17{,}850$ miles; and circumference: $17{,}850 \text{ miles} \cdot 2 \cdot \frac{22}{7} = 112{,}200$ miles.

91 "Longitudo autem...220...miliaris" (VI.554–59). The distance of the concave surface of the sun's sphere from the center of the earth equals the least distance of the sun (*DB*: line VI.552) minus the radius of the sun (line VI.553), or $3{,}910{,}716\frac{550}{660}$ miles $-$ 17,850 miles $= 3{,}892{,}866\frac{550}{660}$ miles. The distance of the convex surface of the sun's sphere from the center of the earth equals the greatest distance of the sun (*AD*: line VI.551) plus the radius of the sun, or $4{,}250{,}779\frac{110}{660}$ miles $+$ 17,850 miles $= 4{,}268{,}629\frac{110}{660}$ miles. Therefore the thickness of the sun's sphere is the distance of the sun's convex less the distance of the sun's concave, or $4{,}268{,}629\frac{110}{660}$ miles $-$ $3{,}892{,}866\frac{550}{660}$ miles $= 375{,}762\frac{220}{660}$ miles.

92 "Ambitus autem...220...miliaris" (VI.559–60). The radius of the far edge of the sun's sphere (line VI.557) is $4{,}268{,}629\frac{110}{660}$ miles. This multiplied by $2 \cdot \frac{22}{7}$ gives $26{,}831{,}383\frac{220}{660}$ miles.

93 "Ambitus quoque...70227 miliaria" (VI.560–64). Radius of sun's eccentric (*AC*: line VI.549) is 4,080,748 miles. This multiplied by $2 \cdot \frac{22}{7}$ gives 25,650,416 miles. The latter divided by $365\frac{1}{4}$ gives 70,227 miles (plus a small fraction, $\frac{17}{1461}$, neglected in the computation).

94 "Corpus...eius" (VI.564–65). As explained subsequently in the text, $11^3 : 2^3 = \frac{1331}{8} = 166\frac{3}{8}$.

95 "6 partes et 30 minuta" (VI.574–75). The *Almagest* figure is 6 units, no minutes (*Almagest* 10. 7: Ptolemy, Manitius, vol. 2, p. 190). This is the only time when Campanus has used a parameter of this kind that differs from its counterpart in the *Almagest*. The most probable explanation is simple error on his part, either through misreading or because of a faulty manuscript. A possible source of confusion is the fact that Saturn's epicycle-radius *is* given in the *Almagest* as 6;30 parts.

This value for the eccentricity of Mars is repeated by Johannes de Lineriis (Price, *Equatorie of the Planetis*, p. 127), a striking proof of his close dependence on the *Theorica*. More surprisingly, it seems also to be repeated by "Chaucer" (see ibid., pp. 69, 126–27).

96 "39 partes et 30 minuta" (VI.575–76). *Almagest* 10. 8 (Ptolemy, Manitius, vol. 2, p. 198).

97 "linea *AD*...14 partes" (VI.576–80). $AD = AC + CD = 60 + 6;30 = 66;30$. $DB = AC - CD = 60 - 6;30 = 53;30$. $ED = AD + AE = 66;30 + 39;30 = 106$. $DF = DB - AE = 53;30 - 39;30 = 14$.

98 "pars una et 10 minuta" (VI.586). Like all the other planetary radii, this is taken from al-Farġānī (22. 4: Alfraganus, Carmody, p. 39).

99 "secundum...una" (VI.581–88). Line *AD* for the sun (line VI.524) $+ r_{\odot} + r_{♂} = 1309;45,50 r_{\oplus} + 5;30 r_{\oplus} + 1;10 r_{\oplus}$; therefore line *DF* for Mars $= 1316;25,50 r_{\oplus}$.

100 "Eruntque...36 secunda" (VI.589–93). The magnitudes in earth-radii are obtained by multiplying those preceding (lines VI.571–80) by $\frac{1316;25,50}{14}$. All the results are computed correctly to the nearest second. They are tabulated in Table 5, section 2, p. 361.

101 "Linea...350...miliaris" (VI.594–602). The magnitudes in miles are obtained, as before, by multiplying the magnitudes in earth-radii by $3{,}245\frac{5}{11}$. All are correctly computed. They are tabulated in Table 5, section 3, p. 361.

102 "Semidiameter...23800 miliaria" (VI.602–4). 1;10 (radius of Mars in earth-radii: line VI.586) multiplied by $3{,}245\frac{5}{11}$ equals $3{,}786\frac{240}{660}$ miles. The latter multiplied by $2 \cdot \frac{22}{7}$ equals 23,800 miles exactly.

103 "Longitudo...310...miliaris" (VI.604–8). The least possible distance of Mars (*DF*: line VI.601) less Mars's radius (lines VI.602–3) gives the near edge of Mars's sphere: thus, $4{,}272{,}415\frac{350}{660}$ miles $- 3{,}786\frac{240}{660}$ miles $= 4{,}268{,}629\frac{110}{660}$ miles.

The greatest possible distance of Mars (*ED*: line VI.600) plus the radius of Mars gives the far edge of Mars's sphere; thus, $32{,}348{,}289\frac{180}{660}$ miles $+ 3{,}786\frac{240}{660}$ miles $= 32{,}352{,}075\frac{420}{660}$ miles. Hence, the thickness of Mars's sphere equals the far edge less the near edge, or $32{,}352{,}075\frac{420}{660}$ miles $- 4{,}268{,}629\frac{110}{660}$ miles $= 28{,}083{,}446\frac{310}{660}$ miles.

104 "Ambitus autem...203355904 miliaria" (VI.608–9). The far edge of Mars's sphere (lines VI.606–7) is $32{,}352{,}075\frac{420}{660}$ miles. This multiplied by $2 \cdot \frac{22}{7}$ equals 203,355,904 miles.

105 "687 dies" (VI.611). If we divide the crude parameter given by Campanus (line VI.79) for the mean daily motion of Mars's epicycle, 0;31,26°, into 360°, we get 11,27;10,... days, or 687 days to the nearest day (cf. section IV, note 73). Al-Farġānī (17. 3: Alfraganus, Carmody, p. 31), too, says, "anno persico et 10^{m} et 22^{d} fere"; i.e., $(365+10 \cdot 30+22)$ days $= 687$ days.

106 "Ambitus quoque...una die" (VI.609–13). The radius of Mars's deferent (*AC*: line VI.595) is $18{,}310{,}352\frac{265}{660}$ miles. This multiplied by $2 \cdot \frac{22}{7}$ gives $115{,}093{,}643\frac{440}{660}$ miles. The latter divided by 687 is indeed $167{,}530\frac{47}{60}$ miles to the nearest sixtieth.

107 "780 dies" (VI.615). If we divide the crude parameter given by Campanus (line VI.67) for Mars's daily mean motion on the epicycle, 0;27,42°, into 360°, we get 12,59;47,... days, or 780 days to the nearest day (cf. section IV, note 73). Al-Farġānī (17. 1: Alfraganus, Carmody, p. 31), too, says, "2 annis et 1^{m} et 20^{d} fere"; i.e., $(2 \cdot 365+30+20)$ days $= 780$ days.

108 "Ambitus...97141 miliaria" (VI.613–17). Radius of epicycle (*AE*: line VI.597) is $12{,}054{,}315\frac{35}{660}$ miles. This multiplied by $2 \cdot \frac{22}{7}$ gives $75{,}769{,}980\frac{220}{660}$ miles. The latter divided by 780 gives 97,141 miles to the nearest sixtieth.

109 "Corpus...et 216" (VI.617–19). $1;10:1 = \frac{7}{6}$, and $7^3 : 6^3 = \frac{343}{216}$, or $1\frac{127}{216}$.

110 "2 partes et 45 minuta" (VI.625–26). *Almagest* 11. 1 (Ptolemy, Manitius, vol. 2, p. 215).

111 "11 partes et 30 minuta" (VI.626–27). *Almagest* 11. 2 (Ptolemy, Manitius, vol. 2, p. 223).

112 "linea *AD*...45 partes et 45 minuta" (VI.627–31). $AD = AC+CD = 60+2;45 = 62;45$. $DB = AC-CD = 60-2;45 = 57;15$. $ED = AD+AE = 62;45+11;30 = 74;15$. $DF = DB-AE = 57;15-11;30 = 45;45$.

113 "4 partes et 34 minuta" (VI.637). Al-Farġānī (22. 4: Alfraganus, Carmody, p. 39) says, "diametrum iouis est quater diametrum terre et dimidium ac 16^{a} unius quantum diametrum terre"; i.e., $d_{♃} = (4+0;30+0;3,45)d_{\oplus}$. This is indeed 4;34 to the nearest minute.

114 "linea *DF*...una" (VI.638–40). Line *ED* for Mars (line VI.593)$+r_{♂}+r_{♃} = 9967;15,36r_{\oplus}+1;10r_{\oplus}+4;34r_{\oplus}$; hence line *DF* for Jupiter $= 9972;59,36r_{\oplus}$.

115 "linea *AC*...40 secunda" (VI.640–44). The figures in the text are derived from the preceding ones by multiplying by $\frac{9972;59,36}{45;45}$ in each case. All are correctly computed to the nearest second. They are tabulated in Table 6, section 2, p. 362.

116 "De miliaribus...360...miliaris" (VI.645–53). The magnitudes in miles are obtained by multiplying the magnitudes in earth-radii by $3{,}245\frac{5}{11}$. All computations are exact. They are tabulated in Table 6, section 3, p. 362.

117 "Semidiameter...93160 miliaria" (VI.653–55). 4;34 (radius of Jupiter in earth-radii: line VI.637) multiplied by $3{,}245\frac{5}{11}$ equals $14{,}820\frac{600}{660}$ miles. The latter multiplied by $2 \cdot \frac{22}{7}$ is exactly 93,160 miles.

118 "Longitudo...520...miliaris" (VI.655–59). Least possible distance of Jupiter (*DF*: line VI.652) less Jupiter's radius (lines VI.653–54) gives the near edge of Jupiter's sphere; thus, $32{,}366{,}896\frac{360}{660}$ miles $-$ $14{,}820\frac{600}{660}$ miles $= 32{,}352{,}075\frac{420}{660}$ miles. Greatest possible distance of Jupiter (*ED*: line VI.651) plus the radius of Jupiter gives the far edge of Jupiter's sphere; thus, $52{,}529{,}881\frac{340}{660}$ miles $+$ $14{,}820\frac{600}{660}$ miles $= 52{,}544{,}702\frac{280}{660}$ miles. The far edge of Jupiter's sphere less the near edge equals the thickness of Jupiter's sphere; thus, $52{,}544{,}702\frac{280}{660}$ miles $-32{,}352{,}075\frac{420}{660}$ miles $= 20{,}192{,}626\frac{520}{660}$ miles.

119 "Ambitus autem...440...miliaris" (VI.659–61). Radius of far edge of Jupiter's sphere (line VI.657) is $52{,}544{,}702\frac{280}{660}$ miles. This multiplied by $2 \cdot \frac{22}{7}$ is $330{,}280{,}986\frac{440}{660}$ miles.

120 "4334 dies" (VI.663). If we divide the crude parameter given by Campanus (lines VI.79–80) for the mean daily motion of Jupiter's epicycle, 0;4,59°, into 360°, we get 1,12,14;26,... days, or 4,334 days to the nearest day, whereas al-Farġānī 17. 3 (Alfraganus, Carmody, p. 31) has "11 annis et 10ᵐ et 16ᵈ"; i.e., $(11 \cdot 365 + 10 \cdot 30 + 16)$ days $= 4{,}331$ days. Alfraganus's figure is confirmed by computation with the accurate parameter for Jupiter's mean motion in longitude taken from the *Almagest*, which leads to 1,12,10;57,... $\approx$ 4,331 days. The discrepancy results from the fact that computation with only two significant figures is not accurate enough to produce a result correct to the nearest day when the motion is so slow. This (combined with the comparable result for Saturn, line VI.711) proves that Campanus's crude mean daily motion figures were the basis of all the "dieta" calculations (cf. section IV, note 73, and p. 29, above).

121 "Ambitus quoque...61564 miliaria" (VI.661–65). Radius of Jupiter's deferent (*AC*: line VI.646) is $42{,}448{,}389\frac{20}{660}$ miles. This multiplied by $2 \cdot \frac{22}{7}$ is $266{,}818{,}445\frac{220}{660}$ miles. The latter divided by 4,334 is 61,564 miles to the nearest mile.

122 "399 dies" (VI.667). If we divide the crude figure given by Campanus (line VI.68) for Jupiter's mean daily motion in anomaly, 0;54,9°, into 360°, we get 6,38;55,... days, or 399 days to the nearest day (cf. note 120, above). Al-Farġānī (17. 1: Alfraganus, Carmody, p. 31), too, says, "in anno et mense et 4ᵈ [MS *K, recte*; Carmody prints the scribal error "28ᵈ"] fere"; i.e., $(365+30+4)$ days $= 399$ days.

123 "Ambitus autem...128171 miliaria" (VI.665–68). Radius of Jupiter's epicycle (*AE*: line VI.648) is $8{,}135{,}941\frac{430}{660}$ miles. This multiplied by $2 \cdot \frac{22}{7}$ is $51{,}140{,}204\frac{440}{660}$ miles. The latter divided by 399 is 128,171 miles to the nearest mile.

124 "Corpus...27000" (VI.668–71). $4;34 : 1 = \frac{274}{60} = \frac{137}{30}$, and $137^3 : 30^3 = \frac{2,571,353}{27,000} = 95\frac{6,353}{27,000}$.

125 "tres partes et 25 minuta" (VI.676). *Almagest* 11. 5 (Ptolemy, Manitius, vol. 2, p. 238).

126 "6 partes et 30 minuta" (VI.676–77). *Almagest* 11. 6 (Ptolemy, Manitius, vol. 2, p. 246).

127 "linea *AD*...5 minuta" (VI.677–79). $AD = AC+CD = 60+3;25 = 63;25$. $DB = AC-CD = 60-3;25 = 56;35$. $ED = AD+AE = 63;25+6;30 = 69;55$. $DF = DB-AE = 56;35-6;30 = 50;5$.

128 "4 partes et 30 minuta" (VI.683–84). Al-Farġānī 22. 4 (Alfraganus, Carmody, p. 39).

129 "Erit...una" (VI.686–87). Line *ED* for Jupiter (lines VI.643–44)$+r_{♃}+r_{♄} =$ $16,185;40,40r_{\oplus}+4;34r_{\oplus}+4;30r_{\oplus}$; hence, line *DF* for Saturn $= 16,194;44,40r_{\oplus}$.

130 "linea *AC*...16 secunda" (VI.687–91). The figures in the text are obtained by multiplying the preceding ones in earth-radii by $\frac{16,194;44,40}{50;5}$. All are correctly computed to the nearest second. They are tabulated in Table 7, section 2, p. 363.

131 "De miliaribus...640...miliaris" (VI.692–700). The magnitudes in miles are obtained by multiplying the magnitudes in earth-radii by $3,245\frac{5}{11}$. All computations are exact. They are tabulated in Table 7, section 3, p. 363.

132 "Semidiameter...91800 miliaria" (VI.700–702). 4;30 (radius of Saturn in earth-radii: lines VI.683–84) multiplied by $3,245\frac{5}{11}$ equals $14,604\frac{360}{660}$ miles. The latter multiplied by $2 \cdot \frac{22}{7}$ is exactly 91,800 miles.

133 "Longitudo...480...miliaris" (VI.702–6). Least possible distance of Saturn (*DF*: line VI.699) less Saturn's radius (lines VI.700–701) gives the near edge of Saturn's sphere; thus, $52,559,306\frac{640}{660}$ miles $-$ $14,604\frac{360}{660}$ miles $= 52,544,702\frac{280}{660}$ miles. Greatest possible distance of Saturn (*ED*: line VI.698) plus the radius of Saturn gives the far edge of Saturn's sphere; thus, $73,373,142\frac{400}{660}$ miles $+$ $14,604\frac{360}{660}$ miles $= 73,387,747\frac{100}{660}$ miles. The far edge of Saturn's sphere less the near edge equals the thickness of Saturn's sphere; thus, $73,387,747\frac{100}{660}$ miles $-$ $52,544,702\frac{280}{660}$ miles $= 20,843,044\frac{480}{660}$ miles.

134 "Ambitus autem...440...miliaris" (VI.706–9). Radius of far edge of Saturn's sphere (lines VI.704–5) is $73,387,747\frac{100}{660}$ miles. This multiplied by $2 \cdot \frac{22}{7}$ is $461,294,410\frac{440}{660}$ miles.

135 "10800 dies" (VI.711). If we divide the crude parameter given by Campanus (line VI.80) for the mean daily motion of Saturn's epicycle, 0;2°, into 360°, we get exactly 10,800 days. Al-Farġānī 17. 3 (Alfraganus, Carmody, p. 31), on the other hand, has "in 29 annis et 5^{m} et 15^{d}"; i.e., $(29 \cdot 365+5 \cdot 30+15)$ days $= 10,750$ days, a figure that is confirmed by computation with the accurate parameter for Saturn's mean motion taken from the *Almagest*. Cf. note 120, above.

136 "Ambitus quoque...36647 miliaria" (VI.709–12). Radius of Saturn's deferent (*AC*: line VI.693) is $62,966,224\frac{520}{660}$ miles. This multiplied by $2 \cdot \frac{22}{7}$ gives 395,787,-$698\frac{440}{660}$ miles. The latter divided by 10,800 is 36,647 miles to the nearest mile.

137 "378 dies" (VI.714). If we divide the crude figure given by Campanus (line

VI.68) for Saturn's mean daily motion on the epicycle, 0;57,8°, into 360°, we get 6,18;3,... days, or 378 days to the nearest day (cf. note 120, above). Al-Farġānī (17. 1: Alfraganus, Carmody, p. 31), too, says, "in anno et 13[d] fere"; i.e., (365+13) days = 378 days.

138 "Ambitus autem...113431 miliaria" (VI.712–16). Radius of Saturn's epicycle (*AE*: line VI.695) is $6{,}821{,}341\frac{290}{660}$ miles. This multiplied by $2 \cdot \frac{22}{7}$ gives $42{,}877{,}003\frac{220}{660}$ miles. The latter divided by 378 is 113,431 miles to the nearest mile.

139 "Corpus...et 8" (VI.716–18). $4;30 : 1 = \frac{9}{2}$, and $9^3 : 2^3 = \frac{729}{8} = 91\frac{1}{8}$.

140 "ut prius dictum est" (VI.722). Lines IV.322–23.

141 "prouenit autem...superficie descripti" (VI.722–27). The formula for the surface-area of a sphere, s, is

$$\begin{aligned} s &= 4\pi r^2 \\ &= 2r \cdot 2\pi r \\ &= 4a, \end{aligned}$$

where a is the area of the great circle and $a = r \cdot c/2$ (c being the circumference, equal to $2\pi r$). To find the surface-area of the earth, al-Farġānī 8. 4 (Alfraganus, Carmody, p. 14) uses the formula $s = d \cdot c$ (i.e., $2r \cdot 2\pi r$).

142 "si duplum...unius miliaris" (VI.727–31). $r_♄$ (distance of far edge of Saturn's sphere: lines VI.704–5) is $73{,}387{,}747\frac{100}{660}$ miles. $c_♄$ (circumference of Saturn's sphere: line VI.708) is $461{,}294{,}410\frac{440}{660}$ miles. $s_♄ = 2rc_♄ = 67{,}706{,}715{,}144{,}825{,}054\frac{38}{99}$ square miles.

143 "in sua...propiori" (VI.739–40). We must translate "on its outer and inner sides" here to make sense, though everywhere else "longitudo longior" and "longitudo propior" have the technical meaning "at the position on the deferent farthest from [or nearest to, respectively] the center of the universe" (cf. section IV, note 62). Campanus means that other astronomers have neglected the fact that the heavenly bodies have magnitude and have treated them as points. Such a procedure would shrink each "sphere" slightly, and each such shrinkage would be reflected in the size of each sphere beyond it; so the resulting accumulated error would not be quite negligible. Amongst "quidam" we may include al-Farġānī, for instance. But the discrepancies among the various ancient and medieval lists of planetary distances derived from Ptolemy's system are a result more of inaccuracies of rounding than of neglect of the magnitudes of the bodies.

144 "que sunt...signorum" (VI.744–45). Stars outside the ecliptic, which move not on a great but on small circles, have as a result a smaller absolute velocity than the $35\frac{1}{12}$ miles per day for stars in the ecliptic.

145 "35 miliaria...pars" (VI.745). $461{,}294{,}410\frac{440}{660}$ miles (inner circumference of sphere of the fixed stars, which is equal to the outer circumference of Saturn's sphere: line VI.708) divided by $100 \cdot 365\frac{1}{4} \cdot 360$ is indeed very little less than $35\frac{1}{12}$ miles.

146 "quemadmodum...luna" (VI.752–53). See lines IV.526–32 and IV.570–71.
147 "in principio istius operis" (VI.756–57). Line II.74.
148 "in una...saturnum" (VI.757–58). The particular planet represented in Figure 22, p. 346, is Saturn.
149 "4 circulos concentricos" (VI.761). The number "four" includes the "largest circle that I can fit," which has already been drawn.
150 "quemadmodum...predictis" (VI.763–64). See lines III.89–153 for the fullest description.
151 "in uenere...sagittarii" (VI.765–68). For the positions of the apogees given here, which are all taken from the Toledan tables, see lines VI.38–41 and note 7, above.
152 "tantum...diuisionibus" (VI.772–74). See section V, note 68.
153 "quemadmodum...mercurio" (VI.778). Lines V.500–503.
154 "60 partes" (VI.780). This number is chosen for convenience (cf. section IV, note 80). Each of these units corresponds to two of the Ptolemaic units, for which see lines VI.408–17 or Table 3, section 1, p. 359, for Venus and lines VI.571–80 or Table 5, section 1, p. 361, for Mars. These sections of the tables give in detail the measurements which are related in the following portion of the text.
155 "21 partes...partis" (VI.787–88). $21+\frac{1}{3}+\frac{1}{4} = 21\frac{7}{12}$, and twice this is 43;10 (the radius of Venus's epicycle; see line VI.412).
156 "Remanebitque...partis" (VI.796–97). Campanus has made the mistake of merely adding up the measurements given and subtracting them from 60. This gives him, for Venus: $60-(1\frac{1}{4}+30+21\frac{7}{12}) = 60-52\frac{10}{12} = 7\frac{1}{6} \approx 7$; and, for Mars: $60-(6\frac{1}{2}+30+19\frac{3}{4}) = 60-56\frac{1}{4} = 3\frac{3}{4}$. He should, however, have subtracted, not $DK+CA+AE$, but $DC+CA+AE$; in other words, the total is too great by $\frac{1}{2}DK$. Subtraction of the correct terms brings the final results to $7\frac{19}{24}$ and 7, respectively. Is this mistake another indication that Campanus never actually constructed his instrument? See p. 33, above.
157 "sicud...mercurii" (VI.799). Lines V.535–50.
158 "96 partes" (VI.801). This number, too, is chosen for convenience (see section IV, note 80). Each unit corresponds to one of the Ptolemaic units (see lines VI. 622–31 or Table 6, section 1, p. 362, for Jupiter, and lines VI.674–79 or Table 7, section 1, p. 363, for Saturn).
159 "Residuum...partis" (VI.810–11). Campanus has made the same mistake as he did for Venus and Mars (see lines VI.796–97 and note 156). The amended remainders here should be $21\frac{3}{4}$ for Jupiter and $26\frac{1}{12}$ for Saturn.
160 "sicud...mercurii" (VI.820). Lines V.526–33.
161 "ita quod...concauari" (VI.825–26). For this and what follows, compare the more detailed description at lines IV.551–610.
162 "medium...solis" (VI.863). The mean centrum of Venus is the mean anomaly of the sun because both have the same apogee (♊ 17;50°) and the same mean motion. For the definition of "argumentum solis," see lines III.159–60.

163 "Ad cetera" (VI.864). I.e., for the mean centra (longitudes) of the other planets, and for the mean motions in anomaly of all the planets.

164 "quemadmodum…docuimus" (VI.865–66). Lines V.629–42, and see section V, note 87.

For the questions raised by the abrupt ending (cf. lines V.647–50), see pp. 29–30, above.

Appendixes
Bibliography
Indexes

Appendix A

Interpolation in Text Giving Ptolemaic Magnitudes for the Lunar Model

The following passage, which is interpolated by the δ group of manuscripts after line IV.414, merely states in words certain data contained in Table 1 (p. 356). The first two sentences complete the description of the system in which *AC*, the radius of the moon's deferent, is 60 units. The remainder then correctly lists the magnitudes of the various parts of the moon's system when *AD*, the maximum distance of the center of the epicycle, is taken as 60 units. The latter is the norm adapted by Ptolemy in the *Almagest* (for the reason, see section IV, note 63).

Linea uero *AB* que est diameter deferentis lune erit 120 partes sine aliqua fractione, est enim dupla ad lineam *AC* semidiametrum eiusdem deferentis que posita est continere 60 tales partes. Linea uero *FB* equalis est *AE* linee quoniam utraque earum est semidiameter epicicli lune cuius quantitas data est et notata.
5 Declaratis itaque magnitudinibus premissis secundum partes de quibus *AC* linea que semidiameter est ecentrici lune continet 60, nunc restat magnitudines declarare easdem secundum partes de quibus linea *AD* que est a centro terre ad longitudinem longiorem centri ecentrici continet 60. Dico ergo quod de partibus predictis linea *AC* que est semidiameter ecentrici ut predictum est erit 49 partes
10 et 41 minuta. Linea uero *CD* que est distantia centri deferentis lune a centro orbis signorum ut prius erit 10 partes et 19 minuta. Linea uero *AE* que est semidiameter epicicli lune erit de eisdem partibus 5 partes et 15 minuta. Linea uero *DB* que est longitudo propior ecentrici lune a centro terre erit 39 partes et 22 minuta. Tota etiam linea *ED* que est longior longitudo centri corporis lune a centro terre que
15 unquam esse potest erit 65 partes et 15 minuta. Linea uero *DF* que est longitudo propior centri corporis lune a centro terre erit 34 partes et 7 minuta. Linea uero *AB* que diameter est deferentis lune erit de predictis partibus 99 partes et 22 minuta. Sic igitur declarantur magnitudines predicte secundum partes de quibus *AD* linea continet 60.

2 est enim dupla: enim cum duplica *Z*
3 partes *om. Z*
5 itaque: utique *DNZ*
6 lune: linee *Z*
9 est semidiameter: demidiameter est *Z*
11 ut prius: dictum est *mg. N*
13 lune: linee *Z*
14 lune: linee *Z*

Appendix B

Eulogy on Frederick II by Petrus de Vineis

The text given here (see the Commentary to p. 367, lines I.4–110) is based on that in "Petri de Vineis, Cancellarii quondam Friderici II Imp. Rom. Epistolarum libri VI...post...Simonis Schardii JC. editionem anni MDLXVI denuo cum Haganoensi exemplari collatum recognitum...auctum...per Germanum Philalethen" (Amberg: Johannes Schönfeldius, 1609), *Epistola* 44, pp. 450–52. Occasional corrections have been introduced on the basis of the text in Dietrich von Nieheim, *Viridarium Imperatorum et Regum Romanorum*, ed. Alphons Lhotsky and Karl Pivec, Monumenta Germaniae Historica, Staatsschriften des späteren Mittelalters, vol. 5.1 (Stuttgart, 1956), pp. 70–71.

Questionis ardue petita responsio, in quantum respondenti permittitur, enodatur. Grandis namque progressus materie, infinitis terminande limitibus, rancoris propinat iudicia, et ex tele diffuse contextu, que de preconio summi Cesaris hostes cedentis orditur, ne quid ex contingentibus omittatur, manus scribentis tremescit et stupet. Quis enim posset nisi amplo famine prepotens tanti principis insignia promere, in cuius pectus confluunt quicquid uirtutes habent, quem nubes pluerunt iustum, et super eum celi desuper rorauerunt.[1] Non Plato, non Tullius, non filii tenebrarum, qui ex ore sedentis in throno in generatione sua prudentiores lucis filiis nuncupantur.[2] Hunc siquidem terra, pontus adorant, et ethera satis applaudunt, utpote qui mundo uerus imperator a diuino prouisus culmine, pacis amicus, caritatis patronus, iuris conditor, iustitie conseruator, patientie filius, mundum perpetua ratione gubernat. Hic est de quo Ezechielis uerba proclamant: Aquila grandis magnarum alarum, longo membrorum ductu, plena plumis et uarietate[3] multiplici. Hic est de quo loquitur Ieremias: Replebo te hominibus quasi botro, et super te celeusma cantabitur.[4] Talis ergo presidio principis protectus totus mundus exultet, talem namque totus orbis uocabat in dominum, talem requirebat iustitie defensorem, qui in potentia strenuus, in strenuitate preclarus, in claritate

1. Cf. Isa. 45:8: "Rorate, caeli, desuper, et nubes pluant iustum."

2. Cf. Luke 16:8: "...quia filii huius saeculi prudentiores filiis lucis in generatione sua sunt."

3. Ezek. 17:3.

4. Jer. 51:14: "Replebo te hominibus quasi brucho, et super te celeuma cantabitur.'

benignus, in benignitate sapiens, in sapientia prouidus, in prouidentia foret humanus. In eo denique insita forma boni, tanquam liuore carens, climata ligat
20 et elementa coniungit, ut conueniant flammis frigora, iungantur arida liquidis, planis associentur aspera, et directis inuia maritentur. Sub eius namque temporibus destruuntur fomenta malitie, uirtus securitatis inseritur. Itaque gladii conflantur in uomeres,[5] pacis federe suffocante timorem, et eius metus instinctu, quicquid libertas negligit, et licentia immoderate presumit, uictorie censura casti-
25 gat. O miranda diuina clementia, fastum compescere prompta, perituro mundo de tam mundo principe tam consulte quam utiliter prouidisti, qui ex omni parte beatus, strenuus, in toto cuiuslibet turbationis orbe pacator iustissimus, sine cura populi solus esse nesciret. Quem supremi manus opificis formauit in hominem, ut rerum habenas flecteret et cuncta sub iuris ordine limitaret. O utinam diuina
30 prouisio, per parenthasim dierum nostrorum numerum resecans, et Cesaree manui fulcimenta contribuens, annos Augusti regnantis augeret. O nature felicitas, quanti priuilegii prerogatiua principem ditasti felicem, concedens alii quod deficit in te ipsa. Hunc trames rationis antistitem, hunc exigebat iustitia defensorem, qui congruam seruans utrobique temperiem conatus cupiditatis infringeret, et
35 eius morsus illicitos refrenaret. Cui iam uirtutum incipiunt mysteria inuidere, ea uidelicet inuidie specie, que non ardore liuoris emulantis destruit animum, sed in suauitatis odorem flatibus incitat caritatis. Viuat igitur, uiuat sancti Friderici nomen in populo, succrescat in ipso feruor deuotionis a subditis, et fidei meritum mater ipsa fidelitas in exemplum subiectionis inflammet.

5. Cf. Isa. 2:4: "...conflabunt gladios suos in uomeres."

Appendix C

Text of Jābir ibn Aflaḥ's Method of Finding the Planetary Arc of Retrogradation

The following extract from Book 8 of Jābir's treatise is taken, except where otherwise indicated, from MS Madrid, Biblioteca Nacional, 10006, fol. 132r. The printed text (Apianus, *Instrumentum Primi Mobilis*, p. 131) is a very poor one; in the short extract below, it omits no fewer than three passages, one quite long. The diagram is copied from that in the manuscript.

Jābir here purports to be criticizing Ptolemy's method of finding the lengths of the arc of retrogradation in the planets. Like many of his criticisms, it shows poor understanding of the *Almagest*, and is in fact completely irrelevant to Ptolemy's method. It would, however, be a legitimate criticism of someone who found the arc lengths from the distances of station from apogee without making the correction explained by Campanus at lines VI.261–340 (see that passage, and Commentary to section VI, note 29). There seems little doubt that Campanus got the idea for his correction from this passage of Jābir. The two round numbers given by Jābir agree quite well with analogous figures that we can derive from Campanus's data. Jābir says that the difference (resulting from Jābir's proposed correction) in the length of half the epicyclic arc of retrogradation for Mars at apogee is about 4° (leading to a time difference of 9 days). In my note 33 to section VI, I found, by calculation from the Toledan tables, $p = 22;32$ and $p' = 18;30$. The difference is 4;2°, and the period of half-retrogradation one can then derive from p' agrees exactly with what Campanus gives (line VI.239), indicating that he did indeed find a correction for the length of arc very close to Jābir's 4°. The other number given by Jābir is about $1\frac{1}{4}$ days for the difference in the time taken by Venus to traverse half the arc of retrogradation at perigee. If we take Campanus's figure (ll. VI.223–24) for the distance of Venus's station from apogee of the epicycle when the epicycle is in the perigee, namely, 5^s 18;21°, subtract it from 180° to get the distance of station from perigee of the epicycle, and divide the result by 0;36,59°, the daily motion of Venus in anomaly, we get an uncorrected figure of about 18;54 days: this value is less than Campanus's corrected result (line VI.238: "20 dies 4 hore") by 1;16 days, or almost exactly $1\frac{1}{4}$ days.

Et similiter occurrit super eum estimatio in hoc, ut duplicaret partes orbis reuolutionis que sunt inter locum stationis prime et inter propinquitatem pro-

piorem uisibilem. Est ergo illud secundum eius estimationem locus stationis secunde, hoc autem non certificatur nisi ita ut sit centrum orbis reuolutionis in
5 hora habitudinis que nominatur extremitas noctis super unum duorum transituum mediorum ecentrici. Verum quando est centrum orbis reuolutionis in sectione ecentrici in qua est longitudo longior aut in sectione in qua est longitudo propior, scilicet duabus sectionibus quas determinant duo transitus medii, non ergo certificatur illud propter motum propinquitatis propioris uisibilis. Quod est, quia nos
10 ponemus [Fig. 27] orbem reuolutionis circulum *ABG* circa centrum *D* et centrum orbis signorum punctum *E* et centrum motus equalis punctum *H* et longitudinem longiorem punctum *Z*, et continuabo centrum orbis signorum cum centro orbis reuolutionis per lineam *ADGE*. Et sit stella in statione sua prima supra punctum *B*, et sit centrum orbis reuolutionis in eo quod est inter punctum longitudinis lon-
15 gioris et transitum medium. Sequitur ergo propter illud ut punctum *G*, quod est

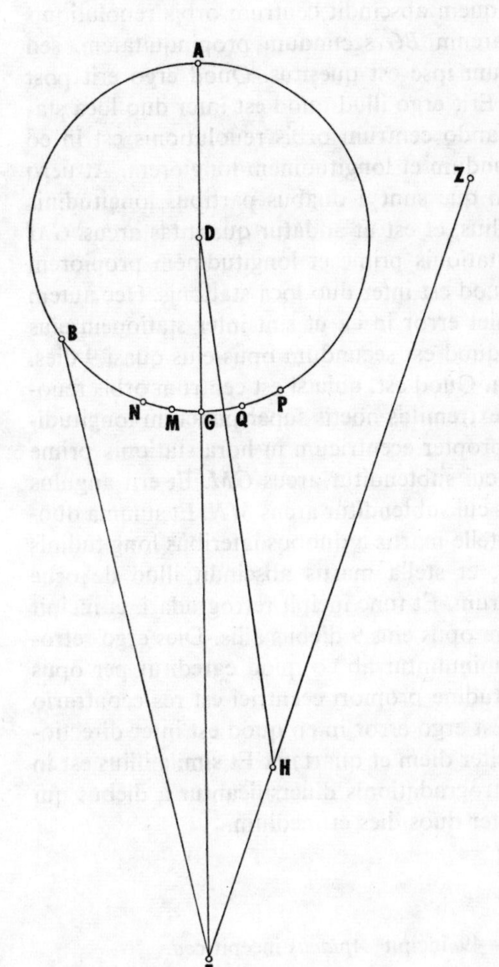

Fig. 27

Figure (after Jābir) to illustrate Jābir's discussion of the method of finding the length of the arc of retrogradation for the planets. *E* represents the earth, *H* the equant point, and *Z* the apogee of the deferent. The epicycle *ABG* is drawn with its center at *D*, which lies between the apogee of the deferent and the point of mean distance on the deferent. *B* is the point on the epicycle in which the planet is at its first station. *G* is the perigee of the epicycle at the moment depicted. Jābir says, correctly, that by the time the planet in its motion on the epicycle has reached the perigee of the epicycle, that perigee is no longer at *G* but at another point *M* nearer *B*, since the epicycle (and hence the epicyclic apogee) has also moved with respect to *E* during that time (the points *G* and *M* here correspond to the points *Z* and *T* in Campanus's Figure 19, p. 316, q.v.). Thus the length of half of the arc of retrogradation is not *BG* but *BM*, and the difference between the two is *GM*. Arc *MN* is constructed equal to arc *GM*, and then arc *GN* represents the difference between double the distance of the station from perigee and the whole arc of retrogradation. The two points *Q* and *P* are not mentioned in the text, but presumably they correspond to *M* and *N*, respectively, for the situation when the epicycle center lies between mean distance and perigee on the deferent. Jābir says that then one has to *add GM* (i.e., *GQ*) to *GB* to get the true arc of half-retrogradation. Compare the lower part of Figure 19 and the accompanying explanation.

longitudo propior uisibilis, permutetur per motum centri orbis reuolutionis ad contrarium motus stelle. Moueatur ergo usque quod perueniat stella in habitudine que nominatur extremitas noctis per arcum *GM*. Erit ergo punctum *M* ipsum punctum quod diuidit illud quod est inter duo loca stationis in duo media, quo-
20 niam quando stella peruenit ad punctum stationis secunde, perambulat punctum longitudinis propioris arcum equalem arcui *GM* qui sit arcus *MN*, et est longitudo puncti stationis secunde a puncto *N* equalis longitudini puncti stationis prime a puncto *M*. Sequitur ergo propter illud ut punctum *M* diuidat quod est inter duo loca stationis in duo media, non punctum *G* sicut ipse putauit. Oportet ergo prop-
25 ter illud ut duplum arcus *BG* addat in hoc loco super longitudinem que est inter duo loca stationis per duplum arcus *GM*, quod est arcus *GN*. Sequitur ergo propter illud in hoc loco, scilicet quando est centrum orbis reuolutionis inter longitudinem longiorem et transitum medium primum, ut minuatur de arcu *BG*, qui est longitudo stationis prime a puncto longitudinis propioris uisibilis, arcus *GM*,
30 qui est superfluitas diuersitatis arcus quem abscindit centrum orbis reuolutionis in tempore in quo abscindit stella arcum *BG* secundum propinquitatem, sed secundum ueritatem arcum *BM*, uerum ipse est quesitus. Quod ergo erit post diminutionem eius duplabimus illud. Erit ergo illud quod est inter duo loca stationis. Et similiter sequitur etiam quando centrum orbis reuolutionis est in eo
35 quod est inter transitum medium secundum et longitudinem longiorem. At uero quando est in una duarum sectionum que sunt a duabus partibus longitudinis propioris, tunc sequitur contrarium illius, et est ut addatur quantitas arcus *GM* super arcum qui est inter duo loca stationis prime et longitudinem propiorem uisibilem. Et duplatio illius est illud quod est inter duo loca stationis. Hec autem
40 res si defuerit in stella martis, perueniet error in ea ut sint inter stationem eius primam secundum ueritatem et illud quod est secundum opus eius quasi 9 dies. Et similiter est in statione eius secunda. Quod est, quia si est centrum orbis reuolutionis in habitudine que nominatur extremitas noctis super punctum longitudinis longioris, est angulus diuersitatis propter ecentricum in hora stationis prime
45 ipse angulus centri orbis reuolutionis cui subtenditur arcus *GM*. Et erit angulus diuersitatis in statione secunda angulus cui subtenditur arcus *MN*. Et summa duorum horum angulorum in diuersitate stelle martis a duobus lateribus longitudinis longioris ecentrici est quasi 4 partes, et stella martis abscindit illud de orbe reuolutionis sue in propinquitate 9 dierum. Et tunc incipit retrogradari, et incipit
50 dirigi ante inceptionem suam secundum opus eius 9 diebus aliis. Dies ergo retrogradationis eius secundum ueritatem minuuntur ab eo quod egreditur per opus eius 18 diebus, et quando est in longitudine propiori ecentrici est res econtrario illius, et hoc apparens est in uenere. Est ergo error in eo quod est inter directionem eius et eius retrogradationem circiter diem et quartam. Et simile illius est in
55 retrogradatione secunda. Dies ergo retrogradationis diuersificantur a diebus qui proueniunt secundum opus suum circiter duos dies et medium.

25 addat *Apianus* addatur *cod.*
40 sint *Apianus* sit *cod.*
49 incipit[2] *Apianus* incepit *cod.*

Bibliography

Abbott, T. K. *Catalogue of the Manuscripts in the Library of Trinity College, Dublin.* Dublin, 1900.

Alfraganus. *Il 'libro dell'aggregazione delle stelle'.* Edited by Romeo Campani. Collezione di Opuscoli Danteschi inediti o rari, nos. 87–90. Citta di Castello, 1910.

———. *Differentie in quibusdam collectis scientie astrorum.* Edited by F. J. Carmody. Mimeographed. Berkeley, 1943.

Anaritius. *In decem libros priores Elementorum Euclidis commentarii.* Edited by Maximilian Curtze as Supplementum to Euclid, *Opera omnia.* Leipzig, 1899.

Apianus, Petrus. *Astronomicum Caesareum.* Ingolstadt, 1540.

———. *Instrumentum Primi Mobilis.... Accedunt iis Gebri filii Affla Hispalensis astronomi vetustissimi pariter et peritissimi libri IX de Astronomia, ante aliquot secula Arabice scripti et per Giriardum Cremonensem latinitate donati....* Nuremberg, 1534.

Aquinas, St. Thomas. *In octo libros Physicorum Aristotelis expositio.* Sancti Thomae Aquinatis Opera Omnia, iussu Leonis XIII P.M. edita, vol. 2. Rome, 1884.

Archimedes. *Opera omnia cum commentariis Eutocii.* Edited by J. L. Heiberg. 2d ed. 3 vols. Leipzig, 1910–15.

Artaud de Montor. *See* Montor, Artaud de

Aschbach, Joseph von. *Geschichte der Wiener Universität.* 3 vols. Vienna, 1865–88.

Bacon, Roger. *Opera hactenus inedita Fr. Rogeri Baconis.* Edited by Robert Steele. 16 fascs. Oxford, 1905/09–1940. Fasc. 13, *Questiones supra libros octo Physicorum Aristotelis.* Edited by F. Delorme. Oxford, 1935.

———. *The "Opus majus" of Roger Bacon.* Edited by J. H. Bridges. 3 vols. Oxford and London, 1897–1900.

———. *Opus tertium.* In *Opera quaedam hactenus inedita,* edited by J. S. Brewer. Vol. 1, pp. 3–310. Rerum Britannicarum Medii Aevi Scriptores. London, 1859.

Baldi, Bernardino. "Vite inedite di matematici italiani." Edited by Enrico Narducci. *Bullettino di bibliografia e di storia delle scienze matematiche e fisiche,* vol. 19 (1886), pp. 591–96.

———. *See also* Steinschneider, Moritz

Bandini, A. M. *Catalogus codicum latinorum Bibliothecae Mediceae Laurentianae.* 5 vols. Florence, 1774–78.

Barnabita, T. B. "Sopra Pietro Peregrino di Maricourt e la sua epistola de magnete." *Bullettino di bibliografia e di storia delle scienze matematiche e fisiche*, vol. 1 (1868), pp. 1–32.

Battānī, al-. *See* Nallino, C. A.

Baur, Ludwig. *See* Grosseteste, Robert

Beaujouan, Guy. *Manuscrits scientifiques medievaux de l'Université de Salamanque et de ses "Colegios mayores."* Bibliothèque de l'école des hautes études hispaniques, fasc. 32. Bordeaux, 1962.

Benjamin, F. S., Jr. "John of Gmunden and Campanus of Novara." *Osiris*, vol. 11 (1954), pp. 221–46.

Bernardus de Virduno. *Tractatus super totam astrologiam*. Edited by P. P. Hartmann, O.F.M. Franziskanische Forschungen, vol. 15. Werl, 1961.

Bīrūnī, al-. *The Chronology of Ancient Nations*.... Translated by C. E. Sachau. London, 1879.

———. *Tafhim, The Book of Instruction in the Elements of the Art of Astrology*. Translated by R. R. Wright. London, 1934.

Björnbo, A. A. "Die mathematischen S. Marcohandschriften in Florenz." *Bibliotheca mathematica*, ser. 3, vol. 12 (1912), pp. 97–132, 193–224.

———. "Studien über Menelaos' Sphärik." *Abhandlungen zur Geschichte der mathematischen Wissenschaften mit Einschluss ihrer Anwendungen*, vol. 14 (1902), pp. 1–154.

Björnbo, A. A., and Vogl, Sebastian. "Alkindi, Tideus und Pseudo-Euklid, Drei optische Werke." *Abhandlungen zur Geschichte der mathematischen Wissenschaften mit Einschluss ihrer Anwendungen*, vol. 26 (1912), no. 3.

Boethius, Anicius Manlius Severinus. *Boetii quae fertur Geometria*. In *De institutione arithmetica*..., pp. 372–428. Edited by Gottfried Friedlein. Leipzig, 1867; reprinted Frankfurt, 1966.

———. *Philosophiae consolatio*. Edited by Ludovicus Bieler. Corpus Christianorum series latina 94, pt. 1. Turnhout, 1957.

Boffito, G., and D'Eril, C. M. *Il trattato dell'astrolabio da Pietro Peregrino di Maricourt: Introduzione e saggio del codice Vaticano Palatino N. 1392 con facsimili*. Pubblicazioni del Collegio alla Querca di Firenze, ser. in-8, no. 21. Florence, 1928.

Boncompagni, Baldassarre. "Della vita e delle opere di Gherardo Cremonese tradutore del secolo duodecimo e di Gherardo da Sabbionetta astronomo del secolo decimoterzo." *Atti dell'Accademia pontificia de'Nuovi Lincei*, vol. 4 (1851), pp. 387–493.

———. *Scritti di Leonardo Pisano*.... 2 vols. Rome, 1857, 1862.

Bond, J. D. "The Development of Trigonometric Methods down to the Close of the Fifteenth Century." *Isis*, vol. 4 (1921–22), pp. 295–323.

Bond, W. H. *See* De Ricci, Seymour

Bouché-Leclercq, A. *L'Astrologie grecque*. Paris, 1899; reprinted Brussels, 1963.

Brahe, Tycho. *Opera omnia*. Edited by J. L. E. Dreyer. 15 vols. Copenhagen, 1913–29.

Braunmühl, A. von. *Vorlesungen über Geschichte der Trigonometrie*. 2 pts. Leipzig, 1900, 1903.

Cadier, Léon, ed. *Les Régistres de Jean XXI, 1276–77*. Bibliothèque des écoles françaises d'Athènes et de Rome, ser. 2, no. 12. Paris, 1898.

Calcoen, R. *Inventaire des manuscrits scientifiques de la Bibliothèque royale de Belgique*. Vol. 1. Brussels, 1965.

Calendar of the Patent Rolls Preserved in the Public Record Office. London, 1891——.

Campani, Romeo. *See* Alfraganus

Campanus. *Tractatus de sphera*. In *Sphera mundi nouiter recognita cum commentariis 7 authoribus*, fols. 153–58. Venice: L. A. de Giunta, 1518.

Carmody, F. J. *Arabic Astronomical and Astrological Sciences in Latin Translation: A Critical Bibliography*. Berkeley, 1956.

———. *Astronomical Works of Thabit b. Qurra*. Berkeley and Los Angeles, 1960.

———. *See also* Alfraganus; Gerardus

Caspar, Max, trans. *Johannes Kepler, Neue Astronomie*. Munich, 1929.

Catalogue des manuscrits de la Bibliothèque royale de Belgique. Brussels, 1839–42.

Catalogue général des manuscrits des bibliothèques publiques de France: Départements. 44 vols. Paris, 1886–1911.

Catalogue of the Manuscripts Preserved in the Library of the University of Cambridge. 6 vols. Cambridge, 1856–67.

Celsus, A. Cornelius. *Quae supersunt*. Edited by F. Marx. Corpus medicorum latinorum, vol. 1, Leipzig and Berlin, 1915.

Chalcidius. *See* Waszink, J. H., ed.

Charisius, Flavius Sosipatrus. *Artis grammaticae libri V*. Edited by K. Barwick. Leipzig, 1925.

Clagett, Marshall. *Archimedes in the Middle Ages*. Vol. 1, *The Arabo-Latin Tradition*. Madison, Wis., 1964.

———. "The Medieval Latin Translation of the *Elements* of Euclid, with Special Emphasis on the Versions of Adelard of Bath." *Isis*, vol. 44 (1953), pp. 16–42.

Cleomedes. *De motu circulari corporum caelestium libri duo*. Edited by Hermann Ziegler. Leipzig, 1891.

Cohen, M. R., and Drabkin, I. E. *A Source Book in Greek Science*. Cambridge, Mass., 1948; reprinted 1958.

Colebrooke, H. T. "On the Notion of the Hindu Astronomers concerning the Precession of the Equinoxes and Motions of the Planets." *Asiatick Researches*, vol. 12 (1816), pp. 209–50; reprinted in *Collected Essays*, vol. 2, pp. 374–416.

Collectio Salernitana.... pubblicati a cura di Salvatore de Renzi. 5 vols. Naples, 1852–59.

Coopland, G. W. *Nicole Oresme and the Astrologers*. Cambridge, Mass., 1952.

Cotta, L. A. *Museo Novarese*. Milan, 1701.

Coxe, H. O. *Catalogi codicum manuscriptorum Bibliothecae Bodleianae*. Vol. 3, *Codices Graecos et Latinos canonicianos complectens*. Oxford, 1854.

Croke, Alexander, ed. *Regimen sanitatis Salernitanum: A Poem...with an Ancient Translation*. Oxford, 1830.

Curtze, Maximilian. "Die Handschrift No. 14836 der Königl. Hof- und Staatsbibliothek zu München." *Abhandlungen zur Geschichte der Mathematik*, vol. 7 (1895), pp. 132–35.

———. *See also* Anaritius

Daniel of Morley. *See* Sudhoff, Karl

Daunou, P. C. F. "Campanus de Novarre, mathematicien." *Histoire littéraire de la France*, vol. 21 (1835), pp. 248–54, 688–89.

———. "Simon de Gènes." *Histoire littéraire de la France*, vol. 21 (1835), pp. 241–48.

Davis, F. N., ed. *Rotuli Roberti Grosseteste, episcopi Lincolniensis, necnon rotulus Henrici di Lexington*. Publications of the Lincoln Record Society, vol. 11. Horncastle, 1914.

Delambre, J. B. J. *Histoire de l'astronomie ancienne*. 2 vols. Paris, 1817; reprinted New York and London, 1965.

———. *Histoire de l'astronomie du Moyen Age*. Paris, 1819; reprinted New York and London, 1965.

della Tuccia, Niccolo. "Cronache di Viterbo e di altre citta." In *Cronache e statuti della Città di Viterbo*. Documenti di storia italiana pubblicati a cura della R. Deputazione sugli studi di storia patria 5. Florence, 1872.

De Ricci, Seymour. *Census of Medieval and Renaissance Manuscripts in the United States and Canada: Supplement*. Edited and augmented by C. U. Faye and W. H. Bond. New York, 1962.

Diels, Hermann. *Die Fragmente der Vorsokratiker*. 10th ed. Edited by Walther Kranz. 3 vols. Berlin, 1961.

Digard, Georges; Fauçon, Maurice; Thomas, Antoine; and Fawtier, Robert. *Les Régistres de Boniface VIII, recueil des bulles de ce pape publiées ou analysées d'après les manuscrits originaux des archives du Vatican*. Vol. 1. Bibliothèque des écoles françaises d'Athènes et de Rome, ser. 2, no. 4. Paris, 1907.

Dozy, R. P. A. *Supplément aux dictionnaires arabes*. 2d ed. 2 vols. Leiden, 1927.

Drake, Stillman, and Drabkin, I. E., trans. and annotators. *Mechanics in Sixteenth-Century Italy*. Madison, Wis., 1969.

Dreyer, J. L. E. "On the Origin of Ptolemy's Catalogue of Stars. II." *Monthly Notices of the Royal Astronomical Society*, vol. 78 (1918), pp. 343–49.

———. *See also* Brahe, Tycho

du Cange, Charles du Fresne, comp. *Glossarium mediae et infimae latinitatis*. New ed., augmented by monks of the order of St. Benedict. Niort, 1883–87.

Duhem, Pierre. "Sur l'*Algorithmus demonstratus*." *Bibliotheca mathematica*, ser. 3, vol. 6 (1905), pp. 9–15.

———. *Le Système du monde: Histoire des doctrines cosmologiques de Platon à Copernic*. 10 vols. Paris, 1913–59.

Eubel, Conrad. *Hierarchia catholica Medii Aevi*. Münster, 1913.

Euclid. *Elementa*. Edited by J. L. Heiberg. 5 vols. Leipzig, 1883–88.

Fabricius, Jo. Albert. *Bibliotheca latina mediae et infimae aetatis*. 6 vols. Hamburg, 1734–46.

Favaro, Antonio. "Intorno alla vita ed alle opere di Prosdocimo de'Beldomandi, matematico padovano del secolo XV." *Bullettino di bibliografia e di storia delle scienze matematiche e fisiche*, vol. 12 (1879), pp. 41–251.

Faye, C. U. *See* De Ricci, Seymour

Fontaine, Jacques, ed. *Isidore de Seville*, "*Traité de la nature*," *suivi de l'épitre en vers du roi Sisebut à Isidore*. Bibliothèque de l'école des hautes études hispaniques, fasc. 28. Bordeaux, 1960.

Franco of Liège. *See* Winterberg, C.

Frati, Lodovico. "Indice dei codici latini conservati nella R. Biblioteca Universitaria di Bologna." *Studi italiani di filologia classica*, vol. 16 (1908), pp. 103–432; vol. 17 (1909), pp. 1–171.

Geber. *See* Apianus, Petrus

Geminus. *Elementa astronomiae*. Edited by Karl Manitius. Leipzig, 1898.

Gerardus. *Theorica planetarum*. Edited by F. J. Carmody. Berkeley, 1942.

Gesner, Conrad. *Bibliotheca universalis, sive catalogus....* Vol. 1. Zürich, 1545. Augmented by J. J. Frisius, Zürich, 1583.

Ginzel, F. K. *Handbuch der mathematischen und technischen Chronologie*. 3 vols. Leipzig, 1906–14; reprinted 1958.

Goldstein, B. R. "The Arabic Version of Ptolemy's *Planetary Hypotheses*." *Transactions of the American Philosophical Society*, n.s., vol. 57 (1967), pp. 3–55.

Grabmann, Martin. *Forschungen über die lateinischen Aristoteles-Übersetzungen des XIII Jahrhunderts*. Beiträge zur Geschichte der Philosophie des Mittelalters, vol. 17, nos. 5–6. Münster i.W., 1916.

———. *Guglielmo di Moerbeke*. Rome, 1946.

Grosseteste, Robert. *De dispositione aeris*. In *Die philosophischen Werke des Robert Grosseteste, Bischofs von Lincoln*, pp. 41–51. Edited by Ludwig Baur. Beiträge zur Geschichte der Philosophie des Mittelalters, vol. 9. Münster i.W., 1912.

———. *De sphaera*. Ibid., pp. 10–32.

———. *Summa philosophiae*. Ibid., pp. 275–643.

———. *See also* Davis, F. N.; Thomson, S. H.

Guiraud, Jean, ed. *Les Régistres d'Urbain IV (1261–64)*. 5 vols. Bibliothèque des écoles françaises d'Athènes et de Rome, ser. 2, no. 13. Paris, 1901–6.

Gunther, R. T. *The Astrolabes of the World*. 2 vols. Oxford, 1932.

———. *Chaucer and Messahalla on the Astrolabe*. Early Science in Oxford, vol. 5. Oxford, 1929.

Haller, Johannes. *Das Papsttum*. 2d ed. 5 vols. Basel, 1951–53.

Halliwell, J. O., ed. *Rara mathematica*. Vol. 1, *Joannis de Sacro-Bosco: Tractatus de arte numerandi*. London, 1839; reprinted 1841.

Halm, C., and others. *Catalogus codicum latinorum Bibliothecae Regiae Monacensis*. 2 vols. in 7 pts. Munich, 1868–81.

Halma, N., ed. *Commentaire de Théon sur les Tables Manuelles de Ptolemée*; *Tables Manuelles Astronomiques de Ptolemée et de Théon*. 3 pts. Paris, 1822–25.

Hartmann, P. P., O.F.M. *See* Bernardus de Virduno

Hartner, Willy. "Mediaeval Views on Cosmic Dimensions and Ptolemy's Kitāb al-Manshūrāt." In *Mélanges Alexandre Koyré*, vol. 2, pp. 254–82. Paris, 1964.

———. "The Mercury Horoscope of Marcantonio Michiel of Venice." In *Vistas in Astronomy*, edited by Arthur Beer. Vol. 1, pp. 84–138. London and New York, 1955.

Haskins, C. H. *Studies in the History of Mediaeval Science*. Cambridge, Mass., 1924; 2d ed. 1927; reprinted New York, 1960.

Hauber, A. "Zur Verbreitung des Astronomen Sūfī." *Der Islam*, vol. 8 (1918), pp. 48–54.

Hebräische Bibliographie: Blätter für neuere und altere Literatur des Judenthums, vols. 1–21. Berlin, 1858–82.

Heiberg, J. L. *See* Archimedes; Euclid; Ptolemy

Hipparchus. *In Arati et Eudoxi Phaenomena Libri Tres.* Edited by K. Manitius. Leipzig, 1894.

Hyginus. *Astronomica.* Edited by Émile Chatelain and Paul Legendre. Paris, 1909.

Ibn Abī Uṣaybiʿa. *Kitāb ʿUyūn al-Anbāʾ fī Ṭabaqāt al-Aṭibbāʾ.* Edited by A. Müller. 2 vols. Königsberg, 1884, 1882.

Ibn al-Qiftī. *Taʾrīkh al-Ḥukamāʾ.* Edited by Julius Lippert. Leipzig, 1903.

Ibn an-Nadīm. *Kitâb al-Fihrist.* Edited by Gustav Flügel. 2 vols. Leipzig, 1871, 1872.

Ibn Ezra, R. Abraham. *El libro de los fundamentos de las Tablas astronómicas de R. Abraham ibn ʿEzra.* Edited by J. M. Millás Vallicrosa. Madrid and Barcelona, 1947.

Icazbalceta, J. G. *Bibliografia Mexicana del Siglo XVI.* Mexico, 1954.

Isidore of Seville. *See* Fontaine, Jacques, ed.

Jābir. *See* Apianus, Petrus

James, M. R. *Bibliotheca Pepysiana: A Descriptive Catalogue of the Library of Samuel Pepys.* Pt. 3, *Mediaeval Manuscripts.* London, 1923.

———. *A Descriptive Catalogue of the Manuscripts in the Library of Corpus Christi College, Cambridge.* 2 vols. Cambridge, 1912 [1909–13].

———. *Lists of Manuscripts Formerly Owned by Dr. John Dee.* Supplement no. 1 to *Transactions,* Bibliographical Society. Oxford, 1921.

John of Gmunden. *See* Benjamin, F. S., Jr.; Klug, Rudolf; Mundy, John

John of Salisbury. *Metalogicon libri IIII.* Edited by C. C. J. Webb. Oxford, 1929.

Kantorowicz, E. H. "The Prologue to *Fleta* and the School of Petrus de Vinea." *Speculum,* vol. 32 (1957), pp. 231–49.

Kāshī, al-. *See* Kennedy, E. S.

Keil, Heinrich. *Grammatici latini ex recensione Henrici Keilii.* 7 vols. plus *Supplementum ex recensione Hermanni Hageni.* Leipzig, 1857–80.

Kennedy, E. S., ed. *The Planetary Equatorium of Jamshīd Ghiyāth al-Dīn al-Kāshī.* Princeton Oriental Studies, vol. 18. Princeton, N.J., 1960.

Kepler, Johannes. *Gesammelte Werke.* Edited by Walther von Dyck and Max Caspar. Munich, 1938——.

Ker, Neil. *Medieval Libraries of Great Britain: A List of Surviving Books.* 2d ed. London, 1964.

Khwārizmī, al-. *See* Suter, H.; Toomer, G. J.; Vogel, Kurt

Kibre, Pearl. "Further Addenda and Corrigenda." *Speculum,* vol. 43 (1968), pp. 78–114.

Klebs, A. C. "Incunabula scientifica et medica." *Osiris,* vol. 4 (1938), pp. 1–359; reprinted separately, Bruges, 1938.

Klug, Rudolf. "Johannes von Gmunden, der Begründer der Himmelskunde auf deutschen Boden." *Sitzungsberichte der Akademie der Wissenschaften in Wien,* phil.-hist. ser., vol. 222 (1943), no. 4.

Kren, Claudia. "Homocentric Astronomy in the Latin West: The *De reprobatione ecentricorum et epiciclorum* of Henry of Hesse." *Isis,* vol. 59 (1968), pp. 269–81.

Lachmann, Karl, ed. *Gromatici Veteres.* In *Die Schriften der Römischen Feldmesser.* 2 vols. Berlin, 1848, 1852; reprinted, Hildesheim, 1967.

Laplace, Pierre Simon de. *Exposition du système du monde.* 4th ed. Paris, 1813.

Leitschuh, F., and Fischer, H. *Katalog der Handschriften der kgl. Bibliothek zu Bamberg*. 3 vols. in 6 pts. Bamberg, 1887–1912.

Lejeune, Albert. *Euclide et Ptolémée: Deux stades de l'optique géometrique grecque*. Université de Louvain, Recueil de travaux d'histoire et de philologie, ser. 3, fasc. 31. Louvain, 1948.

Leonardo of Pisa. *See* Boncompagni, Baldassarre

Levi della Vida, G. "Appunti e questi di storia litteraria araba 5. 'Almanacco.'" *Revista degli studi orientali*, vol. 14 (1934), pp. 265–70.

Litt, Thomas. *Les Corps célestes dans l'univers de saint Thomas d'Aquin*. Louvain and Paris, 1963.

Little, A. G., ed. *Roger Bacon Essays*.... Oxford, 1914.

Löfstedt, Einar. *Coniectanea, Untersuchungen auf dem Gebiete der antiken und mittelalterlichen Latinität*. Uppsala and Stockholm, 1950; reprinted Amsterdam, 1968.

Maass, Ernst, ed. *Commentariorum in Aratum reliquiae*.... Berlin, 1898; reprinted 1958.

Macray, W. D. *Catalogi codicum manuscriptorum Bibliothecae Bodleianae*. Vol. 9, *Codices a Kenelm Digby donatos complectens*. Oxford, 1883.

Macrobius, Ambrosius Theodosius. *Opera*. Edited by J. Willis. 2 vols. Leipzig, 1963.

Madan, Falconer; Craster, H. H. E.; and others. *A Summary Catalogue of Western Manuscripts in the Bodleian Library at Oxford*. 7 vols. in 8 pts. Oxford, 1895–1953.

Maimonides, Moses. *The Guide for the Perplexed*. Translated by M. Friedländer. 2d ed. London, 1904; reprinted New York, 1956.

Manitius, Karl. *See* Geminus; Hipparchus; Ptolemy

Mann, H. K. *The Lives of the Popes in the Early Middle Ages*. 18 vols. London, 1925–32.

Marchand, Prosper. *Dictionaire historique, ou mémoires critiques et littéraires*. The Hague, 1758.

Martianus Capella. *De nuptiis Philologiae et Mercurii libri VIIII*. Edited by Adolf Dick. Leipzig, 1925.

Marx, F. *See* Celsus, A. Cornelius

Marx, J. *Verzeichnis der Handschriften-Sammlung des Hospitals zu Cues bei Bernkastel am Mosel*. Trier, 1905.

Maurolico, Francesco. "Scritti inediti (BN 7473, fols. 1–16)." Edited by Federico Napoli. *Bullettino di bibliografia e di storia delle scienze matematiche e fisiche*, vol. 9 (1876), pp. 23–121.

Mazzatinti, G. [and others]. *Inventari dei manoscritti delle biblioteche d'Italia*. Fasc. 1——. [Various places]. 1887——.

Merlan, P. *From Platonism to Neoplatonism*. 2d ed. The Hague, 1960.

Mette, H. J. *Sphairopoiia, Untersuchungen zur Kosmologie des Krates von Pergamon*. Munich, 1936.

Michel, Henri. *Traité de l'astrolabe*. Paris, 1947.

Migne, J. P. *Patrologiae cursus completus. Series latina*. 221 vols. Paris, 1844–64.

Millás Vallicrosa, J. M. *Estudios sobre Azarquiel*. Madrid and Granada, 1943–50.

———. "El *Liber de motu octave sphere* de Ṯābit ibn Qurra." *Al-Andalus*, vol. 10 (1945), pp. 89–108.

———. *Nuevos estudios sobre historia de la ciencia española.* Barcelona, 1960.

Montor, Artaud de. *The Lives and Times of the Popes.* Translated. 10 vols. New York, 1910.

Moody, E. A., and Clagett, Marshall, eds. *The Medieval Science of Weights: Treatises Ascribed to Euclid, Archimedes, Thabit ibn Qurra, Jordanus de Nemore, and Blasius of Parma.* Madison, Wis., 1952.

Mundy, John. "John of Gmunden." *Isis,* vol. 34 (1943), pp. 196–205.

Muratori, Lodovico Antonio. *Rerum italicarum scriptores.* 25 vols. in 28 pts. Milan, 1723–51.

Murdoch, J. E. "The Medieval Euclid: Salient Aspects of the Translations of the *Elements* by Adelard of Bath and Campanus of Novara." *Revue de synthèse,* vol. 89, ser. 3, nos. 49–52 (1968), pp. 67–94.

Nallino, C. A. *Al-Battānī sive Albatenii "Opus astronomicum."* 3 pts. Pubblicazioni del Reale Osservatorio di Brera in Milano, no. 40. Milan, 1899–1907.

———. "Il Gherardo Cremonese autore della *Theorica planetarum* deve ritenersi essere Gherardo Cremonese da Sabbioneta." *Rendiconti, Reale Accademia dei Lincei, Classe scienze morali, storiche e filologiche,* ser. 6, vol. 8 (1932), pp. 386–404; reprinted in *Raccolta di scritti editi e inediti,* vol. 6, Rome, 1948.

———. "Il valore metrico del grado di meridiano secondo i geografi arabi." In *Raccolta di scritti editi e inediti,* vol. 5, pp. 408–57. Rome, 1944.

Narducci, E. *Catalogo di manoscritti ora posseduti da D. Baldassarre Boncompagni.* 2d ed. Rome, 1892.

Neugebauer, O., ed. *Astronomical Cuneiform Texts.* 3 vols. London, 1955.

———. "Thabit ben Qurra 'On the Solar Year' and 'On the Motion of the Eighth Sphere.'" *Proceedings of the American Philosophical Society,* vol. 106 (1962), pp. 264–99.

———. *Vorlesungen über Geschichte der antiken mathematischen Wissenschaften.* Vol. 1, *Vorgriechische Mathematik.* Berlin, 1934; reprinted 1969.

Niccolo della Tuccia. *See* della Tuccia, Niccolo

Nicomachus of Gerasa. *Introductionis arithmeticae libri II.* Edited by R. Hoche. Leipzig, 1866. Translated by M. L. D'Ooge, with studies by F. E. Robbins and L. C. Karpinski, as *Introduction to Arithmetic.* University of Michigan Studies, Humanistic Series, vol. 16. New York, 1926.

Novum glossarium mediae latinitatis. Edendum Curavit Consilium Academiarum. Copenhagen, 1957——.

Oresme, Nicolas. *Quaestiones super geometriam Euclidis.* Edited by H. L. L. Busard. Janus Suppléments, vol. 3. Leiden, 1961.

P. F. "Bulletin critique et chronique bibliographique." *Bulletin du Cange,* vol. 14 (1939), pp. 66–78.

P. H. M. "De conventu nostri ordinis Viterbiensi." *Analecta Augustiniana,* vol. 11 (1925–26), pp. 227–35.

Panzer, G. W. *Annales typographici ab artis inventae origine ad annum MDXXXVI.* 11 vols. Nuremberg, 1793–1803.

Paoli, C. *I codici Ashburnhamiani della R. Biblioteca Mediceo-Laurenziana di Firenze.* 5 fascs. Ministero dell'Istruzione Pubblica, *Indici e Cataloghi,* vol. 8. Rome, 1887–1917.

Partsch, J. "Die Grenzen der Menschheit: I. Die antike Oikumene." *Berichte über die Verhandlungen der K. sächsischen Gesellschaft der Wissenschaften*, phil.-hist ser., vol. 68 (1916), no. 2.

Pedersen, Olaf. "The Life and Work of Peter Nightingale." In *Vistas in Astronomy*, edited by Arthur Beer. Vol. 9, pp. 3–10. Oxford, 1967.

Peirce, C. S. "Campanus." *Science*, n.s., vol. 13 (1901), pp. 809–11.

Philo. *Philonis Judaei Opera omnia*. 8 vols. Vol. 6, *Quaestiones in Genesim*. Bibliotheca sacra patrum ecclesiae graecorum, pt. 2. Leipzig, 1828–30.

Poulle, Emmanuel. "L'Astrolabe médiéval d'après les manuscrits de la Bibliothèque nationale." *Bibliothèque de l'école des chartes*, vol. 112 (1954), pp. 81–103.

———. *Astronomie théorique et astronomie pratique au Moyen Âge*. Palais de la Découverte, Conference D.119. Paris, 1967.

———. *La Bibliothèque scientifique d'un imprimeur humaniste au XVe siècle: Catalogue des manuscrits d'Arnaud de Bruxelles à la Bibliothèque nationale de Paris*. Travaux d'Humanisme et Renaissance, vol. 57. Geneva, 1963.

———. *Un Constructeur d'instruments astronomiques au XVe siècle, Jean Fusoris*. Bibliothèque de l'école pratique des hautes études, IVe sec., sciences historiques et philologiques, fasc. 218. Paris, 1963.

———. "L'Équatoire de Guillaume Gilliszoon de Wissekerke." *Physis*, vol. 3 (1961), pp. 223–51.

———. "L'Équatoire de la Renaissance." In *Le Soleil à la Renaissance, sciences et mythes: Colloque international*. Brussels, 1965.

———. "Sur un Fragment d'instrument astronomique des Musées de Bruxelles." *Ciel et terre*, vol. 79 (1963), pp. 363–79.

———. "Théorie des planètes et trigonométrie au XVe siècle, d'après un équatoire inédit, le sexagenarium." *Journal des savants*, 1966, pp. 129–61.

Poulle, Emmanuel, and Maddison, Francis. "Un Équatoire de Franciscus Sarzosius." *Physis*, vol. 5 (1963), pp. 43–64.

Powicke, F. M. *King Henry III and the Lord Edward: The Community of the Realm in the Thirteenth Century*. 2 vols. Oxford, 1947.

———. *The Medieval Books of Merton College*. Oxford, 1931.

———. *The Thirteenth Century, 1216–1307*. The Oxford History of England, vol. 4. Oxford, 1953.

Price, D. J., ed. *The Equatorie of the Planetis*. Cambridge, 1955.

Proclus Diadochus. *In Platonis Rem Publicam commentarii*. Edited by W. Kroll. 2 vols. Leipzig, 1899, 1901; reprinted Amsterdam, 1965.

———. *In primum Euclidis Elementorum librum commentarii*. Edited by G. Friedlein. Leipzig, 1873; reprinted Hildesheim, 1967.

Pruckner, Herbert. *Studien zu den astrologischen Schriften des Heinrich von Langenstein*. Leipzig and Berlin, 1933.

Ptolemy. *Almagesti Cl. Ptolemei Pheludiensis Alexandrini Astronomorum principis opus ingens ac nobile*.... Venice: Petrus Liechtenstein, 1515.

———. *Handbuch der Astronomie*. Translated by Karl Manitius. 2 vols. Leipzig, 1912–13; reprinted 1963.

———. *Opera astronomica minora. Opera quae exstant omnia*, pt. 2. Edited by J. L. Heiberg. Leipzig, 1907.

———. *Syntaxis mathematica*. 2 vols. *Opera quae exstant omnia*, pt. 1. Edited by J. L. Heiberg. Leipzig, 1898, 1903.

———. *See also* Goldstein, B. R.

Quétif, Jacques, and Échard, Jacques. *Scriptores ordinis praedicatorum recensiti....* 2 vols. Paris, 1719–21; reprinted 2 vols. in 4 pts., New York, 1959.

Rashdall, Hastings. *The Universities of Europe in the Middle Ages*. Rev. ed. Revised by F. M. Powicke and A. B. Emden. 3 vols. Oxford, 1936.

Reisch, Gregor. *Margarita philosophica nova....* Edited by Johann Grüniger. Strasbourg, 1515.

Renaud, H. J. P. "L'Origine du mot 'Almanach.'" *Isis*, vol. 37 (1947), pp. 44–46.

Reports of the Royal Commission on Historical Manuscripts. London, 1870——.

Riccardi, Pietro. *Saggi di una bibliografia Euclidea*. Memorie della Reale Accademia delle Scienze dell'Istituto di Bologna, ser. 4, vol. 8. Bologna, 1887.

Ricius, Augustinus. *De motu octauae Sphaerae*. Paris, 1521.

Rico y Sinobas, Manuel, ed. *Libros del saber...del rey D. Alfonso X....* 5 vols. Madrid, 1863–67.

Robertus Anglicus. "Le Traité du quadrant." Edited by Paul Tannery. *Notices et extraits des manuscrits de la Bibliothèque nationale*, vol. 35 (1897), pp. 561–640; reprinted in P. Tannery, *Mémoires scientifiques*, vol. 5, pp. 118–97, Paris, 1922.

Rockinger, Ludwig. *Briefsteller und Formelbücher des eilften bis vierzehnten Jahrhunderts*. Quellen und Erörterungen zur bayerischen und deutschen Geschichte, vol. 9, pts. 1–2. Munich, 1863–64; reprinted New York, 1961.

Ross, W. D. *Aristotle*. London, 1923.

Sacrobosco. *See* Halliwell, J. O.; Thorndike, Lynn

Sarton, George. *Introduction to the History of Science*. 3 vols. in 4 pts. Baltimore, 1927–47.

Schöpp, Natalie. *Papst Hadrian V (Kardinal Ottobuono Fieschi)*. Heidelberger Abhandlungen zur mittleren und neueren Geschichte, vol. 49. Heidelberg, 1916.

Schramm, Matthias. *Ibn al-Haythams Weg zur Physik*. Boethius, vol. 1. Wiesbaden, 1963.

Schum, Wilhelm. *Beschreibendes Verzeichniss der Amplonianischen Handschriften-Sammlung zu Erfurt*. Berlin, 1887.

Sedgwick, W. B. "Some Poetical Words of the Twelfth Century." *Bulletin du Cange*, vol. 7 (1932), pp. 223–26.

Sethe, Kurt. *Von Zahlen und Zahlworten bei den alten Ägyptern*. Strasbourg, 1916.

Silverstein, Theodore. *Medieval Latin Scientific Writings in the Barberini Collection*. Chicago, 1957.

Simoni, Cornelio de. "Intorno alla vita ed ai lavori di Andalò di Negro, matematico ed astronomo genovese del secolo decimoquarto, e d'altri matematici e cosmografi genovesi." *Bullettino di bibliografia e di storia delle scienze matematiche e fisiche*, vol. 7 (1874), pp. 313–38.

Smith, D. E. *History of Mathematics*. 2 vols. Boston, 1923–25; reprinted New York, 1958.

———. *Rara arithmetica*. Boston and London, 1908.

Smith, D. E., and Karpinski, L. C. *The Hindu-Arabic Numerals*. Boston and London, 1911.

Solinus. *Collectanea rerum memorabilium*. Edited by Theodor Mommsen. Berlin, 1864.
Steinschneider, Moritz. *Die hebräischen Übersetzungen des Mittelalters und die Juden als Dolmetscher*. Berlin, 1893; reprinted Graz, 1956.
———. "Karaitischen Handschriften." *Hebräische Bibliographie*, vol. 11 (1871), p. 43.
———. "Über das Wort Almanach." *Bibliotheca mathematica*, n.s., vol. 2 (1888), pp. 13–16.
———. "Vite di matematici arabi tratte da un'opera inedita di Bernardino Baldi." *Bullettino di bibliografia e di storia delle scienze matematiche e fisiche*, vol. 5 (1872), pp. 427–534.
Stevenson, Henry, and Rossi, J. B. de. *Codices Palatini Latini Bibliothecae Vaticanae....* Vol. 1. Rome, 1886.
Stillwell, Margaret Bingham. *Incunabula in American Libraries: A Second Census of Fifteenth-Century Books Owned in the United States, Mexico and Canada*. New York, 1940.
Sudhoff, Karl. "Daniels von Morley *Liber de naturis inferiorum et superiorum*, nach der Handschrift Cod. Arundel 377 des Britischen Museums zum Abdruck gebracht." *Archiv für die Geschichte der Naturwissenschaften und der Technik*, vol. 8 (1918), pp. 1–40.
Suter, H. "Die astronomischen Tafeln des Muḥammed ibn Mūsā al-Khwārizmī in der Bearbeitung des Maslama ibn Aḥmed al-Madjrīṭī und der latein. Übersetzung des Athelhard von Bath." *Skrifter, Kongelige Danske Videnskabernes Selskab*, ser. 7, historical and philosophical sec., vol. 3 (1914), no. 1.
———. "Die Mathematiker und Astronomen der Araber und ihre Werke." *Abhandlungen zur Geschichte der mathematischen Wissenschaften mit Einschluss ihrer Anwendungen*, vol. 10 (1900).
Symon de Phares. *See* Wickersheimer, Ernest
Ṯābit ibn Qurra. *See* Carmody, F. J.; Millás Vallicrosa, J. M.; Neugebauer, O.
Tabulae codicum manuscriptorum praeter graecos et orientales in Bibliotheca Palatina Vindobonensi asservatorum. 11 vols. Vienna, 1864–1912; reprinted 1965.
Theon of Smyrna. *Expositio rerum mathematicarum ad legendum Platonem utilium*. Edited by Eduard Hiller. Leipzig, 1878.
Thesaurus linguae latinae, editus auctoritate et consilio Academiarum Quinque Germanicarum. Leipzig, 1900——.
Thiele, Georg. *Antike Himmelsbilder*. Berlin, 1898.
Thomas-Stanford, Charles. *Early Editions of Euclid's "Elements*." London, 1926.
Thomson, S. H. *The Writings of Robert Grosseteste, Bishop of Lincoln 1235–1253*. Cambridge, 1940.
Thorndike, Lynn. *A History of Magic and Experimental Science*. 8 vols. New York, 1923–58.
———. *Latin Treatises on Comets between 1238 and 1368 A.D.* Chicago, 1950.
———. "Notes on Some Astronomical, Astrological and Mathematical Manuscripts of the Bibliothèque Nationale, Paris." *Journal of the Warburg and Courtauld Institutes*, vol. 20 (1957), pp. 112–72.
———. "Notes upon Some Medieval Astronomical, Astrological and Mathematical Manuscripts at Florence, Milan, Bologna and Venice." *Isis*, vol. 50 (1959), pp. 33–50.

———. "Some Little Known Astronomical and Mathematical Manuscripts." *Osiris,* vol. 8 (1949), pp. 41–72.

———. "Some Medieval and Renaissance Manuscripts on Physics." *Proceedings of the American Philosophical Society,* vol. 104 (1960), pp. 188–201.

———. ed. and trans. *The "Sphere" of Sacrobosco and Its Commentators.* Chicago, 1949.

———. "The Study of Mathematics and Astronomy in the Thirteenth and Fourteenth Centuries as Illustrated by Three Manuscripts." *Scripta mathematica,* vol. 23 (1957), pp. 67–76.

Thorndike, Lynn, and Kibre, Pearl. *A Catalogue of Incipits of Mediaeval Scientific Writings in Latin.* Cambridge, Mass., 1937. Revised and augmented, 1963.

Tiraboschi, Girolamo. *Storia della letteratura italiana.* 9 vols. in 16 pts. Milan, 1822–26. Many other editions were published in Italy in the late eighteenth and early nineteenth centuries.

Tobler-Lommatzsch, Altfranzösisches Wörterbuch. Wiesbaden, 1954——.

Tomasini, Jacob. *Bibliothecae patavinae MSS publicae et privatae.* Udine, 1639.

Toomer, G. J. "Ptolemy." In *Dictionary of Scientific Biography.* New York, in press.

———. Review of *The Astronomical Tables of al-Khwārizmī* by O. Neugebauer. *Centaurus,* vol. 10 (1964), pp. 203–12.

———. "A Survey of the Toledan Tables." *Osiris,* vol. 15 (1968), pp. 5–174.

Tropfke, Johannes. *Geschichte der Elementar-Mathematik.* Vol. 4, *Ebene Geometrie,* edited by Kurt Vogel. 3d ed. Berlin, 1940.

Vacca, Giovanni. "Campano da Novara." *Enciclopedia Italiana di Scienze, Lettere ed Arti,* vol. 8 (Milan, 1930), p. 594.

Valentinelli, J. *Bibliotheca manuscripta ad S. Marci Venetiarum.* Venice, 1868–73.

Varāhamihīra, *Brihat Samhita.* Edited and translated by V. Subrahmany Sastri. 2 vols. Bangalore, 1947.

Vogel, Kurt. *Mohammed ibn Musa Alchwarizmi's "Algorismus."* Milliaria, no. 3. Aalen, 1963.

Wadding, Lucas. *Scriptores Ordinis Minorum.* Rome, 1690; reprinted Frankfurt, 1967. Revision begun by J. M. Fonseca in 1731 as *Annales Minorum, seu Trium Ordinum a S. Francisco institutorum,* and continuing, with some 25 vols. to date.

Wallerand, G. *Henri Bate de Malines: Speculum divinorum et quorundam naturalium.* Les Philosophes Belges, vol. 11. Louvain, 1931.

Wartburg, Walther v. *Französisches etymologisches Wörterbuch.* Bonn, Leipzig, Tübingen, and Basel, 1928——.

Waszink, J. H., ed. *Timaeus a Calcidio translatus commentarioque instructus....* Plato Latinus, vol. 4. London and Leiden, 1962.

Weidler, J. F. *Historia astronomiae sive de ortu et progressu astronomiae.* Wittenberg, 1741.

Weissenborn, H. "Die Übersetzungen des Euklid aus dem Arabischen in das Lateinische durch Adelard von Bath nach zwei Handschriften der königliche Bibliothek in Erfurt." *Zeitschrift für Mathematik und Physik,* hist. sec., vol. 25 (1880), pp. 141–66.

———. *Die Übersetzungen des Euklid durch Campano und Zamberti.* Halle, 1882.

Wickersheimer, Ernest. *Recueil des plus célèbres astrologues et quelques hommes faict par Symon de Phares du temps de Charles VIIIe, publié d'après le manuscrit unique de la Bibliothèque nationale*. Paris, 1929.

Winterberg, C. "Der Traktat Franco's von Lüttich *De quadratura circuli.*" *Abhandlungen zur Geschichte der Mathematik*, vol. 4 (1882), no. 2.

Wislocki, W. *Catalogus codicum manuscriptorum Bibliothecae Universitatis Jagellonicae Cracoviensis*. 2 vols. Cracow, 1877–81.

Witelo. *Opticae libri decem*. In *Opticae thesaurus* (with Alhazen's *Optics*). Edited by Friedrich Risner. Basel, 1572.

Wolfson, H. A. *Crescas' Critique of Aristotle*. Cambridge, Mass., 1929.

Wright, J. K. *The Geographical Lore of the Time of the Crusades*. American Geographical Society Research Series, no. 15. Washington, D.C., 1925; reprinted New York, 1965.

Zinner, Ernst. *Die Geschichte der Sternkunde von den ersten Anfängen bis zur Gegenwart*. Berlin, 1931.

———. *Leben und Wirken des Johannes Müller von Königsberg genannt Regiomontanus*. Munich, 1938.

———. *Verzeichnis der astronomischen Handschriften des deutschen Kulturgebietes*. Munich, 1925.

Index of Manuscripts Cited

The following index contains the references to all manuscripts mentioned in the introduction, commentary, and appendixes. They are listed alphabetically by the city or town where they are located. References are to page or to page and note (*n*) numbers. Manuscripts collated for the text of the *Theorica* are so indicated by the addition in parentheses of the siglum each has been assigned.

Index of Technical Terms

The following index includes all the technical terms in the *Theorica* connected with astronomy or the instrument. A number of common technical terms of elementary geometry and arithmetic have been omitted, as follows: abicere, addere, additio, agregare, agregatio, angulus, arcus, augmentum, cadere, colligere (in unum), coniungere (= "to add"), contineri ab, continuare (inter, cum), crescere, cubus, decrescere, demere, demonstrare, demonstratio, describere, determinare, differentia (= "difference"), dimidiare, dimidium, diminuere (de), diminutio, distantia (= "linear distance"), diuidere, ducere, duplare, egredi, equalis (= "equal"), equalitas (= "equality"), excedere, excessus, extendere, includere, inequalis (= "unequal"), intercipere, latus, linea (recta), lineare, minuere, multiplicare, multiplicatio, probare, productum, protendere, protrahere, prouenire, quadrare, reducere ad, relinquere, remanere, remouere, resecare, secare, secundum spatium, subtendere, subtractio, subtrahere, superaddere, superesse, unitas. For words which recur many times in the text, references are given to only a few representative passages, and the recurrence is indicated by *al.* (elsewhere), *fr.* (frequently), or *passim.* References are to section and line number of the text, and if the note (*n*) to that line contains relevant information, the note number within that section is also recorded. As a comparison between the two documents, certain words or phrases found in the *Theorica planetarum Gerardi* (see p. 36, above) are indicated with a "G" together with the section number, as found in Carmody's text, where the usage occurs (e.g., s.v. "accedere," G.1 refers to section 1 of Carmody's edition of Gerardus's *Theorica*).

General Index

This index contains all references to all proper names in the introduction, text, commentary, and appendixes, with the following omissions: purely bibliographical references, names of authors, places and works occurring in the descriptions of manuscripts (pp. 59–120), and trivial occurrences of "Campanus" and of the names of the planets. It is also an index of subjects, but the subjects listed and the subject references entered have both been restricted to those considered of possible interest. References are to pages. For the introduction, note numbers have been added, being indicated by *n* followed by the number in italics. References to the text are indicated by the addition to the page number of *l* followed by the line number on that page.

THE UNIVERSITY OF WISCONSIN PUBLICATIONS IN MEDIEVAL SCIENCE

Marshall Clagett, *General Editor*

1

The Medieval Science of Weights (Scientia de ponderibus): Treatises Ascribed to Euclid, Archimedes, Thābit ibn Qurra, Jordanus de Nemore, and Blasius of Parma
Edited by Ernest A. Moody and Marshall Clagett
448 pages

2

Thomas of Bradwardine: His "Tractatus de proportionibus." Its Significance for the Development of Mathematical Physics
Edited and translated by H. Lamar Crosby, Jr.
216 pages

3

William Heytesbury: Medieval Logic and the Rise of Mathematical Physics
By Curtis Wilson
232 pages

4

The Science of Mechanics in the Middle Ages
By Marshall Clagett
742 pages

5

Galileo Galilei: "On Motion" and "On Mechanics"
Edited and translated by I. E. Drabkin and Stillman Drake
204 pages

6

Archimedes in the Middle Ages. Volume I: The Arabo-Latin Tradition
By Marshall Clagett
752 pages

7

The "Medical Formulary" or "Aqrābādhīn" of al-Kindī
Translated with a study of its materia medica by Martin Levey
424 pages

8

Kūshyār ibn Labbān: "Principles of Hindu Reckoning"
A Translation with introduction and notes by Martin Levey and Marvin Petruck of the *Kitāb fī uṣūl ḥisāb al-hind*
128 pages

9

Nicole Oresme: "De proportionibus proportionum" and "Ad pauca respicientes"
Edited with introductions, English translations, and critical notes
by Edward Grant
488 pages

10

The "Algebra" of Abū Kāmil, "Kitāb fī al-jābr wa'l-muqābala," in a Commentary by Mordecai Finzi
Hebrew text, translation, and commentary, with special reference
to the Arabic text, by Martin Levey
240 pages

11

Nicole Oresme: "Le Livre du ciel et du monde"
Edited by Albert D. Menut and Alexander J. Denomy, C.S.B.
Translated with an introduction by Albert D. Menut
792 pages

12

Nicole Oresme and the Medieval Geometry of Qualities and Motions: A Treatise on the Uniformity and Difformity of Intensities Known as "Tractatus de configurationibus qualitatum et motuum"
Edited with an introduction, English translation, and commentary
by Marshall Clagett
728 pages

13

Mechanics in Sixteenth-Century Italy: Selections from Tartaglia, Benedetti, Guido Ubaldo, and Galileo
Edited and translated by Stillman Drake and I. E. Drabkin
440 pages

14

John Pecham and the Science of Optics: "Perspectiva communis"
Edited with an introduction, English translation, and critical notes
by David C. Lindberg
320 pages

15

Nicole Oresme and the Kinematics of Circular Motion: "Tractatus de commensurabilitate vel incommensurabilitate motuum celi"
Edited with an introduction, English translation, and commentary
by Edward Grant
438 pages

16

Campanus of Novara and Medieval Planetary Theory: "Theorica planetarum"
Edited with an introduction, English translation, and commentary
by Francis S. Benjamin, Jr., and G. J. Toomer